# QUANTUM MECHANICS

## TA-YOU WU

World Scientific

*Published by*

World Scientific Publishing Co. Pte. Ltd.

5 Toh Tuck Link, Singapore 596224

*USA office:* 27 Warren Street, Suite 401-402, Hackensack, NJ 07601

*UK office:* 57 Shelton Street, Covent Garden, London WC2H 9HE

**British Library Cataloguing-in-Publication Data**
A catalogue record for this book is available from the British Library.

**QUANTUM MECHANICS**



ISBN-13 978-9971-978-47-1
ISBN-10 9971-978-47-4
ISBN-13 978-981-238-286-3 (pbk)
ISBN-10 981-238-286-0 (pbk)

# Foreword

It is great news that Prof. Ta-You Wu is publishing his lectures on quantum mechanics. To describe my reaction to this news, let me first quote from Robert Serber's "The Early Years" (in *Oppenheimer*, Scribner's, 1969).

*The years 1925 to 1929 were great years in physics. They saw the development of the quantum theory: the Schrödinger equation, the Dirac equation, field theory and quantum electrodynamics. That it was so completely a European development illustrates the weakness and provincialism of theoretical physics in the United States at the time. Within fifteen years the situation was drastically changed, and American theoretical physics was becoming comparable to the best. A very important element in this change was the influence of Robert Oppenheimer. The alumni of the great school of theoretical physics he established in Berkeley, California, played a large part in the subsequent development of American physics and also in enabling us to meet the demands of World War II.*

If the introduction of the new quantum theory into the United States was important, then its introduction into China was dramatic: It was only in the 1920's that college physics was beginning to be taught in China. Wu himself had learned college physics from Y. T. Yao (Jao Yu-tai). He then studied in Ann Arbor for his Ph.D. and returned to China to teach at Peking University 1934–1937. There were only a handful of physicists in China in those years who knew quantum mechanics and Wu was the one who exposed the young students to the new field. Among them were H. C. Cheng, C. F. Hsueh, A. T. Kiang, S. T. Ma, S. T. Shen and others.

Then the war came and Wu taught at the Southwest Associated University in Kunming 1938–1945. Sheldon S. L. Chang, Kun Huang, Su-Shu Huang, C. F. Yin and I were among his students in undergraduate and graduate courses from 1941–1944.

Wu's courses were always well prepared, comprehensive and lucid. He was very responsible to the students and allowed us to question him at great lengths after classes. Sometimes we were afraid he would miss the last primitive horse carriage that he was supposed to take to Gang-Tou village where he lived.

I have described elsewhere my gratefulness to him for introducing me to the subject of group theory in physics. As I grow older, I am even more convinced that what shapes a scientist's early taste and style is *the* most important element in his later work.

I recognize in this book on Quantum Mechanics a number of items that Wu taught us in 1941–1942 in my senior year. In particular, the Hamilton-Jacobi theory and the intricacies of the many-electron system were topics that we learned that year. Both made deep impressions on me. The chapters in the present book on the Klein-Gordon and Dirac equations and on field theory were not covered in that course of 1941–1942.*

This book is the result of Professor Wu's fifty years of teaching experience in the basic course on quantum mechanics. It has a unique emphasis on the physics of quantum mechanics, embracing both fundamental principles and complex phenomena. In which other textbook could one find discussions of the Raman effect, autoionization and configuration interaction together with a presentation of Heisenberg's initial ideas? Exposure to these subjects, I believe, shapes a student's taste in what physics is all about, and that, in the long run, is more important than the technique and technology that one usually focuses on in a graduate course.

*Stony Brook*                                                               C. N. Yang

* The author has decided to remove these chapters from the present volume for a possible future volume.

# Foreword

Nowadays, any serious student with an interest in science has to be acquainted with quantum mechanics. Yet quantum mechanics is much more than a useful tool; it is an intellectual achievement of the highest order. Thus the proper teaching of quantum mechanics should not consist merely of its mathematical techniques and physical facts; because the subject is so well-developed there are, however, a number of topics that must be covered. An ideal solution would be to have all of that, plus the historical perspective and at least some of the philosophical implications. Of course, this is not easy. Although there are many textbooks on quantum mechanics, none precisely fulfills such a need. That is why this new book by Professor Ta-You Wu is so welcome.

After a brief review of classical physics, Professor Wu discusses in some detail the pre-quantum mechanical period, the early development of matrix mechanics and wave mechanics, and then proceeds to the general theory of quantum mechanics. He particularly emphasizes the historical evolution of the basic ideas. His incisive description of the contributions of the giants of early times, Planck, Einstein, Bohr, Pauli, Heisenberg, Schrödinger, Dirac evokes a sense of how science is done by real people. The discovery of truth rarely follows a straight line; each new segment often contains only a partial truth which is then complemented by the next step taken by someone else. This is especially true in the case of quantum mechanics, as it is brought out clearly in Professor Wu's book.

In the general theory of quantum mechanics, Professor Wu discusses with particular care the postulates involved, with special emphasis on measurement theory and the related tools of the density matrix and Green's functions. He then starts discussions on applications to atomic structure. The chapters on multi-electron atoms, perturbative theory and time-dependent systems, would be very useful to graduate students interested in physics and chemistry. For those doing more advanced work in atomic and molecular structure, these chapters form a good introduction to Professor Wu's standard treatise on vibrational spectra and the structure of polyatomic molecules.

The remaining third of the book is mainly devoted to relativistic theories, from the Klein-Gordon equation through Dirac's theory to field quantization.* Again, the emphasis is on the mathematical formalism as well as the physical implications. A genuine regret is that the book ends at the threshold of quantum field theory, which makes the reader look forward to a sequel to the present volume.

*New York*                                                              T. D. Lee

* The author has decided to remove these chapters from the present volume for a possible future volume.

# Preface

In 1929, at Nankai University, Tientsin, China, I tried to initiate myself into wave mechanics through A. March's *Die Grundlagen der Quantenmechanik* and into matrix mechanics through G. Birtwistle's *The New Quantum Mechanics*. At that time, few books were available. At Ann Arbor in 1931–2, I learned quantum mechanics with Goudsmit and attended Heisenberg's lectures more or less based on his *The Physical Principles of the Quantum Theory* and Uhlenbeck's course of lectures on Dirac's relativistic equation of the electron. I began to teach quantum mechanics in China in 1934 and, after the War, off and on in the United States and Taiwan. In 1956, my notes were mimeographed for my class in Taiwan and, in 1979, these notes were published in Chinese in two volumes on nonrelativistic and relativistic quantum mechanics. These earlier notes and the texts in Chinese found favor with some colleagues who, from time to time, encouraged me to publish them in English. With some misgivings I have prepared the present volume.

My course on quantum mechanics usually begins with a summary of those important developments in physics in the period 1900–1925 that first trumpeted the successes of the Quantum Theory and then set the stage for a revolution that shook the foundations of classical physics. Then the initial ideas and developments of the matrix mechanics of Heisenberg and Born, of the wave mechanics of de Broglie and Schrödinger and of the transformation theory of Dirac are presented. Some elementary examples are treated to familiarize the students with several technical aspects of quantum mechanics. This is followed by the probability interpretation of Born, the uncertainty principle of Heisenberg and the complementarity concept of Bohr. With the completion of the development of the mathematical structure and the physical ideas, I recapitulate and present quantum mechanics in a postulational manner.

The present volume also traces this repetitious path because, from my experience, the average student finds such a course easier to follow than to begin with Dirac's book. The historical order of developments is emphasized by dating important contributions and referring to the original papers. It is believed that a knowledge of who did what and when gives one a feeling of "intimacy" which adds to his appreciation of physics.

After a brief treatment of the perturbation theory for stationary-state systems and of time-dependent systems, the last fourth of the book is devoted to the application of quantum mechanics to atomic problems. The reasons for this preoccupation with atoms are: firstly, that quantum mechanics had its earliest and complete success in atomic problems; secondly, many atomic problems and their treatment are common to other branches of physics; and lastly, my own early interest has been in this area. The last brief chapter on molecules is included for the last reason.

I would like to take this opportunity to record my appreciation for those from whom I learned quantum mechanics: my teachers, Professors Otto Laporte, Samuel A. Goudsmit, George E. Uhlenbeck and David M. Dennison of Michigan, in 1931–3; Professors Werner Heisenberg and Gregory Breit, in 1932; Professor Enrico Fermi and J. H. Van Vleck, in 1933, at the Michigan Summer Physics Symposium. I am thankful for the opportunities that brought me into association with Professor Isidor I. Rabi at Columbia in 1947–9; Professor Gerhard Herzberg at the National Research Council, Ottawa, in 1949–63; Professor Paul A. M. Dirac at the N. R. C. in the summer of 1956; and the Institute for Advanced Study, Princeton, in 1958–9. I am also grateful for the good fortune of having in my classes in China such illustrious physics students as (the late) S. T. Ma in 1934–7; C. N. Yang, Kun Huang, S. L. Chang, (the late) S. S. Huang in 1941–2; and T. D. Lee in 1945–6. Doctors Yang and Lee have each written a warm Foreword for the present volume, for which their old teacher has the most heartfelt appreciation.

I wish to thank Drs. Paul S. T. Lee and J. P. Hsu who have kindly read the typescripts and pointed out many errors.

*Taipei*
*December, 1985*                                                        Ta-You Wu

# Contents


Preface     vii

**0   CLASSICAL PHYSICS—A RESUMÉ**     1

    1. Classical dynamics     2
       (1) Basic concepts: space, time and mass     2
       (2) Dynamical laws: simultaneous space-time *and* momentum-energy knowledge     4
       (3) Canonical equations and transformation theory of dynamics     5
       (4) Hamilton-Jacobi equation     10
    2. Electromagnetic theory     19
       (1) Electromagnetic fields     19
       (2) Symmetry properties in space reflection and time reversal     23
    3. Statistics and probability     25
    Appendix 0: Hamilton's principle and canonical equations     26

**1   PRE-QUANTUM MECHANICAL PERIOD (1900–1925)**     30

    1. Planck's quantum theory (1900)     31
    2. Einstein's theory of the photon (1905) and the wave-particle duality     37
    3. Quantum theory applied to vibrations of atoms (1906–7)     44
    4. Bohr's theory—the concept of stationary states (1913)     46
    5. Bohr's theory—refinements (1915)     49
    6. Einstein's transition probabilities (1916)     52
    7. The correspondence principle (1918)     55
    8. Theory of dispersion—Ladenburg, Kramers and Heisenberg (1921–5)     59
    9. Atomic spectra—empirics (up to 1925)     64
       (1) Empirics: spectral terms, series, multiplets     65
       (2) Anomalous Zeeman effect: Paschen-Back effect     70
    10. Electron spin (1925)     73

(1) Spin-orbit interaction ..... 73
(2) Remarks ..... 74
(3) Stern-Gerlach's experiment ..... 76
(4) Landé's $g$-formula for anomalous Zeeman effect ..... 76
(5) Paschen-Bach effect for strong magnetic fields ..... 77
(6) Quantum numbers $L, S, J$ ..... 77
11. Pauli's exclusion principle (1925) ..... 79
12. The birth of quantum mechanics ..... 82

**2   MATRIX MECHANICS** ..... **85**

1. Heisenberg's initial ideas (1925) ..... 85
2. Matrix algebra ..... 88
Exercises ..... 104
3. Transformation of representations: the theory of Dirac ..... 106
Exercises ..... 108
4. Matrix mechanics—principles ..... 110
(1) Harmonic oscillator ..... 114
Exercises ..... 116
5. Angular momenta and selection rules ..... 118
Exercises ..... 122
6. Perturbation theory ..... 123
(1) Non-degenerate systems ..... 123
(2) Degenerate systems ..... 126
7. Dirac's theory ..... 126

**3   WAVE MECHANICS: EARLY DEVELOPMENTS** ..... **131**

1. de Broglie's ideas (1923–1924) ..... 131
2. Schrödinger's wave mechanics (1926) ..... 135
(1) Digression to Hamilton-Jacobi theory ..... 136
(2) Digression to the relation between wave and ray optics ..... 138
(3) Schrödinger equation ..... 140
3. The Schrödinger theory ..... 142
(1) Generalization to a system of $N$ particles ..... 142
(2) Linearity and superposition principle ..... 142
(3) Wave packet ..... 143
(4) Time independence of $\int \Psi^* \Psi \, dq$ ..... 144
(5) Commutation relation and equivalence between matrix and wave mechanics ..... 144
(6) Fourier transformation and representation transformation ..... 146
(7) Conditions on $\psi$ ..... 148
(8) Relativistic wave equation ..... 148

4. The eigenvalue problem                                                   149
   (1) The Sturm-Liouville problem                              150
   (2) Simple harmonic oscillator                               154
   (3) Central field problem, parity                            157
   (4) Angular momenta                                          160
Appendix A.  Hermite polynomials                                            161
Appendix B.  Harmonic oscillation in Fock representation                    164
Appendix C.  Associated Legendre polynomials                               167
Appendix D.  Angular momentum operators and
                 spherical harmonics                 172
Exercises                                                                   175
5. The hydrogen atom                                                        178
   (1) Discrete states, $E < 0$                                 179
   (2) Continuum states, $E > 0$                               181
   (3) Normalization of continuum wave functions               183
Appendix E.  Associated Laguerre polynomials                               185
Exercises                                                                   189

**4  PROBABILITY POSTULATE AND UNCERTAINTY
PRINCIPLE**                                                                 **191**

Introduction                                                               191
1. Schrödinger equation in integral form                                   193
   (1) Particle in a central field $V(r)$                      193
   (2) Scattering of a particle in a potential field          194
2. Probability interpretation                                              198
3. Uncertainty principle                                                   199
Exercises                                                                   206

**5  QUANTUM MECHANICS: GENERAL THEORY**                                   **209**

1. Complementarity principle                                               209
2. Quantum mechanics: mathematical preliminaries                           212
   (1) Vector space, vectors                                  213
   (2) Linear operators, hermitian operators                  216
   (3) Transformation of representations, unitary operators   219
   (4) Non-commutative hermitian operators                    220
   (5) Unitary transformations                                225
   (6) Displacement operator                                  227
   (7) Time translation operator $U(t)$                        230
Exercises                                                                   231
3. Quantum mechanics in postulational form                                 232
Postulate I                                                                 232
Postulate II                                                                232

Postulate III    233
Postulate IV    233
Postulate V    235
Postulate VI    239
4. Representation and measurement    239
    (1) Representation    239
    (2) Measurement    241
5. Relation between Heisenberg's and Schrödinger's equations of motion    242
6. Density matrix    245
    (1) Pure and mixed state    246
    (2) Density operator and density matrix    247
    (3) Trace    248
    (4) Normalization    248
    (5) $\rho^2$ and the condition for a pure state    248
    (6) Physical meaning of density matrix and mixed state    249
    (7) Transformation property of $\rho$    251
    (8) Liouville equation in quantum mechanics    252
    (9) Density matrix and irreversibility of macroscopic processes    254
7. The Copenhagen school and Einstein's philosophy    254

**6    PERTURBATION THEORY: STATIONARY STATE PROBLEM    259**

1. Perturbation theory for non-degenerate systems    260
    (1) Anharmonic oscillator    261
    (2) Stark effect in alkali metals atoms    261
    (3) Sommerfeld's relativistic correction    262
2. Perturbation theory for degenerate systems    265
    (1) Stark effect in (non-relativistic) hydrogen atoms    267
Appendix F. Parabolic coordinates: Stark effect    269
Exercises    271

**7    TIME-DEPENDENT SYSTEMS    274**

1. Theory of the Raman effect and inverse Stark effect    274
2. Perturbation theory of transitions of Dirac    279
    (1) Method of variation of constants    279
    (2) Einstein's $B_n^m$, $A_n^m$ coefficients    283
    (3) Theory of dispersion    286
    (4) Rearrangement collisions    289
    (5) Transition probability and cross section    291

    3. Unitary (time-translation) operator   293
    4. Method of integral equation and Green's function   296
        (1) The unperturbed system $H_0$   297
        (2) The perturbed system $H = H_0 + V$   299
    5. Green's function as propagator   300
    6. Uncertainty relation for energy and time   304

**8   THE HYDROGEN ATOM   307**

    1. Electron spin   307
        (1) Operators and eigenvectors   307
        (2) Spin-orbit interactions   309
        (3) Transformation between $(m_l, m_s)$- and $(j, m)$-representations   313
    2. Selection rules   314
    3. Fine structure   317
    4. Zeeman effect   321
        (1) Strong magnetic field, Paschen-Bach effect   322
        (2) Weak magnetic field   323
        (3) Arbitrary magnetic field   323
        (4) Non-crossing theorem   324
    5. Scattering of an electron by the hydrogen atom   325
        (1) Born approximation   325
    Exercises   327

**9   TWO- AND MANY-ELECTRON ATOMS   329**

    1. Symmetry property   329
    2. Two-electron atom: energy   332
        (1) Perturbation theory   333
        (2) Ritz variational method   337
        (3) Hartree-Fock method   339
        (4) Hylleraas' method   341
    3. Two-electron atom: configuration and $\{L, S\}$-coupling $^{2S+1}L$   344
        (1) $L, S$ states, $e^2/r_{12}$ energies   344
        (2) Spin-orbit interactions: multiplet structure   350
    4. Two-electron atom: $\{j, j\}$-coupling   353
    5. Two-electron atom: intermediate coupling   357
    6. Configuration interactions   359
        (1) Autoionization   360
        (2) Auger effect   364
    7. Many-electron atoms   364
        (1) Slater's method ($\{L, S\}$-coupling)   365

(2)  Hartree-Fock method                                              370
(3)  Selection rules for many-electron atoms                          372
(4)  Configuration interactions                                       373
8. The Thomas-Fermi potential and many-electron atoms                 376
(1)  The statistical potential                                        376
(2)  Energy of an electron in an atom or ion                          378
(3)  WKB method of solving Eq. (IX-218)                               380
(4)  $f$ states in the actinium series of atoms                       382
(5)  The Rydberg correction to the $f$ states of a heavy atom         383

**10  QUANTUM MECHANICS OF MOLECULES**                                **386**

1. Introduction                                                       386
2. The electronic, vibrational and rotational motion of a
   molecule                                                           388
(1)  $H_2^+$ electronic states with fixed nuclei                      390
(2)  Nuclear vibration                                                391
(3)  Rotational motion                                                392
3. The hydrogen molecule—the theory of Heitler and
   London                                                             393
4. Coupling between vibrational and rotational motions                397
5. Vibrational motion of a polyatomic molecule                        401
6. Rotational motion of a symmetrical top molecule                    405
7. Nuclear spin in molecules                                          407

**NAME INDEX**                                                        **411**

**SUBJECT INDEX**                                                     **414**

# QUANTUM MECHANICS

# Chapter 0

# Classical Physics—a Resumé

In the following, we mean by classical physics the whole system of physics completed before the advent of the quantum theory in 1900 and the theory of relativity in 1905. Classical physics may be grouped into three disciplines, namely, dynamics, electromagnetism and thermal and statistical physics. Dynamics may be said to have begun with Galileo (1564–1642) and the whole system of dynamics was established by Newton (1642–1727). It reached its peak of development with Hamilton (1805–1865). The science of electromagnetism may be said to begin with Ampére (1775–1836) and Faraday (1791–1867), and the theory includes that of optics, completed by Maxwell (1831–1879). The production of electromagnetic waves in the laboratory by Hertz in 1888, the electron theory of Lorentz and the identification of the electron by J. J. Thomson in 1897 seem to have rendered the theory of electromagnetism "complete".

The subjects of heat, the conservation of energy and thermodynamics deal with certain properties of matter in bulk. They were established by Joule (1818–1889), Helmholtz (1821–1894), Clausius (1822–1888) and Lord Kelvin (1824–1907). The kinetic theory of gases may be traced to D. Bernoulli (1700–1782); the main developments are due to Maxwell and Boltzmann (1844–1906) in the 1860's and 1870's. Kinetic theory furnishes the link between the dynamics of molecules and the macroscopic concepts of temperature and thermal energy. Probability is introduced into physics when a property, such as the kinetic energy, is averaged over a large number of molecules. The statistical mechanics of Maxwell and Boltzmann in the early 1870's attempts to bridge the molecular view governed by dynamics and the macroscopic view by thermodynamical variables by means of statistical concepts (such as those of ensembles) and the ergodic hypothesis. This last hypothesis has later been replaced by a postulational formulation.

Toward the end of the nineteenth century, the success of the existing system of physics was very great indeed. Newton's laws of motion, together with Newton's theory of universal gravitation were successful in celestial mechanics; together with the kinetic theory they established the dynamics of fluids. The theory of the electromagnetic field seemed to account for all known phenomena of electromagnetism and light; and thermodynamics and statistical mechanics seemed to account for all thermal properties of matter-in-bulk. The system of physics seemed then to be *almost* complete. Physics was then in such a state that it prompted a prominent physicist to say that the main structure of physics was already there, that what was left was to go to the next decimal place for higher accuracy.

But only "almost"! Toward the end of the century, there came the results of two groups of investigations, namely those of Michelson and Morley in an attempt to detect the effect, if any, of the Earth's motion in its orbit on the velocity of light, and those of Paschen and others on the spectral distribution of the radiation emitted by a "black body" at temperature $T$. The former led to the Lorentz transformation and finally to the special theory of relativity of Einstein in 1905. The latter led to the quantum theory of Planck in 1900.

Then natural radioactivity was discovered by Becquerel in 1896. The law of radioactive decay was discovered by Rutherford and Soddy (1900–03), which turned out later to be a probability law of a nature akin to that of quantum mechanics.

The story of the development of the quantum theory in the period 1900–1925 will be told in Chapter 1, and the development of the quantum mechanics from 1925 onward in the rest of this book. But before this, let us summarize below the essence of classical physics, especially those parts which have been taken as "basic" but have to be fundamentally changed in quantum mechanics.

## 1.  Classical dynamics

### ( 1 )  Basic concepts: space, time and mass

In classical dynamics, the primary concepts are those of space (the measurement of length), time (the measurement of time-interval) and mass. From these, other dynamical concepts are defined. Thus velocity $\mathbf{v} = d\mathbf{r}/dt$; acceleration $\mathbf{a} = d\mathbf{v}/dt$; momentum $\mathbf{p} = m\mathbf{v}$; force $\mathbf{f} = m\mathbf{a}$; kinetic energy $T = \frac{1}{2}mv^2$; potential energy $V$ such that $\mathbf{f} = -\nabla V$, etc.

### (i)  Time

The time concept in Newton's dynamics (and in all of classical physics) is an "absolute time" that "flows continually and uniformly, and independently of anything else." As a consequence of this *absolute time*, the time interval between two events at one point in space will be the same for all observers in motion relative to one another, and if two events taking place at two different places are simultaneous for one observer $A$, they will also be simultaneous to other observers moving relative to $A$.

## (ii)  Space

In classical physics, the measurement of length is an unambiguous process. To measure a moving object, one lays a meter stick alongside of it and reads off the two end marks *simultaneously*. By the concept of absolute time simultaneity is also absolute so that the length of a moving object is also absolute, i.e., the same for all observers in motion relative to the object.

In fact, in classical physics, space and time are independent of each other.

Next we come to the concept of space itself. All our experience in our daily life accustoms us to a space having the properties of Euclidean geometry, and it is therefore natural to take an abstract space with Euclidean geometry as a *physical* space for the description of physical phenomena.

## (iii)  Mass

The concept of mass appears explicitly in Newton's Second Law $d(mv)/dt = $ **f**. It is a constant.

## (iv)  Measurements

In classical physics, it is tacitly assumed that, *in principle*, it is possible to measure any physical quantity as accurately as one pleases, that any errors coming from the perturbations introduced by the measuring process itself can be made as small as one pleases, in fact reduced to zero in the limit. For example it is assumed that one can in principle determine the position and the momentum of a particle at the same time to any degree of accuracy desired.

In the theory of relativity, the speed of light is finite, and the basic principle is that all physical laws have the same form for all inertial frames (i.e. for all observers in uniform relative motion to one another). But Einstein furnished the fundamental key to the Lorentz transformation by defining the concepts of length and time by means of their measurements by rigid rods and synchronized clocks. Absolute time and with it, absolute simultaneity, no longer have meaning in physics. Space and time are no longer independent, but $x$, $y$, $z$, $ct$ are together transformed according to the Lorentz transformations among inertial frames, as the four coordinates of 4-dimensional (pseudo-) Euclidean space.

In the general theory of relativity, Einstein introduced the revolutionary idea that the physical space (as distinguished from the abstract Euclidean space) may have a geometry which is influenced and determined by the presence of matter, or energy, and is therefore non-Euclidean.

It is entirely beyond the scope of the present work to go into the theory of relativity. Suffice it to remark here that the concepts of space and time have undergone fundamental reexaminations and changes through the theory of relativity. We shall remark here in passing that the concept of space (co-ordinates $r$) has also undergone fundamental change—representation by hermitian operators—in quantum mechanics, while the concept of time has remained "classical" (as in the special theory of relativity). The asymmetry is a serious unsatisfactory point in the present structure of our quantum mechanics and relativity theory.

### *(2) Dynamical laws: simultaneous space-time and momentum-energy knowledge*

In classical dynamics, the fundamental principle is the law of motion of Newton. For simplicity, take a particle of mass $m$ under a force $\mathbf{F}$. The law of motion is a differential equation of the second order

$$m\ddot{\mathbf{r}} = -\nabla V(r) \tag{0-1}$$

or, in Hamilton's canonical form, a pair of equations of the first order

$$\dot{p} = -\frac{\partial H}{\partial q}, \qquad \dot{q} = \frac{\partial H}{\partial p}, \tag{0-2}$$

where $q = (x, y, z)$, $p = (m\dot{x}, m\dot{y}, m\dot{z})$, and

$$H = \frac{p^2}{2m} + V(q).$$

Now the solution of either (0-1) or (0-2) presupposes a knowledge of the initial values of $\mathbf{r}$ and $\mathbf{p} = m\dot{\mathbf{r}}$ at a time $t_0$. The possibility of the simultaneous knowledge of $x$, $y$, $z$ and $p_x$, $p_y$, $p_z$ is therefore a basic assumption of classical dynamics.

The solution of (0-1) is a relation of the form of a trajectory

$$\mathbf{r} = \mathbf{r}(t - t_0),$$

and a first integral of (0-1) is the energy integral

$$\frac{1}{2m}p^2 + V = E.$$

We say that classical dynamics allows a description of a system *both* in terms of

$$\text{space-time, i.e.,} \quad \mathbf{r} = \mathbf{r}(t - t_0) \tag{0-3}$$

*and* in terms of

$$\text{momentum-energy, i.e.,} \quad p^2 = 2m(E - V). \tag{0-4}$$

But, as we shall see in the present book, this basic premise of classical dynamics—the possibility of simultaneous exact knowledge of a coordinate $q$ and its conjugate momentum $p$—is denied in quantum mechanics! And so is the simultaneous space-time and energy-momentum description! In quantum mechanics, it is not possible to describe the successive states of motion of a particle by a deterministic differential equation of motion of the form (0-1). We must emphasize, in passing, that changes made in quantum mechanics are in the basic concepts of coordinate and momentum themselves as well as the nature of the structure of physical theory.

*(3) Canonical equations and transformation theory of dynamics*

For a dynamical system described by the generalized coordinates $q = (q_1, \ldots, q_n)$, the Lagrangian function is defined by

$$L(q, \dot{q}, t) = T(q, \dot{q}) - V(q, t), \qquad \dot{q}_k = \frac{dq_k}{dt}, \tag{0-5}$$

where $T$, $V$ are the kinetic and the potential energy. The equations of motion of Newton can be expressed in the following form

$$\frac{d}{dt}\frac{\partial L}{\partial \dot{q}_k} - \frac{\partial L}{\partial q_k} = 0, \qquad k = 1, 2, \ldots, n, \tag{0-6}$$

by direct transformation from equations (0-1). These are a system of $n$ differential equations of the second order for the $q_k$'s. In principle, there are $n$ first integrals

$$f_k(q, \dot{q}, t) = c_k, \qquad k = 1, 2, \ldots, n, \tag{0-7}$$

and the solution of (0-6) is

$$q_k = q_k(q^0, \dot{q}^0, t), \qquad \dot{q}_k = \dot{q}_k(q^0, \dot{q}^0, t - t_0), \tag{0-8}$$

where $q^0 = (q_1^0, \ldots, q_n^0)$, $\dot{q}_k^0 = (\dot{q}_1^0, \ldots, \dot{q}_n^0)$ are the initial values of $q$, $\dot{q}$ at time $t = t_0$.

But the Lagrange equation (0-6) are the Euler-Lagrange differential equation of the following equation in the calculus of variations

$$\delta \int_{t_0}^{t} L\, dt = 0, \tag{0-9}$$

where the variations $\delta q_k$'s are zero at the two limits $t$ and $t_0$ of the independent variable $t$, and are arbitrary between the two limits. The variational form (0-9) is a more general formulation due to Lagrange (1736–1813) of the earlier

Maupertuis principle of least action (1744)[a]

$$\Delta \int_{t_0}^{t} 2T \, dt = 0. \tag{0-10}$$

The next great development came with Hamilton (1805–1865). The momentum $p_k$ conjugate to $q_k$ is defined by

$$p_k = \frac{\partial L}{\partial \dot{q}_k}, \qquad k = 1, 2, \ldots, n. \tag{0-11}$$

These equations define a transformation from

$$q_1, \ldots, q_n, \dot{q}_1, \ldots, \dot{q}_t, t \quad \text{to} \quad q_1, \ldots, q_n, p_1, \ldots, p_n, t, \tag{0-12}$$

which is known as Legendre transformation (but is perhaps due to Euler.) If a (Hamiltonian) function $H(q, p, t)$ is defined by

$$H(q, p, t) + L(q, \dot{q}, t) = \sum p_k \dot{q}_k \tag{0-13}$$

with $p_k = \dfrac{\partial L}{\partial \dot{q}_k}$, then it can be proved generally that

$$\dot{q}_k = \frac{\partial H}{\partial p_k}, \qquad k = 1, 2, \ldots, n \tag{0-14}$$

and

$$\frac{\partial L}{\partial q_k} + \frac{\partial H}{\partial q_k} = 0, \tag{0-15}$$

$$\frac{\partial L}{\partial t} + \frac{\partial H}{\partial t} = 0. \tag{0-16}$$

(0-14) may be taken to define the inverse transformation to (0-11) and (0-12). The proof of (0-14)–(0-16) presupposes

$$\mathscr{H}_{\dot{q}} L \equiv \frac{\partial \left( \dfrac{\partial L}{\partial \dot{q}_1} \cdots \dfrac{\partial L}{\partial \dot{q}_n} \right)}{\partial (\dot{q}_1, \ldots, \dot{q}_n)} \neq 0, \qquad \frac{\partial (\dot{q}_1 \cdots \dot{q}_n)}{\partial (p_1 \cdots p_n)} = [\mathscr{H}_{\dot{q}} L]^{-1}$$

and is straightforward. That

[a] The difference in notations, $\delta$ and $\Delta$, has the following origin: In (0-9), we compare the values of integrals taken along various trajectories having the same initial and the same end-point in $q$-space and having the same time of transit $t - t_0$. In (0-10), we examine the values of integrals taken along various trajectories having the same initial and the same end-point in $q$-space, but, instead of having the same time of transit $t - t_0$, obeying the energy conservation law for all the varied paths, i.e., $\Delta t \neq 0$ at the end-point.

$$\frac{\partial(p_1,\ldots,p_n)}{\partial(\dot{q}_1,\ldots,\dot{q}_n)} \neq 0 \tag{0-17}$$

is guaranteed by the kinetic energy $T(q,\dot{q})$ being a positive definite quadratic in the $\dot{q}_n$'s.

$$2T = \sum_{i,j}\frac{\partial^2 T}{\partial\dot{q}_i\partial\dot{q}_j}\dot{q}_i\dot{q}_j = \sum\frac{\partial^2 L}{\partial\dot{q}_i\partial\dot{q}_j}\dot{q}_i\dot{q}_j > 0. \tag{0-18}$$

On using (0-13) in Hamilton's principle (0-9)

$$\delta\int_{t_0}^{t}\left[\sum p_k\dot{q}_k - H(q,p)\right]dt = 0, \tag{0-19}$$

one obtains

$$\int_{t_0}^{t}\sum_k\left\{\left(\dot{q}_k - \frac{\partial H}{\partial p_k}\right)\delta p_k - \left(\dot{p}_k + \frac{\partial H}{\partial q_k}\right)\delta q_k\right\}dt + \sum p_k\delta q_k\bigg|_{t_0}^{t} = 0. \tag{0-20}$$

The coefficients of $\delta p_k$ vanish by virtue of (0-14), and those of $\delta q_k$ vanish because the $\delta q_k$'s are arbitrary.[b] Hence one obtains the canonical equations

$$\dot{q}_k = \frac{\partial H}{\partial p_k}, \qquad \dot{p}_k = -\frac{\partial H}{\partial q_k}, \qquad k = 1,2,\ldots n, \tag{0-21}$$

which form a system of $2n$ equations of the first order. The solution of (0-21) is the set of $2n$ equations

$$\begin{aligned}
q_k &= q_k(q^0, p^0, t - t_0),\\
p_k &= p_k(q^0, p^0, t - t_0), \qquad k = 1,\ldots,n
\end{aligned} \tag{0-22}$$

where $q^0 = (q_1^0,\ldots q_n^0)$, $p^0 = (p_1^0,\ldots p_n^0)$ are the initial values of $q$ and $p$ at $t = t_0$.

We consider canonical transformations from

$$q,p,t \quad \text{to} \quad Q,P,t$$

such that the equations (0-21) have the same form

$$\dot{Q}_k = \frac{\partial\bar{H}}{\partial P_k}, \qquad \dot{P}_k = -\frac{\partial\bar{H}}{\partial Q_k}, \qquad k = 1,\ldots,n \tag{0-23}$$

where

$$\bar{H} = \bar{H}(Q,P,t) \tag{0-24}$$

does not have to be the same as $H$.

[b] See Appendix 0.

From the form (0-19), it is seen that a *sufficient* condition for canonical transformation is

$$\left(\sum p_k\, dq_k - H\, dt\right) = \lambda\left(\sum p_k\, dQ_k - \bar{H}\, dt\right) + dS, \qquad (0\text{-}25)$$

where $\lambda$ is a constant and $dS$ is an exact differential of a function $S$, $S = S(q, Q, t)$ say. In this case, the following equations

$$p_k = \frac{\partial S}{\partial q_k}, \qquad p_k = -\frac{\partial S}{\partial Q_k}, \qquad \bar{H} = H + \frac{\partial S}{\partial t} \qquad (0\text{-}26)$$

define a canonical transformation for *any* function $S(q, Q, t)$.

A particular function $S$ is the principal function of Hamilton. It is the function obtained by integrating the Lagrangian $L(q, \dot{q}, t)$ along the trajectory of the motion of the system between an initial instant $t_0$ and a final instant $t$

$$S = \int_{t_0}^{t} L\, dt. \qquad (0\text{-}27)$$

From this, one obtains the variation $\delta S$

$$\delta S = -\int_{t_0}^{t} \sum_k \left(\frac{d}{dt}\frac{\partial L}{\partial \dot{q}_k} - \frac{\partial L}{\partial q_k}\right)\delta q_k\, dt + \sum \frac{\partial L}{\partial \dot{q}_k}\delta q_k\bigg|_{t_0}^{t}.$$

By virtue of Lagrange equations (0-6) and (0-11), one obtains,

$$\delta S = \sum p_k \delta q_k \bigg|_{t} - \sum p_k \delta q_k \bigg|_{t_0}$$

$$\equiv \sum p_k \delta q_k - \sum p_k^0 \delta q_k^0. \qquad (0\text{-}28)$$

The principal function $S$ defined by (0-27) is a function of the $4n + 1$ variables $q,\, p,\, q^0,\, p^0,\, t - t_0$; but by means of the $2n$ equations (0-22), it is possible to express $S$ as a function of the $2n + 1$ variables $q,\, q^0,\, t - t_0$ so that (0-28) can be written

$$\delta S(q, q^0, t - t_0) = \sum \left(\frac{\partial S}{\partial q_k}\delta q_k + \frac{\partial S}{\partial q_k^0}\delta q_k^0\right)$$

$$= \sum \left(p_k \delta q_k - p_k^0 \delta q_k^0\right).$$

Thus

$$p_k = \frac{\partial S}{\partial q_k}, \qquad p_k^0 = -\frac{\partial S}{\partial q_k^0}, \qquad k = 1, 2, \ldots n. \qquad (0\text{-}29)$$

Comparison with (0-26) shows that the principal function $S$ in (0-27) defines a canonical transformation from the $q$, $p$, at time $t$ to the $q^0$, $p^0$ at time $t_0$ of the motion of the system. It follows from this that the $q$, $p$ of a system undergo a continual series of canonical transformation during the motion of the system, or, in other words, the motion of a system consists in a continual unfolding of canonical transformations. This statement is of course an immediate consequence of the definition of canonical transformation; (0-28), or (0-29), is only an analytical expression of it.

By regarding the $q_k^0$ in $S(q, q^0, t - t_0)$ as another set of coordinates $Q_k$, (0-28) may be taken as another form of the sufficient condition for canonical transformation

$$\delta S = \sum p_k \delta q_k - \sum P_k \delta Q_k. \tag{0-30}$$

The transformation equations themselves are

$$p_k = \frac{\partial S}{\partial q_k}, \qquad P_k = -\frac{\partial S}{\partial Q_k}, \qquad \bar{H}(Q, P, t) = H(q, p, t) + \frac{\partial S}{\partial t}. \tag{0-31}$$

The condition (0-30) can be expressed in equivalent forms.

If $u, v$ are functions of $q, p$, the classical Poisson bracket expression $(u, v)_{q,p}$ is defined as

$$(u, v)_{q,p} = \sum_k \left( \frac{\partial u}{\partial q_k} \frac{\partial v}{\partial p_k} - \frac{\partial u}{\partial p_k} \frac{\partial v}{\partial q_k} \right). \tag{0-32}$$

It can be shown that a necessary and sufficient condition for (0-30) is

$$(Q_i, Q_j)_{q,p} = 0, \qquad (P_i, P_j)_{q,p} = 0, \qquad (Q_i, P_j) = \delta_{ij}. \tag{0-33}$$

Thus (0-33) is a sufficient condition for the transformation $q, p \to Q, P$ to be canonical.

In terms of the Poisson bracket expression, the canonical equations (0-21) can be put in the form

$$\dot{q}_k = (q_k, H), \qquad \dot{p}_k = (p_k, H). \tag{0-34}$$

Canonical transformations have the following properties:
(1) By definition, they leave the form of the canonical equations (0-21) invariant;
(2) The following integrals (Poincaré's absolute integral invariants)

$$I_1 = \int\!\!\int \sum dp_k \, dq_k,$$

$$I_2 = \int\!\!\int\!\!\int\!\!\int \sum dp_j \, dp_k \, dq_j \, dq_k,$$

$$\tag{0-35}$$

$$\cdot \quad \cdot \quad \cdot \quad \cdot \quad \cdot \quad \cdot \quad \cdot$$

$$I_n = \int \cdots^{2n} \cdots \int dp_1 \ldots dp_n \, dq_1 \ldots dq_n,$$

are invariant, the domain of integration $J_1$ being any two-dimensional manifold $p_k q_k$, etc.;

(3) The integrals (Poincaré's relative integral invariants)

$$J_1 = \int_\Gamma \sum p_k \, dq_k,$$

$$J_2 = \int_\Gamma \int_{\Gamma'} \sum p_j \, dq_j p_k \, dq_k$$

(0-36)

are invariant, the integration in $J_1$ being along any closed curve $\Gamma$ in the $2n$-dimensional $q, p$ space, etc.

(4) Canonical transformations leave the Poisson bracket expressions $(u, v)$ invariant, i.e.,

$$(u, v)_{q, p} = (u, v)_{Q, P}.$$

(0-37)

(5) Canonical transformations have the group property, i.e., if $q, p \to Q, P$ and $Q, P \to w, J$ are two canonical transformations, then $q, p \to w, J$ is also a canonical transformation.

(6) If the Hamiltonian $H$ is invariant under the time reversal transformation, i.e.,

$$q' \equiv \Theta q = q, \quad p' \equiv \Theta p = -p, \quad t' \equiv \Theta t = -t,$$

$$\Theta H(q, p, t) = H(q, p, t),$$

(0-38)

the transformation $\Theta$ in (0-38) is canonical, since equations (0-21) go over into

$$\frac{dq'_k}{dt'} = \frac{\partial H}{\partial p'_k}, \qquad \frac{dp'_k}{dt'} = -\frac{\partial H}{\partial q'_k}.$$

(0-39)

The conditions (0-24) and (0-33) become

$$\sum p_k \, dq_k - H \, dt = -\left(\sum p'_k \, dq'_k - H \, dt'\right) + dS,$$

(0-40)

$$(q'_i, q'_j)_{q, p} = 0, \qquad (p'_i, p'_j)_{q, p} = 0, \qquad (q'_i, p'_j) = -\delta_{ij}$$

(0-41)

and (0-37) becomes

$$(u, v)_{q, p} = -(u, v)_{q', p'}.$$

(0-42)

Note the negative sign in (0-40, -41, -42).


*(4) Hamilton-Jacobi equation*

The principal function $S(q, q^0, t - t_0)$ as defined by (0-27) seems to presuppose the knowledge of the solution (0-22) of the Lagrange equations. If that is the case, it would indeed be pointless to reobtain the solution of the dynamical problem from the $2n$ equations (0-29). But the theory of Hamilton and Jacobi shows that it is not necessary to obtain $S$ by integrating (0-27), but instead, one obtains $S$ from partial differential equations.

From $S = S(q, q^0, t - t_0)$ and (0-29), one obtains

$$\frac{dS}{dt} = \frac{\partial S}{\partial t} + \sum \frac{\partial S}{\partial q_k} \dot{q}_k = \frac{\partial S}{\partial t} + \sum p_k \dot{q}_k.$$

Using (0-27) and (0-13), one arrives at

$$\frac{dS}{dt} = L = \sum p_k \dot{q}_k - H(q, p, t).$$

Hence Hamilton (1834) obtained the equation for $S$

$$H\left(q, p = \frac{\partial S}{\partial q}, t\right) + \frac{\partial S}{\partial t} = 0. \tag{0-43}$$

Jacobi (1837) completed the development by the following theorem: If a complete integral of (0-43) containing $n + 1$ arbitrary constants is known,

$$S(q_1, \ldots, q_n, \alpha_1, \ldots, \alpha_n, t - t_0) + \alpha_{n+1}, \tag{0-44}$$

then the integrals of the canonical equations (0-21) are

$$p_j = \frac{\partial S}{\partial q_j}, \qquad \beta_j = -\frac{\partial S}{\partial \alpha_j}, \qquad j = 1, 2, \ldots n. \tag{0-45}$$

From (0-29), it is seen the $2n$ equations (0-45) define a canonical transformation from $q$, $p$, $t$ to $\alpha$, $\beta$, $t$ such that

$$\left(\sum p_k \dot{q}_k - H\right) dt = \left(\sum \beta_k \dot{\alpha}_k - \bar{H}\right) dt + dS, \tag{0-46}$$

where

$$\bar{H}(\alpha, \beta, t - t_0) = H\left(q, p = \frac{\partial S}{\partial q}, t - t_0\right) + \frac{\partial S}{\partial t}. \tag{0-47}$$

But by hypothesis, $S$ is a complete integral of (0-43). Hence

$$\bar{H}(\alpha, \beta, t - t_0) = 0. \tag{0-48}$$

Application of the canonical equations (0-23) then leads to

$$\alpha_j = \text{const.}, \qquad \beta_j = \text{const.}, \qquad j = 1, \ldots n. \tag{0-49}$$

Equations (0-45) then give the $q_j$ and $p_j$'s in terms of the constants $\alpha$, $\beta$ and $t - t_0$.

Of the infinitely many sets of constants $\alpha_1, \ldots \alpha_n$, there is one set of $\alpha$'s which are the initial values of $q$'s,

$$\alpha_j = q_j^0, \qquad j = 1, \ldots, n. \tag{0-50}$$

In this special case, $S(q, \alpha, t - t_0) = S(q, q^0, t - t_0)$, the principal function of Hamilton in (0-27).

The Hamilton-Jacobi equation (0-43) is simplified somewhat in the case that the Hamiltonian $H$ is not an explicit function of time.

Hamilton defined the characteristic function $S_0$ by the integral

$$S_0 = \int_{t_0}^t 2T\, dt, \tag{0-51}$$

which is related to the principal function $S$ of (0-27) by

$$\begin{aligned}
S &= \int_{t_0}^t L\, dt = \int_{t_0}^t (2T - h)\, dt \\
&= S_0 - h(t - t_0),
\end{aligned} \tag{0-52}$$

where $h$ is the total constant energy of the system

$$T + V = h. \tag{0-53}$$

From

$$\begin{aligned}
L = \frac{dS}{dt} &= \sum \frac{\partial S_0}{\partial q_k} \dot{q}_k + \frac{\partial S_0}{\partial t} - h \\
&= 2T + \frac{\partial S_0}{\partial t} - h
\end{aligned}$$

and

$$L = T - V = 2T - h,$$

it follows that

$$\frac{\partial S_0}{\partial t} = 0 \tag{0-54}$$

so that the Hamilton-Jacobi equation (0-43) takes the form

$$H\left(q, p = \frac{\partial S_0}{\partial q}\right) = h. \tag{0-55}$$

If a complete integral of this equation containing $n$ constants of integration (of which one, say, $\alpha_1$, is taken to be the constant $h$)

$$S_0 = S_0(q, h, \alpha_2, \ldots, \alpha_n) \tag{0-56}$$

is such that

$$\frac{\partial\left(\dfrac{\partial S_0}{\partial q_2},\ldots,\dfrac{\partial S_0}{\partial q_n}\right)}{\partial(\alpha_2,\ldots,\alpha_n)} \neq 0, \tag{0-57}$$

then it can be proved that

$$\frac{\partial\left(\dfrac{\partial S_0}{\partial q_1},\dfrac{\partial S_0}{\partial q_2},\ldots,\dfrac{\partial S_0}{\partial q_n}\right)}{\partial(h,\alpha_2,\ldots,\alpha_n)} \neq 0 \tag{0-58}$$

and one can define a transformation from the $q$, $p$ to $\beta$, $\alpha$ by the $2n$ equations

$$p_k = \frac{\partial S_0}{\partial q_k}, \qquad \beta_1 = \frac{\partial S_0}{\partial h}, \qquad \beta_j = \frac{\partial S_0}{\partial \alpha_j},$$

$$k = 1,\ldots,n, \qquad j = 2,\ldots,n. \tag{0-59}$$

This transforms $H(q,p)$ into

$$\bar{H}(\beta,\alpha) = h. \tag{0-60}$$

The Hamilton-Jacobi equation (0-55) can be solved completely if $H$ is of the form

$$H = \sum_{k=1}^{n} H_k(q_k,p_k). \tag{0-61}$$

In this case, the Hamilton-Jacobi equation can be solved by the method of separation of variables by letting

$$S = \sum S_k(q_k,a_k). \tag{0-62}$$

(0-55) then separates into $n$ equations

$$H_k\!\left(q,\frac{dS_k}{dq_k}\right) = \alpha_k,$$

with

$$\sum_k \alpha_k = h.$$

For periodic systems, one defines the action variables

$$J_k = \oint p_k\,dq_k$$

$$= \oint \frac{dS_k}{dq_k}\,dq_k, \tag{0-63}$$

where $\oint$ indicates integration over a closed curve in the $q_k$, $p_k$ space. According to (0-36), $J_k$ are constant during the motion of the system. We regard the $J_1,\ldots,J_n$ as a new set of momenta, and $w_1,\ldots,w_n$ their conjugate coordinates, and transform from $(q_k, p_k)$ to $(w_k, J_k)$ by a function $S^*(q,J)$

$$p_k = \frac{\partial S^*}{\partial q_k}, \qquad w_k = \frac{\partial S^*}{\partial J_k}, \qquad k = 1,\ldots,n \tag{0-64}$$

$$H(q,p) = \bar{H}(w,J), \tag{0-65}$$

$$\dot{w}_k = \frac{\partial \bar{H}}{\partial J_k}, \qquad \dot{J}_k = -\frac{\partial \bar{H}}{\partial w_k}. \tag{0-66}$$

But the $J_k$'s are constant in time so that (0-65) becomes

$$\bar{H} = \bar{H}(J,\ldots,J_n) \tag{0-67}$$

and (0-66) give

$$J_k = \text{const.},$$

$$w_k = v_k t + \delta_k \quad \text{where} \quad v_k = \frac{\partial \bar{H}}{\partial J_k}. \tag{0-68}$$

The $w_k$'s being linear functions of time are called the angle variables.

The system $H$ is periodic if the $v$'s are commensurable. In general the system is said to be multiply periodic (or, conditionally periodic, according to a terminology due to O. Stande, 1887). If some of the $v$'s are equal, the system is said to be partially degenerate; if all the $v$'s are the same, the system is said to be completely degenerate. (K. Schwarzschild, 1916.)

The change in $w_k$ when $q_j$ has undergone one period is

$$\begin{aligned}
\Delta_j w_k &= \oint \frac{\partial w_k}{\partial q_j} dq_j = \oint \frac{\partial^2 S^*}{\partial q_j \partial J_k} dq_j \qquad \text{by (0-64)} \\
&= \frac{\partial}{\partial J_k} \oint \frac{\partial S^*}{\partial q_j} dq_j \\
&= \frac{\partial}{\partial J_k} J_j \qquad \text{by (0-64, 63).}
\end{aligned} \tag{0-69}$$

This means that $w_k$ changes by 1 when $q_k$ alone makes one period and any other $q_j$ may have returned to its initial value without having completed its period, for if it had, $w_j$ would also have changed by 1.

Thus the $q_k$'s are periodic in their respective $w_k$'s with the period 1. One may express $q_k$ in a generalized Fourier series

$$q_k = \sum_{\{\tau\}} c_{\{\tau\}}^{(k)} e^{2\pi i(\tau_1 w_1 + \cdots + \tau_n w_n)}$$

$$= \sum_{\{\tau\}} c_{\{\tau\}}^{(k)} e^{2\pi i\{(\tau_1 v_1 + \cdots + \tau_n v_n)t + (\tau_1 \delta_1 + \cdots + \tau_n \delta_n)\}}, \tag{0-70}$$

where $c_{\{\tau\}}^{(k)}$ are functions of $\{\tau_1, \tau_2, \ldots \tau_n\}$, the $\tau$'s being integers.

The condition for the Hamilton-Jacobi equation (0-55) to be separable is the following $\frac{1}{2}n(n-1)$ equations (Levi Civita, 1904)

$$\begin{vmatrix} 0 & \dfrac{\partial H}{\partial q_j} & \dfrac{\partial H}{\partial p_j} \\[2ex] \dfrac{\partial H}{\partial q_k} & \dfrac{\partial^2 H}{\partial q_j \partial q_k} & \dfrac{\partial^2 H}{\partial p_j \partial q_k} \\[2ex] \dfrac{\partial H}{\partial p_k} & \dfrac{\partial^2 H}{\partial q_j \partial p_k} & \dfrac{\partial^2 H}{\partial p_j \partial p_k} \end{vmatrix} = 0, \quad j \neq k, \quad j,k = 1,2,\ldots,n. \tag{0-71}$$

A dynamical system can still be completely solvable according to a criterion, in the following form. If the canonical equations (0-21) have $n$ functionally independent first integrals

$$f_j(q,p) = c_j, \qquad c_j = \text{const.} \tag{0-72}$$

and if the Poisson bracket expression of any pair $f_j$, $f_k$ vanishes identically

$$(f_j, f_k) = 0, \qquad j, k = 1, 2, \ldots, n, \tag{0-73}$$

then one defines as action variables $J_k$

$$J_k = f_k(q,p) = \text{const.} \tag{0-74}$$

and defines a canonical transformation from the $q$, $p$ to a new set $w$, $J$ by a function $S(q,J)$[c]

$$p_k = \frac{\partial S}{\partial q_k}, \tag{0-75a}$$

$$w_k = \frac{\partial S}{\partial J_k}. \tag{0-75b}$$

In the new $w$, $J$ coordinates the Hamiltonian is $\bar{H}(w,J)$, and the canonical equations are (0-66). As the $J_k$'s are constants, it follows that $\bar{H} = \bar{H}(J)$ and

---

[c] The relations $(J_j, J_k) = 0$ from (0-73) satisfy the condition for the transformation $(q,p) \to (w,J)$ to be canonical. See (0-41).

$$J_k = \text{const.},$$

$$w_k = v_k t + \delta_k, \qquad v_k = \frac{\partial \bar{H}}{\partial J_k} = \text{const.} \tag{0-76}$$

and $w_k$'s are angle variables (see (0-68)).

To obtain the $S$ of (0-75), one has

$$\begin{aligned} dS &= \sum \frac{\partial S}{\partial q_k} dq_k \\ &= \sum p_k(q, J) \, dq_k \qquad \text{from (0-75a)} \end{aligned}$$

and

$$S(q, J) = \sum_k \int_0^{q_k} p_k(q, J) \, dq_k. \tag{0-77}$$

From (0-75b), one solves for $q_k$ in terms of $w$, $J$

$$q_k = q_k(w, J), \qquad k = 1, \ldots, n. \tag{0-78}$$

Substituting this into (0-75a), one gets

$$p_k = p_k(q, J) = p_k(w, J), \qquad k = 1, \ldots, n. \tag{0-79}$$

These $2n$ equations giving $q$, $p$ in terms of $J_k$ and $w_k = v_k t + \delta_k$ are the solution of the problem.

The above theory applies to systems which are multiply periodic. We shall choose as an illustrating example the problem of an electron, charge $-e$, in the field of a nucleus of charge $Ze$. It turns out that this problem can be solved both by the method of separation of variables (0-61), and the method of first integrals (0-72, 73, 74). In fact, of course, the two methods in this case are the same.

In spherical polar coordinates $(r, \vartheta, \psi)$,

$$L = \frac{m}{2}(\dot{r}^2 + r^2 \dot{\vartheta}^2 + r^2 \sin^2 \vartheta \dot{\psi}^2) + \frac{Ze^2}{r},$$

$$p_r = m\dot{r}, \qquad p_\theta = mr^2 \dot{\vartheta}, \qquad p_\psi = mr^2 \sin^2 \vartheta \dot{\psi}.$$

The Hamilton-Jacobi equation is

$$\frac{1}{2m}\left[ \left(\frac{\partial S_0}{\partial r}\right)^2 + \frac{1}{r^2}\left(\frac{\partial S_0}{\partial \vartheta}\right)^2 + \frac{1}{r^2 \sin^2 \vartheta}\left(\frac{\partial S_0}{\partial \psi}\right)^2 \right] - \frac{Ze^2}{r} = E, \tag{0-80}$$

where $E$ is the total energy ($H = \bar{H} = E$). As $\psi$ is a cyclic coordinate,

$$p_\psi = \frac{\partial S_0}{\partial \psi} = \text{const.} \tag{0-81}$$

and (0-80) is separated, by setting

$$S_0(r, \vartheta) = S_r(r) + S_\vartheta(\vartheta)$$

into

$$\left(\frac{dS_\vartheta}{d\vartheta}\right)^2 + \frac{1}{\sin^2 \vartheta} p_\psi^2 = M^2,$$

$$\left(\frac{dS_r}{dr}\right)^2 + \frac{M^2}{r^2} = 2m\left(E + \frac{Ze^2}{r}\right). \tag{0-82}$$

The "separation constant" $M^2$ can be identified as follows. The Hamilton-Jacobi equation in plane-polar coordinate $r$, $\varphi$ (in the plane of motion) is

$$\frac{1}{2m}\left[\left(\frac{\partial S_0}{\partial r}\right)^2 + \frac{1}{r^2}\left(\frac{\partial S_0}{\partial \varphi}\right)^2\right] - \frac{Ze^2}{r} = E.$$

Since $\varphi$ is a cyclic coordinate,

$$p_\varphi = \frac{\partial S_0}{\partial \varphi} = \text{constant}$$

and comparison with (0-82) shows that $M$ in (0-82) is the angular momentum $p_\varphi$ of the electron,

$$p_\varphi = M. \tag{0-83}$$

From the expressions of the kinetic energy in $r$, $\vartheta$, $\psi$ and in $r$, $\varphi$, one obtains

$$2T = \dot{r}p_r + \dot{\vartheta}p_\vartheta + \dot{\psi}p_\psi,$$

$$2T = \dot{r}p_r + \dot{\varphi}p_\varphi.$$

Hence

$$p_\varphi \, d\varphi = p_\vartheta \, d\vartheta + p_\psi \, d\psi. \tag{0-84}$$

One defines the action variables

$$J_r = \oint p_r \, dr, \qquad J_\vartheta = \oint p_\vartheta \, d\vartheta, \qquad J_\psi = \oint p_\psi \, d\psi,$$

$$J_\varphi = \oint p_\varphi \, d\varphi = J_\vartheta + J_\psi, \tag{0-85}$$

where the integration is over a period of the motion, and has

$$M^2 = p_\varphi^2 = \frac{1}{4\pi^2}(J_\vartheta + J_\psi)^2. \tag{0-86}$$

From (0-82), one has

$$J_r = \oint \left[ 2mE + \frac{2mZe^2}{r} - \left( \frac{J_\vartheta + J_\psi}{2\pi r} \right)^2 \right]^{1/2} dr,$$

where the integration is taken over $r$ from the smaller root $r_{min}$ to the larger root $r_{max}$ of the radicand and back to $r_{min}$. The result is

$$J_r = -(J_\vartheta + J_\psi) + \pi Z e^2 \sqrt{\frac{2m}{-E}},$$

or

$$E = -\frac{2\pi^2 m Z^2 e^4}{(J_r + J_\vartheta + J_\psi)^2}. \tag{0-87}$$

This is the new Hamiltonian

$$E = \bar{H}(J_r, J_\vartheta, J_\psi)$$

in which the angle variables do not appear. From the canonical equations one obtains (0-76)

$$v_r = v_\vartheta = v_\psi, \tag{0-88}$$

i.e., the system is completely degenerate. The degeneracy arises from (i) the central system of $V(r)$, and in addition, (ii) from the Coulomb form $V \propto 1/r$.

One obtains the same results above by using for the three first-integrals (0-72) the integrals (0-83), (0-81)

$$p_\varphi = \text{const.}, \qquad p_\psi = \text{const.} \tag{0-89}$$

and the energy integral (E-80). The proof of (0-74) and the rest may be left as a simple exercise.

We shall end this resumé of classical dynamics by the following remarks bearing on the parts played by classical dynamics in the quantum theory and quantum mechanics.

(1) In classical dynamics, the *action variables* $J_k$ in (0-63) and (0-85) can take on any value. The revolutionary idea in Bohr's Theory (1913) of the hydrogen atom consists in postulating that the $J$'s can take on only values which are integral multiples of Planck's constant $h$

$$J_\xi = n_\xi h, \qquad \xi = r, \vartheta, \psi. \tag{0-90}$$

[See Chap. 1, Secs. 4 and 5 below].

(2) In classical dynamics, the $J$'s, for such systems as (0-61), are not only invariants during the motion of a system (Poincaré's relative integral invariants), but invariant under very slow changes of some external field. They are

adiabatic invariants (Einstein, 1911; Ehrenfest, 1913, 1916; Burgers 1917.) This concept of adiabatic invariants is of interest in connection with the frequency condition in Bohr's theory (Chap. 1, Sec. 4).

(3) Canonical transformations in classical dynamics are somewhat similar to unitary transformations in quantum mechanics. For example, when Poisson brackets are replaced by the quantum Poisson brackets in quantum mechanics

$$(u, v)_{cl.} \to [u, v]_{qu.} \equiv \frac{1}{i\hbar}(uv - vu), \tag{0-91}$$

equations (0-34) become

$$\dot{q} = [q, H], \qquad \dot{p} = [p, H] \tag{0-92}$$

in quantum mechanics. [See Eqs. (II-85) in Chap. 2.]

Also, the invariance of $(u, v)_{cl.}$ under canonical transformation (0-37) has its analogue in the invariance of $[u, v]_{qu.}$ under unitary transformation.

(4) If the condition for time reversal (0-42) is carried over to quantum mechanics, i.e., the quantum Poisson brackets $[u, v]$ changes sign upon time reversal, it is seen from (0-91) that formally $i$ changes sign upon time-reversal. This is formally the same as the Wigner time-reversal which consists of the combined operations of $t \to -t$ and $i \to -i$ and under which the Schrödinger equation is invariant

$$\left(i\hbar\frac{\partial}{\partial t} - H\right)\Psi = 0 \to \left((-i)\hbar\frac{\partial}{\partial(-t)} - H\right)\Psi^* = 0. \tag{0-93}$$

(5) The Hamilton-Jacobi equation (0-55)

$$H\left(q, p = \frac{\partial S_0}{\partial q}\right) = E$$

plays a very important part in Schrödinger's initial approach to the problem of obtaining a "wave equation" for the de Broglie wave. [See Chap. 3, Sec. 2 below.]

## 2. Electromagnetic theory

### (1) Electromagnetic fields

Historically, the basic concepts in magnetostatics are the strength $p$ of magnetic poles (approximated by long thin magnetic needles) and Coulomb's law

$$F = \frac{\mu_0}{4\pi}\frac{p_1 p_2}{r^2}, \tag{0-94}$$

and in electrostatics are the electric charge $e$ and Coulomb's law (1785)

$$F = \frac{1}{4\pi\varepsilon_0}\frac{e_1 e_2}{r^2}. \tag{0-95}$$

The constants $\mu_0$ and $\varepsilon_0$ for vacuum are obviously tied up with the physical concepts of $p$ and $e$. They are independent as long as electrostatic and magnetostatic phenomena are independent.

The link between electricity and magnetism is furnished by the discovery of Oersted (1820), Biot and Savart (1820) and Ampére (1825). An infinitely long conductor carrying an electric current $I$ produces a magnetic field $H$ at a distance $R$ from the conductor, given by

$$\gamma H = \frac{I}{4\pi R}, \tag{0-96}$$

where $\gamma$ is a constant introduced to provide for the connection between electric and magnetic phenomena. The $H$ defined by Ampére's law (0-96) is related to $B$, the magnetic induction, which is by definition the force acting on a unit pole, by[d]

$$B = \mu_0 H. \tag{0-97}$$

The three constants $\mu_0$, $\varepsilon_0$, $\gamma$ are no longer all independent; it can be shown on entirely dimensional considerations that,

$$\frac{\gamma}{\sqrt{\mu_0\varepsilon_0}} \text{ is of the dimension of velocity.}$$

It was the great triumph of Maxwell's theory of the electromagnetic field to show that

$$\frac{\gamma}{\sqrt{\mu_0\varepsilon_0}} = c, \text{ the velocity of light.} \tag{0-98}$$

It is the freedom provided by three constants joined by one single relation (0-98) that makes it possible to define different systems of units, namely, the Gauss system with $\gamma = c$ and $\mu_0\varepsilon_0 = 1$, and the m.k.s.a. system with $\gamma = 1$, $\mu_0\varepsilon_0 = 1/c^2$, $4\pi\varepsilon_0 = 10^7/c^2$, $\mu_0 = 4\pi \times 10^{-7}$.

Maxwell summarized the laws of electromagnetic field in the following

---

[d] One may express Ampére's law (0-96) in the form

$$B = \frac{\mu_0}{4\pi}\frac{I}{R}$$

which defines $B$, and then define $H$ by (0-97). This definition of $B$ is consistent with the definition of $B$ as the force on a unit pole according to (0-94).

equations (in m.k.s.a. units)

Faraday's law:
$$\text{curl } \mathbf{E} = -\frac{\partial \mathbf{B}}{\partial t}, \tag{0-99a}$$

Ampére's law:
$$\text{curl } \mathbf{H} = \mathbf{j} + \frac{\partial \mathbf{D}}{\partial t}, \quad j = \rho v, \tag{0-99b}$$

Coulomb's law:
$$\text{div } \mathbf{B} = 0, \tag{0-99c}$$

Coulomb's law:
$$\text{div } \mathbf{D} = \rho, \tag{0-99d}$$

$$\mathbf{B} = \mu \mathbf{H} \quad (\mu_0 = 4\pi \times 10^{-7} \text{ henry/m for vacuum}),$$

$$\mathbf{D} = \varepsilon \mathbf{E} \quad (\varepsilon_0 = \frac{1}{36\pi} \times 10^{-9} \text{ farad/m for vacuum}), \tag{0-100}$$

$$\frac{1}{\sqrt{\mu\varepsilon}} = v \quad (v = c = 3 \times 10^8 \text{ m/sec. for vacuum}).$$

Equation of continuity:[e]

$$\frac{\partial \mathbf{p}}{\partial t} + \text{div } \mathbf{j} = 0. \tag{0-101}$$

Lorentz force $\mathbf{f}$ on unit volume of charge and current distribution:

$$\mathbf{f} = \rho \mathbf{E} + [\mathbf{j} \times \mathbf{B}]. \tag{0-102}$$

On introducing the scalar potential $\phi(\mathbf{r}, t)$ and the vector potential $\mathbf{A}(\mathbf{r}, t)$ by

$$\mathbf{B} = \text{curl } \mathbf{A}, \qquad \mathbf{E} = -\text{grad } \phi - \frac{\partial \mathbf{A}}{\partial t}, \tag{0-103}$$

one obtains from (0-99), for constant $\mu$, $\varepsilon$,

$$\Box \phi = -\frac{1}{\varepsilon}\rho + \frac{\partial}{\partial t}\left(\text{div } \mathbf{A} + \mu\varepsilon\frac{\partial \phi}{\partial t}\right), \tag{0-104a}$$

$$\Box A = -\mu \mathbf{j} + \text{grad}\left(\text{div } \mathbf{A} + \mu\varepsilon\frac{\partial \phi}{\partial t}\right), \tag{0-105a}$$

where

$$\Box = \nabla^2 - \mu\varepsilon\frac{\partial^2}{\partial t^2}.$$

[e] (0-99a) and div curl $\equiv 0$ imply (0-99c). (0-99b), (0-101) and div curl $\equiv 0$ imply (0-99d). Thus of the 4 equations (a, b, c, d) only (a), (b) are independent, if the conservation of charge is regarded as a fundamental law.

If one chooses $A$ and $\phi$ such that they satisfy the so-called Lorentz condition (1892)

$$\operatorname{div} \mathbf{A} + \mu\varepsilon\frac{\partial\phi}{\partial t} = 0, \tag{0-106}$$

then

$$\Box\phi = -\frac{1}{\varepsilon}\rho \tag{0-104}$$

$$\Box\mathbf{A} = -\mu\mathbf{j} \tag{0-105}$$

which give $\phi$ and $\mathbf{A}$, and hence the field $\mathbf{E}$ and $\mathbf{B}$, arising from a charge and current distribution.

The equations (0-103) do not define a unique $\phi$ and $\mathbf{A}$ by a given (experimentally measured, in classical theory) $\mathbf{E}$ and $\mathbf{B}$. The following gauge transformation (H. Weyl, 1929) of $\phi$ and $\mathbf{A}$ by a scalar function $\Lambda(\mathbf{r}, t)$

$$\mathbf{A}' = \mathbf{A} + \operatorname{grad}\Lambda, \qquad \phi' = \phi - \frac{\partial\Lambda}{\partial t} \tag{0-107}$$

leaves $\mathbf{E}$ and $\mathbf{B}$ invariant. The requirement that the Lorentz condition (0-107) be invariant under (0-106) is met by the relation[f]

$$\Box\Lambda = 0. \tag{0-107a}$$

Some of the important consequences of the Maxwell theory are the following:

In general, a system of electric charges and currents generate electromagnetic waves in accordance with the differential equations (0-104, 105) and (0-103). In the simple case of a homogeneous medium ($\mu$ and $\varepsilon$ are constants) with $\rho = \mathbf{j} = 0$, one obtains from (0-99a, b) the wave equations

$$\left(\nabla^2 - \varepsilon\mu\frac{\partial^2}{\partial t^2}\right)\left\{\begin{matrix}\mathbf{E}\\ \mathbf{H}\end{matrix}\right\} = 0, \tag{0-108}$$

where $\dfrac{1}{\sqrt{\varepsilon\mu}} = v$ is the phase velocity. Introducing the wave vector $\mathbf{k}$, and

$$|k| = \frac{2\pi}{\lambda} = \frac{\omega}{v} = \omega\sqrt{\varepsilon\mu} \tag{0-109}$$

we have, for the solutions of (0-108), the plane waves

---

[f] The expression of the equations (0-99)–(0-107) in tensor, and therefore in Lorentz covariant, form will be summarized in Appendix H in Chap. 15 below.

$$E = E_1 e^{i(k \cdot r - \omega t)} + E_2 e^{i(k \cdot r + \omega t)},$$
$$H = H_1 e^{i(k \cdot r - \omega t)} + H_2 e^{i(k \cdot r + \omega t)}, \tag{0-110}$$

where $E_1, E_2, H_1, H_2$ are constants (vectors). From

$$\operatorname{div} E = \operatorname{div} H = 0,$$

we obtain

$$i(k \cdot E_1)e^{i(k \cdot r - \omega t)} + i(k \cdot E_2)e^{i(k \cdot r + \omega t)} = 0,$$
$$i(k \cdot H_1)e^{i(k \cdot r - \omega t)} + i(k \cdot H_2)e^{i(k \cdot r + \omega t)} = 0,$$

whose solutions are

$$(k \cdot E_1) = (k \cdot E_2) = (k \cdot H_1) = (k \cdot H_2) = 0, \tag{0-111}$$

i.e., $E$ and $H$ are both perpendicular to the direction of propagation $k$. From (0-110) and (0-99a, b), for $E_2 = H_2 = 0$, we obtain

$$[k \times H_1] = -\omega \varepsilon E_1, \qquad [k \times E_1] = \omega \mu H_1$$

showing that $E_1$ and $H_1$ are perpendicular to each other. The Poynting vector is then

$$P = E \times H.$$

In view of

$$\sqrt{\varepsilon}\, E = \sqrt{\mu}\, H,$$
$$P = v \frac{\varepsilon E^2 + \mu H^2}{2} = vu, \tag{0-112}$$

where $u$ is the energy density of the electromagnetic field.

The above simple example is intended only to show that electromagnetic waves and energy propagate in a continuous manner through space.

Another simple consequence of classical electromagnetic theory is that an accelerated electric charge $e$ radiates energy at the rate

$$Q = \iint P \cdot dS = \frac{e^2}{4\pi \varepsilon_0 c^3} (\ddot{x})^2. \tag{0-113}$$

This energy is spread continuously over the wave front of the electromagnetic wave.

*(2) Symmetry properties in space reflection and time reversal*

The Maxwell field equations (0-99) and (0-101, 102) are referred to a coordinate system OXYZ. Let us consider the mirroring operation $M$ (with respect to the OXY plane), such that

$$M_{XZ}x = x, \qquad M_{XZ}y = -y, \qquad M_{XZ}z = z. \qquad (0\text{-}114)$$

This $M$ is equivalent to the inversion operation $P$

$$P_x x = -x, \qquad P_y y = -y, \qquad P_z z = -z \qquad (0\text{-}115)$$

followed by a rotation about the OY axis through $\pi$, i.e.,

$$M_{XZ} = R_y(\pi)P. \qquad (0\text{-}116)$$

To study the effect of $M$ on equations expressing physical laws, it is sometimes convenient to study the effect of $P$, for a rotation in an isotropic space cannot affect the expressions of physical laws.

Under space inversion (or mirroring) we take, by convention, the electric charge to be a scalar, i.e.,

$$Pe = e. \qquad (0\text{-}117a)$$

Since force $\mathbf{f}$ is a polar vector and $\mathbf{f} = e\mathbf{E}$, the electric field $\mathbf{E}$ must be a polar vector

$$P\mathbf{f} = -\mathbf{f}, \qquad P\mathbf{E} = -\mathbf{E}. \qquad (0\text{-}117b)$$

The magnetic induction $\mathbf{B}$ can be seen to be an axial (or, pseudo) vector, i.e.,

$$P\mathbf{B} = \mathbf{B}. \qquad (0\text{-}117c)$$

These relations, together with

$$\begin{aligned}
P\mathbf{j} &= -\mathbf{j}, \qquad P\rho = \rho, \\
P\mathbf{A} &= -\mathbf{A}, \qquad P\phi = \phi,
\end{aligned} \qquad (0\text{-}117d)$$

show that the whole system of equations (0-99)–(0-107) are invariant under the inversion $P$, and hence also of any mirroring operation.

Consider the operation $\Theta$ of reversing the direction of time, i.e.,

$$\Theta t = -t. \qquad (0\text{-}118)$$

It is clear that

$$\begin{aligned}
\Theta\mathbf{j} &= -\mathbf{j}, \qquad \Theta\rho = \rho, \qquad \Theta\mathbf{f} = \mathbf{f}, \\
\Theta\mathbf{E} &= \mathbf{E}, \qquad \Theta\mathbf{D} = \mathbf{D}, \qquad \Theta\phi = \phi.
\end{aligned} \qquad (0\text{-}119a)$$

But from the Biot-Savart or Ampére law, reversing an electric current reverses $\mathbf{H}$ and $\mathbf{B}$, and

$$\Theta\mathbf{H} = -\mathbf{H}, \qquad \Theta\mathbf{B} = -\mathbf{B}, \qquad \Theta\mathbf{A} = -\mathbf{A}. \qquad (0\text{-}119b)$$

With these relations, it is seen that the whole system of equation (0-99)–(0-107) are invariant under time reversal.

In classical physics, at least in the older literature, such questions as the invariance of the electromagnetic field equations under spatial inversion and time reversal were not raised, presumably because physicists felt instinctively certain that basic physical laws should not depend on the choice of a right-handed or a lefthanded coordinate system, or on the sign of the parameter $t$. Indeed, all the experience with the electromagnetic phenomena has borne out this belief so strongly that a left-right symmetry in all natural laws has been taken for granted. Therefore it was a revolutionary proposal by T. D. Lee and C. N. Yang in 1956 that the law of weak interactions (that govern such processes as the beta decays and $\pi^+ \to \mu^+ + \nu_\mu$ and $\mu^- \to e^- + \tilde{\nu}_e + \nu_\mu$) may not have the symmetry (i.e., invariance under $P$, now called the parity operation) and it was in fact a surprise when this proposal was experimentally confirmed by C. S. Wu and others. It was then realized by physicists that this *parity conservation* is not an inviolable law of nature.

## 3. Statistics and probability

Classical thermodynamics deals with the properties of matter-in-bulk in thermodynamic equilibrium. The basic concepts are macroscopic concepts such as temperature, pressure, entropy, free energy, etc. The kinetic theory of gases also deals with the properties of matter-in-bulk, but starts with the molecular point of view and the laws of dynamics. In dealing with a large number, say $10^{23}$, of molecules, the concepts of average values and distribution functions are introduced. Thus in 1859, Maxwell obtained the distribution law of velocities of the molecules of a gas in a state of thermodynamic equilibrium. On certain probability assumptions, the distribution law is obtained that the probability of finding in a gas a molecule whose $x$-component of velocity lies between $u$ and $u + du$ is

$$\phi(u)\, du = \sqrt{\frac{m}{2\pi}}\, \beta\, e^{-\beta(mu^2/2)}\, du, \tag{0-120}$$

where $m$ is the mass of a molecule and $\beta$ is a constant.[g] This function satisfies the condition

$$\int_{-\infty}^{\infty} \phi(u)\, du = 1.$$

---

[g] This constant $\beta$ cannot be determined on dynamical and probability considerations alone. It is only when a further theoretical hypothesis (namely, the connection between energy and temperature) is made that $\beta$ is found to be $1/kT$.

Let us consider the nature of the probability distribution of (0-120). One starts with a dynamical system of a large number $N$, say $10^{23}$, of molecules in an enclosure with perfectly rebounding walls, and let us assume that any two molecules interact with each other according to known laws. If at a given instant $t_0$ the coordinates and momenta of all the $N$ molecules are specified, then *in principle* the motion of the whole system is governed by Newton's laws of motion, and *in principle* it is *possible* to calculate the velocities of the individual molecules, although this is an impracticable job even with the high speed electronic computers now available. It is for overcoming this impracticability that probability ideas are introduced. The emphasis here is that the velocities of a gas are *in principle* knowable; probability concepts are introduced not because the velocities are absolutely intrinsically not knowable.

We may say that in the whole of classical physics, the probability concept has been introduced in this sense; the reason is of course that the basic laws underlying phenomena in classical physics are dynamics and electromagnetic theory, which are themselves deterministic and causal.

The first instance of something basically different is the law of radioactive decay (Rutherford, 1900)

$$dN = -\lambda N \, dt, \quad \text{or} \quad N = N_0 e^{-\lambda t}. \tag{0-121}$$

This law was later found to hold for all the unstable "elementary" particles produced by the high energy accelerators.

This law is interpreted as a basic impossibility of knowing when a given nucleus will decay; one can only know the probability that it may decay in an interval of time $\Delta t$.

In other words, the "probability" does not arise from our having to deal with large numbers, but is of an intrinsic origin.

It is this intrinsic probability has plays a basic part in quantum mechanics, in contrast to classical physics.


### Appendix 0

### Hamilton's principle and canonical equations

In obtaining the canonical equations (0-21) from the Hamilton principle (0-19)

$$\delta \int_{t_0}^{t} \{\sum p_k \dot{q}_k - H(q,p,t)\} \, dt = 0, \tag{0A-1}$$

one usually finds in the literature the argument that the variations $\delta q_k$ and $\delta p_k$ are arbitrary except at the two end points at which $\delta q_k = \delta p_k = 0$. In the form (0-9),

$$\delta \int_{t_0}^{t} L(q, \dot{q}, t)\, dt = 0, \qquad (0A\text{-}2)$$

the $\delta q_k$ are arbitrary but the $\delta \dot{q}_k$ are not, as seen from the following considerations. One passes from a point $q, \dot{q}$ on a trajectory to a point $q + \delta q, \dot{q} + \delta \dot{q}$ on a varied trajectory; the time $t$ is an independent variable not subject to variations, i.e., the condition for all varied trajectories to be traversed in the same time $t - t_0$ precludes the $\delta \dot{q}$'s from being independent. Since the $p$'s are connected with the $\dot{q}$'s (0-11), the $\delta p$'s cannot be arbitrary. Thus treating $\delta p$'s as arbitrary in (0A-1) is a point that merits some detailed discussions.

We shall prove the following theorem: The variational problem (0A-2) in which the $\delta q$'s are arbitrary (except at the two end points), and the variational problem

$$\delta \int_{t_0}^{t} \left\{ L(q, u, t) + \sum \frac{\partial L}{\partial u_k}(\dot{q}_k - u_k) \right\} dt = 0, \qquad (0A\text{-}3)$$

where $\dot{q}_k = \dfrac{dq_k}{dt}$, and $\delta q_k, \delta u_k$ are arbitrary except $\delta q_k = \delta u_k = 0$ at the two end points, are equivalent if

$$\sum_{j,k} \frac{\partial^2 L}{\partial u_j \partial u_k} \xi_j \xi_k > 0 \qquad (0A\text{-}4)$$

for arbitrary $\xi_j, \xi_k$.

To prove this, take, for simplicity, the case $k = 1$ and 2. The Euler equations of (0A-3) are

$$\frac{\partial^2 L}{\partial u_1^2}(\dot{q}_1 - u_1) + \frac{\partial^2 L}{\partial u_1 \partial u_2}(\dot{q}_2 - u_2) = 0,$$

$$\frac{\partial^2 L}{\partial u_2 \partial u_1}(\dot{q}_1 - u_1) + \frac{\partial^2 L}{\partial u_2^2}(\dot{q}_2 - u_2) = 0, \qquad (0A\text{-}5)$$

$$\frac{d}{dt}\frac{\partial L}{\partial u_1} - \frac{\partial L}{\partial q_1} + \frac{\partial^2 L}{\partial u_1 \partial q_1}(\dot{q}_1 - u_1) + \frac{\partial^2 L}{\partial u_2 \partial q_2}(\dot{q}_2 - u_2) = 0,$$

$$\frac{d}{dt}\frac{\partial L}{\partial u_2} - \frac{\partial L}{\partial q_2} + \frac{\partial^2 L}{\partial u_1 \partial q_2}(\dot{q}_1 - u_1) + \frac{\partial^2 L}{\partial u_2 \partial q_2}(\dot{q}_2 - u_2) = 0. \qquad (0A\text{-}6)$$

The condition for the positive-definiteness of

$$\frac{\partial^2 L}{\partial u_1^2}\xi^2 + 2\frac{\partial^2 L}{\partial u_1 \partial u_2}\xi\eta + \frac{\partial^2 L}{\partial u_2^2}\eta^2 > 0 \tag{0A-7}$$

is

$$\frac{\partial^2 L}{\partial u_1^2} > 0, \qquad \begin{vmatrix} \dfrac{\partial^2 L}{\partial u_1^2} & \dfrac{\partial^2 L}{\partial u_1 \partial u_2} \\[2ex] \dfrac{\partial^2 L}{\partial u_2 \partial u_1} & \dfrac{\partial^2 L}{\partial u_2^2} \end{vmatrix} > 0. \tag{0A-8}$$

The condition (0A-8) leads, from (0A-5), to

$$\dot{q}_1 - u_1 = 0, \qquad \dot{q}_2 - u_2 = 0, \tag{0A-9}$$

and, with these in (0A-6), to

$$\frac{d}{dt}\frac{\partial L}{\partial \dot{q}_k} - \frac{\partial L}{\partial q_k} = 0, \qquad k = 1, 2, \tag{0A-10}$$

which are the Euler equations of (0A-2). This proof can be generalized to $q = (q_1, \ldots, q_n)$.[h]

On defining in (0A-2)

$$p_k \equiv \frac{\partial L}{\partial u_k}, \tag{0A-13}$$

and

$$H(q, p, t) \equiv -L(q, u, t) + \sum p_k u_k, \tag{0A-14}$$

then (0A-3) can be put in the form

$$\delta \int_{t_0}^{t} \left\{ \sum p_k \dot{q}_k - H(q, p, t) \right\} dt = 0. \tag{0A-15}$$

---

[h] It can be shown that (0A-8) is the necessary and sufficient condition for the integral (0A-2) to be a (relative) minimum along the path given by the Lagrange equations. For the general case with $n$ coordinates $q_1, \ldots, q_n$, the condition (0A-8) is that all the principal minors of the determinant

$$\begin{vmatrix} \dfrac{\partial^2 L}{\partial \dot{q}_1 \partial \dot{q}_1} & \dfrac{\partial^2 L}{\partial \dot{q}_1 \partial \dot{q}_2} & \dfrac{\partial^2 L}{\partial \dot{q}_1 \partial \dot{q}_3} & \cdots \\[2ex] \dfrac{\partial^2 L}{\partial \dot{q}_2 \partial \dot{q}_1} & \dfrac{\partial^2 L}{\partial \dot{q}_2 \partial \dot{q}_2} & \cdots & \cdots \\[2ex] \cdots & \cdots & \cdots & \cdots \end{vmatrix} \tag{0A-11}$$

be positive. This condition comes from the requirement

$$\sum_{j,k}\frac{\partial^2 L}{\partial \dot{q}_j \partial \dot{q}_k}\xi_j \xi_k > 0. \tag{0A-12}$$

This positive definiteness condition is guaranteed by the positive-definiteness of the kinetic energy of the system.

Since $\delta u_k$ in (0A-3) are arbitrary, and since the $p_k$'s are given by (0A-13) in terms of the $u_k$'s, the $\delta p_k$'s are arbitrary in (0A-15). The Euler equations of (0A-5) are

$$\dot{p}_k + \frac{\partial H}{\partial q_k} = 0, \qquad \dot{q}_k - \frac{\partial H}{\partial p_k} = 0$$

which are (0-21).

# Chapter 1

# Pre-Quantum Mechanical Period, 1900–1925

Toward the end of the period of the Quantum Theory (1900–1925), two independent, revolutionary ideas were born, one from Louis de Broglie in 1923–4 and one from Werner Heisenberg in 1925. These ideas sparked some explosive developments which by 1928 had already led to a complete system, both in its mathematical formalism and its physical and philosophical interpretation. The suggestion of "matter wave" by de Broglie led to the development of "wave mechanics" by Erwin Schrödinger in 1926 (January–June), the probability interpretation of Max Born in 1926 (June–July). The initial idea of Heisenberg (the epoch-making paper completed in July, 1925) led Born and Pascal Jordan to formulate matrix mechanics (September, 1925), culminating in the complete work by Born, Heisenberg and Jordan in November, 1925. The first paper of Heisenberg also stimulated Paul A. M. Dirac (September, 1925) to embark on an independent development of a more formal and general theory of non-commutative algebra of $q$-numbers, not explicitly connected with matrices. Then followed the important development of the transformation theory of quantum mechanics by Dirac, Jordan and others (1926). The mathematical equivalence of matrix mechanics and wave mechanics was shown by Schrödinger (March, 1926). An important missing link in wave mechanics, namely, the physical meaning of the wave function, was supplied by Born (June, July, 1926); the probability interpretation turned out to be one of the new fundamental concepts of quantum mechanics. Then came another epoch-making theory of the uncertainty principle by Heisenberg (March, 1927). The concept of complementarity and the philosophical interpretation of quantum mechanics were then completed (1927) into what is known as the Copenhagen School, represented by Bohr, Heisenberg and others. In 1928, Dirac gave the relativistic theory of the electron which brought together quantum mechanics (at least for the one-electron system) and the special theory of relativity, accounted for the electron spin, and opened up the field of quantum electrodynamics.

The half century following this period of rapid developments unparalleled in the history of science saw the applications of quantum mechanics to atomic, molecular, solid state, nuclear and elementary particle physics, and the problem of quantized fields. Generations of physicists have since grown up who have learned and used quantum mechanics more or less in the same way as they have applied mathematics. Lest we believe quantum mechanics has come about "easily" and "naturally", we shall begin our presentation of quantum mechanics with a brief summary of the important developments of physics in the period of the Quantum Theory, 1900–1925, before the dawn of the new epoch.

## References

For full historical accounts of the developments of the quantum theory and detailed references to the original literatures, see

E. T. Whittaker, *History of the Theories of Aether and Electricity*, Vol. II, Modern Theories, 1900–1926 (Thomas Nelson & Sons, London, 1953).
M. Jammer, *The Conceptual Development of Quantum Mechanics* (McGraw-Hill, New York, 1966).
A. Pais, *'Subtle Is the Lord …': The Science and the Life of Albert Einstein* (Clarendon Press, Oxford, 1982).

## 1. Planck's quantum theory (1900)

In the last two decades of the nineteenth century, theoretical and experimental physicists were interested in obtaining the law of the spectral distribution of the radiation of a blackbody as a function of temperature. If we denote by $E = \psi_\lambda \, d\lambda (\psi_\nu \, d\nu)$ the energy density in the range $\lambda$ and $\lambda + d\lambda$, ($\nu$ and $\nu + d\nu$) of radiation in equilibrium with a blackbody at a temperature $T$, the problem was to obtain the function

$$\psi_\lambda(\lambda, T) \qquad \text{or} \qquad \psi_\nu(\nu, T). \tag{I-1}$$

(i) The first step in this process is the empirical law for the total energy density

$$\psi = \int \psi_\lambda \, d\lambda \qquad \text{or} \qquad \int \psi_\nu \, d\nu \tag{I-1a}$$

obtained by Stefan (1879),

$$\psi = \frac{4\sigma}{c} T^4; \qquad \sigma = 5.67 \times 10^{-5} \text{ erg/cm. sec. degree}^4. \tag{I-2}$$

The temperature dependence in this law was derived (1884) on thermodynamical consideration by Boltzmann (1844–1906). This law was verified by the accurate measurements of Paschen, Lummer and Pringsheim in 1897.

(ii) The second step is the discovery, experimental and theoretical, of the displacement law by W. Wien (1864–1928) in 1894, which states that the product of the temperature $T$ and the wavelength $\lambda_m$ at which $\psi_\lambda(\lambda, T)$ is maximum is a constant for all temperatures, i.e.,

$$\lambda_m T = 0.2884 \quad \text{(for various } T\text{).} \tag{I-3}$$

This law narrows the form of $\psi_\lambda$ (or $\psi_\nu$) to

$$\psi_\lambda = \frac{1}{\lambda^5} f(\lambda T), \qquad \text{or} \qquad \psi_\nu = \nu^3 g\left(\frac{\nu}{T}\right) \tag{I-4}$$

in which the product $\lambda T$ (or $\nu/T$) appears as the argument of the as yet undetermined function $f$ (or $g$).

(iii) Wien (1896) made an assumption (unjustifiable) that the wavelengths $\lambda$ of radiation emitted by molecules are a function of the velocity $v$ of the molecules alone, or conversely, $v^2$ is a function of $\lambda$. On using Boltzmann's distribution $\exp(-mv^2/2kT)$, and the general form (I-4), Wien obtained

$$\psi_\lambda = \frac{\gamma}{\lambda^5} \exp\left(-\frac{\delta}{\lambda T}\right), \qquad \text{or} \qquad \psi_\nu = \alpha \nu^3 \exp\left(-\frac{\beta \nu}{T}\right) \tag{I-5}$$

which were found by Paschen and Wanner to be in fair agreement with experimental measurements at the short wavelength end of the spectrum (or, for small values of $\lambda T$), but to be in violent disagreement for the long wavelength end.

(iv) The "last straw", so to speak, was the publication by Lord Rayleigh (J. W. Strutt, 1842–1919) in June, 1900, and by J. H. Jeans (1877–1946) in 1905, of the Rayleigh-Jeans law

$$\psi_\lambda \, d\lambda = \frac{8\pi}{\lambda^4} d\lambda(kT), \qquad \text{or} \qquad \psi_\nu \, d\nu = \frac{8\pi\nu^2}{c^3} d\nu(kT). \tag{I-6}$$

Note that each formula consists of two factors: $(8\pi/\lambda^4)\,d\lambda((8\pi\nu^2/c^3)\,d\nu))$ is the number of modes of oscillation per unit volume of the radiation field in the wavelength range between $\lambda$ and $\lambda + d\lambda$ (the frequency range between $\nu$ and $\nu + d\nu$), and $kT$ is the average electric and magnetic energy per oscillational degree of freedom on the basis of the principle of equipartition in classical statistical mechanics. Thus, in classical physics, the law (I-6) is "inevitable".

Equation (I-6) is in fact in good agreement with the experimental result for the long wavelength end of the spectrum (or, large values of $\lambda T$). But it fails completely for short $\lambda$ (or, small values of $\lambda T$), and it leads to a divergent result for the total energy.

(v) Max Planck (1858–1947), a master of thermodynamics, studied the problem from the point of view of entropy. If $U$ is the average energy of an oscillator, and $U = U(T)$, and $S$ is the entropy of an oscillator, the second law of thermodynamics gives for constant volume,

$$\left(\frac{\partial S}{\partial U}\right)_V = \frac{1}{T}. \tag{I-7}$$

If Wien's law (I-5) is assumed to be correct, then comparison of (I-5) and (I-6) leads to

$$v^3 \exp\left(-\frac{\beta v}{T}\right) \propto U v^2,$$

or

$$U = \gamma v \exp\left(-\frac{\beta v}{T}\right), \qquad \gamma, \beta = \text{constants}. \tag{I-8}$$

From (I-7), one gets

$$\frac{\partial^2 S}{\partial U^2} = \frac{\text{const.}}{U}.$$

If Rayleigh-Jeans law (I-6) is assumed to be correct, then similar consideration would have led to

$$\frac{\partial^2 S}{\partial U^2} = \frac{\text{const.}}{U^2}.$$

As Wien's and Rayleigh-Jeans laws are correct only at the small $\lambda T$ and large $\lambda T$ limits respectively, Planck made the "guess" that $U$ satisfied

$$\frac{\partial^2 S}{\partial U^2} = \frac{a}{U(b + U)}, \qquad a, b = \text{constants}. \tag{I-9}$$

Using (I-7), Planck obtained

$$\frac{1}{T} = \frac{\partial S}{\partial U} = c \ln\left(\frac{U + b}{U}\right), \qquad c = -\frac{a}{b}$$

and

$$U = \frac{b}{\exp\left(\dfrac{1}{cT}\right) - 1}.$$

But $U$ must have the form (I-8). Hence

$$U = \text{const.} \frac{v}{\exp\left(q\dfrac{v}{T}\right) - 1}, \qquad q = \text{const.}$$

which, together with (I-6), leads to the formula

$$\psi_v \, dv = \frac{8\pi v^2 \, dv}{c^3} \frac{pv}{\exp\left(q\dfrac{v}{T}\right) - 1}.$$

The "empirical" formula was presented by Planck on October 19, 1900, to the Physical Society of Berlin meeting. It was immediately found to be in very good agreement with the experimental results of H. Rubens (1865–1922) and with later measurements of many other physicists.

Planck then spent the next eight weeks or so on intense, strenuous work to justify the "guess" (I-9). He put aside the thermodynamical approach and made recourse to Boltzmann's statistical interpretation of entropy which he had previously criticized. If $U$ is the average energy of an oscillator and $N$ is the number of oscillators, it was *assumed* that $UN$ is distributed into $N$ oscillators not in a continuously divisible manner as in classical physics, but in "discrete elements" of amount $\varepsilon$, i.e., the oscillators emit or absorb energy in multiples of $\varepsilon$. Let $P = UN/\varepsilon$, an integer. The number of ways of distributing $P$ indistinquisable energy elements $\varepsilon$ into $N$ indistinquisable oscillators is[a]

$$W = \frac{(N + P - 1)!}{(N - 1)!P!}. \tag{I-10}$$

With the aid of the Stirling approximation, for $N$, $P$ not too small (say $\gg 1$), we have

$$\ln(N!) = N \ln N - N$$

and the entropy $S_N$ for the system of $N$ oscillators is

$$S_N = k \ln W$$

$$\simeq kN\left[\left(1 + \frac{U}{\varepsilon}\right)\ln\left(1 + \frac{U}{\varepsilon}\right) - \frac{U}{\varepsilon}\ln\frac{U}{\varepsilon}\right]$$

and by using (I-7), for the entropy of one oscillator $S = S_N/N$, Planck obtained

$$U = \frac{\varepsilon}{\exp(\varepsilon/kT) - 1}.$$

The general form (I-4) requires then $\varepsilon \propto v$, or

---

[a] The following simple way of obtaining this expression is due to J. Ehrenfest and H. Kamerlingh Onnes (1914). Let $P$ balls be distributed into $N$ cells formed by $N - 1$ bars along a line. The formula then follows.

$$\varepsilon = h\nu,$$

and

$$\psi_\nu = \frac{8\pi\nu^2}{c^3}\frac{h\nu}{e^x - 1}, \qquad x = \frac{h\nu}{kT}$$

$$= \frac{8\pi\nu^2}{c^3}\left(\frac{x}{e^x - 1}\right)kT \tag{I-11a}$$

which is of the form of the "empirical" formula (I-10). This is known as Planck's law. In $\lambda$, this takes the form

$$\psi_\lambda = \frac{8\pi}{\lambda^4}\left(\frac{x}{e^x - 1}\right)kT. \tag{I-11b}$$

It is readily seen to pass to Wien's and Rayleigh-Jeans law for $x \gg 1$ and $x \ll 1$ respectively. It is also seen to satisfy Wien's displacement law (I-3) and Stefan-Boltzmann's law (I-2), and from these two laws, (I-11), or (I-12), leads to

$$\frac{d\psi_\lambda}{d\lambda} = 0, \qquad \text{or} \qquad \left(1 - \frac{x}{5}\right)e^x = 1, \qquad (x = 4.965 \text{ or } 0)$$

$$\psi = \int_0^\infty \psi_\lambda \, d\lambda = \frac{8\pi^5 k^4}{15(hc)^3} \quad (= 7.562 \times 10^{-15} \text{ erg/cm}^3 \text{ degree}^4).$$

From these one gets (the values accepted at present)

$$k = 1.38 \times 10^{-16} \text{ erg/degree}$$

$$h = 6.626 \times 10^{-27} \text{ erg sec.}[b]$$

Planck's law (I-11) or (I-12), has been verified by all subsequent accurate measurements by Rubens and others.

The Planck law has, since 1900, been obtained on various arguments. Planck's December, 1900, theory as outlined above implied both discrete absorption and discrete emission by oscillators. On basically similar assumption, a much simpler derivation of Planck's law has been given by H. A. Lorentz (1910) and Planck (Oct. 1911). The method is from statistical mechanics. If

---

[b] In Planck's original work of 1900 (and before that), he did not use (I-6), but obtained from classical electrodynamics the relation

$$\psi_\nu = \frac{8\pi\nu^2}{c^3}U$$

connecting the mean energy of an oscillator and the energy density of the radiation field in which the oscillators were immersed. Therefore the original theory, as pointed out by Einstein (1906), mixed the continuous theory of electrodynamics with the discrete theory of energy elements $\varepsilon$.

the oscillator can have only the discrete energies $nh\nu$, $n = 0, 1, 2, \ldots$, the mean energy is

$$U = \frac{\sum nh\nu e^{-nx}}{\sum e^{-nx}} \qquad x = \frac{h\nu}{kT}$$

$$= \frac{h\nu}{e^x - 1},$$

which leads to Planck's law (I-11a).

Then Planck (end of 1911, January, 1912) proposed another theory (called the Second theory) which assumed continuous absorption but discrete emission by oscillators. The assumptions were not tenable and we shall not discuss it here, but only mention the result for the mean energy $U$

$$U = \frac{1}{2}h\nu + \frac{h\nu}{e^x - 1},$$

which has the "zero point energy" $\frac{1}{2}h\nu$ at absolute $T = 0$—a feature substantiated later by quantum mechanics and also by the observed (R. S. Mulliken, 1924) isotope shift in the vibrational band $(0'-0)$ of two isotope species $a$, $b$ of the BO diatomic molecule

$$\Delta\nu = \tfrac{1}{2}h[(\nu_a' - \nu_a) - (\nu_b' - \nu_b)].$$

This Second Theory was soon abandoned by Planck and a Third Theory assuming both absorption and emission to be continuous was proposed (1914). This last theory was also criticized and abandoned.

The accepted theory at present is that of S. N. Bose (1924) who proposed a new statistics (i.e., counting) for photons instead of the classical Boltzmann counting. Bose considered the problem of finding the number of ways of distributing $N$ indistinquishable photons $\varepsilon = h\nu$ in $G_\nu$ cells, and this number was that calculated in (I-10),

$$\frac{(N_\nu + G_\nu - 1)!}{N_\nu!(G_\nu - 1)!}.$$

The total number of having $N_1$ photons $h\nu_1$ in $G_1$, $N_2$ photons $h\nu_2$ in $G_2$, etc. is

$$W_{\text{Bose}} = \prod_\nu \frac{(N_\nu + G_\nu - 1)!}{N_\nu!(G_\nu - 1)!}.$$

Using Stirling's approximation and neglecting 1 compared with $N_\nu$ and $G_\nu$, maximizing $\ln W_{\text{Bose}}$ under the condition

$$\sum N_\nu \varepsilon_\nu = E = \text{const.},$$

one gets

$$N_v = \frac{G_v}{e^{\beta h v} - 1}, \qquad \beta = \text{const.}$$

From Rayleigh's number of modes of oscillations per unit volume of medium, in the range between $v$ and $v + dv$, one gets, for a volume $V$,

$$G_v = \frac{8\pi V v^2}{c^3} dv,$$

and for the radiation energy density $\psi_v \, dv$,

$$\psi_v \, dv = \frac{1}{V} N_v \varepsilon_v = \frac{1}{V} N_v h v$$

$$= \frac{8\pi v^2 \, dv}{c^3} \frac{h v}{e^x - 1}, \qquad x = \frac{h v}{kT}$$

which is Planck's law (I-11a).

This "Bose statistics" was immediately extended by Einstein from photons (for which the total number of photons $\sum_v N_v$ is not a constant since they are constantly being transformed by absorption and emission) to gas molecules. In quantum mechanics, it is found that the "statistics", the spin of the particle concerned, and the symmetry with respect to an exchange of two indistinguishable particles are closely related to one another.

### References

Planck's October, 1900 paper making the "guess" (I-9) is published in *Verhandlungen der Deutschen Physik Gesells* **2**, 202–4 (1900).

Planck's December, 1900 paper giving the Planck law (I-11) is found in *ibid.* **2**, 237–245 (1900); *Annalen d. Physik* **4**, 553–563 (1901).

S. N. Bose, *Z. Phys.* **26**, 178 (1924). Bose sent his manuscript to Einstein for comments. Einstein recognized the importance of the new ideas and translated the manuscript and submitted it for publication. He extended the new statistics to gas molecules for which the condition $\sum N_v = \text{const.}$ must be introduced. At this time Einstein recognized the relevance of Bose's ideas to de Broglie's ideas of particles and "matter waves".

## 2.   Einstein's theory of the photon (1905) and the wave-particle duality

Planck assumed that an oscillator emits or absorbs energy in multiple units of $\varepsilon$ ($= h v$). Einstein in 1905 proposed an even far more revolutionary theory in which radiation exists also in discrete quanta even when propagating in (free) space. The arguments by which he arrived at this conception of radiation are briefly as follows.

Let $S_v(\psi_v, v)\, dv$ be the entropy, per unit volume, of radiation in the spectral range between $v$ and $v + dv$, so that, in a cavity of volume $V$, the entropy $S$ is

$$S = V \int_0^\infty S_v\, dv.$$

The condition for equilibrium is given by

$$\delta \int_0^\infty S_v\, dv = 0$$

with the condition

$$\delta \int_0^\infty \psi_v\, dv = 0,$$

i.e.,

$$\int_0^\infty \left( \frac{\partial S_v}{\partial \psi_v} - a \right) \delta\psi_v\, dv = 0, \qquad a = \text{const.}$$

For this to be true for arbitrary $\delta\psi_v$, we have

$$\frac{\partial S_v}{\partial \psi_v} = a = \text{constant, independent of } v.$$

Hence

$$dS = V \int_0^\infty \frac{\partial S_v}{\partial \psi_v}\, d\psi_v\, dv$$

$$= V \frac{\partial S_v}{\partial \psi_v}\, d\psi = \frac{\partial S_v}{\partial \psi_v}\, d\Psi,$$

where $d\Psi = V \int d\psi_v\, dv$ is the heat reversibly added to the cavity, and hence

$$dS = \frac{d\Psi}{T}.$$

From these two relations, one obtains

$$\frac{\partial S_v}{\partial \psi_v} = \frac{1}{T} \qquad \text{independent of the form of } \psi_v.$$

Einstein now assumed the validity of Wien's law (I-5) (for the spectral region of large $v/T$), and obtained

$$\frac{\partial S_v}{\partial \psi_v} = -\frac{1}{\beta T} \ln \frac{\psi_v}{\alpha v^3}.$$

On integrating,

$$S_v = -\frac{\psi_v}{\beta v}\left(\ln \frac{\psi_v}{\alpha v^3} - 1\right).$$

Let $\mathscr{S}_v = VS_v$, $\Psi_v = V\psi_v$. Then

$$\mathscr{S}_v = -\frac{\Psi_v}{\beta v}\left(\ln \frac{\Psi_v}{\alpha v^3 V} - 1\right).$$

If the volume changes from $V_0$ to $V$, the entropy changes by

$$(\mathscr{S}_v - \mathscr{S}_v^0)\,dv = \frac{\Psi_v\,dv}{\beta v}\ln\left(\frac{V}{V_0}\right)$$

$$= \frac{R}{N}\ln\left(\frac{V}{V_0}\right)^{N\Psi_v\,dv/\beta vR},$$

where $R$ = the gas constant, $N$ = Avogadro's number.

Now in the kinetic theory of gas, the probability of one molecule's being in a volume $V$, if it is originally in a volume $V_0$, is $V/V_0$, and the probability of $n$ mutually independent molecules being in $V$, if they originally occupy $V_0$, is $(V/V_0)^n$. Einstein proposed to interprete $(N/R)(\Psi_v\,dv/\beta v)$ as a number $n$, and with $N/R = k$, the Boltzmann constant, and $\beta = h/k$ on comparing (I-5) with Planck's law (I-11a), Einstein obtained

$$\Psi_v\,dv = nhv. \tag{I-12}$$

This means that radiation itself is in discrete quanta, the discreteness not only appearing in the emission or absorption process by oscillators as assumed in Planck's theory.

This theory of assuming a discrete nature for electromagnetic radiation is imcompatible with the well-established wave properties of light, namely the diffraction, interference and polarization phenomena. For example, O. Lummer and E. Gehrcke (1902) obtained interference between two (split) beams of light (from a mercury lamp) having a path difference of the order of one meter—a phase difference of some $2.6 \times 10^6$ waves. It is difficult to visualize light as propagating in discrete units (called by G. N. Lewis "photons" in 1926).

In this 1905 paper, Einstein suggested, as applications of his theory of light quanta, the phenomena of the photoionization of gases by x-rays and of the photoelectric effect (discovered by H. Hertz in 1886–7). The photon theory predicted that the maximum energy of the photoelectrons is independent of the intensity of the radiation, but is determined only by its frequency, according to the relation

$$\tfrac{1}{2}mv^2 = h\nu - e\phi.$$

This relation has been verified experimentally, in particular, by very accurate measurements (1916) by R. A. Millikan (1868–1953).

Einstein's theory was criticized by P. Ehrenfest (1911–1914) and Natanson (1911), as not leading to Planck's law, but only to Wien's law. The criticisms were based on the analysis of the distinction between the "indistinquishable and discrete photons" of Einstein and the "energy steps" in Planck's theory.

A deeper understanding of the implications of Planck's law and the wave-particle duality was achieved by Einstein (1909) by studying the Planck's law from the point of view of the statistical theory of fluctuations. Consider radiation in an isothermal enclosure of temperature $T$, and denote by $\bar{E}$ ($= \psi_\nu \, d\nu$ or $\psi_\lambda \, d\lambda$) the mean energy density (in the range $\nu$ and $\nu + d\nu$, etc). $\bar{E}$ is assumed to be given by Planck's law (I-11a)

$$\bar{E} = \frac{8\pi h\nu^3 \, d\nu}{c^3} \frac{1}{e^x - 1}, \qquad x = \frac{h\nu}{kT}.$$

In statistical mechanics, the average value of $E$ of a system is

$$\bar{E} = \frac{\displaystyle\int E e^{-\beta E} \, d\Omega}{\displaystyle\int e^{-\beta E} \, d\Omega}, \qquad \beta = \frac{1}{kT},$$

where the integration is over all phase space. The "fluctuation" $\varepsilon$ is defined by

$$\varepsilon = E - \bar{E}$$

and the mean-square fluctuation $\overline{\varepsilon^2} = \overline{E^2} - \bar{E}^2$ can be immediately seen to be

$$\overline{\varepsilon^2} = kT^2 \frac{d\bar{E}}{dT}.$$

With Planck's law for $\bar{E}$, Einstein found

$$\overline{\varepsilon^2} = \frac{8\pi h\nu^3 \, d\nu}{c^3}\left[\frac{h\nu}{e^x - 1} + \frac{h\nu}{(e^x - 1)^2}\right]$$

$$= \bar{E}h\nu + \frac{c^3 \bar{E}^2}{8\pi \nu^3 \, d\nu}. \tag{I-13}$$

Now this result can be compared with the $\overline{\varepsilon^2}$ calculated with Wien's law (I-5) for $\bar{E}$ and Rayleigh-Jeans' law (I-6) for $\bar{E}$

Wien's law:
$$\bar{E} = \alpha v^3 \, dv \, e^{-x}, \qquad \overline{\varepsilon_d^2} = \bar{E} h v;$$

Rayleigh-Jeans law:
$$\bar{E} = \frac{8\pi v^2 \, dv}{c^3} kT, \qquad \overline{\varepsilon_c^2} = \frac{c^3 \bar{E}^2}{8\pi v^2 \, dv}.$$

Thus Planck's law is the result of the sum of two mean-squares of fluctuations: one $\varepsilon_c$ arising from wave interferences (as shown subsequently rigorously by H. A. Lorentz in 1912) and therefore traceable to wave properties of radiation, and one $\varepsilon_d$ that can be interpreted as the fluctuation of the number $n$ of photons of energy $hv$, $\varepsilon_d = (n - \bar{n})hv = \delta hv$, $\overline{\varepsilon_d^2} = \overline{\delta^2}(hv)^2 = \bar{n}(hv)^2 = \bar{E}hv$. Thus Einstein's analysis showed that Planck's law implies the simultaneous presence of both the wave and particle properties of radiation.

Einstein's theory of light quanta has been "completed" by associating a momentum $p$ to the light quantum. In 1908, Planck, from considerations based on special relativity gave a "unified" definition of momentum according to which the energy flux divided by $c^2$ is the momentum density. Each light quantum moving in vacuum with a velocity $c$ must then carry a momentum $p = hv/c$.

Einstein's theory of light quanta can now be expressed by the relations

$$E = hv, \qquad p = \frac{hv}{c} = \frac{h}{\lambda}, \tag{I-14}$$

stating that electromagnetic waves of frequency $v$ and wavelength $\lambda$ have "particle" properties of energy $E$ and momentum $p$ given by these relations.

To depart from the above chronological accounts of the developments, let us take up the Compton effect (1923) which gives strong, direct support to the relations (I-14).

A. H. Compton (1892–1962) found that when x-rays of wave length $\lambda$ were scattered by (light) atoms, the scattered radiations in directions making an angle $\vartheta$ with the incident beam had two peaks in the spectrum, one at $\lambda$ and one at a longer wavelength $\lambda'$ given by

$$\lambda' - \lambda = 0.0242(1 - \cos \vartheta)(\text{Å}).$$

Compton gave a theory of this result on Einstein's theory of the photon. On using (I-14) and assuming energy and momentum conservation laws of dynamics,

$$hv = hv' + (m - m_0)c^2, \qquad m = \frac{m_0}{\sqrt{1 - \left(\dfrac{v}{c}\right)^2}}$$

$$\frac{hv}{c} = \frac{hv'}{c} \cos \theta + mv \cos \phi,$$

$$0 = \frac{hv'}{c} \sin \vartheta - mv \sin \phi,$$

where $m_0$, $v$, $\phi$ are the rest mass, velocity and scattering angle $\phi$ of the recoil electron, he obtained

$$\lambda' - \lambda = \frac{h}{m_0 c}(1 - \cos \vartheta),$$

in agreement with the empirical result above, since $h/m_0 c = 0.0242$ (the "Compton wave length").

This theory of the Compton effect was furthur supported by subsequent experiments. Bothe and Geiger (1924–5) using coincidence counters found that the recordings of the recoil electrons and the photoelectrons from the scattered x-rays $(\lambda')$ were "simultaneous", and with more modern experimental techniques, Hofstadter and McIntyre (1950) found that the recordings are less than $1.5 \times 10^{-8}$ sec. apart. Compton and Simons (1925) using Wilson's cloud chamber technique found that the directions of the scattered x-rays and the recoil electron did satisfy the above conservation equations. With more modern technique, Cross and Ramsey (1950) found that the theoretical value of $\phi$ and the observed value of the recoil electron agree within 1 degree.

All these results are very important, not only as direct support of Einstein's theory (I-14), but as a refutation of another theory of a startling nature at that time (1924). In attempting to reconcile the discontinuity of quantum transitions with the continuity of waves, J. C. Slater (1924) suggested that in atomic processes energy may not be conserved microscopically but only statistically conserved. This idea was formulated by Bohr, Kramers and Slater in a paper (1924) which dealt with the problems of dispersion, refraction and reflection of radiation. The experiments of Bothe, etc. showed definitely this theory was not tenable. The spirit of the theory, namely the readiness to give up established theories for new ideas, was, however, very important, for immediately after that, Heisenberg joined Kramers (at Copenhagen) to work on the theory of dispersion and conceived the ideas that developed into the matrix mechanics.

Coming back to Einstein's theory, represented in the relations (I-14), we must emphasize their profoundly difficult nature and must not be disarmed by their simplicity or by our becoming familiar with them. On their right-hand sides, we have light or x-rays which we know to be waves from the measurements of their wave lengths by diffraction gratings or crystal lattices, (M. Von Laue, 1912; W. L. Bragg and W. H. Bragg, 1913) and from their polarization by means of certain crystals (for light) or scattering (e.g. the experiments of C. G. Barkla on x-rays, 1904–6). Now "wave-length" and "polarization" are characteristic properties of waves. The left-hand sides of (I-14) for photons are defined by the characteristic properties of particles, namely, the energy and momentum. But these properties are now expressed by, or related to, the conceptually different properties of waves.

Then the electron diffraction experiments of C. J. Davisson & L. H. Germer (1927), G. P. Thomson (1927) and E. Rupp (1928) which showed electrons to exhibit wave properties, with wavelengths given by the inverses of equations (I-14), i.e., $\lambda = h/p$, and the experiments of I. Estermann and O. Stern (1930) on hydrogen molecules and helium atoms on impinging on the surface of LiF crystals. These experiments strongly supported the de Broglie theory that a particle of energy $E$ and momentum $p$ has wave properties:

$$v = \frac{E}{h}, \qquad \lambda = \frac{h}{p}. \tag{I-15}$$

Thus the Einstein theory of photons and the de Broglie theory of "matter waves" make the wave-particle duality complete. This can be expressed by combining the relations (I-14) and (I-15), which have similar appearance but fundamentally different meanings, into the "Einstein-de Broglie relations"

$$E \rightleftharpoons h v, \qquad p \rightleftharpoons \frac{h}{\lambda}. \tag{I-16}$$

This duality dilemma is not of a "technical" kind, but strikes very deep into the conceptual foundation of classical physics.

## References

A. Einstein: The classic paper appears in *Annalen d. Physik* **17**, 132–148 (1905); English translation in *Am. J. Phys.* **33**, 367–374 (1965).

P. Ehrenfest, *Annalen d. Physik* **36**, 91 (1911); *Proc. Amst. Acad.* **17**, 870–3 (1914), compare Einstein's and Planck's hypotheses.

A. Einstein: The paper using the theory of fluctuations and bringing the wave-particle duality to the fore appears in *Physik Zeit.* **10**, 185–193 (1909); *Verhandlungen der Deutschen Physik Gesells* **11**, 482–500 (1909).

A. Einstein, *Mitt. Physik Ges.* Zürich **16**, 47 (1916); *Physik. Zeits.* **18**, 121 (1917). In these papers, Einstein completed the particle picture of the light quantum by ascribing to it a momentum $p = hv/c = h/\lambda$.

G. N. Lewis, *Nature* **118**, 874 (1926). The name *photon* was coined in this article.

A. H. Compton, *Bull. Nat. Res. Coun.* Vol. **4**, No. 20 (Oct., 1922); *Phys. Rev.* **21**, 483–502 (1923).

W. Bothe and H. Geiger, *Z. Phys.* **26**, 44 (1924); **26**, 635 (1925).

A. H. Compton and A. W. Simon, *Phys. Rev.* **26**, 289–299 (1925).

J. C. Slater, *Nature* **113**, 307 (1924); N. Bohr, H. A. Kramers and J. C. Slater, *Phil. Mag.* **47**, 785 (1924).

R. Hofstadter and J. A. McIntyre, *Phys. Rev.* **78**, 24–28 (1950).

W. G. Cross and N. F. Ramsey, *Phys. Rev.* **80**, 929–936 (1950).

### 3.  Quantum theory applied to vibrations of atoms (1906–7)

In the preceding sections, the quantum theory is concerned only with radiations (of blackbody, light and x-rays). In 1907 Einstein applied the quantum theory to the vibrational motion of the atoms (about their equilibrium position) of solids, and succeeded in accounting for the observed deviation of the specific heat (or atomic heat) of solids at constant pressure at very low temperature from the value $C = 3R$ according to Dulong and Petit's law which was a consequence of the principle of equipartition of energy of classical statistical mechanics.

For simplicity, Einstein assumed one single frequency $v$ for all atoms, and obtained from (I-7),

$$E = 3N\bar{\varepsilon} = 3RT\frac{x}{e^x - 1}, \qquad x = \frac{hv}{kT}$$

$$C_v = \frac{dE}{dT} = 3R\frac{x^2 e^x}{(e^x - 1)^2},$$

which approaches the classical value $3R$ for $x \ll 1$, i.e., at high temperature, and at low temperature

$$C_v \to 3Rx^2 e^{-x},$$

which agrees with the observed decrease of $C_v$ for all solids at low $T$. The predicted decrease is however too rapid, compared with the observed behavior $C_v \propto T^3$ near the absolute $T = 0$.

Despite this unsatisfactory feature, Einstein's theory is of great significance for the following reason: by extending the concept of the quantum from radiation fields to the vibrational motion of atoms (in solids), Einstein started a new approach to the physics of solids. Such later concepts as the phonon and the polaron referring to the lattice vibrations and their interactions with electrons have their origin in Einstein's theory.

An improvement, at a "practical" level, over Einstein's theory is the theory of Debye (1912). A solid is treated as a continuous body having a continuous spectrum of elastic waves of frequencies $v$ from 0 to a maximum $v_m$ determined by the number of vibrational degrees of freedom of the atoms of the body. If $V$ is the volume of one gram-atom of the solid, $N$, Avogadro's number, $c_l$, $c_t$ the velocities of longitudinal and transverse elastic waves in the solid, then $v_m$ is determined by

$$3N = V \int_0^{v_m} 4\pi \left( \frac{1}{c_l^3} + \frac{2}{c_t^3} \right) v^2 \, dv,$$

(see (I-6) for the number of modes of oscillations per unit volume of a medium.)

If for simplicity one assumes $c_l$, $c_t$ to be independent of the frequency, then, on using (I-7) for the average energy $\bar{\varepsilon}$ for vibration of frequency $v$, one obtains for the energy $E$ and the atomic heat $C_v$ of one gram-atom of solid,

$$E = \frac{9N}{v_m^3} \int_0^{v_m} \frac{kTx}{e^x - 1} v^2 \, dv, \qquad x = \frac{hv}{kT}$$

$$= \frac{9RT^4}{\Theta} \int_0^{\Theta/T} \frac{x^3}{e^x - 1} \, dx, \qquad \Theta = \frac{hv_m}{k}$$

$$C_v = 9R \left[ 4 \left( \frac{T}{\Theta} \right)^3 \int_0^{\Theta/T} \frac{x^3}{e^x - 1} \, dx - \left( \frac{\Theta}{T} \right) \frac{1}{e^{\Theta/T} - 1} \right]. \quad \text{(I-17)}$$

The behavior of $C_v$ is

$$C_v \to \begin{cases} 3R \text{ for } \dfrac{\Theta}{T} \ll 1, \quad \text{i.e., high } T \\[2ex] \dfrac{12\pi R}{5} \left( \dfrac{T}{\Theta} \right)^3 \text{ for } \dfrac{\Theta}{T} \gg 1, \quad \text{i.e., for low } T. \end{cases}$$

The $C_v \propto T^3$ behavior is in excellent agreement with the observed data for all solids, and the values of $\Theta$ obtained from the $C_v$ data are in very good agreement with the values obtained from the elasticity data of many solids. Considering the basic assumptions of the theory (namely, a continuous instead of a discrete structure, and a spectrum with a $v_m$ determined in the manner shown above), one may be awed by these agreements.

An improvement over Einstein's simple theory, in a more fundamental manner, is the theory of Born and von Kármán (1912–3). The starting point is the dynamical motion of the atoms in a lattice structure, each atom interacting with its neighbors with a potential $U$ quadratic in the displacements of the atoms from their equilibrium positions. In essence, the theory is that of the normal vibrations of a system of particles from their equilibrium in classical dynamics, together with the quantum expression (I-7) for the mean energy of vibration of frequency $v$. The actual calculations of course depend on the form of the potential $U$, i.e., the symmetry property of the lattice, as well as the numerical coefficients therein, and the theory cannot be expressed in a general form such as (I-17). But in the few cases where calculations have been made, the results are in good agreement with the experimental data.

## References

A. Einstein, *Annalen d. Physik* **22**, 180–190, 800 (1907).

P. Debye, *ibid.* **39**, 789–839 (1912); comparison with experiments, E. Schrödinger, *Physik. Zeit.* **20**, 420, 450, 474, 497, 523 (1919).

M. Born, and T. Von Kármán, *Physik Zeit.* **13**, 297–309 (1912); **14**, 15–19 (1913). Review article by E. W. Montroll in *Handbook of Physics*, E. U. Condon and H. Odishaw, eds. (McGraw-Hill, New York, 1967).

### 4.  Bohr's theory—the concept of stationary states

In Rutherford's laboratory at Manchester in 1909, H. Geiger (1882–1947) an assistant, and E. Marsden (1890–1970) a student of Rutherford, at Rutherford's suggestion, carried out experiments on the scattering of $\alpha$-particles (by atoms in thin plate of matter) and found scatterings through angles greater than 90°. Rutherford proposed in 1911, the "nuclear model" of the atom—considered by some to be the greatest of all his discoveries (among them, the artificial disintegration of the nucleus of nitrogen in 1919.) The model, however, was confronted with a fundamental difficulty on the basis of classical electrodynamics, namely, that the electrons orbiting around a nucleus will radiate electromagnetic waves and the atom will not be stable.

Niels Bohr (1885–1962), a young Danish physicist visiting Rutherford at Manchester in 1912, worked on the problem of atomic structure and in 1913 published the epoch-making theory of the hydrogen atom and its spectrum. The theory is so familiar that we need only briefly review the main points.

(i) The first revolutionary idea of Bohr is the postulate of "stationary states" which are exempt from the law of classical electrodynamics, i.e., an atom in a stationary state does not radiate energy. The condition for such states is, in the case of an electron in a circular orbit of radius $a$,

$$\text{angular momentum:} \quad pa = n\left(\frac{h}{2\pi}\right), \tag{I-18}$$

where $p = \mu v = \mu \omega a$ is the momentum of the electron of mass $\mu$. This equation, together with the equation

$$\frac{Ze^2}{a^2} = \frac{\mu v^2}{a},$$

leads to the expression for the energy of state $n$

$$E_n = -\frac{2\pi^2 \mu Z^2 e^4}{n^2 h^2}. \tag{I-19}$$

(ii) The second postulate in Bohr's theory is the introduction of the concept of discrete transitions between stationary states, and the frequency condition

$$h\nu_{mn} = E_m - E_n, \tag{I-20}$$

for the frequency of the radiation emitted (absorbed) when the atom goes from state $m$ to state $n$ for $E_m > E_n$ $(E_m < E_n)$. Thus[c]

$$v_{mn} = Z^2 R_\infty \left( \frac{1}{n^2} - \frac{1}{m^2} \right), \qquad \text{(in cm}^{-1}) \qquad\qquad\text{(I-21)}$$

$$R_\infty = \frac{2\pi^2 \mu e^4}{ch^3} = 109721.6 \text{ cm}^{-1}, \qquad (c = \text{velocity of light}).$$

This theory of Bohr immediately achieves a great success when the formula (I-21) reproduces the empirically found Balmer series (1885) of spectral lines of hydrogen for $Z = 1$, $n = 2$ and $m = 3, 4, 5, \ldots$. The series for $n = 1$, $n = 3$, $n = 4$, $n = 5$ and $m \geq n + 1$, $n + 2$, ... predicted by (I-21) have all been observed; they are the Lyman series (1914), the Paschen series (1908), the Brackett series (1922) and the Pfund series (1924) respectively. In the last decade, transitions such as $m = 246$, $n = 245$ have been observed in the radio wave region from nebulae.

The reality of stationary states has received direct evidence from experiments other than spectroscopic. J. Franck and G. Hertz in 1914 studied the collision between electrons with mercury vapor atoms, and found that the resonance line at 2536 Å was emitted when the electron energy exceeded a critical value. This and many other experiments on the "critical potentials" can be simply explained on the existence of stationary states in atoms and molecules and their excitation by electron impact.

But these and other subsequent successes of the Bohr theory must not be allowed to let us forget the revolutionarily new nature of the two postulates which introduce discontinuities into a fundamentally continuous classical physics, namely, the discrete states and the discrete transitions. Consider the postualte (I-20). It may be pointed out that in the prevalent view before Bohr, an atom as a system of electric charges emits radiations by the oscillations of all the charges at the same time. But Bohr postulated that an atom produces one spectral line at a time.[d]

The concepts of stationary states and their definitions by quantization conditions in terms of quantum numbers are amazing in the following respect: these concepts have remained valid in quantum mechanics even though the Bohr theory has been superseded at a basic level. Even in elementary particle physics, the most basic knowledge that we have about them is their energy states defined by quantum numbers, most of which are still of an empirical nature. We shall therefore consider the concept of stationary states at greater length.

[c] When the relative motion of the electron and the nucleus, of mass $M$, is taken into account, $\mu$ is to be replaced by the reduced mass $\mu(1 + (\mu/M))^{-1}$.
[d] It is of historical interest to note that this idea was expressed by A. W. Conway (Dublin, in 1907) even before the Rutherford model.

The first progress following Bohr's 1913 theory was the generalization of the quantum condition (I-18) by A. Sommerfeld (1915) and W. Wilson (1915). For a multiply periodic system of $n$ degrees of freedom, there are to be $n$ quantization conditions in terms of the action variables (0-63)

$$\oint p_i\, dq_i = n_i h, \qquad i = 1, 2, \ldots, n, \tag{I-22}$$

where the integrals are over the periods of the individual coordinates $q_i$.

The action variables $J_k$ have been shown to have a very important property by Burgers (1917), namely, $J_k$ is invariant when the system is subject to a very slow change of an external perturbation. This property of "adiabatic invarience" in a way makes the quantization condition (I-22), now in the form

$$J_i = n_i h \tag{I-23}$$

easier to accept, for then if $J_i$ would change, they would change suddenly, since they do not change slowly. Of course we must emphasize that this is no argument for Bohr's two basic postulates.

There are, however, systems that are not multiply periodic. In classical dynamics, such a system is the well-known 3-body problem, and no general solution has been found. The failure to extend the Bohr theory to helium is not one of obtaining inaccurate numerical results, but is deep rooted in the failure of the quantization conditions (I-22) themselves. In fact, without multiply-periodic properties, one does not know how to formulate the quantization conditions. The answer to this fundamental problem is furnished by quantum mechanics.


## References

The $\alpha$-scattering experiments on which Rutherford's model is based is described in
   H. Geiger and E. Marsden, *Proc. Roy. Soc.* London **A 132**, 495 (July, 1909).
The nuclear model of the atom is proposed in E. Rutherford, *Phil. Mag.* **21**, 669 (May,
   1911).
N. Bohr, *Phil. Mag.* **26**, 1–25, 476–502, 857–875 (1913).
A. Sommerfeld, *Münchener Berichte*, 425–458, 459–500 (1915). The generalized quan-
   tization condition. Also simultaneously and independently.
W. Wilson, *Phil. Mag.* **29**, 795 (1915); **31**, 156 (1916).
The quantum condition seems to have been anticipated by Planck in a paper presented
at the First Solvay Congress, Oct. 30–Nov. 3, (1911). For a harmonic oscillator,
$E = (1/2m)p^2 + \frac{1}{2}k^2 q^2$, the quantum condition

$$\iint dp\, dq = nh,$$

where the integral is taken over the area of the ellipse of the $E(p, q) = $ const. curve, leads to

$$E = nh\nu, \qquad 2\pi\nu = \frac{k}{\sqrt{m}}.$$

M. Born, *The Mechanics of the Atom* (G. Bell & Sons, London, 1927) gives an account of Hamilton-Jacobi theory, action and angle variables, multiply periodic systems, etc.

## 5. Bohr's theory: refinements

The first obvious extension of the Bohr theory is to the same problem in 3- and 2-dimensions. We have treated this problem in Section 0-4, (0-80)–(0-88).

On applying the quantization condition (I-22) to $J_r$, $J_\vartheta$, $J_\psi$, $J_\varphi$ of (0-85), we obtain

$$J_r = n_r h, \qquad J_\vartheta = n_\vartheta h, \qquad J_\psi = mh, \qquad J_\varphi = kh. \tag{I-23}$$

$k$ is the angular momentum, in units of $h/2\pi = \hbar$, of the electron in its orbital motion, $m$ is the component, in $\hbar$, of $k$ about the $z$-axis, so that

$$-k \leq m \leq k, \tag{I-24}$$

i.e., $k$ has $2k + 1$ components $-k$, $-k + 1$, ..., 0, ..., $k - 1$, $k$, so that the cosine of the angle between $k$ and the $z$-axis can take only $2k + 1$ discrete values

$$\cos(kz) = \frac{m}{k}, \tag{I-25}$$

i.e., the orientation of the orbital plane is quantized. If we introduce the principal quantum number $n$

$$n = n_r + n_\vartheta + m, \tag{I-26A}$$

$$= n_r + k \qquad \text{(using (I-23) and (0-85))} \tag{I-26B}$$

the energy $E$ for both the 3-dimensional and the 2-dimension treatment is

$$E_n = -\frac{2\pi^2 \mu Z^2 e^4}{h^2 n^2} \tag{I-27}$$

which is the same as (I-19) for the circular motion.

In classical dynamics, we know the general orbits of an electron in the Coulomb field $V(r) = -Ze^2/r$ to be ellipses

$$\frac{1}{r} = \frac{1 + \varepsilon \cos\phi}{a_n(1 - \varepsilon^2)},$$

where $a_n$ is the semi-major axis and $\varepsilon$ the eccentricity. From (0-82), (0-83),

(0-85), (I-23), (I-26B) one can show

$$1 - \varepsilon^2 = \frac{k^2}{n^2} \tag{I-28}$$

and from the independence of the total energy $T + V$ on $\phi$ in the orbit, one obtains

$$a_n = \frac{h^2 n^2}{4\pi^2 \mu Z e^2} = n^2 a. \tag{I-29}$$

From (I-26B), it is seen that for a given $n$, $k$ can take on values $k = 1, 2, \ldots, n$, corresponding to ellipses of decreasing eccentricity ($k = n$ gives the circular orbit of Bohr's original theory). With (I-29), we may write (I-27) in the form

$$E_n = -\frac{Ze^2}{2an^2}. \tag{I-30}$$

(I-27) shows that $E$ depends only on $n$ and not individually on $m$ and $k$. These degeneracies arise from

(a) the central symmetry of the field $V(r)$, leading to the independence on $m$, and

(b) the inverse-square form of $dV/dr$, leading to the independence on $k$.

These remarks are important, as we shall see in the Zeeman effect and the "screened field" of alkali metal atoms. Also we have exactly the same degeneracies in the Schrödinger equation for the hydrogen-like atoms.

One of the early great triumphs of the Bohr theory was the experimental verification of the predicted quantization of orientation (I-25) by O. Stern and W. Gerlach (1921). The experiment consisted in sending a narrow beam of silver atoms through an inhomogeneous magnetic field having a gradient $\partial\mathcal{H}/\partial z$ perpendicular to the direction of motion of the atomic beam. Now a system having an angular momentum $k\hbar$ has a magnetic moment $(e/2\mu c)k\hbar$, and the gradient $\partial\mathcal{H}/\partial z$ acting on this moment produces a force $m(e\hbar/2\mu c)$ $(\partial\mathcal{H}/\partial z)$. If $k = 1$, then $m = 0, \pm1$ and the beam will be split into 3 components. What is observed is that the beam is split into *two* components. That it is *two* instead of *three* can be accounted for when the electron spin is known (1925); but the *splitting* by itself supports the prediction of the effect of "spatial quantization".

Another important success of Bohr's theory is the extension by Sommerfeld (in 1915–6) to include the effect of the relativistic changes of mass with velocity of the electron into account, and he obtained for the energy $E$ the formula

$$\frac{E}{m_0 c^2} = \left[ 1 + \frac{Z^2 \alpha^2}{(n_r + \sqrt{k^2 - \alpha^2 Z^2})^2} \right]^{-1/2} - 1, \tag{I-31}$$

where $k$ is given in (I-23), and $\alpha$ is the "fine-structure" constant

$$\alpha = \frac{2\pi e^2}{hc} \cong \frac{1}{137}, \tag{I-32}$$

which is a dimensionless universal constant. For $Z^2\alpha^2 \ll 1$, (I-31) can be expressed in a series

$$E = -\frac{Z^2 Rhc}{n^2}\left[1 + \frac{Z^2\alpha^2}{n^2}\left(\frac{n}{k} - \frac{3}{4}\right) + \frac{Z^4\alpha^4}{n^4}(\dots)\dots\right], \tag{I-33}$$

where $R$ is given in (I-21). This makes $E$ dependent on $k$ through the term in $Z^2\alpha^2$, which gives rise to a fine-structure to a spectral line, such as the $H_\alpha$ ($n = 3 \to n = 2$). This fine structure was measured by F. Paschen (1916), and is in good agreement with the theory.

We may remark in passing here that, while the above result is based on a theory (classical dynamics and without the spin-orbit interaction) which has been superseded by quantum mechanics (Dirac's theory (1928) which is relativistic and automatically includes the electron spin), the formula (I-31) gives the same numerical energy values as Dirac's theory, although the quantum numbers $n_r$, $k$ have to be replaced by other quantum numbers with different meanings. This is a case of rare coincidence in the history of physics— an inaccurate theory could have led to the "same" mathematical result as an exact theory.[e]

Sommerfeld's theory of relativistic correction is of interest in another connection. In the equation of motion of the electron, this correction appears as, and is mathematically equivalent to, a small term $C/r^2$, where $C$ is a constant, in addition to the Coulomb potential $-(Ze^2/r)$. The effect of this term is to bring about a dependence of the energy $E$ on $k$ as well as on $n$, as seen from (I-31) or (I-33). The dependence of $E$ on $k$ is small, as $\alpha^2$ is small.

Consider in this connection the valence electron of an alkali metal atom. For such an electron, the effective potential arising from the nucleus and the electrons forming the "core" varies with its distance $r$ from the nucleus as $-(Ze/r)$ for very small $r$, and as $-(e/r)$ for very large $r$, so that the deviation from a Coulomb field is very large. The dependence of $E$ on $k$, for a given $n$, being no longer a relativistic effect, is large. Calculations by means of the Bohr-Sommerfeld condition (I-22) (see $J_r$ below (0-86) for example) show that in such a "screened potential $V(r)$," the energy $E(n,k)$ for a given $n$, is lowest for $k = 1$, corresponding to the most eccentric and therefore deeply penetrating ellipse, and highest for $k = n$, a circular and nonpenetrating,

---

[e] Another case is connected with the theory of thermoelectric effects—Seebeck (1822), Peltier (1834) and Thomson (1854) effects. Lord Kelvin treated these effects incorrectly by neglecting all the irreversible processes, but arrived at the correct result later obtained by Sommerfeld (1928). The reason for this situation was made clear only by the reciprocity relations of Onsager (1931). Cf. A. Sommerfeld, *Thermodynamics and Statistical Mechanics* (Addison-Wesley, Reading, 1964).

orbit. These relations remain valid in the quantum mechanical theory, and are helpful in our understanding of the building-up of the atoms in the periodic table. (see Sec. 11 below).

To conclude this section, we wish to emphasize the concept of "quantum numbers" as defining stationary states. In the case of the Kepler problem, the quantum number $k$, $m$ in (I-23), (I-24) have simple physical meanings. But as we shall see in Sec. 9, in the case of the complex spectra of atoms other than hydrogen, many "quantum numbers" have first been introduced in "empirical" ways on the basis of "empirical" selection rules, before their physical meanings are understood. For example, an "inner" quantum number $j$ is introduced (Sommerfeld, 1920) for the doublet levels of alkali metal atoms; the $S$ in the multiplicity $2S + 1$ and the $L$ of complex spectra; the "even" and "odd" states in Laporte's rule, etc. Their meanings have subsequently been found, with the introduction of the electron spin, angular momentum composition, the concept of parity, etc. But then, in the case of elementary particle physics, "quantum numbers" have been introduced in the systematization of empirical results (selection rules for the transformations, or decays, of particles) whose meaning cannot be (yet) expressible in ordinary physical terms. Thus for elementary particles, we have such "quantum numbers" as "strangeness", "color", "flavor", etc. The use of quantum numbers to define "states" in quantum mechanics has its origin in Bohr's theory of 1913,

## References

A. Sommerfeld, *Munchener Berichte*, 425–458, 459–500, (1915); *Annalen d. Physik* **51**, 1–94, 125–167 (1916) discuss generalized quantization conditions, elliptic orbits, fine structure; *Physik Zeit.* **27**, 491 (1916) deals with the Zeeman effect.

K. Schwarzschild, *Berlin Sitzungsb.* 548 (April, 1916).

P. Epstein, *Physik Zeit.* **17**, 148 (1916); *Annalen d. Physik* **6**, 489 (1916), These two authors treat the problem of the Stark effect by the use of parabolic coordinates.

J. Franck and G. Hertz, *Verhandlungen der Deutschen Physik Gesells* **16**, 457, 517 (May, June, 1914).

M. Born, *The Mechanics of the Atom* (G. Bell & Sons, London, 1927) gives an account of Hamilton-Jacobi equation, angle and action variables, multiply periodic systems, etc.

O. Stern, *Z. Phys.* **7**, 249 (1921).

O. Stern and W. Gerlach, *ibid.* **8**, 110 (1921); **9**, 349, 352 (1922).

## 6. Einstein's transition probabilities (1916)

Bohr's frequency postulate (I-20)

$$h\nu_{mn} = E_m - E_n \tag{I-34}$$

gives only the frequencies of spectral lines, but does not say what transitions $m \to n$ an atom in a state $m$ will make, and in particular what makes it go into a state $n$ and ignore other states intermediate between $m$ and $n$. Such questions are not answered in Bohr's theory. In classical physics, one is accustomed to deterministic and causal views, and in Bohr's theory one naturally has the feeling of incompleteness.

In 1916, Einstein gave a "simple" derivation of Planck's law (I-11), but the theory introduced a new kind of probability concept into physics that was destined to play a fundamental role in quantum mechanics. We shall give a brief account of the theory first.

Consider two states $m$ and $n$ of a system (an atom or molecule) in a radiation field with radiation energy density $\psi(v)\,dv$, where $hv = E_m - E_n > 0$. Einstein assumed that the system in state $m$ has a probability $B_n^m\psi(v)$, per second, of making a transition to a lower state $n$, and the system in state $n$ has a probability $B_n^m\psi(v)$, per second, of making a transition to a higher state $m$. $B_n^m$ is readily understandable, as an "absorption coefficient". $B_n^m$ seems less obvious; it is a kind of "negative absorption", and is called the "stimulated" or "induced" emission coefficient, in contrast with $A_n^m$ which Einstein postulated for a spontaneous transition probability (per second) independent of the presence of the radiation field $\psi(v)$. It is sometimes stated in the literature that the concept of stimulated emission $B_n^m$ is new, but this is not really so, since it is also present in classical electrodynamics, as a feedback effect from an oscillator to the radiation field (see J. H. Van Vleck, *Phys. Rev.* **24**, 330 (1924)). The concept of spontaneous emission $A_n^m$ is however really new; it is similar to the radio-active decay in that it is an intrinsic probability pertaining to an individual atom or nucleus and not arising from our ignorance on account of our having to do with a large number of systems, as in the classical kinetic theory of gases. We shall see in quantum mechanics that such a probability concept plays a basic role. It is most interesting to note that such an "intrinsic probability" was formally introduced by Einstein who in his later years (especially after the development of quantum mechanics in 1925–6) could not accept it in place of classical determinism.

For the equilibrium between the atom and the radiation field $\psi(v)$, the equation expressing "detailed balancing" for the $m \to n$ transitions is

$$N_n B_n^m \psi(v) = N_m B_n^m \psi(v) + N_m A_n^m, \tag{I-35}$$

with

$$N_m = N_n \exp\left(-\frac{E_m - E_n}{kT}\right), \tag{I-36}$$

$$v = (E_m - E_n)/h. \tag{I-37}$$

From the limiting condition that one arrive at the Rayleigh-Jeans law for $x = hv/kT \ll 1$, one obtains Planck's law

$$\psi(v) = \frac{8\pi v^2}{c^3} \frac{hv}{e^x - 1}, \tag{I-38}$$

and the relations[f]

$$B_n^m = B_h^m, \qquad A_h^m = \frac{8\pi h v^3}{c^3} B_h^m. \tag{I-39}$$

The $B_h^m$, $B_h^m$, $A_h^m$ coefficients themselves are not obtainable from the Bohr theory, nor from this theory of Einstein. The $B$'s can be obtained with the perturbation method in quantum mechanics—a system perturbed by a radiation field $\psi(v)$. $A_h^m$, however, must be obtained on the quantized field theory of Dirac (1927).

The application of Einstein's theory to Bohr's theory now furnishes some partial answers to the questions posed at the beginning of the present section, that is, the transitions $m \to n$ are governed by probabilities, and the intensities of the principal series lines of an alkali metal atom in its absorption spectrum, for example, are proportional to $B_n^m \psi(v)$. The next deep question is whether there are some connections between these coefficients and the classical theory of absorbing and radiating systems. In 1918, Bohr formulated the "correspondence principle" for the consideration of such questions that will be discussed in the following section.

The coefficients $B_h^m$, $A_h^m$ have been made use of, in the spirit of the correspondence principle, in translating the theory of dispersion in classical electrodynamics into a quantum theory form. The line of work of Ladenberg, Kramers and Heisenberg turned out to have a key role in Heisenberg's conception of the ideas that lead to matrix mechanics. These will be the subjects of Sec. 8 below.

In passing, we may mention here that the concept of stimulated emission, namely, a system in an excited state $m$ can be induced to make a transition $m \to n$ by a strong radiation field $\psi(v)$ with $v = v_{mn}$, is the basis for the MASER (Microwave Amplification of Stimulated Emission of Radiation) and LASER (Light Amplification of Stimulated Emission of Radiaion) discovered some fifty years after Einstein's 1916–7 papers. The theory of laser is briefly as follows. When a large number of atoms (or molecules) excited by some processes into a metastable state (of long lifetime) $m$ are irradiated by (even very weak) $\psi(v)$, the "simultaneous" stimulated transitions $m \to n$ of the atoms emit a strong, coherent radiation of frequency $v_{mn}$.

The relation (I-36) is the result of Boltzmann statistics. If the system obeys the Bose-Einstein or the Fermi-Dirac statistics, the relation (I-36) has to be modified as follows. Let

$$N_n = N[e^{-\gamma + \beta E_n} - a]^{-1}, \qquad \beta = \frac{1}{kT}, \tag{I-36A}$$

---

[f] We have for simplicity assumed the states $m$, $n$ to be non-degenerate. Otherwise $N_n$, $N_m$ will have to be replaced by $g_n N_n$, $g_m N_m$, where $g_n$, $g_m$ are the weights.

where

$$a = \left\{ \begin{array}{c} 1 \\ 0 \\ -1 \end{array} \right\} \quad \text{for} \quad \left\{ \begin{array}{l} \text{B.E.} \\ \text{Boltzmann} \\ \text{F.D.} \end{array} \right\} \quad \text{statistics.}$$

Then (I-35) has to be replaced by

$$N_n B_n^m \varphi_v(v) = N_m B_n^m \psi(v) + N_m A_n^m (1 + \frac{N_n}{N} a), \tag{I-35A}$$

(I-38, 39) remaining unchanged. See T. Y. Wu, *Ann. Rep.*, Inst. of Phys., Academia Sinica, **13** (1983).

### References

A. Einstein, *Physik Zeit.* **18**, 121 (1917); *Mitt. d. Phys. Ges.* Zurich, No. 18 (1916); *Verhandlungen der Deutschen Physik Gesells* **18**, 318 (1916).

### 7.  The correspondence principle (1918)

Both Bohr's original theory of 1913 and Einstein's theory of transition probabilities of 1916 bear no relation to the classical theory of absorption and radiation of a system of electric charges. Bohr had a basic belief, all through his life, that even though the quantum and the classical theory are fundamentally irreconciliable, there should still be some "asymptotic" links between them since both do contain valid parts. Bohr introduced the concept of the correspondence principle in 1918–1923 which has since served as a guiding spirit in later developments leading, especially, to quantum mechanics.

The principle postulates a "correspondence", a relation, between the classical and the quantum theory in the sense of some limiting conditions. With the aid of this idea, we are to trace some connection between the frequencies, intensities and polarization of spectral lines on the quantum theory, and these quantities on the classical theory.

Firstly, let us consider the correspondence rule for frequencies. In classical theory, the coordinate $q_k$ of a multiply periodic system can be expressed by a Fourier series (0-70),

$$q_k = \sum a_k(\tau) \exp(2\pi(\tau \cdot v)t + \delta_k), \tag{I-40}$$

where $a_k$ and $\delta_\tau$ are functions of the $\tau$'s. The frequencies in classical dynamics are given by (0-68), (0-66)

$$v_{cl} = v_i = \frac{\partial H}{\partial J_i} \tag{I-41}$$

or, the $\tau_i$-th harmonic

$$v_{cl} = \tau_i v_i = \tau_i \frac{\partial H}{\partial J_i}. \tag{I-41a}$$

For simplicity, let us take the singly periodic system.

In the quantum theory the frequencies are

$$v_{qu} = v_{mn} = \frac{E_m - E_n}{h} \equiv v(m, n) = v(m, m - \tau) \tag{I-42}$$

or

$$v_{m,m-\tau} = \frac{E_m - E_{m-\tau}}{h},$$

where

$$J = mh \tag{I-43}$$

and

$$\Delta J = (m - n)h = \tau h \tag{I-44}$$

so that

$$v_{qu} = \frac{\Delta E}{\left(\dfrac{\Delta J}{\tau}\right)} = \tau \frac{\Delta E}{\Delta J}. \tag{I-45a}$$

The correspondence principle (for frequencies) is now expressed as follows. The frequency $v_{qu}$ of a transition $m - n = \tau$, when $\tau \ll m, n$, corresponds to the $\tau$-th harmonic of the frequency $v = \partial H / \partial J$, or

$$\lim_{\Delta m \ll m} v_{qu} = \lim_{\Delta m \ll m} \tau \frac{\Delta E}{\Delta J} = \tau \frac{\partial E}{\partial J} = v_{cl}. \tag{I-46}$$

As an example, take the hydrogen-like atom. From (I-41) and (I-26A), we get

$$v_{cl} = \tau \frac{\partial H}{\partial J_r} = \tau \frac{4\pi^2 \mu Z^2 e^4}{(J_r + J_\vartheta + J_\psi)^3} \tag{I-47}$$

$$= \tau \frac{4\pi^2 \mu Z^2 e^4}{n^3 h^3}. \tag{I-48}$$

From (I-27),

$$v_{qu} = v_{m,m-\tau} = \frac{2\pi^2 \mu Z^2 e^4}{h^3} \left[ \frac{1}{(m - \tau)^2} - \frac{1}{m^2} \right] \tag{I-49}$$

$$\lim_{\tau \ll m} v_{qu} = \tau \frac{4\pi^2 \mu Z^2 e^4}{n^3 h^3} \tag{I-50}$$

so that the correspondence (I-46) is satisfied.

According to the correspondence principle then, if a $\tau$-th harmonic appears in the Fourier series of the electric moment in the classic theory of the system, one may expect a transition $(m, n) = (m, m - \tau)$ in the quantum theory. This is expressed in the form of a "selection rule"

$$\Delta m = m - n = \tau. \tag{I-51}$$

To illustrate the working of the corresponding principle in the problem of the polarization of spectral lines, let us consider the (normal) Zeeman effect (1897). The experimental results are summarized in the following figures.

Longitudinal $\hspace{6em}$ (I-52)

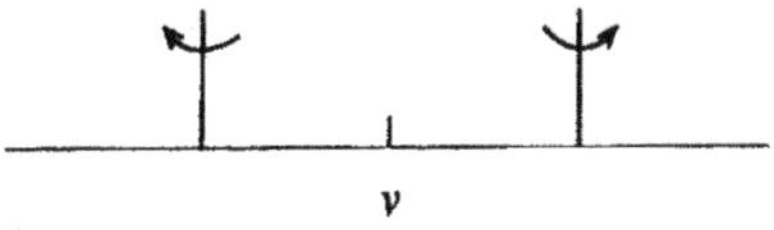

Transverse $\hspace{6em}$ (I-53)

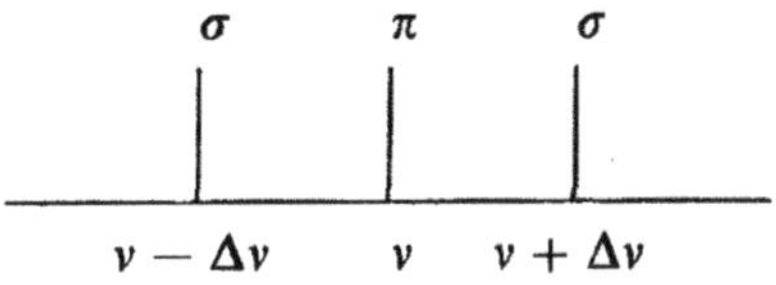

When observed in a direction parallel to the magnetic field $\mathscr{H}$ (longitudinal), there are two circularly polarized spectral lines at $v \pm \Delta v$ from the field-free position $v$, where

$$\Delta v = \frac{e\mathscr{H}}{4\pi\mu c}. \tag{I-54}$$

When observed in directions perpendicular to $\mathscr{H}$ (transverse), the line at $v$ is linearly polarized, with electric field parallel to $\mathscr{H}(\pi)$, and there are two linearly polarized lines at $v \pm \Delta v$, with electric field perpendicular to $\mathscr{H}(\sigma)$.

The quantum theory of the normal Zeeman effect is as follows. The orbital angular moment of the electron is $k\hbar$, and according to classical electromagnetic theory, such an angular momentum is accompanied by a magnetic moment $(e\hbar/2\mu c)\mathbf{k}$.[8] The interaction energy $\Delta E$ of this moment with the magnetic field $\mathscr{H}$ is

$$\Delta E = -\frac{e\hbar}{2\mu c}(\mathbf{k} \cdot \mathscr{H}) = -m\frac{e\mathscr{H}}{2\mu c}\hbar. \tag{I-55}$$

The energy of a state defined by $(n, k, m)$ is then (independent of $k$)

$$E_{n,m} = -\frac{Z^2 Rhc}{n^2} - mh\left(\frac{e\mathscr{H}}{4\pi\mu c}\right). \tag{I-56}$$

---

[8] It must be pointed out that the following treatment does not really apply to the hydrogen-like atoms because of the neglect of the electron spin here. Equation (I-55) can be applied to the singlet $^1p$ state of the alkaline earth atoms, for example. See Sec. 8, (1) below.

The frequency of a transition $n, m \to n', m'$ is then

$$v + \Delta v = v_{nn'} + (m - m')\left(\frac{e\mathcal{H}}{4\pi\mu c}\right).$$

This will agree with the empirical result in the above figure if[h] $\Delta m = m - m'$

$$\Delta m = 0, \pm 1. \tag{I-57}$$

In the classic theory (Lorentz, 1897), the electric moment $er$ has component $ez$ along $\mathcal{H}$ and components $ex, ey$ perpendicular to $\mathcal{H}$. By the Larmor theorem (1897) for the precession with frequency $\vartheta$ of the electron orbit about $\mathcal{H}$, we have

$$z = r \cos \vartheta = \tfrac{1}{2}r(e^{2\pi i v t} + e^{-2\pi i v t}), \tag{I-58}$$

$$x \pm iy = r \sin \vartheta e^{\pm i\psi} = re^{\pm 2\pi i(v \pm o)t}. \tag{I-59}$$

The correspondence principle now leads, on comparing (I-58, 59) with (I-57), indeed to the "selection rule":

$$\Delta m = 0 \qquad \text{for the } z\text{-component of } er, \tag{I-60}$$

$$\Delta m = \pm 1 \qquad \text{for the } x, y \text{ components of } er,$$

which at the same time gives the polarization of the observed radiation summarized in the foregoing figure.

The correspondence principle postulates the following general relation regarding the intensities of spectral lines in the general case: In the quantum theory, the intensities of radiation of frequency $v_{mn}$ are determined by how often the transitions $(m, n)$ take place, i.e., by the probabilities of the transitions $(m, n)$. These probabilities are correlated with the squares of the amplitudes of the Fourier components $\tau = m - n$ of the electric moment (I-40) in the classical theory. The validity of this idea has been verified by H. A. Kramers, and by W. Kossel and A. Sommerfeld.

The correspondence principle has served as a guiding spirit in other areas than spectroscopy. Thus Ladenberg, Kramers and Heisenberg treated the problem of dispersion in the quantum theory, and Heisenberg conceived certain ideas in this work which soon developed into matrix mechanics. All

[h] It is of interest to note that in October and November of 1918, A. Rubinowicz (at Münich) obtained the "selection rule" (I-57), independently of Bohr's correspondence principle, on the following considerations: If in an atomic transition the angular momentum $mh$ changes by $\pm h$ (i.e., $\Delta m = \pm 1$), the conservation of angular momentum requires then that the radiation must carry off an angular momentum $\mp h$. From classical electrodynamics the radiation whose direction is parallel to $\mathcal{H}$ must be circularly polarized, and that whose direction is perpendicular to $\mathcal{H}$ must be linearly polarized. If the angular momentum does not change, $\Delta m = 0$, the radiation must be linearly polarized.

these developments have been under the influence of the correspondence principle.

### References

N. Bohr, "On the Quantum Theory of Line-spectra", *Kong. Danske Vid. Selsk. Skr. Nat. og Math., Afd*, Series 8, **IV**, 1, 1–118 (1918); *Z. Phys.* **6**, 1 (1921); **13**, 117–164 (1922); *Annalen d. Physik* **71**, 228 (1923); *Proc. Phys. Soc.* **35**, 275 (1923); *Proc. Camb. Phil. Soc.* (Supplem.) Part I, 22 (1924).

## 8. Theory of dispersion—Ladenburg, Kramers and Heisenberg (1921–25)

At the end of Sec. 7, we mention that Heisenberg's work, with Kramers in 1924, on the dispersion theory played an important role in Heisenberg's inception of ideas that led to the formulation of the matrix mechanics. To see how startling, new basic ideas can originate from old theories, let us review, briefly, how the classical electrodynamical theory of dispersion evolves, in the spirit of the "correspondence principle", and leads Heisenberg to his new ideas.

In classical electrodynamics, when radiation of frequency $v$ falls on an electric dipole (oscillator of frequency $v_0$) the radiation is absorbed; its reemission of frequency $v$ constitutes "scattering", and the interference of the scattered wave with the incident wave causes "refraction". The effect of the natural frequency $v_0$ on the scattering gives rise to "dispersion". The results of the classical electrodynamics are as follows:

The rate of absorption of energy by a 3-dimensional oscillator ("dispersion electron" in the electron theory) of mass $m$ and natural frequency $v_0$ from an isotropic radiation field of energy density $\psi(v)\,dv$ is

$$\frac{\pi e^2}{m}\psi(v_0). \tag{I-61}$$

If there are $N$ electrons per unit volume, the rate of energy absorption per unit volume is

$$I_{cl} = N\frac{\pi e^2}{m}\psi(v_0). \tag{I-62}$$

The classical electron theory of dispersion can be briefly summarized as follows. The equation of motion of a "dispersion electron" of natural frequency $v_{0j} = \omega_{0j}/2\pi$ under an electric field $\mathbf{E} = \mathbf{E}_0 e^{i\omega t}$ is

$$m(\ddot{r} + \omega_{0j}^2 \mathbf{r} + \gamma\dot{\mathbf{r}}) = e\mathbf{E}_0 e^{i\omega t}. \tag{I-63}$$

The solution in the form $\mathbf{r} = \mathbf{r}_0 e^{i\omega t}$ is given by

$$e\mathbf{r} = \alpha\mathbf{E}_0 e^{i\omega t}, \tag{I-64}$$

where

$$\alpha = \frac{e^2}{m} \frac{1}{\omega_{0j}^2 - \omega^2 + i\gamma\omega}. \tag{I-65}$$

If there are $N_j$ dispersion electrons of frequency $\omega_{0j}$ per unit volume, the induced moment per unit volume is

$$\mathbf{P}_j = \frac{e^2 N_j}{m} \frac{\mathbf{E}}{\omega_{0j}^2 - \omega^2 + i\gamma\omega}. \tag{I-66}$$

We shall neglect the damping term containing $\gamma$, in the following, so that of the complex index of refraction $n = n_r - i n_i$, we take only the real part and obtain[i]

$$n_r^2 = 1 + \frac{e^2 N_j}{\pi m} \frac{1}{v_{0j}^2 - v^2}. \tag{I-67}$$

If there are dispersion electrons of various frequencies $v_{0j}$, ($f_j$ per atom, $N_j$ per unit volume), then (I-65), (I-66) and (I-67) will have to be summed over $j$,

$$\alpha = \frac{e^2}{4\pi^2 m} \sum_j \frac{f_j}{v_{0j}^2 - v^2}, \tag{I-65a}$$

$$\mathbf{P} = \frac{e^2 \mathbf{E}}{4\pi^2 m} \sum_j \frac{N_j}{v_{0j}^2 - v^2}, \tag{I-66a}$$

$$n_r^2 = 1 + \frac{e^2}{\pi m} \sum_j \frac{N_j}{v_{0j}^2 - v^2}. \tag{I-67a}$$

The above classical theory has been simply "translated" into the quantum theory by R. Ladenberg (1921) as follows. If there are $N_j(N_k)$ atoms, per unit volume, in the ground state $j$ (excited state $k$), then the rate of energy absorption from radiation field of energy density $\psi(v)\,dv$ is, by Einstein's theory (Sec. 6),

$$I_{qu} = N_j B_j^k \psi(v) hv,$$

$$= N_j \frac{c^3}{8\pi v^2} A_j^k \psi(v), \tag{I-69}$$

---

[i] On including the damping, the expression $1/(v_{0j}^2 - v^2)$ in (I-67) will have to be replaced by

$$\frac{v_{0j}^2 - v^2}{(v_{0j}^2 - v^2)^2 + \left(\dfrac{\gamma v}{2\pi}\right)^2}. \tag{I-68}$$

The damping makes $n_r$ finite at resonance $v = v_{0j}$, and gives rise to what is known as anomalous dispersion, i.e., $n_r$ decreases with increasing $v$ for $v < v_{0j}$. The same effect applies to all the quantum theory expressions (I-65Q), (I-66Q), (I-67Q) below. The anomalous dispersion near the spectral lines has been experimentally found.

where

$$v = v_{kj} = (E_k - E_j)/h.$$

One identifies $I_{cl}$ with $I_{qu}$, thus leading to the replacement of $f_j$, $N_j$ by $f_{jk}$ and $N_j f_{jk}$

$$f_i \to \frac{mc^3}{2(2\pi e v_{kj})^2} A_j^k \equiv f_{kj},$$

$$N_j \to N_j \frac{mc^3}{2(2\pi e v_{kj})^2} A_j^k \equiv N_j f_{kj}, \tag{I-70}$$

i.e., $f_i$, $N_j$ are replaced by $f_{kj}$, $N_i f_{kj}$ in the quantum theory, which are no longer integers. The $f_{kj}$ are called "the oscillator strengths"; they can be calculated in quantum mechanics.

By means of (I-70), one gets the quantum theory expressions

$$\alpha_j = \frac{e^2}{4\pi^2 m} \frac{f_{kj}}{v_{kj}^2 - v^2}, \tag{I-65Q}$$

$$\mathbf{P}_j = \frac{e^2 \mathbf{E}}{4\pi^2 m} \frac{f_{kj}}{v_{kj}^2 - v^2}, \tag{I-66Q}$$

$$n_r^2 = 1 + \frac{e^2 N_j}{\pi m} \frac{f_{kj}}{v_{kj}^2 - v^2}. \tag{I-67Q}$$

The results of Ladenberg were modified by H. A. Kramers (1924) by taking the state $j$ itself to be an excited state so that there are lower states $i$, $E_j > E_i$ below $j$. The expressions (I-65Q), (I-66Q), (I-67Q) must be summed over all states $k$, $E_k > E_j$, for absorption processes from $j$ to $k$, but must also be summed over all states $i$, $E_j > E_i$, for induced emission processed (negative absorption) from $j$ to $i$. Thus

$$\alpha_i = \frac{e^2}{4\pi^2 m} \left( \sum_k \frac{f_{kj}}{v_{kj}^2 - v^2} - \sum_i \frac{f_{ji}}{v_{ji}^2 - v^2} \right), \tag{I-71}$$

$$n_r^2 = 1 + \frac{N_j e^2}{\pi m} \left( \sum_k \frac{f_{kj}}{v_{kj}^2 - v^2} - \sum_i \frac{f_{ji}}{v_{ji}^2 - v^2} \right). \tag{I-72}$$

These differ from the classical theory expressions (I-67), (I-66) respectively in (i) having the non-integral $f_{kj}$, $f_{ji}$ replacing the integral numbers of dispersion electrons, (ii) having the oscillator frequency $v_{kj}$ replaced by the transition frequencies $v_{kj}$, $v_{ji}$ in the spectrum, and (iii) the presence of the negative absorption terms in the quantum theory.

The result in (I-72) was studied experimentally by Ladenberg and H. Kopfermann (1928) for a line in the spectrum of neon.

Kramers and Heisenberg (January, 1925) extended this work to a multiply periodic system under an incident electric field. For a multiply periodic system, the electric moment $\mathbf{M}_0(t)$ of the unperturbed system may be expressed in the generalized Fourier series (0-70), with (0-68),

$$\mathbf{M}_0 = \frac{1}{2}\sum_{\{\tau\}} \mathbf{C}_{\{\tau\}} e^{2\pi i[(\tau\cdot v)t + (\tau\cdot\delta)]}, \tag{I-73}$$

where

$$\{\tau\} \equiv \tau_1, \tau_2, \ldots, \tau_n, \tag{I-74}$$

$$(\tau\cdot v) \equiv \tau_1 v_1 + \tau_2 v_2 + \cdots + \tau_n v_n, \tag{I-75}$$

$$(\tau\cdot\delta) \equiv \tau_1\delta_1 + \tau_2\delta_2 + \cdots + \tau_n\delta_n, \tag{I-76}$$

the summation being taken over all $\tau_j$ from $-\infty$ to $\infty$, and the $C_{\{\tau\}}$ go to their complex conjugate when the $\tau \to -\tau$, so that $M$ is real. We shall introduce the abbreviations

$$\vartheta \equiv (\tau\cdot v), \qquad \vartheta' \equiv (\tau'\cdot v), \tag{I-77}$$

$$\tau_j^0 \equiv \tau_j + \tau_j', \qquad \vartheta_0 \equiv \vartheta + \vartheta',$$

$$\delta_0 \equiv (\tau^0\cdot\delta),$$

$$\frac{\partial}{\partial J} \equiv \sum_j \tau_j \frac{\partial}{\partial J_j}, \qquad \frac{\partial}{\partial J'} \equiv \sum_j \tau_j' \frac{\partial}{\partial J_j} \tag{I-78}$$

(see (I-41a)). The induced electric moment (along $E$) under an electric field

$$\mathbf{E} = \mathbf{E}_0 e^{2\pi i v t}$$

is then

$$\mathbf{M}(\mathbf{E}) = \mathrm{Re}\left[\sum_{\{\tau\}} e^{2\pi i[(\vartheta_0 + v)t + \delta_0]} \sum_{\{\tau'\}} \left(\frac{C_{\{\tau'\}}}{\vartheta' + v}\frac{\partial C_{\{\tau\}}}{\partial J'} - C_{\{\tau\}}\frac{\partial}{\partial J}\frac{C_{\{\tau'\}}}{\vartheta' + v}\right)\frac{\mathbf{E}_0}{4}\right], \tag{I-79}$$

Re standing for "taking the real part of". This induced moment has all frequencies $\vartheta_0 + v$, i.e.,

$$(\tau_1 + \tau_1')v_1 + (\tau_2 + \tau_2')v_2 + \cdots + (\tau_n + \tau_n')v_n + v \tag{I-80}$$

for all $+$ and $-$ $\tau_j$ and $\tau_j'$. For that component of $\mathbf{M}(\mathbf{E})$ which has the same frequency as the external $\mathbf{E}$, i.e., for $\tau_j' = -\tau_j$ for all $j$, the $\mathbf{M}(\mathbf{E})$ is[j]

---

[j] The components having all other frequencies $\vartheta_0 + v$ are those giving rise ot the Raman effect, as shown a bit earlier by Smekal (1923).

$$\mathbf{M}(v) = \mathrm{Re}\left\{ e^{2\pi i v t} \sum_{\{\tau\}} \frac{\partial}{\partial J}\left(\frac{|C_{\{\tau\}}|^2}{\vartheta - v}\right) \frac{\mathbf{E}_0}{4}\right\},$$

the summation being over $\tau_1, \tau_2, \ldots, \tau_n$ from $-\infty$ to $\infty$. The sum may be rewritten

$$\sum_{\{\tau\}}' \frac{\partial}{\partial J}\left(\frac{|C_{\{\tau\}}|^2}{\vartheta - v} + \frac{|C_{\{\tau\}}|^2}{\vartheta + v}\right) = \sum_{\{\tau\}}' \frac{\partial}{\partial J}\left(\frac{2\vartheta|C_{\{\tau\}}|^2}{\vartheta^2 - v^2}\right), \tag{I-81}$$

where the $\sum'$ is over $\tau_1, \tau_2, \ldots, \tau_n$ such that $\vartheta \equiv (\tau \cdot v) > 0$.

The above expression is the classical theory electric moment for the dispersion.

To translate this into the quantum theory, Kramers and Heisenberg appealed to the correspondence principle in the form (I-46) in which $\partial/\partial J$ is replaced by taking difference, in the sense of (I-71), i.e.,

$$\sum_{\{\tau\}}' \frac{\partial}{\partial J}\left(\frac{2\vartheta|C_{\{\tau\}}|^2}{\vartheta^2 - v^2}\right) \rightarrow \sum_{v_{kj}>0} \frac{2v_{kj}|C_{kj}|^2}{h(v_{kj}^2 - v^2)} - \sum_{v_{ji}>0} \frac{2v_{ji}|C_{ji}|^2}{h(v_{ji}^2 - v^2)}. \tag{I-82}$$

The relation between the amplitude $C_{kj}$ of the electric moment and the Einstein coefficient $B_j^k$, and hence $A_j^k$, is

$$A_j^k = \frac{16\pi^4 v_{kj}^3}{3hc^3}|C_{kj}|^2. \tag{I-83}$$

With the relation (I-70) between $A_j^k$ and $f_{kj}$, Kramers and Heisenberg obtained (I-71) again. They further obtained the following relation

$$\sum_{E_k > E_j} f_{kj} - \sum_{E_i < E_j} f_{ji} = 1 \tag{I-84a}$$

or

$$\frac{2\pi m}{e^2}\left[\sum_k v_{kj} C_{kj}^2 - \sum_i v_{ji} C_{ji}^2\right] = \frac{h}{2\pi}, \tag{I-84b}$$

where $v_{kj} > 0$ and $v_{ji} > 0$. This was found a little earlier by Thomas and by Kuhn, and is known as the Thomas-Kuhn sum rule. We shall not reproduce Kramers-Heisenberg's proof of it here, but shall obtain it in quantum mechanics.[k] This relation is in fact the fundamental commutation relation

$$pq - qp = \frac{\hbar}{i}$$

in disguise [see Chap. 2, Sec. 4].

This work has been most important to Heisenberg, for as we shall see in Chap. 2, Sec. 1, it is this work that led Heisenberg to the new ideas that

---

[k] See Chap. 7, (VII-68).

developed into matrix mechanics, in collaboration with M. Born and P. Jordan.

### References

The classical theory of the scattering of light is due mainly to Lord Rayleigh, *Phil. Mag.* IV, **41**, 107, 274, 447 (1871); *ibid.* V, **12**, 81 (1881); *Nature* **60**, 64, (1899); *Phil. Mag.* V, **47**, 375 (1899), also see J. J. Thomson, *Conduction of Electricity Through Gases*, 2nd ed. 321, (1906).

R. Ladenburg, *Z. Phys.* **4**, 451–471 (1921); **48**, 15 (1928); R. Ladenburg and F. Reiche, *Naturwiss.* **11**, 584–598 (1923).

H. A. Kramers, *Nature* **113**, 673 (May, 1924); **114**, 310, (1924).

M. Born, *Z. Phys.* **26**, 379–395 (June 13, 1924).

H. A. Kramers and W. Heisenberg, *Z. Phys.* **31**, 681–707 (1925).

A. Smekal, *Naturwiss.* **11**, 873–875 (1923).

C. V. Raman, *Indian Jour. Phys.* **2**, 387–398 (1928); C. V. Raman and K. S. Krishnan, *ibid.* 399 (1928).

G. Landsberg and L. Mandelstam, *Naturwiss.* **16**, 557 (1928); *Z. Phys.* **1**, 769 (1928).

W. Thomas, *Naturwiss.* **13**, 627 (1925).

W. Kuhn, *Z. Phys.* **33**, 408 (1925).

### 9.   Atomic spectra; empirics (up to 1925)

The initial signal success of Bohr's theory of the hydrogen atom was followed by many great experimental and theoretical developments in the years 1913–26, namely, Moseley's work on the x-ray spectra of atoms in 1913; Franck and Hertz's experiments on the excitation of atoms by electron collisions in 1914; Sommerfeld's theory of the fine structure of energy states in 1915; Stern and Gerlach's experiments on "spatial quantization" in 1921; Bohr's correspondence principle in 1918; Landé's $g$ formula for the anomalous Zeeman effect in 1923; Pauli's exclusion principle and theory of the periodic table in 1925; and at the end of an era, so to speak, Uhlenbeck and Goudsmit's hypothesis of electron spin in 1925–6. All these developments deal with the spectra and structure of atoms. They lent support to the general scheme of Bohr's theory of the atom; they spurred intensive studies of atomic spectroscopy in that period; but they, by the results they have obtained, also began to give cause to some physicists to doubt the generality and validity of the quantum theory as represented by Bohr's theory. Thus, while it was from the study of atomic structure and spectra that the quantum theory received its greatest support, it was also the study of atomic structure and spectra that initially furnished the crucial tests and the support of the new quantum mechanics. For a better appreciation of the birth of quantum mechanics, we shall give a brief account of the state of atomic physics just before the dawn of quantum mechanics.

### (1) Empirics: spectral terms, series, multiplets

We may perhaps begin with the discovery by J. J. Balmer (1825–98) in 1885 of the series relationship for the frequencies of the lines of hydrogen

$$v = R\left(\frac{1}{4} - \frac{1}{n^2}\right), \qquad n = 3, 4, 5, \ldots \tag{I-85}$$

where the constant $R = 109721.6 \text{ cm}^{-1}$.

In 1883, G. D. Liveing (1827–1924) and J. Dewar (1842–1923) discovered that in the spectra of the alkali metal and alkaline earth atoms, spectral lines form "series" to which H. Kayser (1853–1940) and C. Runge (1856–1927) gave (1890) the names *principal, sharp, diffuse* series. Rydberg (1854–1919) in 1890 and 1896 gave the general series formula for the alkali metal atoms:

Principal series (of doublets) (for fixed $n$, $s$),

$$v = \frac{R}{(n + s)^2} - \frac{R}{(m + p_j)^2}, \qquad \begin{array}{l} m = n + 1, n + 2, n + 3, \ldots \\ j = 1, 2 \end{array}$$

Sharp series (of doublets) (for fixed $n$, $p_{1,2}$),

$$v = \frac{R}{(n + p_j)^2} - \frac{R}{(m + s)^2}, \qquad \begin{array}{l} m = n + 1, n + 2, \ldots \\ j = 1, 2 \end{array} \tag{I-86}$$

Diffuse series (of triplets) (for fixed $n$, $p_{1,2}$),

$$v = \frac{R}{(n + p_j)^2} - \frac{R}{(m + d_k)^2}, \qquad m = n + 1, n + 2, \ldots$$

$R$ is the above constant ($109721.6 \text{ cm}^{-1}$) for all series and all atoms. $s$, $p_1$, $p_2$, $d_1$, $d_2$ are constants in each series; they are called "Rydberg corrections". (I-86) shows that the frequencies of the observed spectral lines of each series can be represented by the difference between a fixed "term" and a series of "terms"; thus, for the principal, sharp and diffuse series, of Na atom, say, (I-86) can be expressed respectively as

$$\text{Principal} \qquad v = T(3\,^2S_{1/2}) - T(n\,^2P_{1/2, 3/2}), \quad n = 3, 4, \ldots,$$

$$\text{Sharp} \qquad v = T(3\,^2P_{1/2, 3/2}) - T(n\,^2S_{1/2}), \quad n = 4, 5, \ldots, \tag{I-86a}$$

$$\text{Diffuse} \qquad v = T(3\,^2P_{1/2, 3/2}) - T(n\,^2D_{3/2, 5/2}), \quad n = 3, 4, \ldots,$$

and another series is known

$$\text{Fundamental} \quad v = T(3\,^2D_{3/2, 5/2}) - T(n\,^2F_{5/2, 7/2}), \quad n = 4, 5, \ldots \tag{I-86b}$$

having low frequencies suggesting the name the "fundamental series" (although there is nothing "fundamental" about it). The terms have numerical values determined empirically from the observed frequencies of spectral lines according to (I-86), i.e., all terms $T(m\,{}^2P)$, $T(m\,{}^2S)$, etc. approach the limit value 0 as $m \to \infty$.

It was noted by W. Ritz (1878–1909) in 1908 that in a spectrum, in addition to the lines that are represented in the various series in (I-86a), there are other lines not included in these series. For example, in the spectrum of Na, such other combinations of terms are also observed,

$$T(4\,{}^2S_{1/2}) - T(n\,{}^2P_{1/2,\,3/2}), \quad n = 4, 5, \ldots$$

$$T(4\,{}^2P_{1/2,\,3/2}) - T(n\,{}^2S_{1/2}), \quad n = 5, 6, \ldots$$

$$T(3\,{}^2D_{3/2,\,5/2}) - T(n\,{}^2P_{1/2,\,3/2}), \quad n = 4, 5, \ldots$$

Thus the frequencies of spectral lines correspond to the differences between two terms in a quadratic scheme

$$\nu_{mn} = T_m - T_n. \tag{I-87}$$

This empirical fact is known as the Ritz combination principle. It plays a role in Bohr's theory, as seen from (I-20) and again in Heisenberg's formulation of his ideas on the new matrix theory.

In the notations in (I-86a) for the *terms*, four quantum numbers $n$, $S$, $L$, $J$ appear as

$$n^{2S+1}L_J. \tag{I-88}$$

### *(i) Quantum numbers $k$ and $L$*

For an individual electron (such as the valence electron of alkali metal atoms), there are the quantum numbers $n$, $k$, where $k\hbar$ is the orbital angular momentum similar to that in the hydrogen atom. For $k = 1, 2, 3, \ldots$, the electron has been given the name $s, p, d, \ldots$ electron and the state $S, P, D, \ldots$,

| $k$ | 1 | 2 | 3 | 4 | 5 | 6 | ... |
|---|---|---|---|---|---|---|---|
| electron | $s$ | $p$ | $d$ | $f$ | $g$ | $h$ | ... |
| state | $S$ | $P$ | $D$ | $F$ | $G$ | $H$ | ... |

$$\tag{I-89a}$$

This naming is purely conventional, the states being related to the spectral series of lines "sharp", "principal", "diffuse", "fundamental" in (I-86a).

For a 2- or many-electron spectrum, the *terms* are similarly designated by $S, P, D, \ldots$ but a quantum number $L$ is assigned to them as follows

| $L$ | 0 | 1 | 2 | 3 | 4 | 5 |
|---|---|---|---|---|---|---|
| State | $S$ | $P$ | $D$ | $F$ | $G$ | $H$ |

$$\tag{I-89b}$$

The meaning of the quantum number $L$ may be taken as the resultant orbital angular momentum of the electrons, according to the adaptation, by A. Landé (1919), of the method of compounding vectors in classical dynamics to the quantum theory; for example,

$$L = k_1 + k_2.$$

But as we shall see below, the empirical results seem to force one to replace $k$ by $l$ $(=k-1)$, so that the meaning of $L$ is also not clear. That $k$ must be replaced by $l$ is shown in quantum mechanics, and $L\hbar$ is given by the quantized sum

$$L = \sum l_i.$$

*(ii)* $S = 0, 1, 2, \ldots; \frac{1}{2}, \frac{3}{2}, \frac{5}{2}, \ldots$

The multiplicity of spectral terms is expressed as $2S + 1$ in terms of a quantum number $S$, whose physical meaning is not clear until the electron spin is introduced in 1925 (see Sec. 10, (v) below). The doublets $^2P$, $^2D$, etc. of the alkali metal atoms correspond to $S = \frac{1}{2}$. For the alkaline earth atoms, there are two terms systems of singlets and triplets with $S = 0$ and 1 respectively. The multiplicities of the spectra of atoms having odd (even) number of valence electrons are even (odd), as shown in the following table

| $S$ multiplicity | No. of electrons | 1 | 2 | 3 | 4 | 5 |
|---|---|---|---|---|---|---|
| 0 | 1 | | singlet | | singlet | |
| $\frac{1}{2}$ | 2 | doublet | | doublet | | doublet |
| 1 | 3 | | triplet | | triplet | |
| $\frac{3}{2}$ | 4 | | | quartet | | quartet |
| 2 | 5 | | | | quintet | |
| $\frac{5}{2}$ | 6 | | | | | sextet |

$$\text{(I-90)}$$

Empirically, in the spectra of *light* atoms (such as He), no lines are observed (or, only with extremely low intensities) corresponding to combinations between terms of different multiplicities, (such as singlet—triplet $^1S$—$^3P$ in He). This may be expressed by the selection rule

$$\Delta S = 0. \qquad \text{(I-91)}$$

For heavy atoms (such as Hg), transitions between what are *designated* as "singlet" and "triplet" states $^1S$, $^3P$ are not only observed, but have strong intensities. In such cases, it means that the designation of a state (term) by a quantum number $S$ does not have very rigorous meaning.[1]

### (iii) j and J

In the doublet spectrum of an alkali metal atom (such as Na), the states (terms) $^2P$, $^2D$, $^2F$, etc. are doublets. In order to differentiate the two levels of a doublet, and also to express the observed fact that in a transition between two doublets only three lines instead of the expected $2 \times 2 = 4$, as shown in (I-92), are found

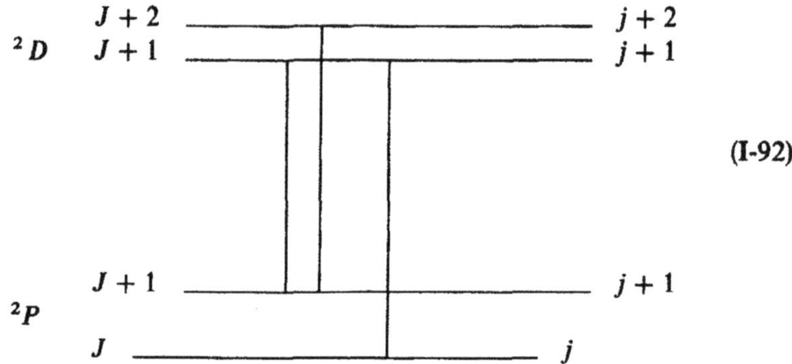


$$\text{(I-92)}$$

Sommerfeld in 1920 introduced a quantum number $j$, so that $j\hbar$ is an angular momentum. Its physical meaning was not clear (until after the introduction of electron spin in 1925); it was thought to be the total angular momentum of the orbital motion of the (valence) electron and the core electrons. In any case, the relative values of $j$ are so assigned that the empirical selection rule

$$\Delta j = 0, \pm 1, \tag{I-93}$$

expresses the observed three lines (see I-92). But this rule fixes $j$ only up to an additive constant.

Later, $j$ was completely fixed by studies of the Zeeman effect. In a magnetic field, the components of $j$ taken on the quantized values

$$-j \leq m \leq j, \tag{I-94a}$$

which are $2j + 1$ in number. Thus a simple counting gives directly the values[m] of $j$. In the following subsection (2) on the anomalous Zeeman effect, we see

---

[1] In quantum mechanics, we express this situation by saying that $S$ is not a "good" quantum number. The reason is that in heavy atoms, the spin-orbit interaction is strong, see Sec. 10, (v) below.

[m] For this reason, the study of the Zeeman effect is much more "useful" in the analysis of atomic spectra than that of the Stark effect.

that for $^2S$, $j = \frac{1}{2}$, and for $^2P$, $j = \frac{1}{2}, \frac{3}{2}$. The meaning of these empirical half-integral values must await the discovery of the electron spin. For the moment, we shall regard $j$ as an empirical quantum number.

But if $j\hbar$ is thought to be resultant of $k$ and another angular momentum, say $r$ (such as that of the core, according to Landé and Sommerfeld), then, by the vector composition rule (Landé, 1919), $j$ should be capable of the values $k - r \leq j \leq k + r$, and if $k < r$; and since $k$ starts from 1, $j$ should have the values $r - 1, r, r + 1$ for $k = 1$; and if $k \geq r$, for $k = 1$, $j$ should have the value $1 - r, 1, 1 + r$, if $r$ is integral. But the *observed* fact is that $^2S$ state is a *single* level having only one value of $j$, which is $j = \frac{1}{2}$ from the Zeeman effect. This forces one to start with $l (= k - 1) = 0$, and $r = \frac{1}{2}$, so that we have for the Na atom, say,

$$^2S_{1/2}, \quad ^2P_{1/2, 3/2}, \quad ^2D_{3/2, 3/2}, \quad \cdots$$

But this $l (= k - 1)$ replacing $k$ and the meaning of $j = l + r$, $r = \frac{1}{2}$, are not understandable!

For a 2- or many electron opectrum, we use the capital $J$ in place of $j$, and the spectroscopic notation is $^{2S+1}L_J$ as in (I-88). The meaning of $J$ will be clear after electron spin is found. (see Sec. 10, (6) below)


*(iv) m and M*

As already shown in (I-94a), in a magnetic field, the angular momentum components $m$ are quantized to the values

$$-j \leq m \leq j.$$

For a 2- or many-electron spectrum, such as He, Mg, the spectroscopic notation for the components of $J$ (along a $z$-axis) is the capital $M$, so that

$$-J \leq M \leq J. \tag{I-94b}$$

For $^3P$ of He, or Mg, the values of $J$ are $J = 0, 1, 2$, and $M = 0; -1, 0, 1; -2, -1, 0, 1, 2$ respectively.


*(v) Parity*

From the analysis of the complex spectrum of iron, Otto Laporte (1924) found that all the observed lines correspond to transitions between the states of one group and the states of another group, but never to transitions among states of the *same* group. The empirical selection rule is known as the Laporte rule. The interpretation of this rule is furnished in quantum mechanics by the concept of symmetry properties of the state (wave functions) with respect to the inversion $(x, y, z) \rightarrow (-x, -y, -z)$. States which are symmetric (antisymmetric) with respect to this inversion are said to be of even (odd) parity. The Laporte rule states that, for electric dipole radiation, the selection rule is

$$\begin{Bmatrix} \text{even} \\ \text{odd} \end{Bmatrix} \text{parity} \leftrightarrow \begin{Bmatrix} \text{odd} \\ \text{even} \end{Bmatrix} \text{parity}. \qquad (\text{I-95})$$

The proof of this selection rule is very simple and will be given in a later chapter.

*(vi)* In (I-86a), there is a number $n$ preceding the notation $^{2S+1}L_J$. In some cases, such as Na, K, He atoms, $n$ may stand for $3s(^2S)$, $3p(^2P)$, or $1s2p(^1P, {}^3P)$, etc., but in other case $n$ may simply be an ordinal number when not enough information is available for a more explicit designation.

Concerning the various quantum numbers $L$, $S$, $J$, $M$, parity, we shall emphasize the following. Of these only $J$ and $M$ have rigorous meaning, while $L$, $S$ are "good quantum numbers" for light atoms and are not so for heavy atoms. For the same reason, the spectroscopic notation $^{2S+1}L_J$ is meaningful for light atoms and only approximate for heavier atoms.

All the statements in this section are to be considered of an empirical nature; their meaning will be clear in quantum mechanics.

### (2) Anomalous Zeeman effect; Paschen-Back effect

After the discovery of the (normal) Zeeman effect in 1896 and its explanation by H. A. Lorentz (1897) on the classical electron theory (see Sec. 7), F. Paschen and E. Back (1912) studied the Zeeman effect of the spectral lines $^2S_{1/2}$—$^2P_{1/2,\,3/2}$ and others of the alkali metal atoms. It is found that for "weak" magnetic fields, the Zeeman pattern is very complicated and is called the anomalous Zeeman effect; and for "strong" magnetic fields, the pattern tends to approach that of the normal Zeeman effect." In Fig. (I-1a), are shown the doublet $^2S_{1/2}$—$^2P_{1/2,\,3/2}$ of Na and also the Zeeman pattern for weak field. When the radiation is viewed along the magnetic field $\mathscr{H}$, the $^2S_{1/2}$—$^2P_{1/2}$, $^2S_{1/2}$—$^2P_{3/2}$ are split into 2 and 4 components respectively; when viewed in directions perpendicular to $\mathscr{H}$, they are split into 4 and 6 components respectively. The frequency shifts of these components from the field-free positions in units of $\mu_B\mathscr{H}/h$ are shown also in the figure.

In Fig. (I-1b) is shown, schematically, the field-free doublet $^2S_{1/2}$—$^2P_{1/2,\,3/2}$ on the left, and the strong field pattern on the right. From the latter, it is seen that, if one neglects the "small" splitting of the two outside components, the pattern is that of a "triplet" of the normal Zeeman effect. This is called the Paschen-Back effect.

---

" A magnetic field $\mathscr{H}$ is said to be weak (strong) if the energy shift $(\mu_B\mathscr{H})$ is small (large) compared with the doublet separation $^2P_{1/2} - {}^2P_{3/2}$, i.e., if $(\mu_B\mathscr{H})$ is small (large) compared with the spin-orbit interaction (see I-102) below.

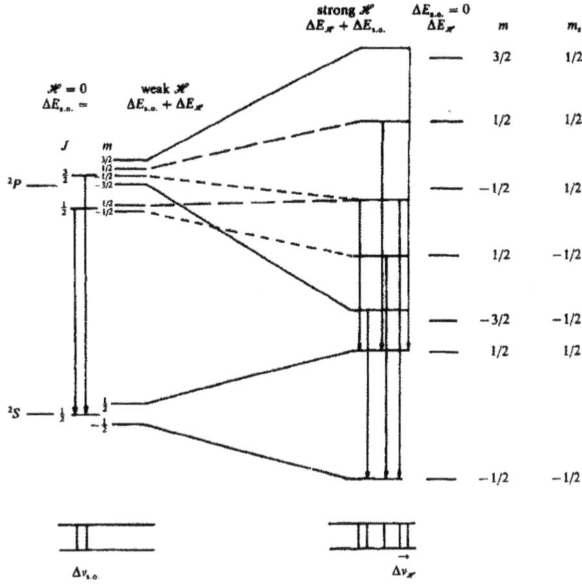

Fig. I-1a. $^2S_{1/2} - {}^2P_{1/2,\,3/2}$ doublet on the left; Paschen-Bach pattern on the right (with $\Delta v$ due to $\mathscr{H}$ drawn on a reduced scale).

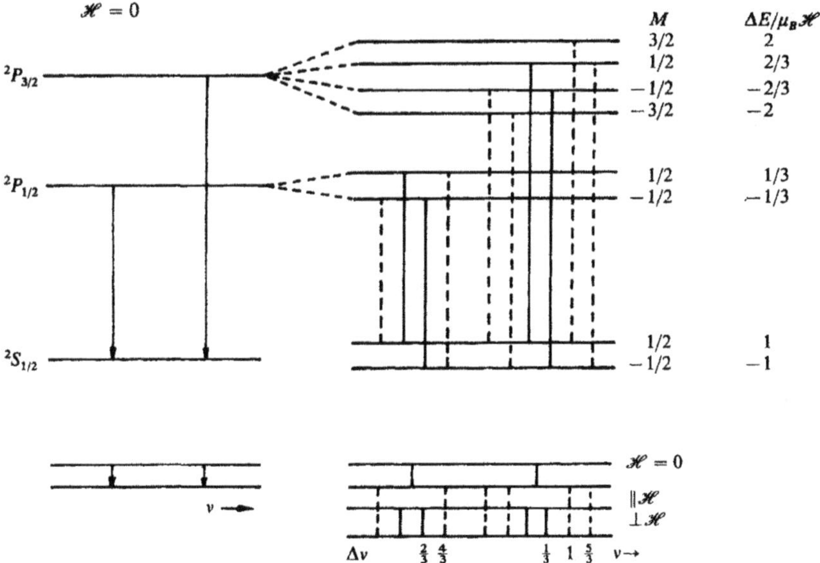

Fig. I-1b. Anomalous Zeeman effect of $^2S_{1/2} - {}^2P_{1/2,\,3/2}$ doublet $\|\mathscr{H}$, $\perp\mathscr{H}$ signify direction of observation $\|$, $\perp$ to $\mathscr{H}$. Dotted lines are circularly polarized; solid lines are linearly polarized along $\mathscr{H}$.

For other multiplets, the anomalous Zeeman effect patterns are different and complicated.

For this anomalous Zeeman effect, A. Landé (1923) obtained an amazing, empirical formula for the energy change $\Delta E_{\mathscr{H}}$ of a state $^{2S+1}L_J$ in a magnetic field $\mathscr{H}$

$$\Delta E_{\mathscr{H}} = Mg\mu_B\mathscr{H} \qquad \left(\mu_B = \frac{e\hbar}{2mc}\right), \tag{I-96}$$

where

$$g = 1 + \frac{J(J+1) + S(S+1) - L(L+1)}{2J(J+1)}, \tag{I-97a}$$

and $M$ is as (I-94),

$$-J \le M \le J. \tag{I-94}$$

For the states $^2S_{1/2}$, $^2P_{1/2,\,3/2}$,

$$\begin{aligned}
^2S_{1/2}, &\qquad g = 2, \\
^2P_{1/2}, &\qquad g = \tfrac{2}{3}, \\
^2P_{3/2}, &\qquad g = \tfrac{4}{3}.
\end{aligned} \tag{I-97b}$$

The calculated $\Delta E_{\mathscr{H}}$ and the Zeeman pattern agree with the observed spectrum shown in Fig. (I-1a).

The "Landé $g$-factor" (I-97) is amazing as the physical meanings of $S$ and $J$ were not known in 1923. Furthermore, for the $^2S_{1/2}$ ($L = 0$) state, the ratio of the magnetic moment to the angular momentum is

$$\frac{Mg\mu_B}{M\hbar} = g\frac{e}{2mc} = 2\left(\frac{e}{2mc}\right) \tag{I-97c}$$

which is twice the classical value $e/2mc$!

Thus the anomalous Zeeman effect presents to physicists a number of questions such as the meaning of $J$, $S$, $g$.


## References


A. Sommerfeld, *Annalen d. Physik* **63**, 221 (1920); **70**, 32 (1923) introduced the quantum number $j$.

O. Laporte, *Z. Phys.* **23**, 135; **24**, 1 (1924) discovered the Laporte rule (parity rule).

F. Paschen and E. Back, *Annalen d. Physik* **39**, 897 (1912); **40**, 960 (1913), discovered the anomalous Zeeman effect for the alkali atoms and the Paschen-Back (strong magnetic field) effect.

A. Landé, *Z. Phys.* **40**, 189 (1923), discusses the Landé $g$-formula.

## 10.  Electron spin (1925)[o]

### (1) Spin-orbit interaction

In the preceding section, we have seen that by the mid 1920's, there are quite a few outstanding problems, namely, the multiplet structure of atomic spectra, the anomalous Zeeman effect, the Landé $g$-formula, and the Stern-Gerlach experiment on the silver atom.

In October, 1925, G. E. Uhlenbeck (1901–    ) and S. A. Goudsmit (1902–1978), both students of P. Ehrenfest at Leiden, proposed that the electron has an intrinsic angular momentum $s\hbar$, with $s = \frac{1}{2}$, and a magnetic moment $\mu_s = g(e/2mc)s\hbar = gs\mu_B$, with $g = 2$, as found by Landé in his $g$-formula (see I-97b, c). Consider an electron moving with a velocity $v$ relative to a nucleus of charge $Ze$. The motion of the nucleus relative to the electron produces at the point where the electron is, a magnetic field $\mathbf{H}$ which according to the Biot-Savart law is

$$\mathbf{H} = \frac{[\mathbf{r} \times Ze\mathbf{v}]}{cr^3} = \frac{Ze[\mathbf{r} \times m\mathbf{v}]}{mcr^3},$$

where $\mathbf{r}$ is the radius vector from the electron to the nucleus. $[-m\mathbf{v} \times \mathbf{r}]$ is the angular momentum of the electron around the nucleus, and according to the quantum theory,

$$[-m\mathbf{v} \times \mathbf{r}] = [\mathbf{r} \times m\mathbf{v}] = \mathbf{l}\hbar.$$

The interaction energy between the magnetic moment $\mu_s$ and the field $\mathbf{H}$ arising[p] from the relative orbital motion of the electron and the nucleus is

$$\Delta E_s = (\mu_s \cdot \mathbf{H})$$

$$= 2g\mu_B^2 \frac{\bar{Z}}{r^3}(\mathbf{l} \cdot \mathbf{s}). \tag{I-98}$$

This bar indicates an averaging over one period of the electron motion. If this is done by quantum mechanics, one gets

$$\frac{\bar{Z}}{r^3} = \frac{Z^4}{a_0^3 n^3 l(l + \frac{1}{2})(l + 1)}, \qquad a_0 = \frac{\hbar^2}{me^2}, \tag{I-99}$$

---

[o] We shall depart from a strict chronological treatment here by first discussing the introduction of the electron spin (October, 1925), before taking up Pauli's exclusion principle which was published early in 1925. In fact Uhlenbeck and Goudsmit read Pauli's paper while working on their electron spin hypothesis. For clarity, we shall first present the theory of electron spin as we know it later, and then come back to the historical developments.

[p] This will be called "the spin-orbit interaction" for brevity.

and

$$(\mathbf{l}\cdot\mathbf{s}) = \tfrac{1}{2}[\,j(j+1) - l(l+1) - s(s+1)]. \tag{I-100}$$

The expression (I-98) was shown by L. H. Thomas (February, 1926) to be incorrect. When the transformation from the laboratory system (in which the nucleus is at rest) to the electron system (in which the electron is at rest) is properly carried out, the expression (I-98) has to be corrected by multiplying it by a factor $\tfrac{1}{2}$.

On putting (I-99), (I-100) into (I-98), one obtains for the two relative orientations of $l$ and $s$ the resultant $j$[q]

$$
\left.
\begin{array}{l}
j = l + s = l + \dfrac{1}{2} \\[2mm]
j = l - s = l - \dfrac{1}{2}
\end{array}
\right\}
\ \Delta E_s = \frac{R\alpha^2 Z^4 hc}{n^3 l(l+\frac{1}{2})(l+1)}
\left\{
\begin{array}{ll}
\dfrac{l}{2} & l = 0, 1, \ldots \\[3mm]
-\dfrac{l+1}{2} & l = 1, 2, \ldots
\end{array}
\right.
\tag{I-101}
$$

where

$$R = \frac{2\pi^2 m e^4}{ch^3}, \qquad \alpha = \frac{e^2}{\hbar c} \simeq \frac{1}{137}.$$

The doublet separation is

$$\Delta\nu_{\text{doublet}} = \frac{R\alpha^2 Z^4}{n^3 l(l+1)}(\text{cm}^{-1}), \qquad l = 1, 2, 3, \ldots \tag{I-102}$$

For $l = 1$, this is the doublet splitting of $^2P_{1/2, 3/2}$. With the appropriate value of "$Z^4$", this should correspond to the separation of the yellow $D$ doublet of Na at 5896, 5890 Å.

### (2) Remarks

The above treatment of the first consequence of the spin hypothesis—the spin-orbit interaction—may give the impression of being "obvious". But this is certainly not so! We shall therefore make the following remarks, to emphasize the by no means obvious nature of the concept of spin and its difficulties in the early development of this hypothesis.

(i) Historically, the need for a fourth quantum number in addition to $n$, $l$, $m$ was (empirically) realized by Pauli (late 1924 and early 1925) in his attempt to understand the periodic table of the atoms. He introduced a fourth

---

[q] The composition of angular momenta in the quantum theory is done by A. Landé, *Verhandlungen der Deutschen Physik Gesells* **21**, 585 (1919).

quantum number[r] $s$, not as one for an internal degree of freedom of rotation in the classical sense of a spinning top, but as a "two-valuedness $s = \pm\frac{1}{2}$ not understandable on classical physics". That Pauli introduced "$s$" and yet had not proposed the "spin" is most probably because of his realization of the difficulties confronting the idea of a "spin". These difficulties are many, such as:

(ii) A genuine rotational degree of freedom in the classical sense, when quantized, will have, not only two, but infinitely many states. For a charged sphere of radius equal to the classical radius of the electron ($e^2/mc^2 = 2.8 \times 10^{-13}$ cm) to spin around to an angular momentum $s\hbar$, the speed of the surface of the sphere will exceed several times that of light. The ratio of the magnetic moment to the angular momentum $\mu_s/s\hbar$ is $g(e/2mc) = e/mc$ and not the value $e/2mc$ according to classical electromagnetic theory.

(iii) Early in 1925, R. de L. Kronig, having learned of Pauli's ideas on the Exclusion Principle through A. Landé, studied electron spin. He calculated the spin-orbit interaction $\Delta E_s$ presumably as in (I-98) (of course, without the later Thomas correction), but then met the difficulty of having to include this additional energy without destroying the seemingly excellent agreement already achieved by Sommerfeld's relativistic fine-structure theory. Kronig discussed his ideas with Pauli and decided not to publish them.

(iv) When Uhlenbeck and Goudsmit later in 1925 independently discovered the idea of electron spin, they were aware of most, if not all, of these difficulties. Their first communication was sent (October, 1925) to the *Naturwisssenschaften* by Ehrenfest who told them that they were still young and could afford to be stupid. To their second communication to *Nature* (January, 1926), Bohr contributed a note expressing his acceptance of the hypothesis. But the hypothesis found general acceptance by physicists, including Pauli, only after Thomas (February, 1926) corrected the expression for the spin-orbit interaction (I-98) by a factor $\frac{1}{2}$.

(v) The difficulty about the spin-orbit interaction's destroying the agreement already achieved by Sommerfeld's relativistic correction is resolved in quantum mechanics. It is found that (1) when the relativistic correction $\Delta E_{\text{rel.}}$ is calculated, not in classical dynamics as in Sommerfeld's theory, but in quantum mechanics, (2) when the spin-orbit energy $\Delta E_s$ is also calculated in quantum mechanics (as done in (I-99)–(I-101) above), and (3) when $\Delta E_{\text{rel.}}$ and $\Delta E_s$ are combined together, the resulting $\Delta E_{\text{rel.}} + \Delta E_s$ amazingly restores the original Sommerfeld formula (I-63) (with some difference in the meaning of

---

[r] It is amazing that Pauli actually used the letter "$s$", almost as if in anticipation of the "spin", which came in October of that year.

the quantum numbers). We shall see this strange coincidence later in the quantum mechanics treatment of this problem.

(vi) The non-classical nature of the electron spin is further brought out by the relativistic wave equation of the electron of Dirac (1928). It is found that the relativistic (i.e., Lorentz covariant) wave equation, without additional assumptions, automatically yields the electron spin, namely, the intrinsic angular momentum $\frac{1}{2}\hbar$, the gyromagnetic ratio ($g = 2$), the spin-orbit interaction.

### *(3) Stern-Gerlach's experiment*

With the electron spin $s\hbar = \frac{1}{2}\hbar$, and the orbital $l\hbar$ (according to quantum mechanics, $l = 0, 1, 2, \ldots$) one obtains for the total angular momentum $j\hbar$, $j = l + s$. For $l = 0$, $j = \frac{1}{2}$, and the components $m = \pm\frac{1}{2}$. For the valence electron of Ag (or, the alkali metal atoms), $l = 0$ and the magnetic moment is $2(e/2mc)s\hbar = \mu_B$, the Bohr magneton. Thus in the Stern-Gerlach experiment, $m\mu_B = \pm\frac{1}{2}\mu_B$ as observed.

### *(4) Landé g-formula for anomalous Zeeman effect*

With the electron spin, the Landé $g$-formula can readily be understood.

For weak magnetic fields[s] $\mathscr{H}$, the $l\mu_B\mathscr{H}$ and $2s\mu_B\mathscr{H}$ are small compared with $\Delta E_s$ of (I-101), so that $l$ and $s$ form a result $j$, and $j$ precesses around $\mathscr{H}$. The interactions of $l\mu_B$ and $gs\mu_B$ with the magnetic field $\mathscr{H}$ are then

$$\Delta E = (l\cos\widehat{l\mathscr{H}} + gs\cos\widehat{s\mathscr{H}})\mu_B\mathscr{H}$$
$$= (l\cos\widehat{lj} + gs\cos\widehat{sj})\cos\widehat{j\mathscr{H}}\,\mu_B\mathscr{H}.$$

Now

$$\cos\widehat{j\mathscr{H}} = \frac{m}{j}, \quad \text{and}$$
$$2lj\cos\widehat{lj} = j^2 + l^2 - s^2$$
$$2sj\cos\widehat{sj} = j^2 + s^2 - l^2$$

and a proper calculation in quantum mechanics requires the replacement of a classical $j^2$ by $j(j + 1)$, etc., one obtains

$$\Delta E_{\mathscr{H}} = mg\mu_B\mathscr{H},$$

---

[s] See footnote at the beginning of Sec. 9, (2).

where

$$g = 1 + \frac{j(j+1) + s(s+1) - l(l+1)}{2j(j+1)} \tag{I-103}$$

which is the Landé $g$ formula (I-97).

### (5) Paschen-Bach effect for strong magnetic fields

For strong $\mathcal{H}$, $l\mu_B\mathcal{H}$, $2s\mu_B\mathcal{H}$ are large compared with the spin-orbit energy $\Delta E_s$. Let the components $l\cos\widehat{l\mathcal{H}}$, $s\cos\widehat{s\mathcal{H}}$ be denoted by $m_l$, $m_s$. Then

$$\Delta E_{\mathcal{H}} = (m_l + 2m_s)\mu_B\mathcal{H}$$

and from (I-98) with the Thomas correction,

$$\Delta E_s = 2\mu_B^2\frac{\bar{Z}}{r^3}(\mathbf{l}\cdot\mathbf{s}) \equiv \zeta(\mathbf{l}\cdot\mathbf{s}) \tag{I-98}$$

$$= m_l m_s \zeta$$

with

$$\Delta E_{\mathcal{H}} \gg \Delta E_s.$$

If we neglect $\Delta E_s$ compared with $\Delta E_{\mathcal{H}}$, we have for the energy change

$$\Delta E = \Delta E_{\mathcal{H}} + \Delta E_s$$

$$\simeq \Delta E_{\mathcal{H}} = (m_l + 2m_s)\mu_B\mathcal{H}. \tag{I-104}$$

As an example, take the $^2S_{1/2}$—$^2P_{1/2,\,3/2}$ doublet of the alkali atoms. For

$$^2P, \quad l = 1, \quad s = \tfrac{1}{2}, \quad m_l + 2m_s = -2, -1, 0, 1, 2$$

$$^2S, \quad l = 0, \quad s = \tfrac{1}{2}, \quad m_l + 2m_s = -1, 0, 1.$$

If the selection rule is

$$\Delta m = \Delta(m_l + m_s) = 0, \pm 1, \tag{I-105}$$

one obtains the pattern of the normal Zeeman effect as seen in Fig. (I-1b).

### (6) Quantum number L, S, J

With the electron spin, we can now have the meaning of the empirical quantum numbers $L$, $S$, $J$ of Sec. 9, (1).

For simplicity, let us take a 2-electron atom (He, Li$^+$, or Mg, Ca having two valence electrons). Let $l_1$, $l_2$, $s_1$, $s_2$ be the orbital and spin angular momenta of the two electrons, in $\hbar$. Let us assume that the spin-orbit energy

$\Delta E_s$ of (I-98) is small,[1] that $l_1$, $l_2$ form a resultant orbital angular momentum $L = l_1 + l_2$, which then takes on the quantized values

$$|l_1 - l_2|, |l_1 - l_2| + 1, \ldots, \qquad \ldots, |l_1 + l_2| - 1, |l_1 + l_2|,$$

that $s_1$, $s_2$ for a resultant spin angular momentum $S = s_1 + s_2$, which takes on the quantized values

$$S = 0, 1. \tag{I-106}$$

Then $L$ and $S$ finally form a total angular momentum

$$J = L + S \tag{I-107a}$$

and

$$|L - S| \leq J \leq |L + S|. \tag{I-107b}$$

For a 2-electron atom, corresponding to (I-106), we have

$$S = 0, \quad J = L$$
$$S = 1, \quad J = L - 1, L, L + 1.$$

As an example, $l = 0, s_1 = \frac{1}{2}, l = 1, s_2 = \frac{1}{2}$, then

$$S = 0, 1$$
$$L = 1$$
$$J = 1, 0, 1, 2$$

and the states in spectroscopic notations are

$$^1P_1, \qquad ^3P_0, \qquad ^3P_1, \qquad ^3P_2.$$

The above angular momentum composition scheme is called $\{L, S\}$ coupling, or Russell-Saunders coupling, scheme. It can be generalized to any number of electrons. But as already pointed out in the preceding footnote, this $\{L, S\}$ coupling scheme

$$L = \sum l_i, \tag{I-108}$$
$$S = \sum s_i,$$
$$J = L + S,$$

is valid for small spin-orbit interaction $\Delta E_s$, and this condition is best satisfied

---

[1] As seen from (I-99), $\Delta E_s$ is proportional to $Z^4$ so that $\Delta E_s$ is small for light atoms. For heavy atoms, $\Delta E_s$ is large. This is the case for Hg, and for the x-ray energy levels of medium and heavy atoms.

in the case of light atoms. It follows that the quantum numbers $S$, $L$ have very definite meaning also only in the case of light atoms.

For heavy atoms such as Hg, $\Delta E_s$ is no longer small, and the $l$ and $s$ of each electron are strongly coupled, forming a resultant total angular momentum $j$, i.e.,

$$j_1 = l_1 + s_1, \qquad j_2 = l_2 + s_2. \tag{I-109a}$$

In such cases, $L$ and $S$ do not have much meaning, and the spectroscopic notation

$$^{2S+1}L$$

also does not have much meaning. Only the total angular momentum $J$, which is now the resultant

$$J = j_1 + j_2 \tag{I-109b}$$

still has a definite meaning.

The coupling scheme (I-109a), (I-109b) is called the $\{j,j\}$ coupling scheme. Examples of this scheme are the x-ray levels for which the inner electrons (near the necleus) move under the much less screened nuclear field $Ze$.

### References

G. E. Uhlenbeck and S. Goudsmit, *Naturwiss.* **13**, 953 (1925); *Nature* **117**, 264 (1926).

M. Abraham, *Annalen d. Physik* **10**, 105 (1903) shows that the gyromagnetic ratio of the magnetic moment to the mechanical angular momentum for orbital motion is $e/2mc$.

L. H. Thomas, *Nature* **117**, 514 (1926).

P. A. M. Dirac, *Proc. Roy. Soc.* London **A 117**, 610; **118**, 351 (1928) discuss the relativistic wave equation of the electron.

### 11.   Pauli's exclusion principle (1925)

The periodic table of the chemical elements was originally mostly of interest and importance to the chemist. But with the discovery of radioactivity, it became of great importance in physics, through the identification of isotopes and the displacement laws of radioactive decays. Then the work of Moseley on the systematics of the x-ray spectra of all the chemical elements known at the time brought together the subjects of the periodic table, x-ray spectroscopy and atomic structure. The application of Bohr's concepts of quantum numbers as learned from the optical spectra to the study of x-ray spectra led to theories on the distribution of electrons in atoms, for example, of J. D. Main Smith (March, 1924) on the chemical side, and of E. C. Stoner (October, 1924) on the physical side. We may say that the work of Stoner is of great importance in Pauli's theory of the periodic table.

The basic empirical fact about the periodic table is the numbers of elements in the periods, namely 2, 8, 8, 18, 18, ... Therefore the first attempt at a theory of the periodic table woule be to find some connection between these numbers and the quantum numbers $n, k, j$ already known in the Bohr and Sommerfeld theory. From the studies of x-ray spectra, it is found that the energy levels of an atom giving rise to its x-ray spectrum can be classified and assigned quantum numbers. In 1922, from all the available x-ray data, Landé proposed a scheme of quantum numbers $n, k, \bar{j}$ for the various x-ray terms in the conventional notations of Barkla (1911) $K, L, M, \ldots$ and Sommerfeld (1915–6). In the following we shall retain Landé's old notations and note that our present day notations $n, l, j$ are related to $n, k, \bar{j}$ by $l = k - 1$ and $j = \bar{j} - \frac{1}{2}$.

| Landé's notation | $K$ | $L_{\mathrm{I}}$ | $L_{\mathrm{II}}$ | $L_{\mathrm{III}}$ | $M_{\mathrm{I}}$ | $M_{\mathrm{II}}$ | $M_{\mathrm{III}}$ | $M_{\mathrm{IV}}$ | $M_{\mathrm{V}}$ |
|---|---|---|---|---|---|---|---|---|---|
| $n$ | 1 | 2 | 2 | 2 | 3 | 3 | 3 | 3 | 3 |
| $k$ | 1 | 1 | 2 | 2 | 1 | 2 | 2 | 3 | 3 |
| $\bar{j}$ | 1 | 1 | 1 | 2 | 1 | 1 | 2 | 2 | 3 |
| $2\bar{j}$ | 2 | 2 | 2 | 4 | 2 | 2 | 4 | 4 | 6 |

$$\text{(I-110)}$$

Stoner (1924) proposed the new idea that the number of electrons in the various x-ray "sub-shells" are $2\bar{j}$, which in our present notation is $(2j + 1)$. Accordingly, the numbers of electrons in the various complete "shells" $n$, for $n = 1, 2, 3, 4$, are $2n^2$,

$$2n^2 = 2, 8, 18(8 + 10), 32(8 + 10 + 14), \ldots \tag{I-111}$$

These numbers are seen to be the periods in the periodic table.

Pauli now asked the question why the electrons did not all go to the lowest energy level $K$ $(n = 1)$.

He proposed to answer this question with the following "principle": In an atom, each electron is specified by four quantum numbers, and no two electrons may have the same set of these four quantum numbers. This is known as the "exclusion principle", when it is generalized to electrons in any system (such as in molecules), and to protons, neutrons and other particles of half-odd-integral spins. This principle is one of those laws of nature that cannot be "derived", but can only be correlated with other laws. Thus later on it was found that this principle was related to the half-odd-integral spin of the particles, to the statistics of the particles, and to the antisymmetry of the wave functions of the system with respect to the interchange of any pair of these particles. They are all laws of nature, and none of them is more understandable so that from it the others can be "derived". We shall simply accept the Exclusion Principle as an empirical law of nature.

For one electron in an atom, we have the quantum numbers $n$, $l$, $m$ (where $l = k - 1$ replaces Sommerfeld's $k$, and $m_l$ is the component of $l$). Pauli introduced a fourth quantum number $s$ which takes on only two values $\pm\frac{1}{2}$. The four quantum numbers are then

$$n, l, m_l, m_s. \tag{I-112}$$

Or, one may choose $n, l, j, m$, where $m$ is the component of $j$, and $-j \leq m \leq j$,

$$n, l, j, m. \tag{I-113}$$

Here, from the x-ray level scheme, it has been known that corresponding to Landé-Sommerfeld's $\bar{j} = 1$, we have $j = l + \frac{1}{2}$ for $l = 0$, and $j = l - \frac{1}{2}$ for $l = 1$; and corresponding to $\bar{j} = 2$, we have $j = l + \frac{1}{2}$ for $l = 1$, and $j = l - \frac{1}{2}$ for $l = 2$. Thus in general, $j$ can take on the value,

$$j = l + \tfrac{1}{2}; \quad l - \tfrac{1}{2} \quad \text{for } l = 1, 2, \ldots$$

or $\qquad\qquad j = l + \tfrac{1}{2}, \qquad\qquad \text{for } l = 0. \tag{I-114}$

The following table gives the number of different sets of $(n, l, m_l, m_s)$ or $(n, l, j, m)$ for various given $n$, or $n$ and $l$.

| Given | No. of different sets of $(n, l, m_l, m_s)$ |
|---|---|
| $n, l, m_l$ | 2 |
| $n, l$ | $2(2l + 1)$ |
| $n$ | $\sum_{l=0}^{n-1} 2(2l + 1) = 2n^2$ |

| | No. of different sets of $(n, l, j, m)$ |
|---|---|
| $n, l, j = l - \frac{1}{2}$ | $2j + 1 = 2l$ |
| $n, l, j = l + \frac{1}{2}$ | $2j + 1 = 2l + 2$ |
| $n, l$ | $2(2l + 1)$ |
| $n$ | $2n^2$ |

$$\tag{I-115}$$

It is seen that with the $(n, l, j, m)$ quantum numbers, the Landé level scheme and the Stoner subshells in (I-110) are completely accounted for. This scheme is particularly suitable for the x-ray levels, since for the electrons in the inner regions, the effective nuclear charge $Z_{eff}$ is large, and the spin-orbit interaction (I-99) and the $j = l \pm \frac{1}{2}$ splitting (I-101, 102) are large.

For the optical spectra of the outer electrons, at least for the light atoms, the spin-orbit interaction is small and the $(n, l, m_l, m_s)$ scheme is valid. But in both cases, the sub-shell $2(2l + 1) = 2, 6, 10, 14, \ldots$ and the complete (closed) shell $2n^2 = 2, 8, 18, 32, \ldots$ are the same.

While the Pauli principle does give the shell structure and the periods of the periodic table, it does not tell the complete story as to how the various subshells and closed shells are built up if we imagine electrons to be fed one by one to a bare nucleus. Here questions of the binding energies of the electrons in various shells and the total energy of the whole atom (or, the atom in successive stages of ionization) are involved. In principle, such questions can be answered in quantum mechanics. But suffice it to say that the Pauli principle satisfactorily accounts for the periodic properties of the chemical elements.

### References


D. I. Mendeléev, *Z. Chem.* **5**, 405 (1869); *Deutsch. Chem. Gesell. Ber.* **4**, 348 (1871); Mendeléev's ideas may be traced to J. A. R. Newlands, *Chem. News* **10**, 94 (Aug., 1864); **12**, 83, 94 (Aug., 1865).

O. Hahn (student of Sir W. Ramsey), *Proc. Roy. Soc.* London **A 76**, 115 (May, 1905); *Chem. News* **92**, 251 (Dec., 1905) found the parent of Th X (radio-Th) to be chemically not separable from Th.

F. Soddy, *Chem. Soc. Ann. Rep.* 285 (1910), proposed the name "isotopes".

J. J. Thomson, *Proc. Roy. Soc.* London **A 89**, 1 (Aug., 1913) found the Neon isotopes 20, 22 by the method of positive (canal) ray analysis.

H. G. J. Moseley, *Phil. Mag.* **26**, 1024 (1913); **27**, 703 (1914).

A. Sommerfeld, *Annalen d. Physik* **63**, 221 (1920) introduced the "inner" quantum number $j$ and the empirical selection rule $\Delta j = 0, \pm 1$.

J. D. Main Smith, *Jour. Soc. Chem. Indus.* (Review) **43**, 325, 437, 490, 548 (1924).

E. C. Stoner, *Phil. Mag.* **48**, 719 (Oct., 1924).

W. Pauli, *Z. Phys.* **31**, 765–785 (1925).


### 12. The birth of quantum mechanics

The decade 1915–1925 following Bohr's theory saw many successes in the field of atomic spectroscopy, as briefly summarized in the preceding sections. But there remained some outstanding problems, both of technical and of more profound nature.

One of the former kind has to do with the quantization condition (I-19). Attempts to extend this theory to the helium atom fail to obtain the energy to an accuracy better than ten percent or so, compared with the accuracy of one part in $10^6$ for hydrogen. The really serious point is that one does not know how to formulate the problem of quantization itself for the helium atom, or for systems that are not multiply periodic.[u] We may recall that in classical dynamics, the helium atom is the famous three-body problem which has not been solved in its generality.

Even the hydrogen atom presents serious difficulties, as shown by O. Klein and by W. Lang (1924) for the problem of a hydrogen atom in crossed uniform electric and magnetic fields.

Of a more profound nature is the gradual realization that, despite the many great successes, something fundamental is amiss in the Bohr theory. The quantization condition, which is one of the foundation pillars of the theory, represents the imposition of entirely foreign, and contradictory, ideas into the classical system of physics. The nature of the shotgun union must not be forgotten even with the efforts of the correspondence principle. There was a prevailing feeling among some physicists that the current quantum theory could only be regarded as a tentative theory, and a logically consistent theory had to be sought.

It must be that by the mid-1920's, the time was ripe. Within less than three years, new ideas were born and developed into a completely new system of physics, by de Broglie in France, Heisenberg and Born in Germany, Dirac in England, and Schrödinger in Switzerland. In Sec. 8, we saw Heisenberg, while working on the quantum theory of dispersion, early in 1925 conceived the ideas that rapidly develop into matrix mechanics in collaboration with

---

[u] In this connection, we may mention the problem of the vibration-rotational band of a linear molecule, such as HCl. The quantization condition

$$\oint P_\phi \, d\phi = jh, \qquad j = 1, 2, 3,$$

(I-116)

leads to the rotational energy

$$E = \frac{h^2}{8\pi^2 I} j^2$$

and the frequencies of a vibration-rotational band

$$\nu = \nu_0 + (\pm j + \tfrac{1}{2})B, \qquad j = \begin{cases} 0, 1, 2, \ldots \\ 1, 2, 3, \ldots \end{cases},$$

$$B = \frac{h}{4\pi^2 I}.$$

This spectrum, however, does not agree with the observed one which is represented by

$$\nu = \nu_0 + \begin{pmatrix} j+1 \\ -j \end{pmatrix} B, \qquad j = \begin{cases} 0, 1, 2, \ldots \\ 1, 2, 3, \ldots \end{cases}$$

(I-117)

It is seen that (I-117) would have resulted had $j$ in (I-116) been replaced by $j + \tfrac{1}{2}$. The use of half-odd-integral quantum numbers is by itself not a serious difficulty since the quantization condition is of a postulational origin anyway. The point is that there is an uncertainty when and where an integer or a half-odd-integer is to be used.

M. Born and P. Jordan. In the fall of 1925, sparked by the initial paper (without reference to matrices yet) of Heisenberg, P. A. M. Dirac independently developed a more symbolic formalism which proved to be an even more general formulation of quantum mechanics. A little earlier, in 1923–4, de Broglie proposed a new idea that a particle has wave properties. This idea sparked the development of wave mechanics by Schrödinger early in 1926, who immediately recognized the mathematical equivalence between his theory and the matrix mechanics.

Then the physical meaning of the new theory was completed by the probability interpretation of Born in mid-1926, and the uncertainty principle of Heisenberg in 1927. The overall physical and philosophical interpretation was integrated by Bohr into what he calls the complementarity principle in 1927. The quantum mechanics is a logically self consistent system of physics and differs in very fundamental ways from classical physics.

Chapter 2

# Matrix Mechanics

In Chapter 1, Sec. 8, we gave a brief account of the work of Kramers and Heisenberg on the quantum theory of dispersion, along the line of Ladenberg, Kramers and Born. The characteristic feature of the quantum theory, a "translation" from the classical theory by means of the correspondence principle, is the replacement of classical dynamical concepts, such as oscillator frequencies, by the frequencies $v_{kj}$ of spectral lines and the oscillator strengths $f_{kj}$ which are related to the Einstein transition coefficients. This observation, together with Einstein's view on the necessity of defining physical concepts by means of criteria for their measurements,[a] led Heisenberg to form his initial ideas of a new basic theory for dealing with atomic phenomena. He presented (July, 1925) a manuscript containing these ideas to M. Born who submitted it to the *Zeits. f. Physik* and began to ponder over Heisenberg's paper, especially the non-commutive multiplication schemes employed. Born, recognized that the quantities in Heisenberg's theory were in fact matrices; he and a young mathematician P. Jordan, an assistant of R. Courant at Göttingen, developed Heisenberg's idea in a rigorous matrix formalism in a paper of September, 1925. Then, Born and Jordan at Göttingen and Heisenberg at Copenhagen cooperated to work on further, and complete the "Matrix Mechanics" system in a paper dated November, 1925. Thus, within a period of a few months, a complete, new system of quantum mechanics, including the canonical transformations and perturbation method, was established.

In the present chapter, we shall give an account of Heisenberg's initial ideas, a digression to matrix algebra and the matrix mechanics itself.

## 1. Heisenberg's initial ideas (1925)

It seems that Heisenberg remembered very well Einstein's insistence on using in physics only a time concept which is defined by an experimental criterion for its measurement. One of his guiding thoughts was to construct a new theory in which the unobservable quantities, such as the electron orbits

---

[a] In the special theory of relativity, Einstein proposed to define a concept in physics, as distinct from philosophy, by defining the procedures for its measurement, thereby defining "time" for each inertial frame and removing absolute time from physics.

and frequencies, did not play a role and instead, the observed quantities such as the frequencies, intensities and polarizations of the spectral lines were employed. From his work with Kramers, on the quantum theory of dispersion, he had found that in the formula for the polarizability of an atom, only the spectral frequencies $v_{kj}$ and the Einstein $A_j^k$ coefficients appear. He set out to construct a mathematical scheme in which these frequencies $v_{kj}$ played an explicit role. He still made use of the correspondence principle as a guide, even though he set out to build a theory by departing from the classical theory.

Now in classical physics, for a system of one degree of freedom, a physical quantity, the coordinate say, $q(t)$ may be represented by a (real) Fourier series

$$q(t) = \sum a_\tau \cos \tau \omega t + \sum b_\tau \sin \tau \omega t$$

$$= \sum_{-\infty}^{\infty} q_\tau e^{i\tau\omega t}, \tag{II-1}$$

where

$$q_\tau = \tfrac{1}{2}(a_\tau - ib_\tau), \qquad q_{-\tau} = \tfrac{1}{2}(a_\tau + ib_\tau)$$

so that

$$q_{-\tau} = q_\tau^*. \tag{II-2}$$

If $q(t)$ is the electric moment, then $q_\tau$ is the amplitude of the component of frequency $\tau\omega$, and

$$|q_\tau|^2 = q_\tau q_{-\tau} \tag{II-3}$$

is proportional to the intensity of radiation of frequency $\tau\omega$.

The $q(t)$ in (II-1) has one important property, namely, when one forms products such as $q^2(t)$, or sums of $q(t) + p(t)$ where $p(t)$ is given by a similar series, or derivatives with respect to time $\dot{q}(t) = \sum i\tau\omega q_\tau e^{i\tau\omega t}$, the frequencies are still $\tau\omega$, i.e., no *new* frequencies appear.

Now in the quantum theory, one does not see the frequencies $\omega$, $2\omega$, $3\omega$, ..., but the frequencies $v_{mn}$ that satisfy the Ritz combination principle (I-87)

$$v_{m,n} = T_m - T_n, \tag{II-4}$$

which form a two-dimensional array instead of a sequence $\omega$, $2\omega$, $3\omega$, ... The relation between the $\tau\omega$ and $v_{m,n}$ is suggested by the correspondence principle, namely, the $\tau$th harmonic $\tau\omega$ in the classical theory *corresponds* to the quantum transitions $m$, $n$ in the sense

$$\tau \to m - n. \tag{II-5}$$

Heisenberg then, in the spirit of the correspondence principle, constructed a two-dimensional set of $q_{mn}^0 e^{i\omega_{mn}t}$ in place of $q_\tau e^{i\tau\omega t}$, and by analogy with (II-2), assumed

$$q_{nm}^0 = q_{mn}^{0*}. \qquad \text{(II-6)}$$

It is natural then to associate $|q_{mn}^0|^2 = q_{mn}^0 q_{nm}^0$ with the intensity of the spectral line $v_{mn}$.

Next, it was necessary to define the time derivative of the two-dimensional array of

$$q_{mn} \equiv q_{mn}^0 e^{i\omega_{mn}t} \qquad \text{(II-7)}$$

as

$$\dot{q}_{mn} = i\omega_{mn} q_{mn}^0 e^{i\omega_{mn}t}. \qquad \text{(II-8)}$$

Heisenberg considered the problem of an harmonic oscillator whose equation of motion in classical dynamics is

$$\ddot{x} + \omega^2 x + \lambda x^2 = 0. \qquad \text{(II-9)}$$

The immediate question is: if $x$ is represented by an array (II-7), what is $x^2$ to be?

In classical theory, we have from (II-1),

$$x(t)x(t) = \sum_\tau \sum_\rho q_\tau e^{i\tau\omega t} q_\rho e^{i\rho\omega t},$$

where the summations are, for both $\tau$ and $\rho$, from $-\infty$ to $+\infty$. On writing $\sigma = \tau + \rho$, this becomes

$$x^2(t) = \sum_\tau \cdot \sum_\sigma x_\tau x_{\sigma-\tau} e^{i\sigma\omega t} \qquad \text{(II-10)}$$

which we rewrite as

$$x^2(t) = \sum_\sigma (x^2)_\sigma e^{i\sigma\omega t}. \qquad \text{(II-10a)}$$

Now multiplying two quantities $x_{mn}$ of the form (II-7), we make the correspondence (II-5) and use the Ritz principle (II-4), so that we make the correspondence

$$\omega_{m,n} + \omega_{p,q} \to \omega_{m,m-\tau} + \omega_{p,p-\rho} \to \omega_{m,m-\tau} + \omega_{m-\tau,m-\tau-\rho}$$

$$\to \omega_{m,m-\tau} + \omega_{m-\tau,m-\sigma} \to \omega_{m,m-\sigma}. \qquad \text{(II-11)}$$

For the product $xx$, we have, corresponding to (II-10) and (II-10a),

$$(x^2) = \sum_{m,\sigma} \left( \sum_\tau x_{m,m-\tau}^0 x_{m-\tau,m-\sigma}^0 \right) e^{i\omega_{m,m-\sigma}t}, \qquad \text{(II-12)}$$

$$(x^2) = \sum_{m,\sigma} (x^2)_{m,m-\sigma} e^{i\omega_{m,m-\sigma}t}. \qquad \text{(II-12a)}$$

From these two expressions, it seems that the correspondence principle suggests the *definition* of the rule of multiplication of the 2-dimensional $(x^2)$:

$$(x^2)_{m,\,m-\sigma} = \sum_j x_{m,\,j} x_{j,\,m-\sigma}, \tag{II-13}$$

where the summation index $\tau$ has been replaced by $m - j$.

Although Heisenberg at that time (before July, 1925) did not know that (II-13) is the rule for matrix multiplication, he was able to obtain many, new interesting results, such as:

(1)  For the harmonic oscillator, i.e., with $\lambda = 0$ in (II-9) the energy is[b]

$$E_n = (n + \tfrac{1}{2})h\nu, \qquad 2\pi\nu = \omega. \tag{II-14}$$

(2)  On considerations of the correspondence principle similar to that in Kramers-Heisenberg's work on the dispersion theory (Chapter I, Sec. 8), Heisenberg obtained the Thomas-Kuhn sum rule in (matrix) form[c]

$$\frac{h}{2\pi} = 2m \sum_{\tau=0}^{\infty} \left\{ |x_{n+\tau,\,n}|^2 \omega_{n+\tau,\,n} - |x_{n,\,n-\tau}|^2 \omega_{n,\,n-\tau} \right\}. \tag{II-15}$$

## 2.  Matrix algebra

In this section we shall present a concise account of those aspects of matrix algebra that are needed in quantum mechanics. The contents of this section are, however, of a purely mathematical nature independent of physics.

**Definition 1.**   A *matrix A* is a two-dimensional array of $m \times n$ elements $A_{mn}$, $m$ being the index for row, $n$ for column,

$$A = \begin{pmatrix} A_{11} & A_{12} & A_{13} & \cdots \\ A_{21} & A_{22} & A_{23} & \cdots \\ A_{31} & A_{32} & A_{33} & \cdots \\ \cdots & \cdots & \cdots & \cdots \\ \cdots & \cdots & \cdots & \cdots \end{pmatrix}. \tag{II-17}$$

If $m = n$, $A$ is called a *square* matrix.

---

[b] For $n = 0$, the oscillator has the "zero-point energy" $\tfrac{1}{2}h\nu$ which the Planck-Bohr-Sommerfeld theory does not show. The existence of this zero-point energy is established experimentally. See Chapter I, Sec. 1.

[c] Heisenberg, without knowing it, has already here found an example of the application of the relation

$$pq - qp = \frac{h}{i}. \tag{II-16}$$

Its discovery will be described in Sec. 3 below.

A *diagonal* matrix is one whose non-diagonal elements are zero,

$$A_{mn} = A_{mm}\delta_m.$$

A *unit* matrix $E$ is a square matrix such that

$$E_{mn} = \begin{cases} 1 \text{ for } m = n \\ 0 \text{ for } m \neq n. \end{cases} \tag{II-18}$$

**Definition 2.** Two matrices $A$ and $B$ are said to be *equal* if

$$A_{mn} = B_{mn} \qquad \text{for all } m, n.$$

**Definition 3.** The *sum* (or *difference*) of two matrices $A$, $B$ is a matrix whose $(m, n)$th element is the sum (or difference) of the $(m, n)$th elements of the two matrices,

$$(A + B)_{mn} = A_{mn} + B_{mn}.$$

It follows that for addition of matrices, the associative law holds

$$(A + B) + C = A + (B + C).$$

**Definition 4.** The *product* of two matrices $A \times B$ is a matrix $C$ whose $(m, n)$th element is obtained according to the rule

$$C_{mn} = (AB)_{mn} = \sum_i A_{mi}B_{in}. \tag{II-19}$$

From this, it follows that[d]

$$EA = AE = A,$$

$$(AB)C = A(BC),$$

$$A(B + C) = AB + AC,$$

$$(A + B)C = AC + BC.$$

But $$AB \neq BA \text{ in general,}$$

that is, the commutative law does not hold for multiplication.

**Definition 5.** The *multiplication* of a matrix $A$ by a number $k$ is the multiplication of each element of $A$ by $k$.

**Theorem 1.** If a matrix $A$ commutes with any arbitrary matrix $B$, i.e., $AB - BA = 0$, then $A$ must be $aE$, where $a$ is a number.

---

[d] The associative law for multiplication is not necessarily valid for matrices of infinitely many rows and columns.

**Definition 6.**   The *inverse*, $A^{-1}$, of a matrix $A$ is a matrix satisfying the relation

$$A^{-1}A = E,$$

which implies

$$AA^{-1} = E.$$

**Definition 7.**   The *transpose* of a matrix $A$, denoted by $\tilde{A}$, is defined by

$$(\tilde{A})_{mn} = A_{nm}. \tag{II-20}$$

**Theorem 2.**

$$(\widetilde{AB}) = \tilde{B}\tilde{A}. \tag{II-21}$$

**Definition 8.**   The *adjoint* of a matrix $A$, denoted by $A^{\dagger}$, is defined by

$$A^{\dagger} = \tilde{A}^{*}, \qquad (A^{\dagger}_{mn} = A^{*}_{nm}) \tag{II-22}$$

where * indicates taking the complex conjugate of all elements.

**Theorem 3.**

$$(AB)^{\dagger} = B^{\dagger}A^{\dagger}. \tag{II-23}$$

**Definition 9.**   The *trace* (also called the *spur*) of a matrix $A$ is the sum of the diagonal elements of $A$,

$$\text{Tr}\,A = \sum_{m} A_{mm}. \tag{II-24}$$

**Theorem 4.**[e]

$$\text{Tr}(AB) = \text{Tr}(BA). \tag{II-25}$$

**Definition 10.**   A matrix $A$ is said to be *hermitian* if its adjoint $A^{\dagger}$ is the matrix $A$ itself.

$$A^{\dagger} = A, \qquad \text{i.e., } A^{*}_{nm} = A_{mn}. \tag{II-26}$$

Hence also the name *self-adjoint*.

**Theorem 5.**   The product $AB$ of two hermitian matrices $A$ and $B$ is hermitian if and only if $AB = BA$.

**Proof:**   (1)

$$(AB)^{\dagger} = B^{\dagger}A^{\dagger} = BA.$$

If $AB = BA$, then $(AB)^{\dagger} = AB$.

---

[e] This theorem is valid for finite matrices. For infinite matrices, it is not true.

(2) As above, $(AB)^\dagger = BA$. If $AB$ is hermitian, i.e., $(AB)^\dagger = AB$, then $AB = BA$.

**Theorem 6.** If $A$, $B$ are hermitian, then $AB + BA$ is hermitian.

**Proof:** From Theorem 3,

$$(AB + BA)^\dagger = B^\dagger A^\dagger + A^\dagger B^\dagger.$$

By hypothesis, the left-hand side is equal to

$$AB + BA = BA + AB$$

$$= (AB + BA). \qquad \text{QED}$$

**Theorem 7.** If $A$, $B$ are hermitian, then $i(AB - BA)$ is hermitian.

**Proof:** From Def. 8 and Theorem 3,

$$(i(AB - BA))^\dagger = -i(B^\dagger A^\dagger - A^\dagger B^\dagger)$$

$$= -i(BA - AB)$$

$$= i(AB - BA). \qquad \text{QED}$$

**Theorem 8.** $AA^\dagger$ is hermitian for arbitrary $A$.

**Proof:**

$$(AA^\dagger)^\dagger = (\widetilde{A^* A^{\dagger *}}) = \tilde{A}^{\dagger *} \tilde{A}^*$$

$$= AA^\dagger. \qquad \text{QED}$$

**Definition 11.** A matrix $A$ is *unitary* if

$$A^\dagger = A^{-1}, \qquad \text{(II-27)}$$

which implies

$$A^\dagger A = AA^\dagger = E.$$

**Theorem 9.** If $A$, $B$ are unitary, $AB$ is also unitary.

**Proof:**

$$(AB)^\dagger = B^\dagger A^\dagger = B^{-1} A^{-1},$$

$$(AB)^\dagger AB = B^{-1} A^{-1} AB = B^{-1} B = E. \qquad \text{QED}$$

**Definition 12.** An *orthogonal* matrix $T$ is such that

$$\tilde{T} = T^{-1}, \quad \text{or} \quad TT^{-1} = T\tilde{T} = E. \qquad \text{(II-28)}$$

From this definition

$$(T\tilde{T})_{mn} = \sum_i T_{mi}\tilde{T}_{in} = \sum_i T_{mi}T_{ni} = \delta_{mn}.$$

Thus an orthogonal matrix $T$ has the same property as the matrix of the coefficients of a linear orthogonal transformation of (orthogonal) axes.

**Definition 13.** If $S$ is a non-singular matrix (i.e., an $S^{-1}$ exists such that $SS^{-1} = E$), and $q$ any matrix, then the transformation from $q$ to $Q$ by

$$Q = S^{-1}qS \tag{II-29}$$

is called a *similarity* transformation.

**Theorem 10.** Let $x = \begin{pmatrix} x_1 \\ x_2 \\ \vdots \end{pmatrix}$, $y = \begin{pmatrix} y_1 \\ y_2 \\ \vdots \end{pmatrix}$ be two column matrices (called *vectors*) and they are related by the transformation matrix $q$

$$y = qx. \tag{II-30}$$

If $x$, $y$ are transformed by a matrix $S^{-1}$ into $X$, $Y$,

$$X = S^{-1}x, \qquad Y = S^{-1}y,$$

then the $X$, $Y$ are related by

$$Y = QX, \qquad Q = S^{-1}qS.$$

**Proof:** We have

$$SX = x, \qquad SY = y,$$
$$SY = qSX,$$
$$Y = S^{-1}qSX.$$

**Theorem 11.** The trace of a matrix is invariant under similarity transformations

**Proof:**

$$\sum_m (S^{-1}AS)_{mm} = \sum_m \left( \sum_{i,j} S^{-1}_{mi}A_{ij}S_{jm} \right)$$
$$= \sum_{i,j} \left( \sum_m S_{jm}S^{-1}_{mi} \right) A_{ij}$$
$$= \sum_{i,j} \delta_{ji}A_{ij} = \sum_i A_{ii}. \qquad \text{QED}$$

**Theorem 12.** The commutation relation $AB - BA = 0$ is invariant under similarity transformations.

**Proof:**

$$(S^{-1}AS)(S^{-1}BS) - (S^{-1}BS)(S^{-1}AS) = S^{-1}(AB - BA)S$$

$$= S^{-1}0S = 0. \qquad \text{QED}$$

**Definition 14.** Eigenvalues and eigenvectors of a matrix.

Let $H$ be a matrix, and $x = \begin{pmatrix} x_1 \\ x_2 \\ \vdots \end{pmatrix}$ a column matrix (vector). Then $Hx$ in

general leads to another vector $y = \begin{pmatrix} y_1 \\ y_2 \\ \vdots \end{pmatrix}$,

$$Hx = y. \tag{II-31}$$

If $z = \begin{pmatrix} z_1 \\ z_2 \\ \vdots \end{pmatrix}$ is such a vector that

$$Hz = \lambda z, \qquad \lambda = \text{constant}$$

then $z$ is said to be an *eigenvector* of $H$, and $\lambda$ an *eigenvalue* of $H$.

The above relation states that $H$ transforms an eigenvector into itself, except for a multiplicative constant $\lambda$. This relation, when written out, is a system of linear equations,

$$\sum_i H_{mi} z_i = \lambda z_m \tag{II-31a}$$

$$
\begin{aligned}
(H_{11} - \lambda)z_1 + H_{12}\phantom{x}z_2 + H_{13}\phantom{x}z_3 + \cdots &= 0, \\
H_{21}\phantom{x}z_1 + (H_{22} - \lambda)z_2 + H_{23}\phantom{x}z_3 + \cdots &= 0, \\
H_{31}\phantom{x}z_1 + H_{32}\phantom{x}z_2 + (H_{33} - \lambda)z_3 + \cdots &= 0, \\
\cdots \qquad\qquad \cdots \qquad\qquad \cdots \qquad \cdots & \\
H_{m1}\phantom{x}z_1 + H_{m2}\phantom{x}z_2 + H_{m3}\phantom{x}z_3 + \cdots &= 0, \\
\cdots \qquad\qquad \cdots \qquad\qquad \cdots \qquad \cdots &
\end{aligned}
\tag{II-31b}
$$

The condition for the system to have non-identically vanishing solution for $z_1, z_2, z_3, \ldots$ is the determinantal equation

$$\det \begin{vmatrix} H_{11} - \lambda & H_{12} & H_{13} & \cdots \\ H_{21} & H_{22} - \lambda & H_{23} & \cdots \\ H_{31} & H_{32} & H_{33} - \lambda & \cdots \end{vmatrix} = 0. \tag{II-32}$$

If the determinant has $n$ rows and $n$ columns, the equation is of order $n$ in $\lambda$

and has $n$ roots

$$\lambda_1, \lambda_2, \ldots, \lambda_n.$$

These roots may or may not be all distinct, depending on the matrix $H$ itself.

If there are some coincident roots, the $H$ is said to have a degeneracy.

Putting any root, say $\lambda_k$, into the system of simultaneous equations, one can solve for the $z_i$ obtaining an eigenvector

$$z^k(z_1^{(k)}, z_2^{(k)}, z_3^{(k)}, \ldots). \qquad k = 1, 2, 3, \ldots$$

Thus there are $n$ eigenvectors for an $n \times n$ matrix $H$, corresponding to the $n$ eigenvalues $\lambda_1, \lambda_2, \ldots$

**Theorem 13.** The eigenvalues of a hermitian matrix are real numbers.

**Proof:** Multiply $\sum_i H_{mi}z_i = \lambda z_m$ by $z_m^*$ and sum[f] over $m$, we have

$$\sum_m \sum_i z_m^* H_{mi} z_i = \lambda \sum_m z_m^* z_m. \tag{II-33}$$

Take the complex conjugate of all quantities. Since $H_{mi}^* = H_{im}$, we have

$$\sum_m \sum_i z_i^* H_{im} z_m = \lambda^* \sum_m z_m^* z_m.$$

$$\therefore \lambda^* = \lambda. \qquad\qquad \text{QED}$$

**Theorem 14.** The eigenvectors of a hermitian matrix form an orthogonal set, i.e.,

$$\sum_i z_i^{*(k)} z_i^{(l)} = 0 \quad \text{for } k \neq l.$$

**Proof:** (1) Assume that $\lambda_k \neq \lambda_l$ for $k \neq l$ (i.e., $\lambda_k, \lambda_l$ are not degenerate). As above,

$$\sum_m \sum_i z_m^{*(k)} H_{mi} z_i^{(l)} = \lambda_l \sum_m z_m^{*(k)} z_m^{(l)},$$

$$\sum_m \sum_i z_m^{(l)} H_{mi}^* z_i^{*(k)} = \lambda_k^* \sum_m z_m^{(l)} z_m^{*(k)}.$$

As $\lambda_k^* = \lambda_k$, $H_{mi}^* = H_{im}$, we have, since $\lambda_l - \lambda_k \neq 0$

$$(\lambda_l - \lambda_k) \sum_m z_m^{*(k)} z_m^{(l)} = \sum_m \sum_i (z_m^{*(k)} H_{mi} z_i^{(l)} - z_i^{*(k)} H_{im} z_m^{(l)}) = 0,$$

$$\sum_m z_m^{*(k)} z_m^{(l)} = 0.$$

(2) Assume now $\lambda_k = \lambda_l$. The above proof fails. For this we introduce a

---

[f] This multiplication is a matrix multiplication. See (II-37a, b)–(II-38) below.

vector $y^{(k)}$ in place of $z^{(k)}$, if[g] $(z^{(k)}, z^{(l)}) \neq 0$,

$$y^{(k)} = c_2 z^{(k)} + c_1 z^{(l)}$$

such that $y^{(k)}$ is orthogonal to $z^{(l)}$, i.e.,

$$c_2^*(z^{(k)}, z^{(l)}) + c_1^*(z^{(l)}, z^{(l)}) = 0.$$

This is called the *Schmidt process*. The method can be extended to the case $\lambda_j = \lambda_k = \lambda_l$, etc.

**Definition 15.** If $U$ is a unitary matrix, then the transformation

$$U^{-1}HU = W, \tag{II-34}$$

is said to be a *unitary transformation*.

**Theorem 15.** The hermitian property of a matrix is invariant under a unitary transformation.

**Proof:** Let

$$U^{-1}HU = W.$$

Given: $U^\dagger = U^{-1}$ and $H^\dagger = H$. To show that $W$ is hermitian, take the adjoint of $W$,

$$W^\dagger = (U^{-1}HU)^\dagger = U^\dagger H^\dagger (U^{-1})^\dagger$$

$$= U^{-1}HU = W. \qquad \text{QED}$$

**Theorem 16.** If in $U^{-1}HU = W$, $H$ and $W$ are hermitian, $U$ may differ from a unitary matrix by a mutiplicative constant.

**Proof:** From $U^{-1}HU = W$, we have

$$HU = UW, \quad \text{and} \quad U^\dagger H = WU^\dagger.$$

Hence

$$HUU^\dagger = UWU^\dagger \quad \text{and} \quad UU^\dagger H = UWU^\dagger.$$

$$HUU^\dagger - UU^\dagger H = 0.$$

Since $H$ is any hermitian matrix, it follows from Theorem 1 that

$$UU^\dagger = aE, \qquad a = \text{a constant}.$$

Let

---

[g] Here we introduce the notation

$$(f, g) \equiv \sum_m f_m^* g_m.$$

$$V = \frac{1}{\sqrt{a}} U, \qquad V^\dagger = \frac{1}{\sqrt{a}} U^\dagger.$$

Then

$$VV^\dagger = E. \qquad\qquad \text{QED}$$

**Definition 16.** If $S^{-1}AS = W$ transforms a matrix $A$ into a diagonal matrix $W$, the transformation is said to be a *principal axes transformation*. To find $S$, we have, from

$$S^{-1}AS = W, \qquad W_{mn} = 0 \quad \text{for } m \neq n, \qquad \text{(II-35a)}$$

we have

$$AS = SW,$$

or

$$\sum_i A_{mi}S_{ik} = S_{mk}W_{kk}, \qquad\qquad \text{(II-35b)}$$

which is a system of simultaneous linear homogeneous equations

$$(A_{11} - W_{kk})S_{1k} + \qquad A_{12}S_{2k} + \qquad A_{13}S_{3k} + \cdots = 0,$$
$$A_{21}S_{1k} + (H_{22} - W_{kk})S_{2k} + \qquad A_{23}S_{3k} + \cdots = 0,$$
$$A_{31}S_{1k} + \qquad A_{32}S_{2k} + (A_{33} - W_{kk})S_{3k} + \cdots = 0,$$
$$\vdots \qquad\qquad\qquad \vdots$$

The condition for non-identically vanishing $S_{1n}$, $S_{2n}$, $S_{3n}$, ... is the determinantal equation

$$\begin{vmatrix} A_{11} - W_{kk} & A_{12} & A_{13} & \cdots \\ A_{21} & A_{22} - W_{kk} & A_{23} & \cdots \\ A_{31} & A_{32} & A_{33} - W_{kk} & \cdots \\ \vdots & \vdots & \vdots & \end{vmatrix} = 0. \qquad \text{(II-36)}$$

On comparing with Definition 14, it is seen that the roots of $W_{kk}$ for $k = 1, 2, 3, \ldots, n$ are the *same* set for all $k$,

$$W_1, \quad W_2, \quad W_3, \quad \ldots, \quad W_n, \qquad\qquad \text{(II-36a)}$$

which are the eigenvalues of $A$. For each root $W_k$, one obtains the eigenvector (a column matrix)

$$S_{1k}, \quad S_{2k}, \quad S_{3k}, \quad \ldots, \quad S_{nk}. \qquad\qquad \text{(II-36b)}$$

Thus: a transformation $S^{-1}AS$, where $S$ is made up of the $n$ eigenvectors as columns of $S$, transforms $A$ into a diagonal matrix $W$, whose elements $W_{kk}$ are the eigenvalues of[h] $A$.

From Theorem 13, if $A$ is a hermitian matrix, then the diagonal elements $W_{kk}$ above are the real eigenvalues of $A$.

The eigenvalues $W_1$, $W_2$, $W_3$, ... may or may not be all distinct. The remarks under Def. 14 apply.

If the transformation matrix $S$ is unitary, then

$$U^{-1}AU = W.$$

Since $U$ is a special case of $S$, the above results hold also for unitary transformations $U$.

Let us transform the column matrix (vector) $y_k$ into $x_k$, and the row matrix $y_k^*$ into $x_k^*$ by

$$x = U^{-1}y, \qquad x^* = y^*U \tag{II-37a}$$

or the inverse,

$$y = Ux, \qquad y^* = x^*U^{-1}. \tag{II-37b}$$

The matrix $y^*Hy$ becomes

$$y^*Hy = x^*U^{-1}HUx$$

$$= x^*Wx,$$

or

$$\sum_{m,n} y_m^* H_{mn} y_n = \sum_{mn} x_m^* W_{mm} \delta_{mn} x_n,$$

$$\sum_{m,n} y_m^* H_{mn} y_n = \sum_{m} x_m^* W_{mm} x_m. \tag{II-38}$$

This shows that the quadratic form $\sum_{m,n} y_m^* H_{mn} y_n$ is transformed into the normal form $\sum_m x_m^* W_{mm} x_m$ (a sum of squares of $x_m^* x_m$). For this reason the transformation to a diagonal form is called a *principal axes transformation*—a geometrical analogue to the transformation (rotation) of the coordinate axes to the principal axes of an ellipsoid in 3-dimensional space.

**Theorem 17.** Two matrices $A$, $B$ can be simultaneously transformed into diagonal matrices if and only if the matrices commute, i.e., $AB - BA = 0$.

**Proof:** (1) The sufficiency part.

Let $U^{-1}AU$ transform $A$ into a diagonal matrix $a$, and $U^{-1}BU$ transform $B$ into $b$. From $AB - BA = 0$, we have

---

[h] The above treatment holds for finite matrices. For infinite matrices (having infinitely many rows and columns), we shall assume that the results obtained above hold.

$$U^{-1}(AUU^{-1}B - BUU^{-1}A)U = 0$$

or

$$ab - ba = 0$$

or

$$a_{mm}b_{mn} - b_{mn}a_{nn} = b_{mn}(a_{mm} - a_{nn}) = 0.$$

$$b_{mn} = 0 \qquad \text{for } m \neq n \text{ if } a_{mm} \neq a_{nn}.$$

If $a_{mm} = a_{nn} = k$ for $m \neq n$, then $a$ is of the form

$$a = kE, \qquad k = \text{const.}$$

so that $b$ can now transformed into a diagonal matrix by $V^{-1}bV$ without changing $a$ any further, since $V^{-1}aV = kV^{-1}EV = kE$.

(2) The necessity part.

If $A$, $B$ are simultaneously transformed into diagonal form $a$, $b$, then

$$ab - ba = 0.$$

By Theorem 12, $S^{-1}aSS^{-1}bS - S^{-1}bSS^{-1}aS = AB - BA = 0$.

**Theorem 18.** Two hermitian matrices have the same common eigenvectors if and only if the two matrices commute ($AB - BA = 0$).

This theorem has the same content as the preceding one. Theorem 18 (or 17) is of the greatest importance in quantum mechanics, as we shall see below.

**Definition 17.** The differentiation of a matrix with respect to a parameter is defined by

$$\frac{dA}{dt} = \begin{vmatrix} \dfrac{d}{dt}A_{11} & \dfrac{d}{dt}A_{12} & \cdots \\[2ex] \dfrac{d}{dt}A_{21} & \dfrac{d}{dt}A_{22} & \cdots \\[2ex] \dfrac{d}{dt}A_{31} & \dfrac{d}{dt}A_{32} & \cdots \end{vmatrix}. \tag{II-39}$$

**Definition 18.** The differentiation of a function $f(X)$ of a matrix $X$ with respect to the argument matrix $X$ is defined by

$$\frac{df(X)}{dX} = \lim_{a \to 0} \frac{f(X + aE) - f(X)}{a}, \tag{II-40}$$

where $E$ is the unit matrix and $a$ is a parameter (number).

**Example 1.**

$$\frac{dX^2}{dX} = 2X.$$

**Example 2.**

$$\frac{df(X)g(X)}{dX} = f\frac{dg}{dX} + \frac{df}{dX}g.$$

Here the order of $f$ and $g$ (matrix functions) must not be changed.

**Theorem 19.** If $H(q, p)$ is a matrix function formed by the processes of addition and multiplication of two matrices $p$ and $q$, and if $p$, $q$ satisfy the relation

$$pq - qp = kE, \tag{II-41}$$

$k$ being a constant (number) and $E$ the unit matrix, then

$$Hq - qH = k\frac{\partial H}{\partial p},$$

$$pH - Hp = k\frac{\partial H}{\partial q}. \tag{II-42}$$

**Proof:** We shall first prove that if $f$, $g$ satisfy these two relations, then $f + g$ and $fg$ also satisfy them:

$$(f + g)q - q(f + g) = fq - qf + gq - qg$$

$$= k\frac{\partial f}{\partial p} + k\frac{\partial g}{\partial p} = k\frac{\partial}{\partial p}(f + g).$$

$$(fg)q - q(fg) = f(gq - qg) + (fq - qf)g$$

$$= fk\frac{\partial g}{\partial p} + k\frac{\partial f}{\partial p}g = k\frac{\partial(fg)}{\partial p}, \quad \text{etc.}$$

In the same way, any function of sums of $f$ and $g$ and products of $f$ and $g$ satisfies the two relations.

Next let $f = q$ (or $p$) and $g = q$ (or $p$). These obviously satisfy the two relations. The proof then follows by induction.

**Theorem 20.** If $H(q, p)$ is a matrix function and

$$\dot{q} = \frac{\partial H}{\partial p}, \qquad \dot{p} = -\frac{\partial H}{\partial q}, \tag{II-43}$$

with

$$pq - qp = k, \qquad \text{a constant,}$$

then

$$\dot{q} = \frac{1}{k}(Hq - qH), \qquad \dot{p} = \frac{1}{k}(Hp - pH), \tag{II-44}$$

and for any function $F(q, p)$,

$$\dot{F} = \frac{1}{k}(HF - FH). \tag{II-45}$$

The proof of this last relation is similar to that of (II-42).

**Theorem 21.**   If $H = H(q, p)$ and

$$k\frac{\partial H}{\partial p} = Hq - qH, \qquad -k\frac{\partial H}{\partial q} = Hp - pH$$

and if

$$\dot{q} = \frac{\partial H}{\partial p}, \qquad -\dot{p} = \frac{\partial H}{\partial q},$$

then $pq - qp$ is a constant in time.

**Proof:**   On putting $H = p$ and $H = q$ in the first two relations, one gets

$$pq - qp = k.$$

$$\frac{dk}{dt} = \dot{p}q - q\dot{p} + p\dot{q} - \dot{q}p$$

$$= -\frac{\partial}{\partial q}(Hq - qH) + \frac{\partial}{\partial p}(pH - Hp)$$

$$= -\frac{\partial}{\partial q}\left(k\frac{\partial H}{\partial p}\right) + \frac{\partial}{\partial p}\left(k\frac{\partial H}{\partial q}\right)$$

$$= -\left(\frac{\partial k}{\partial q}\dot{q} + \frac{\partial k}{\partial p}\dot{p}\right) = -\frac{dk}{dt}.$$

$$\therefore k = \text{const.}$$

**Definition 19.**   A matrix whose rows and columns are not discrete but continuous is called a *continuous matrix*. We shall use Greek letters for labelling rows and columns, thus $A(\alpha, \beta)$ for an element at $\alpha$th row and $\beta$th column.

    A matrix may have both discrete and continuous rows an columns, as show in the following

$$A = \begin{pmatrix} A_{11} & A_{12} & A_{13} & \cdot & \cdot & \cdot & A(1,\alpha) & A(1,\beta) \\ A_{21} & A_{22} & A_{23} & \cdot & \cdot & \cdot & A(2,\alpha) & A(2,\beta) \\ \vdots & \vdots & \vdots & \vdots & \vdots & \vdots & & \\ A(\alpha,1) & & A(\alpha,2) & & & & A(\alpha,\alpha) & A(\alpha,\beta) \\ A(\beta,1) & & A(\beta,2) & & & & A(\beta,\alpha) & A(\beta,\beta) \end{pmatrix}. \tag{II-46}$$

The multiplication of two such matrices is defined by the same rule as for discrete matrices.

Thus

$$(AB)_{m\beta} = \sum_i A_{mi}B_{in} + \int A(m,\gamma)\,d\gamma B(\gamma,\beta), \tag{II-47}$$

the sum being extended to all the discrete part which may be finite or infinite in rows and column, and the integral being taken over the continuous range which may again be finite or infinite in extent.

**Definition 20.** *Dirac's $\delta$ function.* Let $\delta(x - a)$ represent a function having the properties:[i]

$$\delta(x - a) = \begin{cases} 0 & \text{for } x \neq a, \\ \infty & \text{for } x = a, \end{cases} \tag{II-48}$$

$$\int_{-b}^{b} \delta(x - a)\,dx = 1, \qquad -b < a < b, \tag{II-49}$$

so that

$$\int_{-b}^{b} f(x)\delta(x - a)\,dx = f(a), \qquad -b < a < b. \tag{II-50}$$

**Definition 21.** A *continuous matrix* whose elements $\delta(\alpha',\alpha'')$ are the Dirac $\delta$ function $\delta(\alpha' - \alpha'')$ is taken as the *unit continuous matrix*. This unit matrix, as an operator, is defined by the following operational rule for the multiplication of the unit matrix with another (continuous) matrix $A$: Take the $(\alpha', \alpha'')$ element of $\delta A$,

---

[i] A function having these properties had been introduced by G. Kirchhoff in 1882 in connection with the theory of light waves, by O. Heavisde in 1893 into the electromagnetic theory, and in problems such as the vibrations of a string with a singular loading at a point. It was introduced first by Dirac into quantum mechanics in 1926, not as a proper function but as a very useful operator. In fact the two properties (II-48), (II-49) are inconsistent within the Riemann or the Lebesque conception of the integral. The $\delta$ function and its derivative $\delta'$, however, have been put on a mathematically legitimate basis by L. Schwartz by his theory of distributions in 1945. The $\delta$ function plays a very important role in quantum mechanics, not only as a tool in practical problems, but in the early fundamental transformation theory of quantum mechanics of Hilbert (1926–7).

$$(\delta A)(\alpha', \alpha'') = \int \delta(\alpha', \beta)\, d\beta\, A(\beta, \alpha'')$$

$$= \int \delta(\alpha' - \beta)\, d\beta\, A(\beta, \alpha'')$$

$$= A(\alpha', \alpha''),$$

i.e.,

$$\delta A = A. \tag{II-51}$$

**Definition 22.** A diagonal continuous matrix is one whose diagonal elements $A(\alpha', \alpha'') = A(\alpha')$, and

$$A(\alpha', \alpha'') = A(\alpha')\delta(\alpha' - \alpha''). \tag{II-52}$$

Thus the product $AB$, where $B$ is any continuous matrix, is

$$(AB)(\alpha', \alpha'') = \int A(\alpha', \beta)\, d\beta\, B(\beta, \alpha'')$$

$$= \int A(\alpha')\delta(\alpha' - \beta)\, d\beta\, B(\beta, \alpha'')$$

$$= A(\alpha')B(\alpha', \alpha''). \tag{II-53}$$

**Definition 23.** The *derivative $\delta'$* matrix, as an operator,

$$\delta'(\alpha, \beta) = \frac{d\delta(\alpha, \beta)}{d\beta} \tag{II-54}$$

is defined by the following rule of operation

$$(\delta' A)(\alpha', \alpha'') = \int \delta'(\alpha', \beta)\, d\beta\, A(\beta, \alpha'')$$

$$= \int \frac{d}{d\beta}\delta(\alpha', \beta)\, d\beta\, A(\beta, \alpha'')$$

$$= \delta(\alpha' - \beta)A(\beta, \alpha'')|_{-\infty}^{\infty} - \int \delta(\alpha' - \beta)\, d\beta \frac{\partial A(\beta, \alpha'')}{\partial \beta}$$

$$= -\frac{\partial A(\beta, \alpha'')}{\partial \beta}\bigg|_{\beta = \alpha'} = -A'(\alpha', \alpha''), \tag{II-55}$$

the integrated part $\delta(\alpha' - \beta)A(\beta, \alpha'')|_{-\infty}^{\infty}$ vanish since $\delta(\alpha' - \beta)$ vanishes at the limits of $\beta$.

### Notational convention

For convenience, we shall adopt the following convention for labelling the rows and columns of a continuous matrix: Just as in the case of discrete matrices, a continuous matrix, say $Q$, can be transformed into a diagonal matrix of the form (II-52). Let us choose the letter $q'$ to denote the values of the diagonal elements of $Q$, and at the same time use $q'$, $q''$, $q'''$ as indices for the labelling of the elements. Thus

$$Q(q', q'') = q'\delta(q' - q''), \tag{II-56a}$$

and a function $f(Q)$, being also diagonal when $Q$ is diagonal, has matrix elements

$$f(q', q'') = f(q')\delta(q' - q''). \tag{II-56b}$$

When $Q$ is diagonal, another matrix, say $P$, cannot, in general, be transformed at the same time as $Q$ into a diagonal form (see Theorem 17 above). Its matrix elements can, however, be labelled as $P(q', q'')$. This labelling is to remind us that we are using a preferential "representation" in which a matrix $Q$ has been transformed into a diagonal form. If the matrix $P$ does not commute with $Q$, then, in the $Q$ representation, $P$ is not diagonal, and the $P(q', q'')$ does not have the form (II-56b).

This convention is of very great help in clarity in the transformation theory in the following.

**Theorem 22.** If $P$, $Q$ are two (continuous) matrices that obey the relation

$$PQ - QP = k\delta, \tag{II-57}$$

where $k$ is a constant (number), if $Q$ is diagonal, $Q(q', q'') = q'\delta(q' - q'')$, then

$$P(q', q'') = -k\delta'(q', q''). \tag{II-58}$$

**Proof:** Take the $(q', q'')$th element of both sides of (II-57), we have

$$\int P(q', q''') dq''' \, q''' \delta(q''' - q'') - \int q'\delta(q' - q''') dq''' \, P(q''', q'') = k\delta(q' - q'')$$

or

$$P(q', q'')q'' - q'P(q', q'') = k\delta(q' - q'')$$

$$P(q', q'') = \frac{k}{q'' - q'}\delta(q' - q''). \tag{II-59}$$

Next consider the operator

$$(q' - q''')\delta'(q' - q'''),$$

we have, from (II-55),

$$\int (q' - q''')\delta'(q' - q''')\,dq''' \, F(q''', q'') = -\frac{\partial}{\partial q'''}\{(q' - q''')F(q''', q'')\}|_{q'''=q'}$$

$$= F(q', q'').$$

On comparison with (II-51), it is seen that, as operators,

$$(q' - q''')\delta'(q' - q''') = \delta(q' - q''').  \tag{II-60}$$

Hence (II-59) can be written

$$P(q', q'') = -k\delta'(q', q'').  \tag{II-61}$$

From (II-55), we have

$$\int P(q', q''')\,dq''' \, A(q''', q'') = k\frac{\partial A(q''', q'')}{\partial q''}\bigg|_{q'''=q'}.  \tag{II-62}$$

Thus, as operator relation,

$$PA = k\frac{\partial}{\partial q}A,  \tag{II-63}$$

where $k$ is a constant appearing in the relation (II-57)

$$PQ - QP = k\delta.$$

**Exercises**

1.  Show that the absolute value of eigenvalue $a$ of a unitary matrix $U$ is unity, i.e.,

$$Uz = az, \qquad a^*a = 1.$$

2.  Show that a $2 \times 2$ unitary matrix must be of the form

$$U = \begin{pmatrix} \cos\vartheta & \sin\vartheta \\ -\sin\vartheta & \cos\vartheta \end{pmatrix}.$$

3.  Show that the eigenvectors $z^{(k)}$ and $z^{(l)}$ belonging to two different eigenvalues $\lambda_k \neq \lambda_l$ of a unitary matrix $U$ are orthogonal, i.e.,

$$Uz^{(k)} = \lambda_k z^{(k)}, \qquad Uz^{(l)} = \lambda_l z^{(l)},$$

$$(z^{(k)}, z^{(l)}) = 0.$$

4. Show that any matrix $A$ can be decomposed into two hermitian matrices $B, C$ as

$$A = B + iC.$$

5. Show that the product $AB$ of any two unitary matrices $A, B$ is also unitary.

6. If the eigenvalues of a hermitian matrix $A$ are $a_n$, show that

$$U = e^{iA}$$

is a unitary matrix whose eigenvalues are $e^{ia_n}$.

7. Let the eigenvalues and eigenvectors of a hermitian matrix $X$ be $x$ and $\xi$ respectively, i.e.,

$$X\xi_x = x\xi_x.$$

Let $P$ be a (conjugate) hermitian matrix whose eigenvalues and eigenvectors are $p$ and $\eta$,

$$P\eta_p = p\eta_p,$$

and

$$PX - XP = \frac{\hbar}{i}1, \qquad 1 = \text{unit matrix.}$$

If $x'$ is an arbitrary real number, show that $e^{-ipx'/\hbar}\xi_x$ is the eigenvector of $X$ for the eigenvalue $x + x'$.

If $p'$ is an arbitrary real number, show that $e^{-ip'x/\hbar}\eta_p$ is the eigenvector of $P$ with the eigenvalue $p - p'$.

8. Show

$$\frac{1}{2\pi} \int_{-\infty}^{\infty} e^{ikx}\, dx = \lim_{a \to \infty} \frac{1}{2\pi} \int_{-a}^{a} e^{ikx}\, dx$$

$$= \lim_{a \to \infty} \frac{\sin ax}{\pi x}$$

has the property (II-50) of the $\delta$-function and can be taken as one of the representations of the $\delta$-function.

9. Show that the operator

$$\lim_{a \to \infty} \frac{d}{dx}\left(\frac{\sin ax}{\pi x}\right)$$

has the property (II-55) of the $\delta'$ function

$$\int_{-\infty}^{\infty} \delta'(x)f(x)\,dx = -\left(\frac{df}{dx}\right)_{x=0}.$$

10. Show the following relations to be true:

(1) $\delta(ax) = \dfrac{1}{a}\delta(x)$, $a = \text{constant} > 0$,

(2) $\delta[(x-a)(x-b)] = \dfrac{1}{|a-b|}[\delta(x-a) - \delta(x-b)]$, $a \neq b$,

(3) $\delta(x+y)\delta(x-y) = \tfrac{1}{2}\delta(x)\delta(y)$,

(4) $\displaystyle\prod_{i=1}^{n} \delta\left(\sum_{k=1}^{n} A_{ik}x_k\right) = \dfrac{1}{\|A_{ik}\|} \prod_{k=1}^{n} \delta(x_k)$,

where $\|A_{ik}\| = |\det A_{ik}| \neq 0$.

### 3. Transformation of representations: the theory of Dirac

In the last section (following (II-55)), we introduced the concept of "representation". We shall here further explain this concept by means of examples. In Sec. 2, Definition 16, (II-35a), we show how a given matrix $A$ can be transformed by a similarity transformation $S^{-1}AS$ into a diagonal matrix; and in Theorem 17, we saw that only when all the matrices $A$, $B$, $C$, ... commute among themselves, $AB - BA = 0$, $AC - CA = 0$, $BC - CB = 0$, etc., can they be simultaneously (i.e., by the same transformation matrix $S$) transformed into diagonal forms. In Theorem 18, we see that in the case of simultaneously diagonal matrices, they have the same common eigenvectors.

With these theorems, we may illustrate the concept of "representations". We may think of the eigenvectors of a matrix $Q$ as defining a representation. Another matrix $P$ for which $PQ - QP \neq 0$ does not have the same eigenvectors as $Q$, and in the $Q$-representation, $P$ is not diagonal.[j]

Now let $A$ be a matrix, and let it be transformed into the diagonal form, so that its elements in its own representation are, in the notation (II-56a),

$$A(a', a'') = a'\delta(a' - a''). \tag{II-64}$$

[j] To help visualizing this concept, consider an ellipsoid in 3-dimensional space. Its principal axes define a coordinate system in which the equation of the surface of the ellipsoid takes the normal form ($\alpha$) $ax^2 + by^2 + cz^2 = 1$. Another arbitrary ellipsoid whose principal axes do not coincide with this set will have the quadratic form ($\beta$) $Ax^2 + By^2 + Cz^2 + 2Dxy + 2Eyz + 2Fzx = 1$ in this coordinate system. One may transform the form ($\beta$) into the normal form by a linear coordinate transformation, but then the ($\alpha$) form will be thrown out of the normal form.

Let $Q$ be a matrix such that $AQ - QA \neq 0$. In its own $Q$-representation, $Q$ is diagonal

$$Q(q',q'') = q'\delta(q' - q''). \tag{II-65}$$

In the $Q$-representation, $A$ is no longer diagonal, i.e., $A(q',q'')$ are no longer of the form (II-64), (II-65).

Let $U$ be a unitary matrix that transforms $A(q',q'')$ (from the $Q$-representation) to a diagonal matrix (i.e., into the $A$-representation),

$$\sum_{q',q''} U^{-1}(a',q')A(q',q'')U(q'',a'') = A(a',a'')$$

$$= a'\delta(a' - a'') \tag{II-66}$$

and if the $Q$ matrix has both discrete and continuous eigenvalues $q$,

$$\sum \int U^{-1}(a',q')\,dq'\, A(q',q'')\,dq''\, U(q'',a'') = a'\delta(a' - a''). \tag{II-66a}$$

Similarly, one can transform any matrix $F$ from an initial $Q$-representation to the $A$-representation by $U$,

$$\int U^{-1}(a',q')\,dq'\, F(q',q'')\,dq''\, U(q'',a'') = F(a',a''). \tag{II-67}$$

Here $F$ is in general, of course, not diagonal. (II-67) is the general formula, of which (II-66a) is a special case (in which the final representation is that of $F$ itself.)

The meaning of the unitary matrix $U$ and the transformation (II-67) can be seen as follows.

Let $F$ in (II-67) be the $Q$ matrix, so that

$$F(q',q'') = Q(q',q'') = q'\delta(q' - q'').$$

Since $U$ is unitary, $U^\dagger = U^{-1}$, and

$$U^{-1}(a,q) = U^*(q,a),$$

and (II-67) gives

$$\int U^*(q',a')\,dq'\, q'\delta(q' - q'')\,dq''\, U(q'',a'') = Q(a',a''),$$

$$\int U^*(q',a')\,dq'\, q'\delta(q' - q'')\,dq''\, U(q'',a') = Q(a',a'), \tag{II-68a}$$

or

$$Q(a', a') = \int U^*(q', a') U(q', a') q' \, dq'$$

$$= \int |U(q', a')|^2 q' \, dq'. \tag{II-68}$$

Now for the unitary matrix $U$, we have

$$\sum_{q'} U^*(q', a') U(q', a') = E,$$

$$\sum \int U^*(q', a') U(q', a'') \, dq'' \, \delta(a' - a'') = \delta(a' - a''), \tag{II-69}$$

i.e.,

$$U^\dagger U = \delta \qquad \text{(unit matrix)}.$$

We are thus led to the following interpretation of $|U(q', a')|^2 \, dq' = U^{-1}(a', q') U(q', a') \, dq'$. It is the probability, when $A$ has the eigenvalue $a'$, that $Q$'s eigenvalue lies between $q'$ and $q' + dq'$. The matrix elements $U(q', a')$ are the probability amplitudes.

This theory is of basic importance in quantum mechanics.

Jordan (December, 1926) developed the transformation from a somewhat different point of view and in a more general way so that it encompasses and unifies the different view points of Heisenberg, Born-Wiener, Schrödinger and Dirac. For an exposition of Jordan's important work, reference may be made to the article by Kennard.

### Exercises

1. The transformation (II-67) can be written in the form

$$\int F(q', q'') \, dq'' \, U(q'', a') = \int U(q', a'') \, da'' \, F(a'', a'),$$

$$\int U^{-1}(a', q'') \, dq'' \, F(q'', q') = \int F(a', q'') \, da'' \, U^{-1}(a'', q'). \tag{A-1}$$

We shall introduce the mixed representation $(q, a)$, $(a, q)$ by defining

$$F(q', a') = \int U(q', a'') \, da'' \, F(a'', a'),$$

$$F(a', q') = \int F(a', a'') \, da'' \, U^{-1}(a'', q'). \tag{A-2}$$

Show that if the matrix $Q$ satisfies[k]

$$QP - PQ = -\frac{\hbar}{i}\,1,$$

then

$$Q(q',a') = q'U(q',a'), \tag{A-3}$$

and

$$P(q',a') = \frac{\hbar}{i}\frac{\partial}{\partial q'}U(q',a')^*. \tag{A-4}$$

2. Let us take the energy-representation (so that in (II-64), $A$ is the Hamiltonian and $a'$, $a''$ are the energy eigenvalues $E'$, $E''$, ...),

$$H(E',E'') = E'\delta(E' - E''). \tag{A-5}$$

Show that, from (A-3), (A-4), one obtains

$$H\left(q',\frac{\hbar}{i}\frac{\partial}{\partial q'}\right)U(q',E) = E'U(q',E), \tag{A-6}$$

which is the Schrödinger equation if we write $\psi_E(q')$ for $U(q',E)$.

### References for Secs. 1, 2, 3

W. Heisenberg, *Z. Phys.* **33**, 879 (July, 1925). This paper contains the initial ideas described in Sec. 1 above.

A. Cayley, *Jour. f. Math.* **1**, 282 (1855); *Phil. Trans.* **148**, 17 (1858), develops matrix algebra. The name "matrix" was, however, introduced by J. J. Sylvester in 1850.

M. Born and P. Jordan, *Z. Phys.* **34**, 858 (1925). This paper contains the matrix algebra summarized in Sec. 2 above.

P. A. M. Dirac, *Proc. Roy. Soc.* London A **113**, 621 (Dec., 1926). This paper introduces the $\delta$-function and the transformation theory described in Secs. 2 and 3 above.

P. A. M. Dirac, *The Principles of Quantum Mechanics*, 1st. ed. (Clarendon Press, Oxford, 1980). The first edition of this book develops quantum mechanics in an abstract, mathematical form.

P. Jordan, *Z. Phys.* **40**, 809–838 (1927), (Dec., 1926).

E. H. Kennard, *ibid.* **44**, 326–352 (July, 1927).

M. Born and N. Wiener, *J. Math. Phys.* (MIT) **5**, 84–98 (1925–6); also appearing in *Z. Phys.* **36**, 174–187 (Jan., 1926). This work introduces operators as a generalization of matrices.

[k] These results have already been obtained in 1926 by P. A. M. Dirac, *Proc. Roy. Soc.* London A **113**, 621 (Dec., 1926). For a detailed treatment of the transformation theory and the Schrödinger equation for the momentum $p$, see Chap. 5, Sec. 2, (3), especially (V-64).

For references to the mathematical justification of the Dirac $\delta$-function, see M. Jammer, *The Conceptual Development of Quantum Mechanics* (McGraw-Hill, New York, 1966), pp. 313–4; in particular, Laurent Schwartz, *Theorie des Distributions*, 2 Vol. (Hermann et Cie, Paris, 1950–51), and an elementary account in M. J. Lighthill, *Introduction to Fourier Analysis and Generalized Functions* (Cambridge Univ. Press, Cambridge, 1959).

### 4. Matrix mechanics—principles

After recognizing that the two-dimensional arrays of $q_{mn}$ and their multiplication rules (II-13) discovered by Heisenberg are matrices and matrix multiplications, Born and Jordan, and then Born, Heisenberg and Jordan developed the matrix mechanics in the period July–November, 1925. We shall present below the crucial ideas of this new theory.

Born and Jordan started from the classical theory for the action variable $J$ of a periodic system with one degree of freedom

$$J = \oint p\,dq = \int_0^T p\dot{q}\,dt, \qquad T = \frac{1}{\nu} = \frac{2\pi}{\omega}.$$

With the Fourier series

$$p = \sum_n p_n e^{in\omega t}, \qquad \dot{q} = \sum_m im\omega q_m e^{im\omega t},$$

$$J = i\omega \sum_n \sum_m \int_0^T m p_n q_m e^{i(n+m)\omega t}\,dt$$

$$= i\omega \sum_n \sum_k \int_0^T (k-n) p_n q_{k-n} e^{ik\omega t}\,dt$$

$$= -2\pi i \sum_\tau \tau p_\tau q_{-\tau},$$

$$\frac{\partial J}{\partial J} = 1 = -2\pi i \sum_\tau \tau \frac{\partial}{\partial J}(p_\tau q_{-\tau}). \tag{II-70}$$

According to the correspondence principle, the classical $\tau(\partial/\partial J)$ is to correspond to the quantum theoretical

$$\tau \frac{\partial}{\partial J} \leftrightarrow \lim_{\Delta n \ll n} \Delta n \frac{\Delta}{\Delta(nh)} = \frac{\Delta}{h}$$

so that

$$\tau \frac{\partial}{\partial J}(p_\tau q_{-\tau}) \leftrightarrow \frac{1}{h}(p_{n+\tau,n} q_{n,n+\tau} - p_{n,n-\tau} q_{n-\tau,n})$$

and (II-70) becomes the quantum version

$$\sum_{\tau=-\infty}^{\infty} (p_{n,n-\tau}q_{n-\tau,n} - p_{n,n+\tau}q_{n+\tau,n}) = \frac{\hbar}{i}.$$

The left-hand side is the diagonal element of the matrix $pq - qp$, so that

$$(pq - qp)_{nn} = \frac{\hbar}{i}. \tag{II-71}$$

For the non-diagonal matrix elements, there was no hint, and Born-Jordan made the guess (assumption) that they all vanish, so they assumed the relation

$$pq - qp = \frac{\hbar}{i}E, \tag{II-72}$$

where $E$ is the unit matrix. As $p$, $q$ are continuous and infinite matrices, this relation does not contradict the Theorem 4, (II-25), $\mathrm{Tr}\, AB - \mathrm{Tr}\, BA = 0$.

Note that (II-72) has *not* been *derived*; it is a basic postulate in the new theory.

On assuming (II-72), it follows from Theorem 19, (II-42), that any function $F$ of $p$, $q$ consisting of sums and products of $p$, $q$ satisfy the relation

$$\frac{\hbar}{i}\frac{\partial F}{\partial p} = Fq - qF, \qquad -\frac{\hbar}{i}\frac{\partial F}{\partial q} = Fp - pF. \tag{II-73}$$

We assume the following equations of motion,

$$\dot{q} = \frac{\partial H}{\partial p}, \qquad \dot{p} = -\frac{\partial H}{\partial q}, \tag{II-74}$$

where $H$ is the Hamiltonian, constructed as a matrix function of $p$, $q$ in the same functional form as in classical dynamics, except that the $p$, $q$ are now matrices, and their products must be symmetrized to ensure the hermitian property of $H$ (see Sec. 2, Theorem 6). Now let $F$ in (II-73) be this hermitian $H$. From (II-74) and (II-73), we have

$$\dot{q} = \frac{1}{i\hbar}(qH - Hq), \qquad \dot{p} = \frac{1}{i\hbar}(pH - Hp), \tag{II-75}$$

which are now the equations of motion in an alternative form. If we define the "quantum Poisson Brackets" by

$$[A, B] = \frac{1}{i\hbar}(AB - BA), \tag{II-76}$$

(II-75) take the form

$$\dot{q} = [q, H], \qquad \dot{p} = [p, H], \tag{II-77}$$

which have the form of the canonical equations in classical dynamics where the Poisson brackets $(q, H)$, $(p, H)$ replace the quantum Poisson brackets $[q, H]$, $[p, H]$.

Now by induction arguments (similar to that employed in proving (II-45)), it is easily seen that any function $F$ formed by sums or products of $q$ and $p$ also satisfies (II-75), i.e.,

$$\dot{F} = \frac{1}{i\hbar}(FH - HF). \tag{II-78}$$

For $F = H$, this leads to[1] $\dot{H} = 0$, i.e., $H_{mn}$ are all time-independent.

Let us assume that originally, we have constructed $H(q, p)$ of hermitian matrices

$$(p)_{mn} = p^0_{mn} e^{i\omega_{mn}t}, \qquad (q)_{mn} = q^0_{mn} e^{i\omega_{mn}t}$$

with
$$\tag{II-79}$$

$$p^0_{mn} = p^{0*}_{nm}, \qquad q^0_{mn} = q^{0*}_{nm}.$$

Hence $H_{mn}$ must depend on $t$ through the same factor $e^{i\omega_{mn}t}$, i.e.,

$$(H)_{mn} = H^0_{mn} e^{i\omega_{mn}t}. \tag{II-80}$$

But the assumption of the canonical equations (II-74) has lead to $H_{mn} = $ const. Hence $H$ must be diagonal, i.e.,

$$H^0_{mn} = H^0_{mm}\delta_{mn}. \tag{II-81}$$

From (II-75), one obtains, on taking the $(m, n)$th element,

$$i\omega_{mn}(q)_{mn} = \frac{1}{i\hbar}((q)_{mn}H_{nn} - H_{mm}(q)_{mn}),$$

which gives the Bohr frequency condition

$$\hbar\omega_{mn} = H_{mm} - H_{nn}. \tag{II-82}$$

To summarize this formulation of the theory, one starts with the commutation relation (II-72), and the canonical equations (II-74). From these, one arrives at (i) $H = $ constant in time and is a diagonal matrix of the $q$, $p$ given in (II-79), and (ii) the Bohr frequency condition $h\nu_{mn} = H_{mm} - H_{nn}$.

The question may now be raised: If one constructs the Hamiltonian $H$ with the time-dependent hermitian $p$, $q$ of (II-7a), the $H$ is *not* diagonal, since $H_{mn} = H^0_{mn} e^{i\omega_{mn}t}$ in general. But the use of the canonical equations (II-74) implies $H$ being diagonal. This means that before one may use (II-74), $H$ must

---

[1] $\dot{H} = 0$ also follows directly from $\dot{H} = (\partial H/\partial p)\dot{p} + (\partial H/\partial q)\dot{q}$ and (II-74).

be in the diagonal form! Doesn't there seem to be a missing step somewhere?

This question is answered by Born-Heisenberg-Jordan in the following way. One constructs the Hamiltonian $H$ with the $p$, $q$ of (II-79) and $H$ is indeed nondiagonal in general.[m] Let us transform $H$ by a unitary transformation into a diagonal form $W$,

$$U^{-1}HU = W. \tag{II-83}$$

The diagonal elements $W_m$ are real, and are the eigenvalues of $H$ (see Sec. 2, Theorems 13, 15). From $W_m$, one obtains the frequencies

$$h\nu_{mn} = W_m - W_n. \tag{II-82}$$

The transformation matrix $U$ can be found by the method described in Sec. 2, Definition 16, (II-35)–(II-36). The $q$, $p$ are transformed by the same $U$ into

$$Q = U^{-1}qU, \qquad P = U^{-1}pU. \tag{II-84}$$

In terms of $Q$, $P$, $\bar{H}(Q,P) = U^{-1}H(q,p)U = W$, and

$$\dot{Q} = \frac{\partial \bar{H}}{\partial P}, \qquad \dot{P} = -\frac{\partial \bar{H}}{\partial Q},$$
$$= [Q, \bar{H}] \qquad = [P, \bar{H}], \tag{II-85}$$

are the equations of motions.

The matrix mechanics of Born-Heisenberg-Jordan may be summarized as follows:

(1) All physical quantities are represented by hermitian matrices; the observed values of these quantities are to be their eigenvalues.

(2) The hermitian coordinate $q$ and momentum $p$ are to obey the basic postulate

$$pq - qp = \frac{\hbar}{i}1. \tag{II-86}$$

(3) The Hamiltonian $H(q,p)$ constructed as a matrix function of $q$, $p$ (in the same functional form as in classical dynamics, with all products involving the $p$ and $q$ properly symmetrized to ensure the hermitian property of $H$) is transformed by a unitary transformation into diagonal form $W$

$$U^{-1}H(q,p)U = W. \tag{II-87}$$

The eigenvalues $W_1$, $W_2$, ... are then the energy values of the system.

---

[m] One exceptional case is the harmonic oscillator. See below.

(4) The frequencies of the spectral lines are given by

$$h\nu_{mn} = W_m - W_n. \tag{II-88}$$

(5) From the commutation relation above, one obtains the "equations of motion"

$$\frac{\hbar}{i}\frac{\partial H}{\partial p} = Hq - qH, \qquad -\frac{\hbar}{i}\frac{\partial H}{\partial q} = Hp - pH. \tag{II-89}$$

These equations are in arbitrary representation (i.e., $H$ is not diagonal in general.)

In the $H$-representation, i.e., $H$ is diagonal, the equations of motion takes the form

$$\dot{q} = \frac{\partial H}{\partial p}, \qquad \dot{p} = -\frac{\partial H}{\partial q}. \tag{II-90}$$

In this representation the equations of motion lead to the frequency condition.

The Born-Heisenberg-Jordan paper treats many applications of the new matrix mechanics some of which will be given in Secs. 5, 6 below. But even before the appearance of this paper, soon after the Born-Jordan paper, Pauli had applied the new theory to the hydrogen atom and the problem of crossed electric and magnetic fields (November, 1925).

*( 1 ) Harmonic oscillator*

As an example illustrating the applications of matrix mechanics, we follow Born-Jordan in treating the harmonic oscillator.

We start with the hermitian $p$ and $q$

$$q_{mn} = q_{mn}^0 e^{i\omega_{mn}t}, \qquad p = \mu\dot{q} \tag{II-91}$$

and

$$H = \frac{1}{2\mu}p^2 + \frac{1}{2}\mu\omega_0^2 q^2 \tag{II-92}$$

so that[n]

$$\ddot{q} = -\omega_0^2 q. \tag{II-93}$$

From (II-93),

---

[n] (II-93) is obtained by using the canonical equations (II-90). It turns out that the Hamiltonian (II-92) is in fact already diagonal with the $q$, $p$ of (II-91), so that the use of (II-90) is consistent. The harmonic oscillator is therefore an unusual case in which the Hamiltonian does not need diagonalization by canonical transformation.

$$(-\omega_{mn}^2 + \omega_0^2)q_{mn}^0 = 0,$$

$$\therefore q_{mn}^0 = 0 \qquad \text{unless} \qquad \omega_{mn} = \pm\omega_0. \tag{II-94}$$

This states that the electric moment $eq$, and therefore the spectral frequencies $\omega_{mn}$ have the (classical) frequency $\omega_0$ of the oscillator.

From the commutation relation (II-86) and $p = \mu\dot{q}$, we have (dropping the superscript 0 from $q_{mn}^0$ in the following)

$$i\mu\sum_k(\omega_{mk}q_{mk}q_{kn} - q_{mk}\omega_{kn}q_{kn}) = \frac{\hbar}{i}\delta_{mn}.$$

From $q_{mn} = q_{nm}^*$, $\omega_{mk} = -\omega_{km}$, the diagonal element $m = n$ is

$$\sum_k\omega_{mk}|q_{mk}|^2 = -\frac{\hbar}{2\mu}.$$

From (II-94), it is seen that, for a given state $m$, there are two states $k$; $k = j, l$, for which $q_{mk} \neq 0$, namely,

$$\omega_{mj} = \omega_0, \qquad \omega_{ml} = -\omega_0,$$

so that

$$\omega_{mj}(|q_{mj}|^2 - |q_{ml}|^2) = -\frac{\hbar}{2\mu}. \tag{II-95}$$

Here we introduce the concept of a "lowest state", that is, there is a state $m$ for which there exists only a state $l$, $\omega_{ml} = -\omega_0$ but no state $j$, $\omega_{mj} = \omega_0$. This state $m$ will be labelled by 0, i.e., $m = 0$. (II-95) thus can be expressed by

$$-|q_{01}|^2 = -\frac{\hbar}{2\mu\omega_0},$$

$$|q_{10}|^2 - |q_{12}|^2 = -\frac{\hbar}{2\mu\omega_0}, \tag{II-95a}$$

$$|q_{21}|^2 - |q_{23}|^2 = -\frac{\hbar}{2\mu\omega_0}, \text{ etc.}$$

From these, we get

$$|q_{12}|^2 = 2\frac{\hbar}{2\mu\omega_0}, \qquad |q_{23}|^2 = 3\frac{\hbar}{2\mu\omega_0},$$

$$|q_{m,m+1}|^2 = \frac{(m+1)\hbar}{2\mu\omega_0}. \tag{II-96}$$

The Hamiltonian in (II-92) is

$$H_{mn} = \frac{1}{2}\mu \sum_k (\omega_0^2 - \omega_{mk}\omega_{kn})q_{mk}q_{kn}e^{i\omega_{mn}t}.$$

From (II-95), (II-96), for $q_{mk} \neq 0$, we must have

$$\omega_{m,m+1} = -\omega_0, \qquad \omega_{m+1,m} = \omega_0,$$

so that

$$H_{mm} = \mu\omega_0^2(|q_{m,m-1}|^2 + |q_{m,m+1}|^2)$$

$$= (m + \tfrac{1}{2})\hbar\omega_0, \tag{II-97}$$

$$H_{mn} = 0 \qquad \text{for } n \neq m. \tag{II-97a}$$

Thus the $H$ constructed with the $q_{mn}$ of (II-91) is already diagonal, so that, so to speak, by starting with (II-91), one has guessed correctly the solution. That this is not always so is seen from the problem of the anharmonic oscillator for which the Hamiltonian is

$$H = \frac{1}{2\mu}p^2 + \frac{1}{2}k_1 q^2 + \frac{1}{3!}k_2 q^3 + \frac{1}{4!}k_3 q^4.$$

See Sec. 6 below.

**Exercises**

1. *The Harmonic oscillator in Fock representation.* Take the Hamiltonian in (II-92). Let

$$y = x\sqrt{\frac{\mu\omega_0}{\hbar}}.$$

(II-92) then becomes

$$\frac{1}{2}\left(y^2 + \frac{1}{\hbar^2}p^2\right) = \frac{E}{\hbar\omega_0}.$$

From (II-97), it is known that $E$ is diagonal. Hence

$$\frac{1}{2}\left(y^2 + \frac{1}{\hbar^2}p^2\right)_{mn} = \frac{E_m}{\hbar\omega_0}\delta_{mn}.$$

Define

$$b = \frac{1}{\sqrt{2}}\left(y + \frac{i}{\hbar}p\right), \qquad b^\dagger = \frac{1}{\sqrt{2}}\left(y - \frac{i}{\hbar}p\right).$$

(1) Show that $b$ and $b^\dagger$ so defined are adjoint of each other.
(2) From the commutation relation

$$py - yp = \frac{\hbar}{i} 1,$$

show that

$$bb^\dagger - b^\dagger b = 1.$$

(3) Let $\Lambda$ be the diagonal matrix defined by

$$\Lambda_m \delta_{mn} = \left(\frac{E_m}{\hbar\omega_0} - \frac{1}{2}\right)\delta_{mn}.$$

Show that

$$\Lambda_m = \sum_n |b_{nm}|^2 = \sum_n |b_{mn}^\dagger|^2,$$

$$b_{mn}^\dagger(\Lambda_m - \Lambda_n - 1) = 0,$$

$$b_{m,m-1}^\dagger = b_{m-1,n} = \sqrt{m},$$

i.e.,

$$b^\dagger = \begin{pmatrix} 0 & 0 & 0 & 0 & \cdot \\ 1 & 0 & 0 & 0 & \cdot \\ 0 & \sqrt{2} & 0 & 0 & \cdot \\ 0 & 0 & \sqrt{3} & 0 & \cdot \\ 0 & 0 & 0 & \sqrt{4} & \cdot \\ \cdot & \cdot & \cdot & \cdot & \cdot \end{pmatrix},$$

$$\Lambda = \begin{pmatrix} 0 & 0 & 0 & 0 & \cdot \\ 0 & 1 & 0 & 0 & \cdot \\ 0 & 0 & 2 & 0 & \cdot \\ 0 & 0 & 0 & 3 & \cdot \\ \cdot & \cdot & \cdot & \cdot & \cdot \end{pmatrix}.$$

(4) Show that

$$y_{m,m-1} = y_{m-1,m} = \sqrt{\frac{m\hbar}{2\mu\omega_0}}.$$

(5) Show that, while $y$, $p$ are hermitian variables, $b^\dagger$, $b$ are *not* hermitian, and are not "observables" in the sense of Dirac.

(6) Obtain the Thomas-Kuhn sum rule of Chap. I, (I-84b) by means of matrix mechanics,

$$\frac{4\pi m}{\hbar}\left(\sum_{k>m} v_{km}|q_{km}|^2 - \sum_{j<m} v_{mj}|q_{mj}|^2\right) = 1.$$

**References**

W. Heisenberg, *Z. Phys.* **33**, 879–893 (July, 1925).

M. Born and P. Jordan, *ibid.* **34**, 858–888 (Sept., 1925). This paper gives the basic theorems in matrix algebra, the initial step to the commutation relation from the correspondence principle, the harmonic oscillator, anharmonic oscillator and the quantization of the electromagnetic field.

M. Born, W. Heisenberg and P. Jordan, *ibid.* **35**, 557 (Nov., 1925). This paper generalizes Born-Jordan's work to systems of more than one degree of freedom, the theory of canonical transformations, the perturbation theory for time-independent and time-dependent perturbations, degeneracies, angular momenta, selection rules.

W. Pauli, *ibid.* **36**, 336–363 (1926). This paper treats, in matrix mechanics, the hydrogen atom and the problem of crossed electric and magnetic fields—a problem of unsurmountable difficulties in the Bohr theory (see Chap. 1, Sec. 12).

## 5. Angular momenta and selection rules

As another example, we take up the matrix mechanics of angular momenta which is of great importance in general, and in the problem of central fields in particular.

The angular momentum $\mathbf{L}$ and its components $L_x$, $L_y$, $L_z$ are defined by

$$L_x = yp_z - zp_y,$$

$$L_y = zp_x - xp_z, \tag{II-98}$$

$$L_z = xp_y - yp_x,$$

$$\mathbf{L}^2 = L_x^2 + L_y^2 + L_z^2. \tag{II-99}$$

The commutation relations (II-72) between two classical canonical conjugate variables $p_x$, $x$ can be expressed in terms of the quantum Poisson brackets (II-76) in the form

$$[A, B] = \frac{1}{i\hbar}(AB - BA),$$

$$[p_x, x] = -1, \qquad [p_y, y] = -1, \qquad [p_z, z] = -1, \tag{II-100}$$

and all other Q. P. B. vanish

$$[p_x, y] = [x, y] = 0, \text{ etc.} \tag{II-100a}$$

From these relations, one gets the following relations

$$[L_z, x] = y, \qquad [L_z, y] = -x, \qquad [L_z, z] = 0, \tag{II-101}$$

$$[L_z, p_x] = p_y, \qquad [L_z, p_y] = -p_x, \qquad [L_z, \dot{p}_z] = 0, \tag{II-102}$$

$$[L_x, L_y] = L_z, \qquad [L_y, L_z] = L_x, \qquad [L_z, L_x] = L_y, \tag{II-103}$$

$$[\mathbf{L}^2, L_x] = [\mathbf{L}^2, L_y] = [\mathbf{L}^2, L_z] = 0. \tag{II-104}$$

Since $\mathbf{L}^2$, $L_z$ commute, they can be represented simultaneously in diagonal form, i.e., they define a $(\mathbf{L}^2, L_z)$—representation (see Sec. 2, Theorems 17 and 18, and Sec. 3),[°] i.e., $\mathbf{L}^2$ and $L_z$ are diagonal.

We introduce $L_+$, $L_-$ defined by

$$L_\pm = L_x \pm iL_y \tag{II-105}$$

and the notation for matrix element labelling

$$\langle m|A|n \rangle = A_{mn}. \tag{II-106}$$

From (II-99), (II-101) and (II-107), one gets

$$L_z(x + iy) - (x + iy)L_z = (x + iy)\hbar,$$

$$L_+(x + iy) - (x + iy)L_+ = 0,$$

$$L_-(x + iy) - (x + iy)L_- = -2z\hbar,$$

$$L_\pm L_z - L_z L_\pm = \mp L_\pm \hbar, \tag{II-107}$$

$$L_+ L_- = \mathbf{L}^2 - L_z^2 + L_z\hbar, \tag{II-108a}$$

$$L_- L_+ = \mathbf{L}^2 - L_z^2 - L_z\hbar. \tag{II-108b}$$

From the $\langle m\|n \rangle$ element of (II-107),

$$\langle m|L_\pm|n \rangle \langle n|L_z|n \rangle - \langle m|L_z|m \rangle \langle m|L_\pm|n \rangle = \mp \langle m|L_\pm|n \rangle \hbar,$$

it is seen that

$$\langle m|L_\pm|n \rangle = 0 \tag{II-109}$$

unless

$$\langle n|L_z|n \rangle = \langle m|L_z|m \rangle \pm \hbar. \tag{II-110}$$

From (II-108a)

$$\sum_n \langle m|L_+|n \rangle \langle n|L_-|m \rangle = \langle m|\mathbf{L}^2|m \rangle - \left( \langle m|L_z|m \rangle - \frac{\hbar}{2} \right)^2 + \frac{1}{4}\hbar^2.$$

Since $L_+$, $L_-$ are hermitian and $L_+ = L_-^\dagger$, the left-hand side is $\sum_n |\langle m|L_+|n \rangle|^2$ which is $\geq 0$. Hence

---

[°] Although $\mathbf{L}^2$ also commutes with $L_y$, $L_z$, one must not jump to the conclusion that $\mathbf{L}^2$ and $L_x$, $L_y$, $L_z$ can be made diagonal simultaneously, as the $L_x$, $L_y$, $L_z$ do not commute among themselves. This situation arises from "degeneracy", as we shall see below. We choose $\mathbf{L}^2$, $L_z$ to define a representation only by convention.

$$\langle m|\mathbf{L}^2|m\rangle - (\langle m|L_z|m\rangle - \tfrac{1}{2}\hbar)^2 + \tfrac{1}{4}\hbar^2 \geq 0. \tag{II-111a}$$

Similarly, from (II-108b), one gets

$$\langle m|\mathbf{L}^2|m\rangle - (\langle m|L_z|m\rangle + \tfrac{1}{2}\hbar)^2 + \tfrac{1}{4}\hbar^2 \geq 0. \tag{II-111b}$$

From (II-99), it follows that

$$\langle m|\mathbf{L}^2|m\rangle \geq \langle m|L_z^2|m\rangle = |\langle m|L_z|m\rangle|^2$$

so that for a fixed $\langle m|\mathbf{L}^2|m\rangle$, $\langle m|L_z|m\rangle$ must have a maximum values $\langle m|L_z|m\rangle_{\text{max}}$ and minimum $\langle m|L_z|m\rangle_{\text{min}}$ for which

$$\langle m|\mathbf{L}^2|m\rangle - (\langle m|L_z|m\rangle_{\text{min}} - \tfrac{1}{2}\hbar)^2 + \tfrac{1}{4}\hbar^2 = 0,$$
$$\langle m|\mathbf{L}^2|m\rangle - (\langle m|L_z|m\rangle_{\text{max}} + \tfrac{1}{2}\hbar)^2 + \tfrac{1}{4}\hbar^2 = 0. \tag{II-112}$$

The difference must be

$$\langle m|L_z|m\rangle_{\text{max}} - \langle m|L_z|m\rangle_{\text{min}} = 2l\hbar, \tag{II-113}$$

where $2l$ is an integer, so that $l$ may be an integer or a half-odd integer. Using (II-113) in (II-112), one gets

$$2[\langle m|\mathbf{L}^2|m\rangle + \tfrac{1}{4}\hbar^2]^{1/2} = (2l + 1)\hbar$$

or

$$\langle m|\mathbf{L}^2|m\rangle = l(l + 1)\hbar^2, \tag{II-114}$$

and from (II-112)

$$-l\hbar \leq \langle m|L_z|m\rangle \leq l\hbar. \tag{II-115}$$

This relation may be written

$$\langle m|L_z|m\rangle = m\hbar \tag{II-115a}$$

with

$$-l \leq m \leq l. \tag{II-115b}$$

From (II-113) and (II-115a), it is seen that a proper labelling of matrix elements is $\langle l, m|A|l', m'\rangle$ in the $\mathbf{L}^2, L_z$ representation,

$$\langle l, m|\mathbf{L}^2|l', m'\rangle = l(l + 1)\hbar^2\delta_{ll'}\delta_{mm'},$$
$$\langle l, m|L_z|l', m'\rangle = m\hbar\delta_{ll'}\delta_{mm'}. \tag{II-116}$$

To obtain the matrix elements of the other angular momenta, from (II-107) and (II-116), it is seen that the only nonvanishing matrix elements in the $\mathbf{L}^2$, $L_z$ representation are

$$\langle l, m | L_+ | l, m - 1 \rangle \neq 0,$$

$$\langle l, m | L_- | l, m + 1 \rangle \neq 0,$$

and using these in (II-108a), (II-108b), one gets

$$|\langle l, m | L_+ | l, m - 1 \rangle|^2 = (l + m)(l + 1 - m)\hbar^2,$$

$$|\langle l, m | L_- | l, m + 1 \rangle|^2 = (l + m + 1)(l - m)\hbar^2. \tag{II-117}$$

From (II-103)

$$L_y L_z - L_z L_y = -\frac{\hbar}{i} L_x,$$

one obtains

$$\langle l, m | i L_y | l, m - 1 \rangle = \langle l, m | L_x | l, m - 1 \rangle,$$

$$\langle l, m | i L_y | l, m + 1 \rangle = -\langle l, m | L_x | l, m + 1 \rangle. \tag{II-118}$$

Finally,

$$\langle l, m | L_x | l, m - 1 \rangle = i \langle l, m | L_y | l, m - 1 \rangle$$

$$= \tfrac{1}{2}\sqrt{(l + m)(l + 1 - m)}\,\hbar, \tag{II-119a}$$

$$\langle l, m | L_x | l, m + 1 \rangle = -i \langle l, m | L_y | l, m + 1 \rangle$$

$$= \tfrac{1}{2}\sqrt{(l + m + 1)(l - m)}\,\hbar. \tag{II-119b}$$

From (II-116), it is seen that $\mathbf{L}^2$ is degenerate in $m$ (independent of $m$). It is for this reason that $\mathbf{L}^2$ commutes also with $L_x$, $L_y$, even though $L_x$, $L_y$ are not diagonal in the $\mathbf{L}^2$, $L_z$ representation.[p]

The following examples are the $L_z$, $L_x$, $L_y$ matrices in the $\mathbf{L}^2$, $L_z$ representation for $l = 1$ and $l = \tfrac{3}{2}$.

(1) $l = 1$.

$$
L_z = \begin{array}{c} \\ \\ 1 \\ 0 \\ -1 \end{array}
\begin{array}{c} \begin{array}{ccc} m \quad 1 & 0 & 1 \end{array} \\ \hline \left(\begin{array}{ccc} 1 & 0 & 0 \\ 0 & 0 & 0 \\ 0 & 0 & -1 \end{array}\right) \end{array} \hbar,
$$

[p] It is worth noting that quantum mechanics shows a difference from classical dynamics here. In the latter, the total angular momentum of an isolated system (such as the solar system assumed to be isolated from the rest of the stars) is an invariant—an invariable line—in magnitude and direction, so that its components $L_x$, $L_y$, $L_z$ are also constants. But in quantum mechanics, $\mathbf{L}^2$ and one of the $L_x$, $L_y$, $L_z$, say $L_z$, are constant, but the other two, are not diagonal at the same time and therefore are not known with $\mathbf{L}^2$ and $L_z$!

$$L_x = \frac{1}{\sqrt{2}} \begin{pmatrix} 0 & 1 & 0 \\ 1 & 0 & 1 \\ 0 & 1 & 0 \end{pmatrix} \hbar, \qquad L_y = \frac{i}{\sqrt{2}} \begin{pmatrix} 0 & -1 & 0 \\ 1 & 0 & -1 \\ 0 & 1 & 0 \end{pmatrix} \hbar. \qquad \text{(II-120)}$$

(2) $l = \frac{3}{2}$.

$$L_z = \begin{array}{c} \\ \frac{3}{2} \\ \frac{1}{2} \\ -\frac{1}{2} \\ \frac{3}{2} \end{array} \begin{pmatrix} \frac{3}{2} & 0 & 0 & 0 \\ 0 & \frac{1}{2} & 0 & 0 \\ 0 & 0 & -\frac{1}{2} & 0 \\ 0 & 0 & 0 & -\frac{3}{2} \end{pmatrix} \hbar,$$

$$L_x = \begin{pmatrix} 0 & \sqrt{3}/2 & 0 & 0 \\ \sqrt{3}/2 & 0 & 1 & 0 \\ 0 & 1 & 0 & \sqrt{3}/2 \\ 0 & 0 & \sqrt{3}/2 & 0 \end{pmatrix} \hbar,$$

$$L_y = i \begin{pmatrix} 0 & -\sqrt{3}/2 & 0 & 0 \\ \sqrt{3}/2 & 0 & -1 & 0 \\ 0 & 1 & 0 & -\sqrt{3}/2 \\ 0 & 0 & \sqrt{3}/2 & 0 \end{pmatrix} \hbar. \qquad \text{(II-121)}$$

(3) $l = \frac{1}{2}$.

$$L_z = \begin{array}{c} \frac{1}{2} \\ -\frac{1}{2} \end{array} \begin{pmatrix} \frac{1}{2} & 0 \\ 0 & -\frac{1}{2} \end{pmatrix} \hbar, \qquad L_x = \begin{pmatrix} 0 & \frac{1}{2} \\ \frac{1}{2} & 0 \end{pmatrix} \hbar, \qquad L_y = \begin{pmatrix} 0 & -i/2 \\ i/2 & 0 \end{pmatrix} \hbar.$$

$$\text{(II-122)}$$

The $l = \frac{1}{2}$ case applies to the electron spin and the matrices are denoted usually as $\frac{1}{2}\sigma_x, \frac{1}{2}\sigma_y, \frac{1}{2}\sigma_z$, where $\sigma_x, \sigma_y, \sigma_z$ are the Pauli matrices.

### Exercises

1. Show that the $L_+$, $L_-$ defined in (II-109) are adjoint matrices of each other, i.e.,

$$L_+^\dagger = L_-, \qquad L_+ = L_-^\dagger.$$

2. The dispersion $\Delta A$ of $A$ is defined as

$$(\Delta A)^2 = \overline{(A - \bar{A})^2},$$

where

$$\bar{A} = \langle l, m | A | l, m \rangle$$

is the diagonal matrix elements of $A$ in the $\mathbf{L}^2$, $L_z$ representation (see (II-114)).

Calculate $\Delta L_x$, $\Delta L_y$.

3. Calculate $(\Delta L_x)^2$, $(\Delta L_y)^2$ for the case $l = 1$ (see (II-116)), $m = \pm 1$, 0. If a top, of total angular momentum $2\hbar$, precesses about the $z$-axis, show that $(\Delta L_x)^2$, $(\Delta L_y)^2$ are $\frac{1}{2}\hbar^2$ if the $z$-component $L_z$ is $\pm\hbar$, and are $\hbar^2$ if the $L_z$ is 0.

## 6. Perturbation theory

### (1) Non-degenerate systems

The general theory for solving a given problem in matrix mechanics is to transform, by a unitary transformation, the Hamiltonian into a diagonal form

$$UHU^{-1} = W \tag{II-123}$$

(See Sec. 4, (II-82) or Sec. 2, Theorems 13, 15). In many problems, the Hamiltonian $H$ is such that the solution of (II-123) is very difficult. The perturbation theory method consists in comparing $H$ with the $H^0$ of another system for which the solution is known, i.e., $H^0$ is already in a diagonal form, and developing $H - H^0$ in a series of a parameter $\lambda$,

$$H = H^0 + \lambda H^{(1)} + \lambda^2 H^{(2)} + \cdots \tag{II-124}$$

together with the corresponding

$$U = U^0 + \lambda U^{(1)} + \lambda^2 U^{(2)} + \cdots, \tag{II-125}$$

$$W = H^0 + \lambda E^{(1)} + \lambda^2 E^{(2)} + \cdots, \tag{II-126}$$

where to various orders in $\lambda$,

$$\lambda^0: \qquad U^0 H^0 = H^0 U^0, \tag{II-127}$$

$$\lambda: \qquad U^0 H^{(1)} + U^{(1)} H^0 = H^0 U^{(1)} + E^{(1)} U^{(0)}, \tag{II-128}$$

$$\lambda^2: \quad U^0 H^{(2)} + U^{(1)} H^{(1)} + U^{(2)} H^0 = H^0 U^{(2)} + E^{(1)} U^{(1)} + E^{(2)} U^{(0)}. \tag{II-129}$$

By hypothesis, $H^0$ is diagonal, and nondegenerate so that, from (II-127),

$$U^0_{mn}(H^0_{mm} - H^0_{nn}) = 0 \tag{II-130}$$

so that

$$U^0_{mn} = 0 \qquad \text{for } m \neq n.$$

The unitarity of $U^0$ requires that $U^\dagger U = 1$, so that

$$U^{0*}_{mm} U^0_{mm} = 1$$

whose solutions are

$$U^0_{mm} = 1 \text{ or } i.$$

One may take

$$U^0 = 1, \text{ unit matrix.} \tag{II-131}$$

From (II-128), one gets

$$H^{(1)}_{mn} + U^{(1)}_{mn}(H^0_{nn} - H^0_{mm}) = E^{(1)}_{mm}\delta_{mn}$$

from which,

$$E^{(1)}_{mm} = H^{(1)}_{mm}, \tag{II-132}$$

$$U^{(1)}_{mn} = \frac{H^{(1)}_{mn}}{H^0_{mm} - H^0_{nn}} \qquad \text{for } m \neq n. \tag{II-133}$$

The $U^{(1)}_{mm}$ elements are, from $U^\dagger U = 1$, up to $\lambda^0$,

$$U^{(1)} + U^{(1)\dagger} = 0,$$

whose solutions are $U^{(1)}_{mm} = 0$ or $i$. We shall take

$$U^{(1)}_{mm} = 0. \tag{II-134}$$

From (II-129), we obtain by the same procedure as above the following results

$$E^{(2)}_{mm} = H^{(2)}_{mm} + \sum_k{}' \frac{|H^{(1)}_{mk}|^2}{H^0_{mm} - H^0_{kk}}, \qquad (k \neq m), \tag{II-135}$$

$$U^{(2)}_{mm} = 0, \tag{II-136}$$

$$U^{(2)}_{mn} = -\frac{(H^{(1)}_{mm} - H^{(1)}_{nn})H^{(1)}_{mn}}{|H^0_{mm} - H^0_{nn}|^2} + \frac{H^{(2)}_{mn}}{H^0_{mm} - H^0_{nn}}$$

$$+ \sum{}' \frac{H^{(1)}_{mk}H^{(1)}_{kn}}{(H^0_{mm} - H^0_{nn})(H^0_{mm} - H^0_{kk})}, \qquad (k \neq m). \tag{II-137}$$

It is a general feature that the diagonal elements of $H^{(1)}_{mm}$ appear in the $\lambda$ order in $E^{(1)}$, the nondiagonal elements $H^{(1)}_{mn}$ appear in the $\lambda^2$ order in $E^{(2)}$, etc.

It is of interest to note that while $U + \lambda U^{(1)}$ transforms $H^0 + \lambda H^{(1)}$ into $H^0$ + diagonal elements of $H^{(1)}$ only, as seen from

$$(U^0 + \lambda U^{(1)})(H^0 + \lambda H^{(1)})(U^0 - \lambda U^{(1)}) = H^0 + \lambda(H^{(1)} + U^{(1)}H^0 - H^0 U^{(1)})$$

$$+ \lambda^2(\ldots)$$

$$= H^0 + \lambda H^{(1)}_{mm}\delta_{mn},$$

by the use of (II-133) and (II-134). $U^0 + \lambda U^{(1)}$ transforms $H^0$ itself into $H^0$ minus the nondiagonal elements of $H^{(1)}$,

$$(U^0 + \lambda U^{(1)})H^0(U^0 - \lambda U^{(1)}) = H^0 - \lambda H^{(1)}_{mn}, \qquad (m \neq n).$$

*Anharmonic oscillator*

As an example, take the anharmonic oscillator

$$H = \frac{1}{2\mu}p^2 + \frac{1}{2}k_1 x^2 + \frac{1}{3!}k_2 x^3 + \frac{1}{4!}k_3 x^4. \tag{II-138}$$

For convenience, let

$$H^0 = \frac{1}{2\mu}p^2 + \frac{1}{2}k_1 x^2, \qquad H^{(1)} = \frac{1}{6}k_2 x^3, \qquad H^{(2)} = \frac{1}{24}k_3 x^4.$$

The matrix elements of $H^{(1)}$ and $H^{(2)}$ can all be obtained by means of (II-96).

$$x_{m,m+1} = \sqrt{\frac{(m+1)\hbar}{2\mu\omega_0}}, \qquad k_1 = \mu\omega_0^2,$$

$$H^{(1)}_{mn} = \frac{1}{6}k_2 \sum_{i,j} x_{mi}x_{ij}x_{jn},$$

$$H^{(2)}_{mn} = \frac{1}{24}k_3 \sum_{i,j,k} x_{mi}x_{ij}x_{jk}x_{kn}.$$

The relevant elements are

$$H^{(1)}_{mm} = 0,$$

$$H^{(2)}_{mm} = \frac{k_3}{24}A^4\left(m^2 + m + \frac{1}{2}\right),$$

$$H^{(1)}_{m,m+1} = \frac{k_2}{6}A^3\sqrt{3(m+1)^3},$$

$$H^{(1)}_{m,m+3} = \frac{k_2}{6}A^3\sqrt{(m+1)(m+2)(m+3)}, \tag{II-139}$$

$$A = \sqrt{\frac{\hbar}{2\mu\omega_0}}.$$

The energy, up to $\lambda^2$, is

$$W_{nn} = \left(n + \frac{1}{2}\right)\hbar\omega_0 - \frac{5}{6}\frac{k_2^2}{\hbar\omega_0}A^6\left(n^2 + n + \frac{11}{30}\right)$$

$$+ \frac{1}{4}k_3 A^4\left(n^2 + n + \frac{1}{2}\right). \tag{II-140}$$

### *(2) Degenerate systems*

The method has been given by Born, Heisenberg and Jordan. We shall not give it here, but shall give the theory in the wave mechanics.


### 7. Dirac's theory

In Sepember, 1925, Dirac, after reading Heisenberg's first article (*Z. Phys.* **33**, 879, (1925)) but not yet the Born-Jordan paper (*ibid.* **34**, 858, (1925)) which established Heisenberg's theory in matrix form and obtained the commutation relation, developed independently a theory of operators, obtaining also the commutation relation *without* being aware of, or using, matrix algebra. Dirac's first paper was submitted for publication in November (1925). In two subsequent papers (January and July, 1926), Dirac developed a $q$-number algebra (obeying a noncommutative rule of multiplication, in contrast to $c$-numbers which are those used in classical physics.)

Dirac started with Heisenberg's multiplication rule, (II-13), Chap. 2, (*before* it was recognized by Born to be that of matrices),

$$(xy)_{mn} = \sum_k x_{mk} y_{kn}. \tag{II-141}$$

He investigated the form of a quantum operation denoted by $d/dv$, which satisfies the laws

$$\frac{d}{dv}\cdot(x + y) = \frac{d}{dv}\cdot x + \frac{d}{dv}\cdot y, \tag{II-142}$$

$$\frac{d}{dv}\cdot(xy) = \left(\frac{d}{dv}\cdot x\right)y + x\frac{d}{dv}\cdot y. \tag{II-143}$$

On assuming that the "components" of $(d/dv)\cdot x$, namely, $((d/dv)\cdot x)_{mn}$, are linear functions of those of $x$, i.e.,

$$\left(\frac{d}{dv}\cdot x\right)_{mn} = \sum_{i,j} a_{mn;\,ij} x_{ij}, \tag{II-144}$$

one obtains from (II-143) and the rule of Heisenberg (II-141),

$$\sum_{m'n'} a_{mn;m'n'} \sum_k x_{m'k} y_{kn'} = \sum_{k,m',k'} a_{mk;m'k'} x_{m'k'} y_{kn} + \sum_{k,k'n'} x_{mk} a_{kn;k'n'} y_{k'n'}. \qquad \text{(II-145)}$$

On comparing the coefficients of $x_{m'k} y_{k'n'}$ on both sides, one obtains

$$a_{mn,m'n'} \delta_{kk'} = a_{mk',m'k} \delta_{nn'} + a_{kn,k'n'} \delta_{mm'}. \qquad \text{(II-146)}$$

For $k = k'$

$$a_{mn,m'n'} = 0 \qquad \text{unless } n = n', \text{ or } m = m', \text{ or } \begin{cases} n = n', \\ m = m'. \end{cases} \qquad \text{(II-147)}$$

For $k = k', n = n', m \neq m'$

$$a_{mn,m'n} = a_{mk,m'k} \qquad \text{independently of } n, \text{ or } k,$$
$$\equiv b_{mm'} \qquad \text{(by definition).} \qquad \text{(II-148)}$$

For $k = k', n \neq n', m = m'$

$$a_{mn,mn'} = a_{kn,kn'} \qquad \text{independently of } m, \text{ or } k,$$
$$\equiv y_{n'n} \qquad \text{(by definition).} \qquad \text{(II-149)}$$

For $k \neq k, m = m', n = n'$

$$a_{mk',mk} + a_{kn,k'n} = 0,$$

or

$$y_{kk'} + b_{kk'} = 0. \qquad \text{(II-150)}$$

For $k = k', m = m, n = n,$

$$a_{mn,mn} = a_{mk,mk} + a_{kn,kn},$$
$$= + b_{mm} + y_{nn} \qquad \text{(by (II-148, 149)),}$$
$$= - y_{mm} + y_{nn} \qquad \text{by (II-150).} \qquad \text{(II-151)}$$

From this, one obtains

$$\left( \frac{d}{dv} \cdot x \right)_{mn} = \sum_j (a_{mn,mj} x_{mj} + a_{mn,jn} x_{jn}),$$
$$= \sum_j (y_{jn} x_{mj} + b_{mj} x_{jn}) \qquad \text{by (II-149, 148),}$$
$$= \sum_j (x_{mj} y_{jn} - y_{mj} x_{jn}) \qquad \text{by (II-150),} \qquad \text{(II-152)}$$

or

$$\frac{d}{dv} \cdot x = xy - yx. \qquad \text{(II-153)}$$

This is a strange relation; it states that the most general operation on a quantum variable $x$ is the difference between the "Heisenberg products" (II-141) of $x$ with *another* variable $y$ and $y$ with $x$.

To see what the new product difference $xy - yx$ corresponds to in classical physics, Dirac took the classical dynamical system with one degree of freedom, with action and angle variable $J$, $w$ (see Chap. 1, Sec. 5). Let two classical variable $x$, $y$ be periodic functions of $w$, i.e.,

$$x = \sum_\tau x_\tau(J)e^{2\pi i \tau w}, \qquad w = \omega t + \delta$$

$$y = \sum_\tau y_\tau(J)e^{2\pi i \tau w}, \tag{II-153}$$

$\tau = $ being positive and negative integers, $x_\tau$, $y_\tau$ being the amplitudes of the $\tau$th harmonic and functions of $J$.

Dirac introduced the correspondence principle according to which the quantum $x_{n,n-\tau}$ corresponds to the classical $x_\tau$ (see (II-1) and (II-7)). Thus, from (II-152), the quantum

$$x_{m,m-\tau} y_{m-\tau,m-\tau-\sigma} - y_{m,m-\sigma} x_{m-\sigma,m-\sigma-\tau}$$

$$= (x_{m,m-\tau} - x_{m-\sigma,m-\sigma-\tau})y_{m-\tau,m-\tau-\sigma} - (y_{m,m-\sigma} - y_{m-\tau,m-\tau-\sigma})x_{m-\sigma,m-\tau-\sigma} \tag{II-154}$$

$$= \frac{\Delta x}{\Delta J}\Delta Jy - \frac{\Delta y}{\Delta J}\Delta Jx, \tag{II-155}$$

where

$$J = nh$$

is to correspond to the classical

$$\sigma h \frac{\partial x_\tau}{\partial J}y_\sigma - \tau h \frac{\Delta y_\sigma}{\Delta J}x_\tau. \tag{II-156}$$

From (II-153)

$$\sigma y_\sigma e^{2\pi i \sigma w} = \frac{1}{2\pi i}\frac{\partial}{\partial w}(y_\sigma e^{2\pi i \sigma w}), \text{ etc.} \tag{II-157}$$

The right-hand side of (II-154) corresponds to the classical

$$\frac{h}{2\pi i}\sum\left[\frac{\partial}{\partial J}(x_\tau e^{2\pi i \tau w})\frac{\partial}{\partial w}(y_\sigma e^{2\pi i \sigma w})\right.$$

$$\left. - \frac{\partial}{\partial J}(y_\sigma e^{2\pi i \sigma w})\frac{\partial}{\partial w}(x_\tau e^{2\pi i \tau w})\right], \tag{II-158}$$

the sum being over $\tau$ and $\sigma$ such that

$$m - \tau - \sigma = n.$$

By means of (II-153), this can be written

$$\frac{ih}{2\pi}\left(\frac{\partial x}{\partial w}\frac{\partial y}{\partial J} - \frac{\partial y}{\partial w}\frac{\partial x}{\partial J}\right) = \frac{ih}{2\pi}(x, y), \tag{II-159}$$

where $(x, y)$ is the classical Poisson bracket expression for $x$, $y$ with respect to the canonical variables $w$, $J$.

The correspondence principle says that the quantum $xy - yx$ in (II-153) is to correspond to the classical Poisson bracket expression $(x, y)$ as

$$xy - yx \leftrightarrow \frac{ih}{2\pi}(x, y). \tag{II-160}$$

In classical dynamics, the Poisson brackets of the conjugate coordinates $q$, $p$ with respect to any pair, such as $w$, $J$, connected by a canonical transformation have the values

$$(q, q)_{w,J} = (p, p)_{w,J} = 0, \qquad (q, p)_{w,J} = 1. \tag{II-161}$$

Dirac made the postulate that, the quantum variable $qp - pq$ corresponds to the classical $(ih/2\pi)(q, p)$, i.e.,

$$pq - qp = \frac{h}{2\pi i}1, \tag{II-162}$$

which is the commutation relation obtained by Born and Jordan, (II-72).

The basic idea of Dirac in this approach is then as follows: One starts with classical dynamics and formulates the equations of motion in terms of Poisson bracket expressions, namely,

$$\dot{q} = (q, H), \qquad \dot{p} = (p, H). \tag{II-163}$$

The quantum mechanical theory is obtained by replacing the classical Poisson brackets by the quantum Poisson brackets

$$(A, B) \rightarrow \frac{2\pi}{ih}(AB - BA) = [A, B] \tag{II-164}$$

so that the quantum mechanical equations of motion are

$$\dot{q} = [q, H], \qquad \dot{p} = [p, H]. \tag{II-165}$$

The result is the same as in Sec. 4, (II-74)–(II-78). The Dirac's theory, although identical in its results with the matrix mechanics of Heisenberg, Born and Jordan is conceptually simpler. One starts with classical dynamics, and passes over into quantum mechanics by replacing the classical Poisson brackets by the quantum Poisson brackets. This single step constitutes the transition to new basic concepts ($q$-numbers instead of $c$-numbers), the quantization condition in the form of the quantum Poisson bracket

$$[q, p] = 1,$$

for (II-162), (replacing the Bohr-Sommerfeld condition) and also the mathematical structure of the new theory.

## References

P. A. M. Dirac, *Proc. Roy. Soc.* London A **109**, 642 (Nov., 1925).
P. A. M. Dirac, *ibid.* **110**, 561 (Jan., 1926).
P. A. M. Dirac, *ibid.* **111**, 281 (March, 1926).
P. A. M. Dirac, *Proc. Camb. Phil. Soc.* **23**, 412 (July, 1926). This paper gives the $q$-number algebra.
P. A. M. Dirac, *The Principles of Quantum Mechanics*, 1st ed. (Clarendon Press, Oxford, 1930). In this first edition, quantum mechanics is developed in abstract form.

# Chapter 3

# Wave Mechanics: Early Developments

Einstein, in 1905, proposed the quantum theory of radiation in which electromagnetic waves having the characteristic properties of wave length, frequency, diffraction and interference are ascribed the particle properties of energy and momentum. In 1923, Louis de Broglie (1892–    ) proposed the converse, namely, a particle is associated with wave properties thereby completing the particle-wave symmetry, or duality and, for a while, further deepening the dilemma confronting physics. Fortunately the new ideas of de Broglie led to the development of wave mechanics by Schrödinger in 1926, and by 1927, to a new, complete, at least self-consistent, system of physics, including the matrix mechanics of Heisenberg, Born and Jordan described in the preceding chapter.

## 1.  de Broglie's ideas (1923–1924)

de Broglie's ideas—namely, an even closer, more basic relation between "particle" and "wave" than that suggested by Einstein for the photon theory—had been suggested in some observations of W. R. Hamilton (1805–1865) as early as between 1828–1837. Hamilton noted that there is a close formal similarity between the rays in optics and the trajectories in particle dynamics. For a light ray in a medium whose index of refraction $n$ is a function of spatial coordinates $n = n(x, y, z)$, Fermat gives the equation in the form of a variational principle—known as the principle of least time (1657)

$$\delta \int_A^B \frac{ds}{u} = 0, \qquad u = \text{phase velocity},$$

which states that the light ray from a point $A$ to $B$ follows a path such that the time of transit is minimum. As $v\lambda = u$ and $u = c/n$, the equation can be expressed in the form

$$\delta \int_A^B \frac{1}{\lambda(x, y, z)} ds = 0. \tag{III-1}$$

On the other hand, for a particle in a potential field, Maupertuis gave the principle of least action (1744)

$$\delta \int_A^B 2T \, dt = 0 \tag{III-2}$$

which states that the motion of the particle in going from a point $A$ to a point $B$ is such that the integral along the actual trajectory is minimum.[a] For a particle in a potential field $V(x, y, z)$,

$$\int_A^B 2T \, dt = \int_A^B mv \, ds$$

$$= \int_A^B p \, ds = \int_A^B \sqrt{2m(E - V)} \, ds;$$

equation (III-2) takes the form

$$\delta \int_A^B p \, ds = 0. \tag{III-3}$$

Comparison of (III-1) and (III-3) suggests that the light ray in a medium of index of refraction $n(x, y, z)$ and the trajectory of a particle in a potential field $v(x, y, z)$ are the same if

$$n(x, y, z) \propto p(x, y, z)$$

or

$$\frac{1}{\lambda} \propto p. \tag{III-4}$$

This observation of Hamilton was extremely interesting, but the equality of a light ray in optics and a particle trajectory in dynamics was then only formal; and it remained unpursued for about a hundred years until de Broglie makes it the starting point of his new ideas.

de Broglie started with the idea that the wave and the particle properties are more closely related than the formal similarity of the two laws (III-1) and (III-3) suggests. He proposed that to a particle of energy $E$ and momentum $p$, there is associated a wave with frequency $v$ and wavelength $\lambda$ given by

$$v = \frac{E}{h}, \qquad \lambda = \frac{h}{p}, \tag{III-5}$$

on the following considerations. In the special theory of relativity, in an inertial

---

[a] The basis for Maupertuis' principle is not clear, but the variational method had since been developed by Euler (1744), Lagrange (1760), and Hamilton formulated the Hamilton principle in 1834, of which the principle of least action is a special case when the conservation of energy is imposed on the system. The form (III-3) is due to H. Hertz (1892).

system, the energy-momentum 4-vector of a particle of rest mass $m_0$ and velocity $v$ is

$$\left(p, \frac{iE}{c}\right) = \left(\frac{m_0 v_x}{\sqrt{1-\beta^2}}, \frac{m_0 v_y}{\sqrt{1-\beta^2}}, \frac{m_0 v_z}{\sqrt{1-\beta^2}}, \frac{im_0 c}{\sqrt{1-\beta^2}}\right),$$

$$\beta = \frac{v}{c}. \tag{III-6}$$

In the same inertial system, a plane wave is represented by $e^{i(\mathbf{k}\cdot\mathbf{r}-\omega t)}$, or

$$\exp i(k_x x + k_y y + k_z z - \omega t), \tag{III-7}$$

where $k$ is the wave vector $(k_x, k_y, k_z)$ along which the wave propagates, and

$$|k| = \frac{2\pi}{\lambda},$$

$(\mathbf{k}\cdot\mathbf{r} - \omega t)$ is the scalar product of the 4-vectors

$$\left(\mathbf{k}, \frac{i\omega}{c}\right), \qquad (\mathbf{r}, ict).$$

On comparing $(\mathbf{k}, i\omega/c)$ with $(p, iE/c)$, de Broglie made the proportionality relation

$$k \propto p, \qquad \omega \propto E,$$

or

$$k = \frac{p}{\hbar}, \qquad \omega = \frac{E}{\hbar}, \tag{III-8}$$

or

$$\lambda = \frac{h}{p}, \qquad v = \frac{E}{h},$$

which is (III-5). This "completes" the formal relationship between a "wave" (III-7), (III-8) with a "particle" (III-6). While the nature of the wave is not yet clear, the following properties can be deduced.

(i) Let $u$ be the *phase velocity* of the wave[b] (III-7), i.e.,

$$u = v\lambda.$$

For a particle with mass $m$ and momentum $p$,

[b] It seems that the term "phase wave" is first used by de Broglie in 1923.

$$m = \frac{m_0}{\sqrt{1-\beta^2}}, \qquad p = \frac{m_0 v}{\sqrt{1-\beta^2}}, \qquad E = mc^2,$$

(III-5) gives

$$u = \frac{c^2}{v}. \qquad\qquad\qquad \text{(III-9)}$$

Thus the phase velocity $u > c$, as the velocity of the particle $v < c$ according to the relativity theory.

(ii) The *group velocity* $v_g$ of a wave is given (due to Hamilton and Lord Rayleight) by

$$\frac{1}{v_g} = \frac{d}{dv}\left(\frac{1}{\lambda}\right), \qquad u = v\lambda. \qquad\qquad \text{(III-10)}$$

From (III-5), we may rewrite this relation

$$v_g = \frac{dhv}{d\left(\dfrac{h}{\lambda}\right)} = \frac{dE}{dp}.$$

From (III-6),

$$p^2 - \frac{E^2}{c^2} = -m_0^2 c^2$$

so that

$$\frac{dE}{dp} = \frac{c^2 p}{E} = v.$$

Hence the important result that

$$v_g = v, \qquad\qquad\qquad \text{(III-11)}$$

i.e., the group velocity $v_g$ of the wave is equal to the velocity $v$ of the particle.

(iii) The wave (III-7) associated with the electron in the hydrogen atom will lead to the Bohr quantization condition as follows. If we assume that the condition for stationary states is the condition for standing waves of the associated waves

$$2\pi pa = n\lambda, \qquad\qquad\qquad \text{(III-12)}$$

where $a$ is the radius of the electron orbit, we have $\lambda = h/p$ and

$$2\pi pa = nh,$$

which is just the Bohr condition.

de Broglie seemed to have suggested (1924) to A. Dauvillier a diffraction experiment of electrons by crystals to test his theory, but unfortunately the experiment had not been tried. In what later became the Bell Telephone Laboratory, C. J. Davisson (1881–1958) and C. H. Kunsman had been experimenting on the scattering of electrons by nickel (1921), platinum and magnesium (1923), and from the intensity maxima and angles of scattering found evidence for diffraction. There are other experiments too, by C. Ramsauer (1921–3), J. S. Townsend and V. A. Bailey (1922), and R. N. Chaudhuri (1923), on the observed small scattering of slow electrons in rare gases—a phenomenon explainable in terms of diffraction of electrons by the (rather abrupt changes in the) outer regions of the atoms. These experimental results are reviewed by Elsasser (1925) on the basis of de Broglie's wave theory. But the crucial experiments with electrons come in 1927–8, from Davisson and L. H. Germer of Bell Laboratories, G. P. Thomson (son of J. J. Thomson) of England, S. Kikuchi (1928) of Japan and E. Rupp (1928) of Germany, and experiments with molecular beams of hydrogen and helium by I. Estermann, O. Stern and R. Frish (1930–1), with neutron beams by E. Fermi and L. Marshall (1947) and by W. H. Zinn (1947). All these experiments are in agreement with the relation

$$\lambda = \frac{h}{p},$$

which can be regarded as having been firmly experimentally established.

## References

W. R. Hamilton, *Trans. Roy. Irish Acad.* **15**, 69 (1828); **16**, 1, 93 (1830–1); **17**, 1 (1837).

L. de Broglie, *Compt. Rendus* **177**, 507; 548, 630 (1923); summarized in *Phil. Mag.* **47**, 446 (1924); Thesis (1924); *Annales de Physique* **3**, 22–128 (1925).

C. J. Davisson and L. H. German, *Phys. Rev.* **30**, 705 (1927).

G. P. Thomson and A. Reid, *Nature* **119**, 890 (1927); G. P. Thomson, *ibid.* **120**, 802 (1927); *Proc. Roy. Soc.* London A **117**, 600 (1927).

S. Kikuchi, *Proc. Imp. Acad.* Tokyo **4**, 271, 275, 354, 471 (1928).

E. Rupp, *Annalen d. Physik* **85**, 981 (1928) and *Naturwiss.* **16**, 656 (1928) describe electron diffraction by optical grating, *Physik. Zeits.* **29**, 837 (1928); *Z. Phys.* **52**, 8 (1928); *Annalen d. Physik* **1**, 713 (1929).

I. Estermann and O. Stern, *Z. Phys.* **61**, 95 (1930); I. Estermann, R. Frisch and O. Stern, *ibid.* **73**, 348 (1931).

E. Fermi and L. Marshall, *Phys. Rev.* **71**, 166 (1947).

W. H. Zinn, *ibid.* **71**, 752 (1947); E. O. Wollan, E. G. Shull and M. C. Marney, *ibid.* **73**, 527 (1948).

## 2. Schrödinger's wave mechanics (1926)

The story goes that Schrödinger, at the University of Zurich, was asked (1925?) by P. Debye, at the Eidgenössische Technische Hochschule, to give a seminar report on de Broglie's thesis, and the result was the series of papers between January and June of 1926 which gave a formulation of and completed the mathematical structure of what he called "Wave Mechanics".

Schrödinger posed the problem of finding the equation describing the de Broglie wave. His approach was guided by the following considerations. In (III-1) and (III-3), we saw the analogy between geometrical (ray) optics and particle dynamics. It had been known before 1911, through the work of Debye, that the wave equation of wave optics, in the limit of very short wavelength, passes into the equation of the eikonal which is the equation of ray optics; that the eikonal equation is analogous in form to the Hamilton-Jacobi equation of dynamics. Schrödinger then sought a wave equation of a particle which bore a relation to particle dynamics (as represented by the Hamilton-Jacobi equation) as a wave equation in optics bears on ray optics (as represented by the eikonal equation). We shall give an account of this part of Schrödinger's epoch-making work not only for historical interest, but also because we wish to show how a mastery of the old is often needed for the creation of new theories.

## (1) Digression to Hamilton-Jacobi theory

Hamilton (*ca.* 1834) defined the principal function $S(q, q^0, t)$ by the integral

$$S(q, q^0, t) = \int_0^t L(q, \dot{q}) \, dt \qquad \text{(III-13)}$$

of the Lagrangian $L = T - V$ over the actual trajectory of motion of a dynamical system, i.e., by substituting into $L$ the solutions of the equations of motion

$$\frac{d}{dt}\frac{\partial L}{\partial \dot{q}_k} - \frac{\partial L}{\partial q_k} = 0, \qquad k = 1, 2, \ldots, n \qquad \text{(III-14)}$$

$$q_k = q_k(q^0, \dot{q}^0, t), \qquad \dot{q}_k = \dot{q}_k(q^0, \dot{q}^0, t), \qquad \text{(III-15)}$$

where $q_k^0, \dot{q}_k^0$ are the initial values of $q_k, \dot{q}_k$ at $t = 0$. By means of the $2n$ equations (III-15), we may express the $2n$, $\dot{q}_k, \dot{q}_k^0$ in terms of the $2n + 1$, $q_k, q_k^0$ and $t$, and $S$ in (III-13) is expressed as a function of the $q_k, q_k^0, t$. We then may define a transformation

$$p_i = \frac{\partial S}{\partial q_i}, \qquad p_i^0 = \frac{\partial S}{\partial q_i^0}, \qquad i = 1, 2, \ldots, n, \qquad \text{(III-16)}$$

from $(q, p)$ to $(q^0, p^0)$, and the $2n$ equations give

$$\begin{aligned} q_i &= q_i(q^0, p^0, t), \\ p_i &= p_i(q^0, p^0, t), \end{aligned} \qquad \text{(III-17)}$$

which are the solutions of the equations of motion of the system. But this procedure is entirely redundant, since one already assumes the solutions (III-15) to begin with, before one can integrate (III-13).

But Hamilton showed that $S$ can be obtained from a partial differential equation, without the integration in (III-13). We have

$$\frac{dS}{dt} = \sum_i \frac{\partial S}{\partial q_i} \dot{q}_i + \frac{\partial S}{\partial t}$$

and from (III-13),

$$\frac{dS}{dt} = L = \sum_i p_i \dot{q}_i - H.$$

By means of $p_i = \partial S/\partial q_i$ in (III-16), we obtain

$$H\left(q_i, p_i = \frac{\partial S}{\partial q_i}, t\right) + \frac{\partial S}{\partial t} = 0. \tag{III-18}$$

Jacobi (1837) then completed the theory by showing that any complete solution $S(q_1, \ldots, q_n, \alpha_1, \ldots, \alpha_n, t) + \alpha_{n+1}$ containing $n + 1$ arbitrary constants can be used to define a canonical transformation from $(q, p)$ to $(\alpha, \beta)$ by means of

$$p_k = \frac{\partial S}{\partial q_k}, \qquad \beta = -\frac{\partial S}{\partial \alpha}, \qquad k = 1, 2, \ldots, n \tag{III-19}$$

which are the integrals of the equations of motion, namely,

$$\begin{aligned} q_k &= q_k(\alpha, \beta, t), \\ p_k &= p_k(\alpha, \beta, t). \end{aligned} \qquad k = 1, 2, \ldots, n \tag{III-20}$$

If the Hamiltonian $H$ is not an explicit function of time, i.e., $\partial H/\partial t = 0$, then

$$\begin{aligned} \frac{dH}{dt} &= \sum_i \left(\frac{\partial H}{\partial q_i} \dot{q}_i + \frac{\partial H}{\partial p_i} \dot{p}_i\right) + \frac{\partial H}{\partial t} \\ &= 0 \end{aligned}$$

so that

$$H = E, \quad \text{constant.} \tag{III-21}$$

In this case

$$L = T - V$$

$$L = \sum p\dot{q} - H = 2T - H$$

so that (III-13) becomes

$$S = \int_0^t 2T\, dt - Et \tag{III-22}$$

$$= S_0 - Et,$$

where $S_0$ is called *Hamilton's characteristic function.* From this equation—the definition of $S_0$—we have

$$\frac{dS_0}{dt} = 2T = \sum_i p_i \dot{q}_i.$$

From direct differentiation, using (III-16), we have

$$\frac{dS_0}{dt} = \sum_i \frac{\partial S_0}{\partial q_i}\dot{q}_i + \frac{\partial S_0}{\partial t}$$

$$= \sum_i p_i \dot{q}_i + \frac{\partial S_0}{\partial t}.$$

On comparing these two expressions for $dS_0/dt$, it is seen that $S$ is explicitly independent of time. Hence the Hamilton-Jacobi equation (III-18) now has the form

$$H\left(q_i, p_i = \frac{\partial S_0}{\partial q_i}\right) - E = 0. \tag{III-23}$$

A simple example is the problem of a linear harmonic oscillator. Here

$$H = \frac{p^2}{2m} + \frac{1}{2}kq^2.$$

The Hamilton-Jacobi equation is

$$\frac{1}{2m}\left(\frac{dS}{dq}\right)^2 + \frac{1}{2}k^2q^2 = E. \tag{III-24}$$

Another example is the Kepler motion already seen in Chap. 0, Sec. 1(4).

*(2) Digression to the relation between wave optics and ray optics*

Let us start with the basic equation for wave optics

$$\left(\nabla^2 - \frac{1}{u^2}\frac{\partial^2}{\partial t^2}\right)\Phi(r, t) = 0 \tag{III-25}$$

and assume the separation

$$\Phi(r, t) = e^{i\omega t}\phi(r) \tag{III-26}$$

leading to the wave equation (in space) for $\phi(r)$

$$(\nabla^2 + k^2)\phi(r) = 0. \tag{III-27}$$

$k$ is the wave vector related to the wavelength $\lambda$ and phase velocity $u$,

$$|\mathbf{k}| = \frac{2\pi}{\lambda} = \frac{\omega}{u}. \tag{III-28}$$

In a medium of index of refraction $n(r)$, $k$, $\lambda$ and $u$ are functions of $n$ and hence of $r$. Let $k_0$, $\lambda_0$, $u_0 = c$ be their values in vacuum, then

$$|\mathbf{k_0}| = \frac{2\pi}{\lambda_0} = \frac{\omega}{u_0} = \frac{\omega}{c},$$

$$\frac{u_0}{u} = n, \qquad \frac{k}{k_0} = n, \qquad \frac{\lambda_0}{\lambda} = n. \tag{III-29}$$

If $\phi(r)$ is given the assumed form

$$\phi(r) = A(r)e^{ik_0 S(r)}, \tag{III-30}$$

where $A(r)$ is an amplitude, then $S(r)$ satisfies an equation which can be separated into a real and an imaginary part,

$$-k_0^2\left[(\nabla S)^2 - \left(\frac{k}{k_0}\right)^2\right] + (\nabla \ln A)^2 + \nabla^2 \ln A = 0,$$

$$k_0\left[\frac{1}{2}\nabla^2 S + (\nabla \ln A)\cdot(\nabla S)\right] = 0. \tag{III-31}$$

In the limit of very short wavelength, $\lambda_0 \to 0$, or $k_0 \to \infty$, the above two equations reduce to

$$(\nabla S)^2 - n^2(r) = 0. \tag{III-32}$$

Consider surfaces of constant $S$, i.e., $S(x, y, z) = b$. The gradient $\nabla S$, normal to the wave front $S = b$, represents the direction of the light ray. Thus Eq. (III-32) describes the *waves* $\phi(r)$ as *rays* $\nabla S = n(r)$ in the limit of short wavelength. The function $S(r)$ is called the eikonal, introduced by H. Bruns in 1895.[c]

Now, starting from the equation of geometrical optics (III-32), one can obtain a first-order differential equation by the following procedure. On setting

---

[c] In an article in *Annalen d. Physik* **35**, 277 (1911), A. Sommerfeld and J. Runge give a discussion of the relation between the work of Hamilton and that of Bruns, *Leipziger Berichte* **21**, 323 (1895). From this article, it seems that the passage from the wave equation to the eikonal equation in the limit of very short wavelength (from (III-27) to (III-32) above) is due to an incidental remark of P. Debye.

$$\psi(r) = e^{k_0 S}, \tag{III-33}$$

one obtains from (III-32), with $k = nk_0$,

$$(\nabla\psi)^2 - k^2\psi^2 = 0. \tag{III-34}$$

If one formulates the following variational equation

$$\delta \iiint [(\nabla\psi)^2 - k^2\psi^2]\, dr = 0, \tag{III-35}$$

the Euler equation is

$$(\nabla^2 + k^2)\psi = 0, \tag{III-36}$$

which is a wave equation.

### *(3) Schrödinger equation*

After these digressions, let us return to Schrödinger's ideas. The analogous relation between optics and dynamics can be summarized as follows.

| Optics | Dynamics |
|---|---|
| Geometrical ray | Particle trajectory |
| Fermat principle | Maupertuis principle |
| $\delta \int_A^B \dfrac{1}{\lambda}\, ds = 0,$   or | $\delta \int_A^B p\, ds = 0,$   or |
| Eikonal equation | Hamilton-Jacobi equation |
| $(\nabla S)^2 - n^2(r) = 0$   (III-32) | $\dfrac{1}{2m}(\nabla S_0)^2 - (E - V) = 0$   (III-24) |
| Wave equation | Wave equation |
| $u = e^{k_0 S}, \quad k = nk_0$   (III-33) | $\psi = e^{(1/\hbar)S_0}$   (III-37) |
| and | |
| $\delta \iiint [(\nabla u)^2 - k^2 u^2]\, dr = 0$   (III-35) | $\delta \iiint \left[(\nabla u)^2 - \dfrac{2m}{\hbar^2}(E - V)\psi^2\right] = 0$   (III-38) |
| or | or |
| $(\nabla^2 + k^2)u(r) = 0$   (III-36) | $\left[\dfrac{\hbar^2}{2m}\nabla^2 + (E - V)\right]\psi(r) = 0$   (III-39) |

Schrödinger started from the Hamilton-Jacobi equation which for the case of one particle has the form (III-24). This is the dynamical analogue to ray optics (eikonal equation (III-32)). From the Hamilton-Jacobi equation, Schrödinger then obtained the dynamical analogue to wave optics (wave equation (III-36)) by following through the same procedure (III-32)–(III-36), thereby obtaining (III-37), (III-38), (III-39). The last,

$$\left(-\frac{\hbar^2}{2m}\nabla^2 + V\right)\psi(r) = E \tag{III-39}$$

is the *Schrödinger time-independent wave equation*. It can be written in the form

$$H\psi(r) = E\psi(r), \tag{III-39a}$$

where $H$ is the Hamiltonian in which the momentum components $p_x$, $p_y$, $p_z$ are replaced by differential operators

$$p_x \to \frac{\hbar}{i}\frac{\partial}{\partial x}, \qquad p_y \to \frac{\hbar}{i}\frac{\partial}{\partial y}, \qquad p_z \to \frac{\hbar}{i}\frac{\partial}{\partial z}. \tag{III-40}$$

For one particle in three-dimensional motion, (III-39a) is a partial differential equation. It is to be solved subject to certain boundary and other conditions on the function $\psi(r)$. In mathematics, the problem is called an eigenvalue problem, in the sense that only for certain (may be infinitely many) values of the "parameter" $E$, say $E_n$, does the equation (III-39a) possess solutions $\psi_n$ that satisfy the imposed conditions, i.e.,

$$H\psi_n(r) = E_n\psi_n(r). \tag{III-41}$$

The eigenvalue problem (of linear differential or integral equations) is well-known in mathematics, but Schrödinger in the first (January, 1926) of his series of communications postulated the problem of quantization to be an eigenvalue problem of the equation (III-39a). The first application of this idea was to the Kepler problem, and the eigenvalues of the Hamiltonian were found to be the same formula as in Bohr's theory, and hence a signal success. We shall return to the eigenvalue problem in the Sec. 4.

To obtain a time-dependent wave equation, Schrödinger assumed a function

$$\Psi(r,t) = e^{-iE_n t/\hbar}\psi_n(r) \tag{III-42}$$

so that

$$\frac{\partial\Psi}{\partial t} = -\frac{i}{\hbar}E_n\Psi_n. \tag{III-43}$$

Replacing $E_n$ in (III-39) by $-(\hbar/i)(\partial/\partial t)$, Schrödinger obtained the general equation

$$-\frac{\hbar}{i}\frac{\partial\Psi}{\partial t} = H\Psi, \tag{III-44}$$

where $H$ is the Hamiltonian—an operator obtained from the Hamiltonian in classical dynamics by making the replacements (III-40).

We must strongly emphasize that the procedure (III-24)–(III-39) is never to be taken as a deduction of the Schrödinger equation (III-39a) in the sense that it has been deduced from something more basic and already known. We must see that there is no logic involved in the procedure sketched above. The account is given just to show how Schrödinger had arrived at his new equation. The same applies to the equation (III-44); it is a basic postulate in quantum mechanics; and even after the original theory of Schrödinger has been developed much further, it has still remained a postulate.

### 3. The Schrödinger theory

#### (1) Generalization to a system of N particles

Equations (III-39a) and (III-43) are for one particle. For $N$ particles, one constructs the Hamiltonian as in classical dynamics in terms of the $p_x$, $p_y$, $p_z$, $x$, $y$, $z$ of the $N$ particles (properly symmetrized to ensure the hermitian property of any products involving $x$, $p_x$, etc.), and make the replacement (III-40).[d] The Schrödinger equation is then

$$-\frac{\hbar}{i}\frac{\partial\Psi}{\partial t} = H\Psi(q,t), \qquad (\text{III-45a})$$

with

$$\Psi(q,t) = e^{-iE_n t/\hbar}\psi_n(q), \qquad (\text{III-45b})$$

$$(H - E_n)\psi_n(q) = 0, \qquad (\text{III-45c})$$

where $n$ denotes collectively a set of quantum numbers. The wave equation (III-45a) is in the configuration space of the $N$ particles and not in the ordinary 3-dimensional space.

#### (2) Linearity and superposition principle

The Schrödinger equations (III-45a), (III-45c) are linear in the sense that the operators $H$, $\partial/\partial t$ have the following properties: For $c_1$, $c_2$ = constants,

$$\left(H + \frac{\hbar}{i}\frac{\partial}{\partial t}\right)(c_1\Psi_1 + c_2\Psi_2) = c_1\left(H + \frac{\hbar}{i}\frac{\partial}{\partial t}\right)\Psi_1 + c_2\left(H + \frac{\hbar}{i}\frac{\partial}{\partial t}\right)\Psi_2,$$

$$H(c_1\psi_1 + c_2\psi_2) = c_1 H\psi_1 + c_2 H\psi_2. \qquad (\text{III-46})$$

---

[d] When coordinates other than Cartesian are used, the procedure is to start with Cartesian coordinates, make the replacement (III-40) and transform the operators to other coordinates. In (III-45a, b, c), $q$ stands for all the coordinates and $H = H(q,p,t)$ is the operator so obtained.

This linearity is the basis for the superposition principle, namely, that the sum of two solutions is also a solution of a linear equation.

The simplest example is that of a free particle. The Schrödinger equation (III-45a) is now

$$\frac{\hbar}{i}\frac{\partial\Psi}{\partial t} - \frac{\hbar^2}{2m}\nabla^2\Psi(r,t) = 0.\qquad\text{(III-47a)}$$

One solution is the plane wave (monochromatic)

$$\Psi(r,t) = e^{i(\mathbf{k}\cdot\mathbf{r}-\omega t)}.\qquad\text{(III-47b)}$$

The superposition (wave packet)

$$\Psi(r,t) = \frac{1}{(2\pi)^{3/2}}\iiint dk\,A(k)e^{i(\mathbf{k}\cdot\mathbf{r}-\omega t)}\qquad\text{(III-48)}$$

is also a solution.

In the general case, a general solution of (III-45a) is the sum

$$\Psi(q,t) = \sum_h c_n e^{-iE_n t/\hbar}\psi_n(q).\qquad\text{(III-49)}$$

### (3) Wave packet

The integral in (III-48) represents a superposition of phase wave of different wavelengths $\lambda = 2\pi/k$ and different directions of propagation $\mathbf{k}$. From (III-48) and (III-47a), one obtains

$$\hbar\omega - \frac{1}{2m}\hbar^2 k^2 = 0,\qquad\text{(III-50a)}$$

which is the so-called dispersion relation, from which one obtains the phase velocity $u$ and the group velocity $v_g$

$$u = \frac{\omega}{k}\left(=\frac{E}{p}\right),\qquad\text{(III-50b)}$$

$$v_g = \frac{d\omega}{dk} = \frac{\hbar k}{m}\left(=\frac{p}{m}\right),\qquad\text{(III-50c)}$$

the expressions in parentheses being obtained by means of the Einstein-de Broglie relations

$$E = \hbar\omega,\qquad p = \frac{h}{\lambda} = \hbar k.$$

Thus the group velocity $v_g$ of the de Broglie wave is equal to the velocity $p/m$ of the particle—in agreement with de Broglie's result (III-11).

On account of the dispersion $u = \omega/k = (E/h)\lambda$, for a particle of a given energy, the phase velocity depends on $\lambda$. Thus a wave packet disperses as it propagates. This dispersion, or impermanence, of a wave packet is a rather general situation for a particle in a field $V(r)$.[e] This dispersion renders Schrödinger's early theory of the $\psi$ representing an extended electron untenable.

### (4) Time-independence of $\int \Psi^* \Psi \, dq$

From the general Schrödinger equation (III-45a), one obtains

$$\frac{\partial}{\partial t} \int \Psi^* \Psi \, dq = \frac{i}{\hbar} \int [-\Psi^* H \Psi + (H^\dagger \Psi^*) \Psi] \, dq = 0 \qquad \text{(III-51)}$$

since the Hamiltonian $H$ is self-adjoint (hermitian).[f] Thus

$$\int \Psi^* \Psi \, dq = \text{independent of time.} \qquad \text{(III-51a)}$$

If the integral is finite (so that it can be normalized), the above relation is consistent with the probability interpretation of $\Psi$.

The explicit time dependence assumed in (III-45b) leads directly to

$$\Psi_n^* \Psi_n(q, t) = \psi_n^* \psi_n(q),$$

which is independent of time as the general case (III-51). The same remark above applies.

### (5) Commutation relation and equivalence between matrix and wave mechanics

With the relations (III-40)

$$p_x = \frac{\hbar}{i} \frac{\partial}{\partial x}, \qquad \text{etc.,}$$

Schrödinger noted (March, 1926) that

$$(p_x x - x p_x)\psi = \frac{\hbar}{i} \psi,$$

---

[e] The exception is the case of wave packets built up of the wave functions of the harmonic oscillator, as pointed out by Heisenberg, in his epoch-making paper on the uncertainty relation, *Z. Phys.* **43**, 172 (March, 1927).

[f] For self-adjoint operator (to be defined in Sec. 4, (1) (III-71) below),

$$\int \psi^* A \psi \, dq = \int (A^\dagger \psi^*) \psi \, dq = \int (A \psi^*) \psi \, dq.$$

or as an operator equation,

$$p_x x - x p_x = \frac{\hbar}{i},\tag{III-52}$$

which is of the same form as the commutation relation in matrix mechanics (II-72). This formal equivalence between wave and matrix mechanics was then completed by Schrödinger on the following considerations.

Let $u_k(r)$, $k = 1, 2, \ldots$ be a complete set of orthogonal and normalized functions[g] in the sense that any function $f(r)$ can be expanded in terms of them,

$$\int u_m^*(r) u_n(r)\, dr = \delta_{mn},\tag{III-53}$$

$$f(r) = \sum_k C_k u_k(r), \qquad C_k = \int u_k^*(r) f(r)\, dr.\tag{III-54}$$

We shall make use of a relation

$$\sum_j u_j(r) u_j^*(r') = \delta(r - r'),\tag{III-55}$$

which, involving the $\delta$-function, must be regarded as an operator equation.[h]

Let $A$, $B$ be two (hermitian) operators constructed of $q$ and $p$, and let the integrals be denoted by

$$\int u_m^*(r) A u_n(r)\, dr = A_{mn},$$

$$\int u_m^* A B u_n\, dr = (AB)_{mn}, \qquad \text{etc.}\tag{III-56}$$

We wish to show that these integrals have the properties of matrix elements. That

$$(A + B)_{mn} = A_{mn} + B_{mn}$$

follows from the definitions above. For the product $AB$, we have, on introducing (III-55),

---

[g] The definition of "complete set" will be given in Sec. 4, (1), below. The $u_k(r)$ may contain both discrete and continuous functions. The continuous $u_k(r)$ will be normalized by the method of Weyl, to be described in Sec. 5, (3) below. In the following equations, the summation $\sum$ is understood to include integration over the index $k$ of the continuous $u_k(r)$ as well.

[h] (III-55) as an operator equation can be verified as follows. Multiply both sides by $u_k(r')\, dr'$ and integrate over $r'$, and one gets the identity $u_k(r) = u_k(r)$. In the braket notation of Dirac, it takes the form

$$\sum_k |k, r\rangle\langle r', k| = \delta(r - r').\tag{III-55a}$$

The relation, sometimes called the closure relation, is very useful, as we shall see in (III-57) below.

$$\int u_m^* A B u_n \, dr = \iint (u_m^*(r)A)\delta(r - r')(B u_n(r')) \, dr \, dr'$$

$$= \sum_j \iint (u_m^* A)(r)u_j(r)u_j^*(r')(B_n u)(r') \, dr \, dr'$$

i.e.,

$$(AB)_{mn} = \sum_j A_{mj} B_{jn} \tag{III-57}$$

so that the integrals (III-56) satisfy the matrix multiplication rule. Thus did Schrödinger show the mathematical equivalence of matrix and wave mechanics.

### (6) Fourier transformation and representation transformation

The expression (III-48) is a Fourier transformation of $A(k)e^{-i\omega t} = \Phi(k, t)$ into $\Psi(r, t)$. This equation and its inverse transformation can be written in the symmetric form

$$\Psi(\mathbf{r}, t) = \frac{1}{(2\pi)^{3/2}} \iiint d\mathbf{k}\Phi(\mathbf{k}, t)e^{i\mathbf{k}\cdot\mathbf{r}},$$

$$\Phi(\mathbf{k}, t) = \frac{1}{(2\pi)^{3/2}} \iiint d\mathbf{r}\Psi(\mathbf{r}, t)e^{-i\mathbf{k}\cdot\mathbf{r}}. \tag{III-58}$$

We have seen that the Fourier transform of the Schrödinger equation (III-47a)

$$\frac{\hbar}{i}\frac{\partial\Psi}{\partial t} + \frac{\hbar^2}{2m}\nabla^2\Psi = 0, \tag{III-47a}$$

is

$$\frac{\hbar}{i}\frac{\partial\Phi}{\partial t} - \frac{\hbar^2 k^2}{2m}\Phi(k, t) = 0. \tag{III-59}$$

This is the result of a methematical transformation. Now, if we introduce the Einstein-de Broglie relation

$$\hbar k = \frac{h}{\lambda} = p, \tag{III-60}$$

the above equation can be written

$$\left(\frac{\hbar}{i}\frac{\partial}{\partial t} - \frac{p^2}{2m}\right)\Phi(p, t) = 0. \tag{III-61}$$

This is said to be the *momentum representation* of the Schrödinger equation

(III-47a) for a free particle of momentum $p$. The representation for (III-47a), in which $p_x$ is $(\hbar/i)(\partial/\partial x)$, is the coordinate representation, or sometimes referred to as the *Schrödinger representation*.

By means of (III-58), we can readily find the $x$, $p_x$, $x^2$, $p_x^2$, etc. in the coordinate and the momentum representation, as shown in the following table

| Coordinate or Schrödinger representation | | Momentum representation |
|---|---|---|
| $\Psi(\mathbf{r}, t)$ | (III-58) | $\Phi(\mathbf{p}, t)$ |
| $x\Psi(\mathbf{r}, t)$ | | $-\dfrac{\hbar}{i}\dfrac{\partial}{\partial p_x}\Phi(\mathbf{p}, t)$ |
| $\dfrac{\hbar}{i}\dfrac{\partial}{\partial x}\Psi(\mathbf{r}, t)$ | | $p_x\Phi(\mathbf{p}, t)$    (III-62) |
| $x^2\Psi(\mathbf{r}, t)$ | | $-\hbar^2\dfrac{\partial^2}{\partial p_x^2}\Phi(\mathbf{p}, t)$ |
| $-\hbar^2\dfrac{\partial^2}{\partial x^2}\Psi(\mathbf{r}, t)$ | | $p_x^2\Phi(\mathbf{p}, t)$ |

Plane wave equation

$$\frac{\hbar}{i}\frac{\partial\Psi}{\partial t} - \frac{\hbar^2}{2m}\nabla^2\Psi = 0 \quad \text{(III-47a)} \qquad \frac{\hbar}{i}\frac{\partial\Phi}{\partial t} - \frac{p^2}{2m}\Phi = 0 \qquad \text{(III-61)}$$

Simple harmonic oscillator

$$-\frac{\hbar^2}{2m}\frac{d^2\psi(x)}{dx^2} + \frac{m}{2}\omega^2 x^2\psi = E\psi(x) \qquad \frac{1}{2m}p_x^2\phi - \frac{m\omega^2\hbar^2}{2}\frac{d^2\phi}{dp_x^2} = E\phi(p).$$

$$\text{(III-63)}$$

The above transformation is a Fourier transformation, (III-58) with $k = p/\hbar$, but it is also a unitary transformation of the kind that we have discussed in Chap. 2, Sec. 2, following (II-56a), and Sec. 3.[i] In the $x$-representation, $x$ is represented by $x$, but its conjugate momentum $p_x$ is represented by $(\hbar/i)(\partial/\partial x)$. In the $p$-representation, $p_x$ is represented by $p_x$, but its conjugate coordinate $x$ is represented by $-(\hbar/i)(\partial/\partial p_x)$. The commutation relation (III-52) is satisfied in both representations.

---

[i] The transformation between the $x$- and the $p$-representation discussed here is but a special case of the general unitary transformation from any $A$-representation to any $B$-representation. For example, in matrix mechanics, one may start with *any* representation $A$ for the $q$ and $p$ and construct a hermitian Hamiltonian $H$ in that representation $A$, and then transform to the energy-representation in which $H$ is diagonal. The Schrödinger equation (III-45c) starts with the coordinate-representation and one finds the eigenvalues of $H$ by solving the equation for the eigenfunctions of $H$. This is equivalent to transforming to the $H$-representation in matrix mechanics, diagonalizing $H$, obtaining the eigenvalues and eigenvectors.

This completes the equivalence between wave and matrix mechanics.

In a later chapter, these representations, their transitions and the meaning of the $\Psi$ and $\Phi$ in (III-62) will be further discussed.


### (7) Conditions on $\psi$

We have postponed till now the question concerning the conditions to be imposed on the solution $\psi$ of the Schrödinger equation, because the conditions suggests by Schrödinger in the first of his series of communications (1926) have been the subject of many later discussions.[j] It seems that the most basic condition is, for stationary, bound states, the finiteness of the integral (over the configuration space)

$$\int \Psi^*\Psi\, dq = \text{finite}, \tag{III-64}$$

which is a necessary condition for the probability interpretation of $\Psi^*\Psi$.[k,l]


### (8) Relativistic wave equation

The Schrödinger equation for a particle (III-44)

$$\frac{\hbar}{i}\frac{\partial\Psi}{\partial t} = \left(-\frac{\hbar^2}{2m}\nabla^2 + V\right)\Psi$$

is of the first order with respect to time and of the second order with respect to $x$, $y$, $z$, so that it is obviously non-relativistic.

An equation which satisfies the Lorentz transformation covariance has been found, soon after Schrödinger's first communication, by O. Klein, L. de Broglie, Schrödinger himself, V. Fock and W. Gordon. The equation is obtained from the energy-momentum relation for a free particle in the special theory of relativity

$$p_x^2 + p_y^2 + p_z^2 - \left(\frac{E}{c}\right)^2 = -m_0^2 c^2 \tag{III-65}$$

by following the rule (III-40)

$$p_x = \frac{\hbar}{i}\frac{\partial}{\partial x}, \qquad \text{and} \qquad E = -\frac{\hbar}{i}\frac{\partial}{\partial t}, \tag{III-66}$$

---

[j] The early conditions of $\psi$ being real, twice continuously differentiable are too stringent. The condition of integral angular momentum quantum numbers has been discussed most and retained although the argument of single-valuedness of $\psi$ is no longer accepted.

[k] The probability interpretation of $\Psi^*\Psi$ is proposed by M. Born in mid-1926. It remains one of the basic postulates of quantum mechanics. See Chap. 4, Sec. 2.

[l] For states in the continuum, the $\psi$ has to be normalized by a different procedure. See Sec. 4.

the last being suggested by $-E$, $t$ being formally canonical conjugates in classical dynamics. The wave equation is then

$$\left(\nabla^2 - \frac{1}{c^2}\frac{\partial^2}{\partial t^2}\right)\Psi = \left(\frac{m_0 c}{\hbar}\right)^2 \Psi, \qquad \text{(III-67)}$$

which is known as the *Klein-Gordon equation*. Its form is Lorentz invariant. When modified for the Coulomb field and applied to the electron in the hydrogen atom, it does not give the correct result. It is known later that this is because the Klein-Gordon equation does not include the electron spin.

### References

Schrödinger's series of articles, under the title "Quantizierung als Eigenwertproblem", are as follows:
I.   *Annalen d. Physik* **79**, 361–376 (Jan. 27, 1926);
II.  *ibid.* **79**, 489–527 (Feb. 23, 1926);
III. *ibid.* **80**, 437–490 (May 10, 1926);
IV.  *ibid.* **81**, 109–139 (June 21, 1926).
The first communication I obtains the Schrödinger equation (III-39) of the present section from a variational principle applied to the Hamilton-Jacobi equation. The equation is applied to the hydrogen atom, the method of Laplace transformation being used. Bohr's formula is obtained. II goes back to the underlying thoughts of his initial approach—to obtain a "wave mechanics" that bears a relation to particle dynamics (Hamilton-Jacobi equation) as wave optics bears to ray optics (Eikonal equation). Applications to the harmonic oscillator, rigid rotator and vibrating-rotator are discussed. III treats the time-independent perturbation theory, including degenerate systems, and applies it to the Start effect, using parabolic coordinates. IV obtains the time-dependent Schrödinger equation and treats the time-dependent perturbation theory. The suggested interpretation of $\psi^*\psi$ is that it is a weight function in configuration space representing the fluctuations of the space density of the electric charges.
*Annalen d. Physik* **79**, 734–756 (March, 1926). In this paper, Schrödinger points out the mathematical equivalence between the Heisenberg-Born-Jordan matrix mechanics and wave mechanics.

The relativistic wave equation was found by Schrödinger, in IV above, and also by
L. de Broglie, *Compt. Rendus* **183**, 272 (July, 1926)
O. Klein, *Z. Phys.* **37**, 895 (1926)
V. Fock, *ibid.* **38**, 242; **39**, 226 (1926)
W. Gordon, *ibid.* **40**, 117 (1926)

### 4.   The eigenvalue problem

The Schrödinger equation for one particle system

$$H\psi_n(r) = E_n\psi_n(r) \qquad \text{(III-68)}$$

can be solved if the partial differential equation is separable into ordinary differential equations. In such cases, the eigenvalue problem (III-68) is reduced to that of linear second order ordinary differential equation—the so-called Sturm-Liouville problem.

### (1) The Sturm-Liouville problem

The differential equations considered are of the type[m]

$$\Lambda y + \lambda \rho(x) y = 0, \tag{III-69}$$

where $\rho$ is a real function of $x$ and $\rho(x) > 0$ for $b \le x \le a$, $\lambda$ is a parameter and $\Lambda$ a self-adjoint operator

$$\Lambda = \frac{d}{dx}\left(p(x)\frac{d}{dx}\right) - q(x), \tag{III-70}$$

$p(x)$, $q(x)$ being functions of $x$, and the self-adjointness is defined by the condition on $\Lambda$:

$$z\Lambda y - y\Lambda z = \frac{dF}{dx} \tag{III-71}$$

or $\qquad\left(z^*\Lambda y - y\Lambda^* z^* = \dfrac{dF}{dx}\ \text{ for complex } \Lambda\right),$

where $z(x)$, $y(z)$ are two arbitrary functions of $x$, and $dF/dx$ is the total derivative of a function $F(x)$. The Sturm-Liouville problem is to find the eigenvalues $\lambda$ and eigenfunctions of satisfying the boundary conditions for $a \le x \le b$,

$$F(a) = F(b) = 0, \qquad \text{or} \qquad F(b) - F(a) = 0, \tag{III-72}$$

the range $b \le x \le a$ being finite.

The condition (III-71) for self-adjointness can be put in the more explicit form

$$\int_a^b z^*\Lambda y \, dx = \int_a^b (\Lambda^* z^*) y \, dx, \tag{III-71a}$$

or with notation

---

[m] Any equation of the form

$$A(x)y'' + B(x)y' + C(x)y + \lambda D(x)y = 0, \qquad A \ne 0,$$

can be transformed into the form (III-69) by multiplying by $p/A$, where $p = e^{\int (B/A)\,dx}$, and putting $q(x) = -pC/A$, $\rho(x) = pD/A$.

$$(\psi, A\phi) \equiv \int_a^b \psi^* A\phi \, dx, \tag{III-71b}$$

$$(z, \Lambda y) = (\Lambda z, y).$$

Consider two self-adjoint operators $\Lambda$ and $\Pi$. Then by the definition (III-71a),

$$\int_a^b z^*(\Pi\Lambda y)\,dx = \int_a^b (\Pi^* z^*)\Lambda y \, dx = \int_a^b (\Lambda^* \Pi^* z^*) y \, dx$$

or in the notation (III-71b),

$$(z, \Pi\Lambda y) = (\Lambda\Pi z, y). \tag{III-71c}$$

Note that in this definition of self-adjointness, the range of the variable $x$ is important. An operator which is self-adjoint for a certain $a \leq x \leq b$ may not be self-adjoint for a different range $c \leq x \leq d$.

Examples of self-adjoint operators, for the range $-\infty \leq x \leq \infty$ are $x^n$, $p_x = (\hbar/i)(\partial/\partial x)$, $p_x^2$, $i(p_x x - x p_x)$ for functions $z(x)$, $y(x)$ for which the integrals (III-71a) and $\int z^* z \, dx$, $\int z^* y \, dx$, etc. exist (i.e., finite), as can be verified by (III-71a).

A last remark on the definition: Given an operator $\Lambda$, its adjoint, denoted by $\Lambda^\dagger$, is such an operator that

$$\int_a^b (\Lambda^\dagger z)^* y \, dx = \int_a^b z^* \Lambda y \, dx.$$

When $\Lambda^\dagger = \Lambda$, that is, $\Lambda$ is its own adjoint, this relation reduces to (III-71a) above.

The results of the Sturm-Liouville problem can be summarized in the form of a few theorems.

**Theorem 1.** The eigenfunctions of the Sturm-Liouville equation corresponding to different eigenvalues are orthogonal.

**Proof:** Let $\lambda_1$, $\lambda_2$ and $y_1$ and $y_2$ be the eigenvalues and their respective eigenfunctions, with $\lambda_1 \neq \lambda_2$. One has

$$\Lambda y_1 = -\lambda_1 \rho y_1, \qquad \Lambda y_2 = -\lambda_2 \rho y_2,$$

$$\Lambda^* y_1^* = -\lambda_1^* \rho y_1^*, \qquad \Lambda^* y_2^* = -\lambda_2^* \rho y_2^*.$$

From these, one gets

$$\int_a^b (y_2^* \Lambda y_1 - y_1 \Lambda^* y_2^*)\,dx = (\lambda_2^* - \lambda_1)\int_a^b \rho y_2^* y_1\,dx = F(b) - F(a) = 0$$

$$\int_a^b y_2^* y_1 \rho\,dx = 0. \tag{III-73}$$

Here $\rho(x)$ is a "weight" function.                                             QED

**Theorem 2.**   The eigenvalues of the Sturm-Liouville equation are real.

**Proof:**   In the expressions in Theorem 1, put $\lambda_2$ equal to $\lambda_1$ so that

$$(\lambda_1^* - \lambda_1)\int_a^b |y_1|^2 \rho\,dx = 0.$$

Since $\rho(x) > 0$ for $a \leq x \leq b$, the integral $\neq 0$,

$$\therefore \ \lambda \text{ is real (for all } \lambda).$$

**Theorem 3.**   The eigenfunctions of the Sturm-Liouville equation form a complete set.*

**Proof:**   To prove the theorem, we first define *completeness*. Let $\psi_1, \psi_2, \psi_3, \ldots$ be a set of orthonormal functions, i.e.,

$$\int_a^b \psi_m^* \psi_n \rho\,dx = \delta_{mn}. \tag{III-73}$$

Let $f(x)$ be an arbitrary function satisfying the same boundary condition as the $\psi$'s, and let us assume that $f(x)$ be expanded in the $\psi$'s

$$f(x) = \sum_k^\infty C_k \psi_k, \tag{III-74a}$$

$$C_k = \int_a^b \psi_k^* f(x)\rho\,dx \equiv (\psi_k, \rho f), \tag{III-74b}$$

where the summation is to include integration if the $\psi$'s have continuous functions as well as discrete $\psi_k$'s.

Let

$$\Delta_n \equiv \int_a^b f^*(x)f(x)\rho\,dx - \int_a^b \left(\sum_k^n C_k \psi_k\right)^* \left(\sum_j^n C_j \psi_j\right)\rho\,dx \tag{III-75a}$$

$$= \int_a^b f^*(x)f(x)\rho(x)\,dx - \sum_k^n |C_k|^2.$$

---

* Cf. E. C. Kemble, *Fundamental Principles of Quantum Mechanics* (McGraw-Hill, New York, 1937; Dover, New York, 1958).

The set $\psi_1, \psi_2, \ldots$ is said to be complete if

$$\lim_{n \to \infty} \Delta_n = 0. \tag{III-75b}$$

The conditions for the set $\psi_k$ to satisfy this condition (III-75b) are:

(i)
$$\delta\lambda_{n+1} = \delta \left( -\frac{\displaystyle\int_a^b \psi_{n+1}^* \Lambda \psi_{n+1}\, dx}{\displaystyle\int_a^b \psi_{n+1}^* \psi_{n+1} \rho\, dx} \right) = 0 \tag{III-76a}$$

subject to the conditions

$$\int_a^b \psi_{n+1}^* \psi_k \rho\, dx = 0, \qquad k = 1, 2, \ldots, n, \tag{III-76b}$$

(ii)
$$\lim_{n \to \infty} \lambda_n = \infty. \tag{III-77}$$

To prove that (i), (ii) are the conditions for (III-75b), let

$$t_n \equiv f(x) - \sum_k^n C_k \psi_k, \tag{III-78}$$

where $C_k$ is given by (III-74b). Hence

$$(t_n, \rho\psi_i) = \sum_k^\infty C_k^* (\psi_k, \rho\psi_i) - \sum_k^n C_k^* (\psi_k, \rho\psi_i)$$

$$= 0 \text{ if } \psi_i \text{ is in the set } \psi_1, \psi_2, \ldots, \psi_n,$$

$$(\rho\psi_i, t_n) = (t_n, \rho\psi_i)^* = 0, \qquad \text{for } i \leq n.$$

From (III-69),

$$(t_n, \Lambda\psi_i) = -(t_n, \lambda\rho\psi_i) = -\lambda_i(t_n, \rho\psi_i) = 0, \qquad \text{for } i \leq n. \tag{III-79}$$

From (III-71), (III-72),

$$\int_a^b t_n^* \Lambda \psi_i\, dx - \int_a^b \psi_i \Lambda^* t_n^*\, dx = 0$$

or

$$(t_n, \Lambda\psi_i) - (\Lambda t_n, \psi_i) = 0. \tag{III-80}$$

The eigenvalue $\lambda_{n+1}$ is the lowest value of

$$-\frac{(\psi_{n+1}, \Lambda\psi_{n+1})}{(\psi_{n+1}, \rho\psi_{n+1})}.$$

Equation (III-78) shows that $t_n$ contains only $\psi_{n+1}, \psi_{n+2}, \ldots$ so that

$$\lambda_{n+1} < -\frac{(t_n, \Lambda t_n)}{(t_n, \rho t_n)}. \tag{III-81}$$

The condition (ii), $\lim_{n\to\infty} \lambda_{n+1} = \infty$, now requires that either

$$\lim_{n\to\infty} \{-(t_n, \Lambda t_n)\} = \infty, \tag{III-82}$$

or

$$\lim_{n\to\infty} (t_n, \rho t_n) = 0. \tag{III-83}$$

But from (III-78), one forms

$$(f, \Lambda f) = (t_n, \Lambda t_n) + \sum_k^n |C_k|^2 \lambda_k + \sum_k^n \{C_k^*(\psi_k, \Lambda t_n) + C_k(t_n, \Lambda \psi_k)\}.$$

From (III-79), the last two sums vanish, and

$$(t_n, \Lambda t_n) = (f, \Lambda f) - \sum_k^n |C_k|^2 \lambda_k = \sum_{k=n+1}^{\infty} |C_k|^2 \lambda_k.$$

But this does not meet the requirement (III-82). Hence one tries, again using (III-78),

$$(f, \rho f) = (t_n, \rho t_n) + \sum_k^n |C_k|^2.$$

From (III-75a),

$$\Delta_n = (f, \rho f) - \sum_k^n |C_k|^2 = (t_n, \rho t_n)$$

and

$$\lim_{n\to\infty} \Delta_n = \lim_{n\to\infty} (t_n, \rho t_n).$$

Thus the condition (III-83) leads to (III-77), and also to

$$\lim_{n\to\infty} \Delta_n = 0.$$

## *(2) Simple harmonic oscillator*

Let the mass be $\mu$ and the frequency be $\omega$. The Schrödinger equation is

$$-\frac{\hbar^2}{2\mu} \frac{d^2\psi}{dx^2} + \frac{1}{2}\mu\omega^2 x^2 \psi = E\psi. \tag{III-84}$$

On introducing the dimensionless variable $\xi$ and energy $\lambda$

$$\xi = x\sqrt{\frac{\mu\omega}{\hbar}}, \qquad \lambda = \frac{2E}{\hbar\omega}, \tag{III-85}$$

the equation becomes

$$\frac{d^2\psi}{d\xi^2} + (\lambda + \xi^2)\psi = 0. \tag{III-86}$$

Asymptotically for large $\xi$, $\psi$ behaves as $e^{\pm \xi^2/2}$. Assume, for quadratic integrability (III-64),

$$\psi(\xi) = v(\xi)e^{-1/2\xi^2} \tag{III-87}$$

and obtain

$$\frac{d^2v}{d\xi^2} - 2\xi\frac{dv}{d\xi} + (\lambda - 1)v = 0. \tag{III-88}$$

One tries a solution of the form

$$v(\xi) = \xi^l \sum_{k=0}^{\infty} a_k \xi^k. \tag{III-89}$$

On equating to zero the coefficient of the lowest power of $\xi$, one gets the indicial equation[n]

$$l(l - 1) = 0, \qquad \text{i.e.,} \quad l = 0, 1. \tag{III-90}$$

One may take $l = 0$. From (III-89) and (III-88), one obtains the recurrence relation

$$(k + 1)(k + 2)a_{k+2} + (\lambda - 1 - 2k)a_k = 0. \tag{III-91}$$

For large $k$, the ratio of two succesive terms approaches $(2/k)\xi^2$ so that, if (III-89) is an infinite series, it behaves asymptotically as $e^{2\xi^2}$, and $\psi$ in (III-87) as $e^{3/2\xi^2}$. To avoid this non-quadratic integrability, one requires the series to terminate at $\xi^n$ by setting $a_{n+2} = 0$, i.e.,

$$\lambda - 1 - 2n = 0.$$

---

[n]According to the theory of differential equations, a second-order differential equation has two independent solutions; they correspond to the two roots of the indicial equation. But if the two roots differ by an integer, they lead to one and the *same* solution, and the other independent solution $u(\xi)$ satisfies the relation with $v(\xi)$

$$\text{Wronskian} = u\frac{dv}{d\xi} - v\frac{du}{d\xi} = be^{\xi^2}, \qquad b = \text{constant}.$$

If $v(\xi)$ is a polynomial as in (III-89), then $u(\xi)$, or $du/d\xi$, must behave as $e^{\xi^2}$ asymptotically at large $\xi$, and $\psi$ in (III-87) will behave as $e^{1/2\xi^2}$ and be non-quadratically integrable. Hence $u(\xi)$ must be discarded.

i.e., from (III-85),

$$E = (n + \tfrac{1}{2})\hbar\omega, \qquad n = 0, 1, 2, \ldots \tag{III-92}$$

This result is the same as that obtained in matrix mechanics. The existence of the zero point energy $\tfrac{1}{2}\hbar\omega$ is not only verified by spectroscopic data (see Chap. 1, Sec. 1), but is a consequence of the uncertainty principle whose mathematical expression is the commutation relation and which has crept into the Schrödinger equation in disquise through the relation $p_x = (\hbar/i)(\partial/\partial x)$, etc.

The eigenfunctions of the harmonic oscillator are (III-87) with $v(\xi)$, polynomials, given by the recurrence relation

$$a_{k+2} = \frac{2n - 2k}{(k + 1)(k + 2)} a_k. \tag{III-93}$$

Let the polynomial $v(\xi)$ of degree $n$ so defined by denoted by $H_n(\xi)$

$$H_n(\xi) = \sum_k^n a_k \xi^k.$$

The summation begin with $k = 0\,(1)$ according as $n$ is even (odd). Explicitly,

$$H_n(\xi) = (2\xi)^n - n(n - 1)(2\xi)^{n-2} + \frac{1}{2!} n(n - 1)(n - 2)(n - 3)(2\xi)^{n-4} \cdots$$

$$+ \begin{cases} (-1)^{1/2 n} n! \Big/ \left(\dfrac{1}{2}n\right)! & \text{for } n \text{ even,} \\[2mm] (-1)^{1/2(n-1)} n! \xi \Big/ \left(\dfrac{n - 1)}{2}\right)! & \text{for } n \text{ odd.} \end{cases} \tag{III-94}$$

$$H_0 = 1, \qquad\qquad H_3 = (2\xi)^3 - 12\xi,$$

$$H_1 = 2\xi, \qquad\qquad H_4 = (2\xi)^4 - 12(2\xi)^2 + 12, \tag{III-95}$$

$$H_2 = (2\xi)^2 - 2, \qquad H_5 = (2\xi)^5 - 20(2\xi)^3 + 120\xi.$$

$H_n(\xi)$ satisfy the equation (III-88), or

$$\left(\frac{d^2}{d\xi^2} - 2\xi\frac{d}{d\xi} + 2n\right) H_n(\xi) = 0. \tag{III-96}$$

The polynomials $H_n(\xi)$ are called the *Hermite polynomials.* They have the property[o]

$$\int_{-\infty}^{\infty} H_m(\xi) H_n(\xi) e^{-\xi^2}\, d\xi = \begin{cases} 0, & m \neq n \\ 2^n \sqrt{\pi}\, n!, & m = n. \end{cases} \tag{III-97}$$

[o] See Appendix A at the end of this section.

The normalized $\psi_n(\xi)$ of (III-87) are

$$\psi_n(\xi) = \frac{1}{(2^n\sqrt{\pi}\,n!)^{1/2}}\,e^{-1/2\xi^2}H_n(\xi), \tag{III-98}$$

and in the variable $x$ of (III-84), the normalized $\psi_n(x)$ are

$$\psi_n(x) = \sqrt[4]{\frac{\mu\omega}{\hbar}}\,(2^n\sqrt{\pi}\,n!)^{-1/2}e^{-\mu\omega x^2/2\hbar}H_n\left(\sqrt{\frac{\mu\omega}{\hbar}}\,x\right), \tag{III-99}$$

so that

$$\int \psi_n^*(x)\psi_n(x)\,dx = 1.$$

### (3) Central field problem; parity

For a particle in a central field $V(r)$, the Schrödinger equation is, in spherical polar coordinates,

$$-\frac{\hbar^2}{2\mu}\left[\frac{\partial^2}{\partial r^2} + \frac{2}{r}\frac{\partial}{\partial r} + \frac{1}{r^2\sin\vartheta}\frac{\partial}{\partial\vartheta}\left(\sin\vartheta\frac{\partial}{\partial\vartheta}\right) + \frac{1}{r^2\sin^2\vartheta}\frac{\partial^2}{\partial\varphi^2}\right]\psi(r)$$

$$+ V(r)\psi(r) = E\psi(r). \tag{III-100}$$

The equation is "separable," by setting

$$\psi(r) = R(r)\Theta(\vartheta)\Phi(\varphi), \tag{III-100a}$$

leading to three equations

$$\frac{1}{\Phi}\frac{d^2\Phi}{d\varphi^2} = -m^2, \tag{III-101}$$

$$\frac{1}{\sin\vartheta}\frac{d}{d\vartheta}\left(\sin\vartheta\frac{d\Theta}{d\vartheta}\right) + \left(\lambda - \frac{m^2}{\sin^2\vartheta}\right)\Theta = 0, \tag{III-102}$$

$$r\frac{d^2}{dr^2}(rR) + \frac{2\mu}{\hbar^2}(E - V)r^2R = \lambda R. \tag{III-103}$$

The solution of (III-101) is, for periodic conditions

$$\Phi(\varphi) = \frac{1}{\sqrt{2\pi}}e^{im\varphi}, \qquad m = \pm\,\text{integer}. \tag{III-104}$$

The equation (III-102) is well known in the classical theory of potentials. It is convenient to write it in the form

$$\frac{d}{dx}\left[(1 - x^2)\frac{d\Theta}{dx}\right] + \left[\lambda - \frac{m^2}{1 - x^2}\right]\Theta(x) = 0, \qquad x = \cos\vartheta. \tag{III-105}$$

The problem is to find $\Theta(x)$ which are finite at the singular points $x = \pm 1$ of (III-105), and such an analysis leads to the eigenvalues for $\lambda$[p]

$$\lambda = l(l + 1), \qquad l = 0, 1, 2, \ldots \tag{III-106}$$

with

$$-l \le m \le l \tag{III-107}$$

and the eigenfunctions are the associated Legendre polynomials

$$\Theta = P_l^m(x) = (-1)^m (1 - x^2)^{m/2} \frac{d^m P_l(x)}{dx^m}, \qquad m = |m|, \tag{III-108}$$

where $P_l(x)$ are the Legendre polynomials which satisfy the equation

$$\frac{d}{dx}\left[(1 - x^2)\frac{dP_l}{dx}\right] + l(l + 1)P_l = 0. \tag{III-109}$$

The $P_l^m(x)$ have property

$$\int_{-1}^{1} P_l^m(x) P_l^m(x)\, dx = \frac{2}{2l + 1} \frac{(l + m)!}{(l - m)!} \delta_{l,k} \qquad (m = |m|). \tag{III-110}$$

The normalized eigenfunctions of (III-102) are

$$\Theta_{l,m}(x) = (-1)^l \left[\frac{2l + 1}{2} \frac{(l - m)!}{(l + m)!}\right]^{1/2} P_l^m(x). \tag{III-111}$$

Explicitly,

$$\Theta_{0,0} = \frac{1}{\sqrt{2}}, \qquad \Theta_{1,0} = \sqrt{\frac{3}{2}}\cos\vartheta, \qquad \Theta_{1,\pm 1} = \mp\sqrt{\frac{3}{4}}\sin\vartheta,$$

$$\Theta_{2,0} = \sqrt{\frac{5}{8}}(2\cos^2\vartheta - \sin^2\vartheta), \qquad \Theta_{2,\pm 1} = \mp\sqrt{\frac{15}{4}}\cos\vartheta\sin\vartheta,$$

$$\Theta_{2,\pm 2} = \sqrt{\frac{15}{16}}\sin^2\vartheta,$$

$$\Theta_{3,0} = \sqrt{\frac{7}{8}}(2\cos^2\vartheta - 3\sin^2\vartheta)\cos\vartheta,$$

$$\Theta_{3,\pm 1} = \mp\sqrt{\frac{21}{32}}(4\cos^2\vartheta - \sin^2\vartheta)\sin\vartheta,$$

$$\Theta_{3,\pm 2} = \sqrt{\frac{105}{16}}\cos\vartheta\sin^2\vartheta \qquad \Theta_{3,\pm 3} = \mp\sqrt{\frac{35}{32}}\sin^2\vartheta. \tag{III-112}$$

---

[p]See Appendix B at the end of this section. Note that $l$ starts from $l = 0$.

The normalized eigenfunctions of (III-102), (III-103) combined are, from (III-104) and (III-111), denoted by

$$Y_{l,m}(\vartheta, \varphi) = \Phi_m(\varphi)\Theta_{l,m}(\cos \vartheta). \tag{III-113}$$

The $Y_{l,m}$ have the property[q]

$$\sum_{m=-l}^{l} Y_{l,m}^* Y_{l,m} = \frac{2l + 1}{4\pi}. \tag{III-114}$$

## (i) Parity

For a system possessing central symmetry, i.e., the potential energy is a function of $|r|$, the Hamiltonian is invariant under the inversion operation

$$P: \mathbf{r} \to -\mathbf{r}, \qquad (x \to -x, y \to -y, z \to -z) \tag{III-115}$$

and

$$PH = H. \tag{III-116}$$

The result of an inversion operation on the wave function $\psi(r)$ of a one-particle, central field system is as follows

$$P^2\psi = P(P\psi) = \psi,$$

i.e., the eigenvalues of $P^2$ is 1, and

$$P\psi = \pm\psi. \tag{III-117}$$

The state $\psi$ is said to be of even (odd) parity if $P\psi = +1(-1)\psi$. This "parity" is regarded as a "quantum number", although strictly speaking it is the eigenvalues $\pm 1$ of the parity (inversion) operator $P$.

In spherical polar coordinates, (III-115) become

$$Pr = r, \qquad P\vartheta = \pi - \vartheta, \qquad P\varphi = \pi + \varphi \tag{III-1118}$$

and

$$P\Phi_m(\varphi) = e^{im\pi}\Phi_m(\varphi) = (-1)^m\Phi_m,$$

$$P\Theta_{l,m}(\cos \vartheta) = (-1)^{l-m}\Theta_{l,m}(\cos \vartheta), \tag{III-119}$$

$$PR(r) = R(r)$$

so that on $\psi(r)$ of (III-100a),

$$P\psi(r) = P(\Phi_m\Theta_{l,m}R) = (-1)^l\psi(r), \tag{III-120}$$

[q] This property is useful in connection with the problem of closed electron shells in atoms.

i.e., $\psi(r)$ has even (odd) parity according as $l$ is even (odd). Thus parity concept is a basic one in connection with the selection rules for radiation of systems having central or reflection (mirroring) symmetry. (See Chap. 7, Sec. 1 and Chap. 8, Sec. 2)

### (ii) Matrix elements of r (electric moment er)

The electric (dipole) moment has components

$$M_x = er\sin\vartheta\cos\varphi, \qquad M_y = er\sin\vartheta\sin\varphi, \qquad M_z = er\cos\vartheta$$

or

$$M_x \pm iM_y = er\sin\vartheta e^{\pm i\varphi}. \tag{III-121}$$

The non-vanishing matrix elements

$$\iint \Theta_{l,m}\Phi_m^* \begin{Bmatrix} \cos\vartheta \\ \sin\vartheta e^{\pm i\varphi} \end{Bmatrix} \Theta_{l,m'}\Phi_{m'}\, d\sin\vartheta\, d\varphi$$

are

$$\langle l+1, m|\cos\vartheta|l, m\rangle = -\left[\frac{(l-m+1)(l+m+1)}{(2l+1)(2l+3)}\right]^{1/2},$$

$$\langle l+1, m+1|\sin\vartheta e^{i\varphi}|l, m\rangle = \langle l, m|\sin\vartheta e^{-i\varphi}|l+1, m+1\rangle$$

$$= \left[\frac{(l+m+1)(l+m+2)}{(2l+1)(2l+3)}\right]^{1/2}, \tag{III-122}$$

$$\langle l, m+1|\sin\vartheta e^{i\varphi}|l+1, m\rangle = \langle l+1, m|\sin\vartheta e^{-i\varphi}|l, m+1\rangle$$

$$= -\left[\frac{(l-m)(l-m+1)}{(2l+1)(2l+3)}\right]^{1/2}.$$

### (4) Angular momenta

The angular momentum operators, $\mathbf{L}^2$, $L_x$, $L_y$, $L_z$, $L_+$, $L_-$ are defined in Chap. 2 (II-98, 99, 105). In spherical polar coordinates (III-121), one has

$$\frac{\partial}{\partial z} = \cos\vartheta\frac{\partial}{\partial r} - \frac{\sin\vartheta}{r}\frac{\partial}{\partial\vartheta},$$

$$\frac{\partial}{\partial x} = \sin\vartheta\cos\varphi\frac{\partial}{\partial r} + \frac{\cos\vartheta\cos\varphi}{r}\frac{\partial}{\partial\vartheta} - \frac{\sin\varphi}{r\sin\vartheta}\frac{\partial}{\partial\varphi}, \tag{III-123}$$

$$\frac{\partial}{\partial y} = \sin\vartheta\sin\varphi\frac{\partial}{\partial r} + \frac{\cos\vartheta\sin\varphi}{r}\frac{\partial}{\partial\vartheta} + \frac{\cos\varphi}{r\sin\vartheta}\frac{\partial}{\partial\varphi},$$

$$L_z = \frac{\hbar}{i}\frac{\partial}{\partial\varphi},$$

$$L_x = i\hbar\left(\sin\varphi\frac{\partial}{\partial\vartheta} + \cos\varphi\cot\vartheta\frac{\partial}{\partial\varphi}\right),$$

$$L_y = i\hbar\left(-\cos\varphi\frac{\partial}{\partial\vartheta} + \sin\varphi\cot\vartheta\frac{\partial}{\partial\varphi}\right), \tag{III-124}$$

$$L_\pm = \pm\hbar e^{\pm i\varphi}\left(\frac{\partial}{\partial\vartheta} \pm i\cot\vartheta\frac{\partial}{\partial\varphi}\right),$$

$$\mathbf{L}^2 = -\hbar^2\left[\frac{1}{\sin\vartheta}\frac{\partial}{\partial\vartheta}\left(\sin\vartheta\frac{\partial}{\partial\vartheta}\right) + \frac{1}{\sin^2\vartheta}\frac{\partial^2}{\partial\varphi^2}\right].$$

With the notation for the matrix element of $A$

$$\langle l, m|A|l', m'\rangle = \iint Y_{l,m}^{*}\, A\, Y_{l',m'}\, d\cos\vartheta\, d\varphi,$$

one obtains from (III-104), (III-101), (III-102) and (III-106), the following non-vanishing matrix elements

$$\langle l, m|L_z|l, m\rangle = m\hbar,$$
$$\langle l, m|\mathbf{L}^2|l, m\rangle = l(l+1)\hbar^2, \tag{III-125}$$

$$\langle l, m|L_x|l, m+1\rangle = -i\langle l, m|L_y|l, m+1\rangle$$
$$= \tfrac{1}{2}\sqrt{(l-m)(l+m+1)}\,\hbar,$$
$$\langle l, m|L_x|l, m-1\rangle = i\langle l, m|L_y|l, m-1\rangle \tag{III-126}$$
$$= \tfrac{1}{2}\sqrt{(l-m+1)(l+m)}\,\hbar,$$

which are in complete agreement with (II-116) and (II-119a, b) found in matrix mechanics.


## Appendix A
### Hermite polynomials

The Hermite polynomials $H_n(x)$ given by (III-93)–(III-95) satisfy Eq. (III-96)

$$\left(\frac{d^2}{dx^2} - 2x\frac{d}{dx} + 2n\right)H_n(x) = 0. \tag{A-1}$$

It is easy to show that the functions defined as the coefficients of $(1/n!)t^n$ in the expansion of the following generating function

$$\phi(x, t) = e^{x^2} e^{-(t-x)^2} \tag{A-2}$$

$$= \sum_{n=0}^{\infty} \frac{1}{n!} H_n(x) t^n, \tag{A-2a}$$

namely

$$H_n(x) = \left( \frac{\partial^n \phi}{\partial t^n} \right)_{t=0} = (-1)^n e^{x^2} \frac{d^n}{dx^n} e^{-x^2} \tag{A-3}$$

satisfy equation (A-1). From (A-2a) and

$$\frac{\partial \phi}{\partial x} = 2t\phi, \qquad \frac{\partial \phi}{\partial t} + 2(t - x)\phi = 0,$$

one obtains

$$\frac{dH_n}{dx} = 2nH_{n-1}, \qquad n \geq 1, \tag{A-4a}$$

$$H_{n+1} - 2xH_n + 2nH_{n-1} = 0, \qquad n \geq 1. \tag{A-4b}$$

From these two relations, (A-1) follows.

The $H_n(x)$ have the property (see (III-94))

$$H_n(-x) = \pm H_n(x) \qquad \text{for} \qquad n = \begin{cases} \text{even,} \\ \text{odd.} \end{cases} \tag{A-5}$$

From (A-2), one obtains

$$\int_{-\infty}^{\infty} \phi(x, t)\phi(x, s)e^{-x^2} \, dx = \sum_{m=0}^{\infty} \sum_{n=0}^{\infty} s^m t^n \frac{1}{m!\, n!} \int_{-\infty}^{\infty} H_m H_n e^{-x^2} \, dx \tag{A-6}$$

$$= \int_{-\infty}^{\infty} e^{-s^2 - t^2 + 2(s+t)x - x^2} \, dx$$

$$= e^{2st} \int_{-\infty}^{\infty} e^{-(x-s-t)^2} \, d(x - s - t)$$

$$= \sqrt{\pi} \left\{ 1 + 2st + \frac{(2st)^2}{2!} + \frac{(2st)^3}{3!} + \cdots \right\}. \tag{A-6a}$$

One comparing the coefficients of powers of $s^m t^n$ in (A-6) and (A-6a), one gets

$$\int_{-\infty}^{\infty} H_m H_n e^{-x^2} \, dx = \begin{cases} 0, & m \neq n, \\ 2^n \sqrt{\pi}\, n!, & m = n. \end{cases} \tag{A-7}$$

From (A-5) and (A-7), one gets

$$\int_{-\infty}^{\infty} H_m x H_n e^{-x^2}\, dx = \begin{cases} 2^n \sqrt{\pi}\,(n+1)!, & m = n + 1, \\ 2^n \sqrt{\pi}\, n!, & m = n - 1, \\ 0, & m \neq n \pm 1. \end{cases} \tag{A-8}$$

The normalized $H_n$, are from (A-7)

$$\frac{1}{\sqrt{2^n \sqrt{\pi}\, n!}}\, H_n(x). \tag{A-9}$$

For normalized $H_n(x)$, one gets

$$\int_{-\infty}^{\infty} H_{n+1} x H_n e^{-x^2}\, dx = \sqrt{\frac{n+1}{2}},$$

$$\int_{-\infty}^{\infty} H_{n-1} x H_n e^{-x^2}\, dx = \sqrt{\frac{n}{2}}. \tag{A-10}$$

Similarly, by repeated applications of (A-5) and (A-8)–(A-10), one obtains the following integrals

$$\int_{-\infty}^{\infty} H_n x^2 H_n e^{-x^2}\, dx = \frac{2n+1}{2},$$

$$\int_{-\infty}^{\infty} H_{n+2} x^2 H_n e^{-x^2}\, dx = \frac{1}{2}\sqrt{(n+1)(n+2)},$$

$$\int_{-\infty}^{\infty} H_{n+1} x^3 H_n e^{-x^2}\, dx = 3\left(\frac{n+1}{2}\right)^{3/2},$$

$$\int_{-\infty}^{\infty} H_{n+3} x^3 H_n e^{-x^2}\, dx = \sqrt{\frac{(n+1)(n+2)(n+3)}{8}}, \tag{A-10a}$$

$$\int_{-\infty}^{\infty} H_n x^4 H_n e^{-x^2}\, dx = \frac{3}{2}\left(n^2 + n + \frac{1}{2}\right),$$

$$\int_{-\infty}^{\infty} H_m x^{2p} H_n e^{-x^2}\, dx$$

$$= \sqrt{\frac{m!\, n!}{\pi}} \sum_j \binom{2p}{2j} \binom{2j}{j + \frac{n-m}{2}} \frac{1}{2^j \left(\frac{n+m}{2} - j\right)!}\, \Gamma\left(p - j + \frac{1}{2}\right),$$

for $m + n = $ even, the limits of summation being from $j = n - m/2$ to $j = p$, or $n + m/2$, whichever is the smaller.

## Appendix B
### Harmonic oscillation in Fock representation

In Problem 1, Chap. 2, Sec. 4, we introduced the Fock representation, in matrix form, for the harmonic oscillator. Here we shall recast it in wave mechanics, because it is an elegant method and also because it introduces in an elementary way the annihilation and creation operators.

The Schrödinger equation (III-86), in the dimensionless variable $\xi$, is

$$\frac{1}{2}\left(\xi^2 - \frac{d^2}{d\xi^2}\right)\psi = \frac{E}{\hbar\omega}\psi. \tag{B-1}$$

Define the operators

$$b = \frac{1}{\sqrt{2}}\left(\xi + \frac{d}{d\xi}\right), \qquad b^\dagger = \frac{1}{\sqrt{2}}\left(\xi - \frac{d}{d\xi}\right). \tag{B-2}$$

For any function $f(\xi)$, it is seen that

$$b^\dagger b f = \frac{1}{2}\left(\xi^2 - \frac{d^2}{d\xi^2}\right)f - \frac{1}{2}f, \tag{B-3}$$

$$(bb^\dagger - b^\dagger b)f = f,$$

or

$$bb^\dagger - b^\dagger b = 1, \tag{B-4}$$

and

$$\int_{-\infty}^{\infty} f b g \, d\xi = \int_{-\infty}^{\infty} (b^\dagger f) g \, d\xi,$$

$$\int_{-\infty}^{\infty} f b^\dagger g \, d\xi = \int (bf) g \, d\xi. \tag{B-5}$$

Thus $b^\dagger$, $b$ are adjoint of each other, but are *not* self-adjoint.

Let

$$\lambda = \frac{E}{\hbar\omega} - \frac{1}{2}. \tag{B-6}$$

(Note the different notation from (III-85).) Then (B-1) and (B-3) can be written

$$b^\dagger b\psi = \lambda\psi \tag{B-7}$$

and, for normalized $\psi$,

$$\int \psi b^\dagger b\psi \, d\psi = \lambda \int \psi^2 \, d\xi = \lambda.$$

From (B-5), it follows that $\lambda$ is real and positive,

$$\lambda = \int (b\psi)^2 \, d\xi \geq 0, \qquad (B-8)$$

i.e., except for $b\psi = 0$, $\lambda > 0$.

From (B-7) and (B-4), one gets

$$b(b^\dagger b\psi) = \lambda b\psi,$$

$$(b^\dagger b + 1)b\psi = \lambda b\psi,$$

$$b^\dagger b(b\psi) = (\lambda - 1)(b\psi), \qquad (B-9)$$

i.e, $b\psi$ is the eigenfunction of $b^\dagger b$ for the eigenvalue $\lambda - 1$. Similarly,

$$b^\dagger b(b^\dagger\psi) = (\lambda + 1)(b^\dagger\psi). \qquad (B-10)$$

Thus the terminology: $b$ is an *annihilation*, $b^\dagger$ a *creation* operator.

From (B-9), one gets

$$b^\dagger b(b^n\psi) = (\lambda - n)(b^n\psi), \qquad (B-11)$$

where $\lambda - n$ becomes negative for sufficiently large values of $n$. But this contradicts (B-8). To avoid this contradiction, one has to assume that in the sequence $b\psi, b^2\psi, b^3\psi, \ldots$, there is one $\psi_0 = b^s\psi$ such that

$$b\psi_0 = (\lambda - s)\psi_0 = 0, \qquad (B-12)$$

i.e., there is an eigenfunction $\psi_0$ whose eigenvalue is zero,

$$b^\dagger b\psi_0 = 0. \qquad (B-13)$$

The other eigenvalues are, according to (B-10),

$$\lambda = 1, 2, 3, \ldots$$

The energy $E$, from (B-6), is then given by

$$E = \left(n + \frac{1}{2}\right)\hbar\omega, \qquad n = 0, 1, 2, \ldots \qquad (B-14)$$

The $\psi_0$ can be obtained from (B-2) and (B-12), i.e.,

$$\left(\frac{d}{d\xi} + \xi\right)\psi_0 = 0,$$

whose normalized solution is

$$\psi_0 = \frac{1}{\sqrt[4]{\pi}} e^{-1/2\xi^2}. \qquad (B-15)$$

The eigenfunctions for $\lambda_1, \lambda_2, \lambda_3, \ldots$ are obtained from (B-10)

$$b^\dagger \psi_n = N\psi_{n+1}, \tag{B-16}$$

where $N$ is a constant that can be determined as follows. From (B-4) and (B-7),

$$bb^\dagger \psi_n = (b^\dagger b + 1)\psi_n = (n + 1)\psi_n. \tag{B-17}$$

From (B-16),

$$N^2 \int \psi_{n+1}\psi_{n+1}\, d\xi = \int (b^\dagger \psi_n)(b^\dagger \psi_n)\, d\xi = \int (bb^\dagger \psi_n)\psi_n\, d\xi = n + 1.$$

Hence

$$b^\dagger \psi_n = \sqrt{n + 1}\,\psi_{n+1}. \tag{B-18}$$

From this, one obtains

$$\psi_n = \frac{1}{\sqrt{n!}}(b^\dagger)^n\psi_0. \tag{B-19}$$

From (B-2), one has

$$b^\dagger f(\xi) = \frac{1}{\sqrt{2}}\left(\xi - \frac{d}{d\xi}\right)f = -\frac{1}{\sqrt{2}}e^{1/2\xi^2}\frac{d}{d\xi}(e^{-1/2\xi^2}f)$$

with $\psi_0$ in (B-15),

$$\psi_n(\xi) = \frac{(-1)^n}{\sqrt{2^n n!\sqrt{\pi}}}e^{-1/2\xi^2}\frac{d^n}{d\xi^n}(e^{-\xi^2}). \tag{B-20}$$

From (A-3), the Hermite polynomial is

$$H_n(\xi) = (-1)^n e^{\xi^2}\frac{d^n}{d\xi^n}(e^{-\xi^2}). \tag{B-21}$$

Hence

$$\psi_n(\xi) = \frac{1}{\sqrt{2^n n!\sqrt{\pi}}}e^{-1/2\xi^2}H_n(\xi), \tag{B-22}$$

which is (III-98).

The use of the operators $b$, $b^\dagger$ facilitates many calculations. For example,

$$\langle m|\xi|n\rangle = \int \psi_m \xi \psi_n\, d\xi = \frac{1}{\sqrt{2}}\int \psi_m(b + b^\dagger)\psi_n\, d\xi.$$

From (B-18) and a similar relation

$$b\psi_n = \sqrt{n}\,\psi_{n-1}, \tag{B-23}$$

one obtains

$$\xi\psi_n = \frac{1}{\sqrt{2}}(\sqrt{n}\,\psi_{n-1} + \sqrt{n+1}\,\psi_{n+1}) \tag{B-24}$$

so that

$$\langle n - 1|\xi|n\rangle = \sqrt{\frac{n}{2}},$$
$$\tag{B-25}$$
$$\langle n + 1|\xi|n\rangle = \sqrt{\frac{n+1}{2}},$$

which are the same as (A-10).

## Appendix C
### Associated Legendre polynomials

*(1) Legendre coefficients $P_l(\cos\vartheta)$*

Consider two vectors with an angle $\vartheta$ between them, and $|r_<| < |r_>|$. The $P_l(\cos\vartheta)$ are defined by

$$\frac{1}{|r_> - r_<|} = \frac{1}{r_>\sqrt{1 + \xi^2 - 2\xi x}} = \sum_{l=0}^{\infty}\frac{1}{r_>}\xi^l P_l(x), \tag{C-1}$$

i.e., by the generating function $\phi(x,t)$

$$\phi(x,t) = \frac{1}{\sqrt{1 - 2xt + t^2}} = \sum_{l=0}^{\infty} t^l P_l(x). \tag{C-1a}$$

From this, one gets, on differentiating with respect to $t$ and $x$,

$$(n + 1)P_{n+1} - (2n + 1)xP_n + nP_{n-1} = 0, \tag{C-2}$$

$$P'_{n+1} - 2xP'_n + P'_{n-1} - P_n = 0. \tag{C-3}$$

From (C-2) and (C-3), one gets

$$xP'_n - P'_{n-1} - nP_n = 0. \tag{C-4}$$

From (C-3) and (C-4), one gets

$$P'_{n+1} - P'_{n-1} = (2n + 1)P_n. \tag{C-5}$$

From this,

$$P'_{n+1} + P'_n = \sum_{l=0}^{n} (2l + 1)P_l. \tag{C-6}$$

Similarly,

$$P'_{n+1} - xP'_n = (n + 1)P_n, \tag{C-7}$$

$$(1 - x^2)P'_n = nP_{n-1} - nxP_n. \tag{C-8}$$

From (C-8) and (C-4), one gets

$$\frac{d}{dx}\left[(1 - x^2)\frac{dP_l}{dx}\right] + l(l + 1)P_l = 0, \tag{C-9}$$

which is the Legendre equation.

From (C-9), one obtains readily the orthogonality relation

$$\int_{-1}^{1} P_m(x)P_n(x)\,dx = 0 \qquad \text{for } m \neq n, \tag{C-10}$$

and from (C-1a), one obtains[r]

$$\int_{-1}^{1} P_m(x)P_m(x)\,dx = \frac{2}{2m + 1}. \tag{C-10a}$$

The polynomial $P_l(x)$ can be obtained explicitly by putting in (C-9) the sum

$$P_l = \sum_{k=0} a_k x^k,$$

leading to the recurrence relation

$$a_{k+2} = -\frac{(l - k)(l + k + 1)}{(k + 1)(k + 2)}a_k, \tag{C-11}$$

so that the highest power is $x^l$, and

$$P_l(x) = a_l\left[x^l - \frac{(l - 1)l}{2(2l - 1)}x^{l-2} + \frac{(l - 3)(l - 2)(l - 1)l}{2 \cdot 4(2l - 3)(2l - 1)}x^{l-4}\cdots\right],$$

the last term being $x^0(x)$ according as $l =$ even (odd). $a_l$ is chosen by convention

---

[r] In view of (C-10),

$$\int_{-1}^{1}\frac{dx}{1 + t^2 - 2tx} = \sum_{m=0}^{\infty} t^{2m}\int_{-1}^{1}[P_m]^2\,dx.$$

But

$$\int_{-1}^{1}\frac{dx}{1 + t^2 - 2tx} = \frac{2}{t}\left[t + \frac{t^3}{3!} + \frac{t^5}{5!} + \cdots\right], \text{ etc.}$$

to be

$$a_l = \frac{(2l-1)!!}{l!} = \frac{1 \cdot 3 \cdot 5 \dots (2l-1)}{l!}.$$

$P_l(x)$ has the following property:

$$P_l(-x) = \pm P_l(x) \qquad \text{for } l = \begin{cases} \text{even,} \\ \text{odd.} \end{cases} \tag{C-12}$$

$P_l(x)$ has many other equivalent forms, for example, the Rodrigues form

$$P_l(x) = \frac{1}{2^l l!} \frac{d^l}{dx^l} (x^2 - 1)^l. \tag{C-13}$$

A few $P_l(x)$ are:

$$P_l(1) = 1 \qquad \text{for} \qquad l \geq 0,$$

$$P_0(x) = 1, \qquad P_1(x) = x, \qquad P_2(x) = \frac{1}{2}(3x^2 - 1),$$

$$P_3(x) = \frac{1}{2}(5x^2 - 3)x, \qquad P_4(x) = \frac{1}{8}(35x^4 - 30x^2 + 3), \tag{C-14}$$

$$P_5(x) = \frac{1}{8}(63x^4 - 70x^2 + 15)x.$$

*(2) Associated Legendre polynomials*

Consider the equation

$$\frac{d}{dx}\left[(1 - x^2)\frac{d\Theta}{dx}\right] + \left[\alpha^2 - \frac{m^2}{1 - x^2}\right]\Theta = 0, \tag{C-15}$$

where $m = \pm$integers, and $\alpha^2$ is the eigenvalue to be found. This equation is a typical Sturm-Liouville equation, as the range of $x$ is $-1 \leq x \leq 1$. The points $x = \pm 1$ are poles.

For the character of the equation at $x = 1$, let

$$z \equiv 1 - x,$$

so that (C-15) becomes

$$\frac{d^2\Theta}{dz^2} + \frac{2(1 - z)}{z(2 - z)}\frac{d\Theta}{dz} + \left[\frac{\alpha^2}{z(2 - z)} - \frac{m^2}{z^2(2 - z)^2}\right]\Theta = 0. \tag{C-16}$$

Let

$$\Theta = z^\lambda \sum_{k=0} a_k z^k.$$

The indicial equation

$$\lambda(\lambda - 1) + \lambda - \frac{m^2}{4} = 0$$

has two roots

$$\lambda = \pm \frac{|m|}{2},$$

which differ from each other by an integer and therefore lead to one and the same solution. A second independent solution has to be obtained differently (associated Legendre functions of the second kind, not to be discussed here.) We shall take the root $\lambda = +|m|/2$.

Similarly for the pole at $x = -1$, we get $z = 1 + x$ and obtain a factor $(1 + x)^\lambda$ where again $\lambda = |m|/2$. Thus we let

$$\Theta(x) = (1 - x^2)^{|m|/2} u(x). \tag{C-17}$$

The equation for $u(x)$ is

$$(1 - x^2)\frac{d^2 u}{dx^2} - 2(m + 1)x\frac{du}{dx} + [\alpha^2 - m(m + 1)]u = 0, \tag{C-18}$$

where $m$ stands for $|m|$.

Let

$$u(x) = \sum_{k=0} C_k x^k. \tag{C-19}$$

The recurrence relation is

$$(k + 1)(k + 2)C_{k+2} = [k(k - 1) + 2(m + 1)k - \alpha^2 + m(m + 1)]C_k.$$

The ratio of two successive terms is, asymptotically for large $k$,

$$\frac{C_{k+2}}{C_k}x^2 \approx x^2$$

so that the series behaves asymptotically as $1/(1 - x^2)$, and $\Theta$ in (C-17) as $(1 - x^2)^{1/2m-1}$. The integral $\int_{-1}^{1} [\Theta]^2\, dx$ does not converge at $x = \pm 1$ for $m = 0$ or 1. To avoid this, one sets $C_{k+2} = C_{k+4} = \cdots = 0$, i.e.,

$$\alpha^2 = (k + m)(k + m + 1) = l(l + 1), \qquad \text{where} \qquad l \geq m. \tag{C-20}$$

This determines the eigenvalues $\alpha^2$ in (C-15).

To obtain the solution $\Theta(x)$ of (C-15) with $\alpha^2 = l(l + 1)$, one differentiates (C-9) $m$ times, obtaining

$$(1 - x^2)\frac{d^2}{dx^2}\left(\frac{d^m P}{dx^m}\right) - 2(m + 1)x\frac{d}{dx}\left(\frac{d^m P}{dx^m}\right) + (l - m)(l + m + 1)\frac{d^m P}{dx^m} = 0.$$

$$(C\text{-}21)$$

It is seen that $d^m P_l(x)/dx^m$ and $u(x)$ of (C-18) satisfy the same differential equation. Hence, from (C-17),

$$\Theta(x) = (-1)^m(1 - x^2)^{1/2m}\frac{d^m P_l}{dx^m} \equiv P_l^m, \qquad (C\text{-}22)$$

the $(-1)^m$ being introduced by convention.

To obtain the integral

$$F(m) \equiv \int_{-1}^{1} P_l^m P_l^m\, dx = \int_{-1}^{1} (1 - x^2)^m \frac{d^m P_l}{dx^m}\frac{d^m P_l}{dx^m}\, dx, \qquad (C\text{-}23)$$

one obtains, on integration by parts,

$$F(m) = -\int_{-1}^{1} \frac{d^{m-1} P_l}{dx^{m-1}}\frac{d}{dx}\left[(1 - x^2)^m\frac{d^m P_l}{dx^m}\right] dx. \qquad (C\text{-}24)$$

Replacing, in (C-21), $m$ by $m - 1$, and multiplying by $(1 - x^2)^{m-1}$, one gets

$$\frac{d}{dx}\left[(1 - x^2)^m\frac{d^m P_l}{dx^m}\right] = -(l - m + 1)(l + m)(1 - x^2)^{m-1}\frac{d^{m-1} P_l}{dx^{m-1}}$$

and using this in (C-24) leads to a recurrence relation

$$F(m) = (l - m + 1)(l + m)(l - m + 2)(l + m - 1)F(m - 2)$$

$$= \frac{(l + m)!}{(l - m)!} F(0). \qquad (C\text{-}25)$$

From (C-10) and (C-10a), we have $F(0) = 2/2m + 1$. Hence

$$\int_{-1}^{1} P^m(x)P^m(x)\, dx = \frac{2}{2l + 1}\frac{(l + m)!}{(l - m)!}. \qquad (C\text{-}26)$$

The normalized $P^m(x)$ are denoted by

$$\Theta_{l,m}(x) = (-1)^l\sqrt{\frac{2l + 1}{2}\frac{(l - m)!}{(l + m)!}}\, P_l^m(x), \qquad m > 0, \qquad (C\text{-}27)$$

where

$$P_l^m(x) = (-1)^m(1 - x^2)^{1/2m}\frac{d^m P_l(x)}{dx^m}, \qquad m > 0, \qquad (C\text{-}28)$$

$$P_l(x) = \frac{1}{2^l l!} \frac{d^l (x^2 - 1)^l}{dx^l}.$$ (C-29)

The factors $(-1)^m$, $(-1)^l$ are chosen by convention.

For negative values of $m$, we have to define $P_l^m$. This will be done in Appendix D.

## Appendix D
### Angular momentum operators and spherical harmonics

The angular momentum operators are defined in Chap. 2, (II-98, 99, 105). In particular, we have

$$L_+ L_z - L_z L_+ = -L_+ \hbar,$$
$$L_- L_z - L_z L = L_- \hbar,$$ (D-1)

$$L_+ L_- = \mathbf{L}^2 - L_z^2 + L_z \hbar,$$
$$L_- L_+ = \mathbf{L}^2 - L_z^2 - L_z \hbar,$$ (D-2)

and from (III-125), we have[s]

$$\langle \lambda, \mu | \mathbf{L}^2 | \lambda', \mu' \rangle = \lambda(\lambda + 1)\hbar^2 \delta_{\lambda \lambda'} \delta_{\mu \mu'},$$ (D-3)

$$\langle \lambda, \mu | L_z | \lambda', \mu' \rangle = \mu \hbar \delta_{\lambda \lambda'} \delta_{\mu \mu'},$$ (D-4)

i.e.,

$$\mathbf{L}^2 Y_{\lambda \mu} = \lambda(\lambda + 1)\hbar^2 Y_{\lambda \mu},$$ (D-3a)

$$L_z Y_{\lambda \mu} = \mu \hbar Y_{\lambda \mu}.$$ (D-4a)

From $\langle \lambda, \mu | \mathbf{L}^2 | \lambda, \mu \rangle \geq \langle \lambda, \mu | L_z^2 | \lambda, \mu \rangle$, it follows that

$$\lambda(\lambda + 1) \geq \mu^2.$$ (D-5)

Let $L_+$ operate on (D-4a) and using (D-1), we have

$$L_z(L_+ Y_{\lambda \mu}) = (\mu + 1)(L_+ Y_{\lambda \mu}),$$ (D-6)

i.e., $L_+ Y_{\lambda \mu}$ is the eigenfunction of $L_z$ for the eigenvalue $(\mu + 1)\hbar$, and we may write

$$L_+ Y_{\lambda \mu} = a_{\lambda \mu} Y_{\lambda, \mu+1},$$

where $a_{\lambda \mu}$ is a constant that can be obtained from normalization. Let all $Y_{\lambda \mu}$ be

---

[s] In the following, we use $\lambda$, $\mu$ instead of $l$, $m$ as the expressions are general and not restricted to integral values of $l$, $m$. Thus $Y_{\lambda \mu}(\vartheta, \varphi)$ is not limited to integral $l$, $m$ as in $Y_{l,m}$.

already normalized, so that,

$$
\begin{aligned}
|a_{\lambda\mu}|^2 &= (L_+ Y_{\lambda\mu}, L_+ Y_{\lambda\mu}) \\
&= (L_+^\dagger L_+ Y_{\lambda\mu}, Y_{\lambda\mu}) \\
&= (L_- L_+ Y_{\lambda\mu}, Y_{\lambda\mu}), \quad \text{since } L_+^\dagger = L_-, \\
&= ((\mathbf{L}^2 - L_z^2 - L_z \hbar) Y_{\lambda\mu}, Y_{\lambda\mu}) \\
&= \lambda(\lambda + 1) - \mu(\mu + 1) = (\lambda + \mu + 1)(\lambda - \mu).
\end{aligned}
$$

$$
\therefore \; L_+ Y_{\lambda\mu} = \sqrt{(\lambda + \mu + 1)(\lambda - \mu)}\, Y_{\lambda, \mu+1}. \tag{D-7a}
$$

Similarly, we have

$$
L_- Y_{\lambda\mu} = \sqrt{(\lambda - \mu + 1)(\lambda + \mu)}\, Y_{\lambda, \mu-1}. \tag{D-7b}
$$

From (D-5), it is seen that $\mu$ has a maximum and a minimum value, and from (D-7a, b)

$$
-\lambda \le \mu \le \lambda, \tag{D-8}
$$

$$
\mu_{\max} - \mu_{\min} = 2\lambda, \tag{D-9}
$$

so that

$$
L_+ Y_{\lambda\lambda} = 0, \qquad L_- Y_{\lambda, -\lambda} = 0. \tag{D-10}
$$

The above results are general. Let us now apply them to the functions $Y_{l,m}(\vartheta, \varphi)$ in (III-113).

From (III-124) and (D-10) above, we have

$$
e^{i\varphi}\left(\frac{\partial}{\partial \vartheta} + i \cot \vartheta \frac{\partial}{\partial \varphi}\right) Y_{ll} = 0. \tag{D-11a}
$$

The solution is

$$
Y_{ll}(\vartheta, \varphi) = N_{ll} e^{il\varphi} \sin^l \vartheta. \tag{D-12a}
$$

The periodic condition leads to $l = $ integer, and the normalization constant is

$$
N_{ll} = \frac{1}{2^l l!} \sqrt{\frac{(2l + 1)!}{4\pi}}. \tag{D-13}
$$

From (D-7a), and (III-124), $L_- Y_{l,m}$ is

$$
L_- Y_{lm} = \sqrt{(l - m + 1)(l + m)}\, Y_{l, m-1}. \tag{D-14a}
$$

The left-hand side is

$$
L_- Y_{l,m} = \frac{1}{\sin^{m-1} \vartheta} e^{-i\varphi} \frac{\partial}{\partial \cos \vartheta} (\sin^m \vartheta\, Y_{lm}). \tag{D-15a}
$$

On starting from $Y_{ll}$ and lowering $m$ by successive application of $L_-$ by means of (D-15a), one obtains

$$Y_{l,m} = \frac{1}{2^l l!} \sqrt{\frac{2l+1}{4\pi} \frac{(l+m)!}{(l-m)!}} \frac{1}{\sin^m \vartheta} \frac{d^{l-m}(\sin^{2l} \vartheta)}{d\cos \vartheta^{l-m}} e^{im\varphi}. \qquad \text{(D-16a)}$$

From (III-124) and (D-10) we have

$$e^{-i\varphi}\left(\frac{\partial}{\partial \vartheta} - i\cos\vartheta \frac{\partial}{\partial \varphi}\right) Y_{l,-l} = 0, \qquad \text{(D-11b)}$$

whose solution is

$$Y_{l,-l}(\vartheta, \varphi) = \frac{1}{2^l l!} \sqrt{\frac{(2l+1)!}{4\pi}} e^{-il\varphi} \sin^l \vartheta \qquad \text{(D-12b)}$$

and from (D-7b),

$$L_+ Y_{l,m} = \sqrt{(l+m+1)(l-m)}\, Y_{l,m+1}, \qquad \text{(D-14b)}$$

and

$$L_+ Y_{l,m} = -e^{i\varphi} \sin^{m+1}\vartheta \frac{\partial}{\partial \cos\vartheta}\left(\frac{1}{\sin^m \vartheta} Y_{l,m}\right). \qquad \text{(D-15b)}$$

On starting from $Y_{l,-l}$ and raising $m$ by successive application of $L_+$ by means of (D-15b), one obtains

$$Y_{l,m}(\vartheta, \varphi) = \frac{(-1)^m}{2^l l!} \sqrt{\frac{2l+1}{4\pi} \frac{(l-m)!}{(l+m)!}} \sin^m \vartheta \frac{d^{l+m} \sin^{2l}\vartheta}{d\cos\vartheta^{l+m}} e^{im\varphi}. \qquad \text{(D-16b)}$$

The two expressions (D-16a), (D-16b) must be identical, and in fact they are, as seen on carrying out the differentiations.

Now, from (C-13) and (C-22),

$$P_l(x) = \frac{1}{2^l l!} \frac{d^l (x^2-1)^l}{dx^l}, \qquad \text{(C-13)}$$

$$P_l^m(x) = (-1)^m (1-x^2)^{1/2m} \frac{d^m P_l(x)}{dx^m}, \qquad \text{(D-22)}$$

we have

$$P_l^m(x) = \frac{(-1)^{l+m}}{2^l l!} \sin^m \vartheta \frac{d^{l+m} \sin^{2l}\vartheta}{d\cos\vartheta^{l+m}}, \qquad \text{(D-17)}$$

which is used to define $P^m(x)$ for negative values of $m$ as well as $m > 0$. Thus (D-16a), (D-16b) become respectively

$$Y_{l,m}(\vartheta, \varphi) = (-1)^{l-m} \sqrt{\frac{2l+1}{4\pi} \frac{(l+m)!}{(l-m)!}} P_l^m(x) e^{im\varphi}, \qquad \text{(D-18a)}$$

$$Y_{l,m}(\vartheta, \varphi) = (-1)^l \sqrt{\frac{2l+1}{4\pi} \frac{(l-m)!}{(l+m)!}} P_l^m(x) e^{im\varphi}. \qquad \text{(D-18b)}$$

The equivalence of these two expressions leads to the relation

$$P_l^{-m}(x) = (-1)^m \frac{(l-m)!}{(l+m)!} P_l^m(x). \qquad \text{(D-19)}$$

### Exercises

1. Show that the Sturm-Liouville equation (III-69) is equivalent to the variational equation

$$\delta \int_a^b y(\Lambda y)\, dx = 0$$

   with the auxiliary condition

$$\int_a^b \rho(x) y^2\, dx = 1.$$

2. Show that the Schrödinger equation (III-68) is the differential equation of the variational problem

$$\delta \langle H \rangle \equiv \delta \int \psi^* H \psi\, d^3 r = 0$$

   with the auxiliary condition

$$N = \int \psi^* \psi\, d^3 r = 1.$$

3. Show that the Schrödinger equation (III-68) is equivalent to the problem

$$\delta E \equiv \delta \left( \frac{\langle H \rangle}{N} \right) = 0.$$

4. The kinetic and potential energy of a two-dimensional harmonic oscillator add up to

$$\frac{\mu}{2}(\dot{r}^2 + r^2 \dot{\varphi}^2) + \frac{1}{2} \mu \omega^2 r^2.$$

   Introduce the following notation

$$\rho = \sqrt{\frac{\mu\omega}{\hbar}}\, r, \qquad p_\rho = \frac{\hbar}{\omega}\dot\rho, \qquad p_\varphi = \frac{\hbar}{\omega}\rho^2\dot\varphi.$$

Show that the Schrödinger equation is

$$\left[\frac{\partial^2}{\partial\rho^2} + \frac{1}{\rho}\frac{\partial}{\partial\rho} + \frac{1}{\rho^2}\frac{\partial^2}{\partial\varphi^2} + \frac{2E}{\hbar\omega} - \rho^2\right]\Psi(\rho,\varphi) = 0.$$

Show that the eigenvalues and eigenfunctions are

$$E_n = (n+1)\hbar\omega,$$

$$\Psi_{nl}(\rho,\varphi) = \sqrt{\frac{k!}{[(l+k)!]^3}}\,\frac{1}{\sqrt{\pi}}\,e^{\pm il\varphi}\exp\left(-\frac{1}{2}\rho^2\right)L_{l+k}^l(\rho),$$

where

$$k = \frac{1}{2}(n-l)$$

$$= \begin{cases} 0, 1, 2, \ldots, \dfrac{n}{2} & \text{for } n = \text{even integer}, \\[2mm] 0, 1, 2, \ldots, \dfrac{n-1}{2} & \text{for } n = \text{odd integer}. \end{cases}$$

$L_{l+k}(x)$ satisfies the equation

$$x\frac{d^2 L_{k+l}^l}{dx^2} + (l+1-x)\frac{dL_{k+l}^l}{dx} + kL_{l+k} = 0.$$

(See Eq. (E-4) of Chap. 3, Sec. 5, Appendix E.) Show that for $n = $ even integer, $l$ can take $(n/2) + 1$ values, and for $n = $ odd integer, $l$ can take $n + 1$ values. Show that

$$\int_0^\infty \Psi_{n,l}^* \rho^2 \Psi_{n,l}\rho\, d\rho = n + 1,$$

$$\int_0^\infty \Psi_{n+2,l}^* \rho^2 \Psi_{n,l}\rho\, d\rho = \frac{1}{2}\sqrt{(n+2)^2 - l^2},$$

$$\int_0^\infty \Psi_{n,l}^* \rho^4 \Psi_{n,l}\rho\, d\rho = \frac{1}{2}(3n^2 + 6n + 4 - l^2).$$

5. The Schrödinger equation of a three-dimensional isotropic harmonic oscillator is

$$\left[-\frac{\hbar^2}{2\mu}\nabla^2 + \frac{1}{2}\mu\omega^2 r^2 - E\right]\psi(r) = 0.$$

Set

$$\psi(\rho) = R(\rho)\, Y_{l,m}(\vartheta, \varphi),$$

$$\rho = \sqrt{\frac{\mu\omega}{\hbar}}\, r,$$

and

$$R(\rho) = \exp\left(-\frac{1}{2}\rho^2\right)\rho^{l+1}F(\rho).$$

Show that $F(\rho)$ satisfies the equation

$$\frac{d^2F}{d\rho^2} + 2\left(\frac{l+1}{\rho} - \rho\right)\frac{dF}{d\rho} + \left[\frac{2E}{\hbar\omega} - 2l - 3\right]F(\rho) = 0.$$

Show that the eigenvalue $E$ is given by

$$E_n = \left(n + \frac{3}{2}\right)\hbar\omega, \qquad n \geq l$$

and find the eigenfunction $R(\rho)$. Show that the state $n$ is $\frac{1}{2}(n + 1)(n + 2)$-fold degenerate.

6. Show that in the ground state ($n = 0$) of an harmonic oscillator, the expectation values of the kinetic and the potential energy are equal:

$$\int_{-\infty}^{\infty} \psi_0^* \frac{p^2}{2\mu} \psi_0\, dx = \int_{-\infty}^{\infty} \psi_0^* \frac{1}{2}\mu\omega x^2 \psi_0\, dx = \frac{1}{4}\hbar\omega.$$

Discuss this from the point of view of the uncertainty principle.

7. With the notation

$$(\psi_m, Q\psi_n) \equiv \int \psi_m^* Q\psi_n\, dq,$$

self-adjointness may be defined by

$$(\psi_m, Q\psi_n) = (Q\psi_m, \psi_n).$$

Let

$$A_z = \cos\vartheta, \qquad A_\pm = \sin\vartheta\, e^{\pm i\varphi}.$$

Show that

$$L_+ A_+ - A_+ L_+ = 0,$$

$$L_- A_- - A_- L_- = 0,$$

$$L_+ A_- - A_- L_+ = 2A_z, \qquad L_- A_+ - A_+ L_- = -2A_z.$$

Show also that

$$A_- Y_{l,m} = -\sqrt{\frac{(l-m+1)(l-m+2)}{(2l+1)(2l+3)}}\, Y_{l+1,m-1}$$

$$+ \sqrt{\frac{(l+m)(l+m-1)}{(2l-1)(2l+1)}}\, Y_{l-1,m-1},$$

$$A_+ Y_{l,m} = \sqrt{\frac{(l+m+1)(l+m+2)}{(2l+1)(2l+3)}}\, Y_{l+1,m+1}$$

$$- \sqrt{\frac{(l-m-1)(l-m)}{(2l-1)(2l+1)}}\, Y_{l-1,m+1},$$

$$A_z Y_{l,m} = -\sqrt{\frac{(l-m+1)(l+m+1)}{(2l+1)(2l+3)}}\, Y_{l+1,m}$$

$$- \sqrt{\frac{(l-m)(l+m)}{(2l-1)(2l+1)}}\, Y_{l-1,m}.$$

These are in agreement with the results in (III-122).

## 5.  The hydrogen atom

The Schrödinger equation is (III-100), with the Coulomb potential $-Ze^2/r$ for $V(r)$. The angular part of the wave function has been given in (III-113). The radial wave equation (III-103) is now

$$\frac{d^2R}{dr^2} + \frac{2}{r}\frac{dR}{dr} + \left[\frac{2\mu}{\hbar^2}\left(E + \frac{Ze^2}{r}\right) - \frac{l(l+1)}{r^2}\right] R(r) = 0.$$

$$\tag{III-127}$$

$$\mu = \frac{mM}{M+m}.$$

$\mu$ is the reduced mass of the electron and the nucleus. It is convenient, and customary, to introduce the atomic units of Hartree, namely, $a$ for length and $e^2/a$ for energy, where

$$a = \frac{\hbar^2}{me^2} = 5.29 \times 10^{-9} \text{ cm},$$

$$\tag{III-128}$$

$$\frac{e^2}{a} = \frac{me^4}{\hbar^2} = 4.3 \times 10^{-11} \text{ erg} = 27.2 \text{ eV},$$

$a$ being the radius of the first ($n = 1$) Bohr circular orbit, and $e^2/a$ is twice the ionization energy of the hydrogen atom. In the following $r$ and $E$ are dimensionless length and energy in these units. Equation (III-127) becomes

$$\frac{d^2R}{dr^2} + \frac{2}{r}\frac{dR}{dr} + \left[2E + \frac{2Z}{r} - \frac{l(l+1)}{r^2}\right]R(r) = 0. \qquad \text{(III-129)}$$

*(1) Discrete states, $E < 0$*

Let

$$\varepsilon = +\sqrt{-2E}, \qquad \text{real and positive.}$$

For large $r$, the asymptotic solution of (III-129) is

$$R(r) \simeq e^{-\varepsilon r}.$$

Assume the form

$$R(r) = e^{-\varepsilon r}u(r). \qquad \text{(III-130)}$$

The equation for $u(r)$ is then

$$\frac{d^2u}{dr^2} + 2\left(\frac{1}{2} - \varepsilon\right)\frac{du}{dr} + \left[\frac{2(Z - \varepsilon)}{r} - \frac{l(l+1)}{r^2}\right]u = 0. \qquad \text{(III-131)}$$

The point $r = 0$ is a pole. Let

$$u(r) = r^\alpha \sum_{k=0}^{\infty} b_k r^k \equiv r^\alpha W(r). \qquad \text{(III-132)}$$

Substituting this into (III-131), equating to zero the coefficient of the lowest power of $r$, one gets the indicial equation

$$\alpha(\alpha + 1) - l(l+1) = 0,$$

whose roots are

$$\alpha = l, \qquad \text{and} \qquad -(l + 1). \qquad \text{(III-133)}$$

Taking $\alpha = l$, one gets from (III-131) the recurrence relation

$$b_{k+1} = 2\frac{(k + l + 1)\varepsilon - Z}{(k + 1)(k + 2l + 2)}b_k. \qquad \text{(III-134)}$$

If the series $W(r)$ in (III-132) is an infinite series, the ratio of two successive terms is, for large $k$,

$$\lim_{k \to \infty} \frac{b_{k+1}}{b_k}r = \frac{2\varepsilon}{k}r,$$

so that the series has the asymptotic form of

$$W(r) \simeq e^{2\varepsilon r}, \qquad \text{(III-134a)}$$

leading to the asymptotic form $r^l e^{+\varepsilon r}$ for $R(r)$ which is non-quadratically integrable. To avoid this, one may require the series to terminate at a highest power $k = n_r$, i.e., $b_{n_r+1} = b_{n_r+2} = \cdots = 0$, so that $W(r)$ becomes a polynomial. The condition $b_{n+1} = 0$ is, from (III-134),

$$\varepsilon = \frac{Z}{n_r + l + 1}, \qquad \text{or} \qquad E = -\frac{Z^2}{2n^2}\left(\frac{\mu}{m}\right) \qquad \text{(III-135a)}$$

or, in cgs units,

$$E = -\frac{Z^2 \mu e^4}{2\hbar^2 n^2}, \qquad\qquad\qquad \text{(III-135)}$$

where

$$n = n_r + l + 1. \qquad\qquad\qquad \text{(III-136)}$$

(III-135) is the Bohr formula.[1]

---

[1] In the general theory of differential equations, the two roots $x = l$ and $-(l + 1)$, differing by an integer, give one and the same solution $W(r)$. This can be seen by choosing $\alpha = -(l + 1)$ in (III-133), thereby leading to

$$b_{k+1} = 2\frac{(k - l)\varepsilon - Z}{(k + 1)(k - 2l)}b_k. \qquad \text{(III-137)}$$

If $b_0 = 0$, then $b_1 = b_2 = \cdots = b_{2l} = 0$, and

$$b_{2l+1} = 2\frac{l\varepsilon - Z}{(2l + 1)0}\cdot 0 = \text{finite}$$

and

$$b_{2l+1+s} = 2\frac{(l + s)\varepsilon - Z}{(s + 2l + 1)s}b_{2l+s},$$

which is the same as (III-134), with $k + 1$ replaced by $s$, and leads to the same solution $W(r)$ as $\alpha = l$. If $b \neq 0$, then, from (III-137), one finds $b_{2l+1}, b_{2l+2}, \ldots$ all are infinite and $W(r)$ in (III-132) is not a solution at all.

Thus the statement, in most books in the last fifty years, that the root $\alpha = -(l + 1)$ leads to divergent wave function at $r = 0$ and is to be discarded, is not correct.

The other independent solution of (III-131), $v(r)$ must satisfy the Wronskian relation

$$\left(\frac{du}{dr}v - \frac{dv}{dr}u\right)r^2 = Ce^{2\varepsilon r}. \qquad \text{(III-138)}$$

If $u(r)$ is the polynomial defined by (III-134), then the asymptotic behavior of $rv(r)$ (or $r(dv/dr)$) must be that of $e^{2\varepsilon r}$, and in (III-130),

$$rR(r) = re^{-\varepsilon r}v(r)$$

will be like $e^{\varepsilon r}$, and $R(r)$ is then non-quadratically integrable.

It can in fact be shown that the second, independent solution $v(r)$ has the form

$$v(r) = r^{-(l+1)}f(r) + u(r)\ln r,$$

where $f(r)$ is such that $f(0) \neq 0$ and is finite.

For a discussion of the problem of this footnote, see Whittaker and Watson's *A Course on Modern Analysis* (Cambridge Univ. Press, Cambridge, 1927) p. 200. See this reference for the "Whittaker function" solutions of the Eq. (III-131).

In Schrödinger's first communication, the eigenvalue problem is solved by the method of Laplace transformation in which the above problem of this footnote does not arise.

### *Wave functions*

To obtain the eigenfunctions of (III-129), we transform it first by performing a change of variable from $r$ to a new dimensionless one:

$$\rho = \frac{2Z}{n} r, \tag{III-139a}$$

$$\frac{d^2 R}{d\rho^2} + \frac{2}{\rho}\frac{dR}{d\rho} + \left[ -\frac{1}{4} + \frac{n}{\rho} - \frac{l(l+1)}{\rho^2} \right] R(\rho) = 0. \tag{III-139}$$

The solution is, from (III-130), (III-132), (III-134) and (III-135a),

$$R(\rho) = e^{-1/2\rho}\rho^l w(\rho), \tag{III-140}$$

where $w(\rho)$ is a polynomial of degree $n_r$, or from (III-135a),

$$n_r = n - l - 1.$$

Thus $R_{n,l}(\rho)$ has $n_r = n - l - 1$ zeros or nodes (between $r = 0$ and $r = \infty$, not including the origin $r = 0$). The polynomial $w(\rho)$ is given by the recurrence relation (III-134) with $\varepsilon = Z/(n_r + l + 1) = Z/n$, but $w(\rho)$ can be shown to be a known function—the associated Laguerre polynomial. The properties of this function will be given in Appendix E at the end of this section.

The normalized $R_{n,l}(\rho)$ is

$$R_{n,l}(\rho) = \left[ \frac{(n-l-1)!}{2n[(n+l)!]^3}\left(\frac{2Z}{na}\right)^3 \right]^{1/2} e^{-1/2\rho}\rho^l L_{n+l}^{2l+1}(\rho), \tag{III-141}$$

which is normalized according to

$$\int_0^\infty [R_{n,l}(\rho)]^2 r^2\, dr = 1, \qquad \rho = \frac{2Z}{na} r.$$

A few explicit $L_k^j$ are:

$$L_0^0(\rho) = 1, \qquad L_1^0(\rho) = -\rho + 1, \qquad L_1^1(\rho) = -1,$$

$$L_2^0(\rho) = \rho^2 - 4\rho + 2, \qquad L_2^1(\rho) = 2\rho - 4, \qquad L_2^2(\rho) = 2,$$

$$L_3^0(\rho) = -\rho^3 + 9\rho^2 - 18\rho + 6,$$

$$L_3^1(\rho) = -3\rho^2 + 18\rho - 18, \qquad L_3^2(\rho) = -6\rho + 18, \qquad L_3^3(\rho) = -6. \tag{III-142}$$

### *(2) Continuum states $E > 0$*

For $E > 0$ (corresponding to hyperbolic orbits in classical dynamics), the states are in the continuum and the wave functions are said to be continuous wave functions.

Let

$$\varepsilon = i\sqrt{2E} = ik, \qquad k, \text{ real and positive} \tag{III-143}$$

and set

$$R(r) = e^{ikr}u(r). \tag{III-144}$$

Equations (III-131)–(III-134a) remain valid (with $\varepsilon = ik$), and the series (III-132) does not have to terminate to become a polynomial. There is then no such a condition as (III-135a) for discrete state and all positive energies $E$ are allowed.

Now, instead of (III-135a), and (III-139a),

$$\varepsilon = \frac{Z}{n}, \qquad \rho = \frac{2Z}{n}r,$$

we put[u]

$$n = \frac{Z}{\varepsilon} = -i\frac{Z}{k}, \qquad \rho = \varepsilon r = 2ikr. \tag{III-145}$$

The wave function $R_{nl}(\rho)$ in (III-144) is found to be

$$R_{nl}(\rho) = \frac{c}{(2l+1)!}(-i\rho)^l e^{-1/2\rho}F(l+1-n, 2l+2, \rho), \tag{III-146}$$

where the confluent hypergeometric function $F(a, b, x)$ is given by

$$F(a, b, z) = 1 + \frac{a}{b\cdot 1}z + \frac{a(a+1)}{b(b+1)2!}z^2 + \frac{a(a+1)(a+2)}{b(b+1)(b+2)3!}z^3 + \cdots \tag{III-147}$$

In many problems, the asymptotic form for $R_{nl}(\rho)$ for large $\rho$ is useful; it is

$$R_{nl}(r) = \frac{c\exp\left(-\dfrac{\pi Z}{2k}\right)}{\Gamma\left(l+1-i\dfrac{Z}{k}\right)}\frac{1}{kr}\cos\left[kr + \frac{Z}{k}\ln 2kr - \frac{\pi}{2}(l-1) - \sigma_l\right], \tag{III-148}$$

where

$$\sigma_l = \arg\Gamma\left(l+1+i\frac{Z}{k}\right).$$

(III-148) is, asymptotically, a spherical wave, whose wavelength is not a constant, but changes with the distance $r$ from the center through the term

---

[u] From (III-143), $2E = ((\hbar k)^2/\mu)(\hbar^2/\mu e^4) = (p^2/(\mu c)^2)(1/\alpha^2) = (v/c\alpha)^2$, $\alpha = (e^2/\hbar c)$. $Z/k = Z\alpha(c/v)$. For the first Bohr orbit, $v = \alpha c$.

$(Z/k)\ln 2kr$ in the phase.[v] This is the result of the long range nature of the Coulomb potential $V(r)$.

### (3) Normalization of continuum wave functions

The wave function (III-146), being oscillatory in $\rho$, as seen from (III-145) and (III-148), is not quadratically integrable—a result corresponding to the situation that the probability of the electron being found between $r$ and $r + dr$ is not confined to small $r$. In this case, the normalization follows the method of Weyl.[w]

The Sturm-Liouville equation (III-69, 70) is

$$[\Lambda + \lambda\rho]y = \left[\frac{d}{dx}\left(p(x)\frac{d}{dx}\right) - q(x) + \lambda\rho(x)\right]y = 0. \qquad \text{(III-149)}$$

For the continuum, $\lambda$ is continuous.

The eigendifferential $\Delta_\lambda\psi(x)$ of a continuum function $\psi_\lambda$ is defined by

$$\Delta_\lambda\psi = \int_\lambda^{\lambda+\varepsilon} \psi_{\lambda'}(x)\,d\lambda'. \qquad \text{(III-150)}$$

The normalization of $\psi_\lambda$ is defined by

$$\lim_{\varepsilon\to 0}\left[\frac{1}{\varepsilon}\int_0^\infty |\Delta_\lambda\psi|^2\rho(x)\,dx\right] = 1, \qquad \text{(III-151)}$$

or

$$\lim_{\varepsilon\to 0}\left[\frac{1}{\varepsilon}\int_0^\infty dx\rho(x)\left|\int_\lambda^{\lambda+\varepsilon} \psi(x,\lambda')\,d\lambda'\right|^2\right] = 1. \qquad \text{(III-151a)}$$

Let $f(x)$ be an arbitrary, twice differentiable function for which the following integrals converge

$$\int_0^\infty |f(x)|^2\rho\,dx, \qquad \int_0^\infty \frac{1}{\rho}|\Lambda f|^2\,dx. \qquad \text{(III-152)}$$

Let us assume that the complete set of eigenfunctions of (III-149) consists of discrete functions $\psi_n(x)$ and continuum functions $\psi_\lambda(x)$. The expansion of $f(x)$ in the complete set is then

$$f(x) = \sum_n C_n\psi_n(x) + \int d\lambda C(\lambda)\psi_\lambda(x), \qquad \text{(III-153)}$$

---

[v] Compare this with the same effect in the scattering of a particle by a Coulomb field in parabolic coordinates.

[w] H. Weyl, *Math. Annalen* **68**, 220 (1910); cf. E. C. Kemble, ref. on p. 152.

where

$$C_n = \int_0^\infty \psi_n^* f(x)\rho \, dx \equiv (\psi_n, \rho f), \qquad \text{(III-154a)}$$

$$
\begin{aligned}
C(\lambda) &= \lim_{\varepsilon \to 0} \frac{1}{\varepsilon} \int_0^\infty dx \rho f(x) \left[ \int_\lambda^{\lambda+\varepsilon} \psi(x, \lambda') \, d\lambda' \right] \\
&= \lim_{\varepsilon \to 0} \frac{1}{\varepsilon} \left( \int_\lambda^{\lambda+\varepsilon} \psi(x, \lambda') \, d\lambda', \rho f \right).
\end{aligned}
\qquad \text{(III-154b)}
$$

If $f(x)$ is absolutely integrable (i.e., the integral $\int_0^\infty |f(x)|\, dx$ exists) and has an upper bound, $C(\lambda)$ can be replaced by

$$C(\lambda) = (\psi(x, \lambda), \rho f).$$

Similarly, for a function $g(x)$,

$$g(x) = \sum_n B_n \psi_n(x) + \int d\lambda B(\lambda) \psi_\lambda(x),$$

$$B_n^* = (\psi_n, \rho g)^*, \qquad B^*(\lambda) = \lim_{\varepsilon \to 0} \frac{1}{\varepsilon} \left( \int_\lambda^{\lambda+\varepsilon} \psi(x, \lambda') \, d\lambda', \rho g \right)^*.$$

Then

$$(f, \rho f) = \sum_n |C_n|^2 + \int d\lambda |C(\lambda)|^2,$$

$$(g, \rho f) = \sum_n C_n B_n^* + \int d\lambda C(\lambda) B^*(\lambda).$$

As an example to illustrate the above results, let us take the plane wave

$$\psi(x, k) = A e^{ikx}.$$

Then the eigendifferential is

$$\Delta_k \psi = \frac{A}{ix}(e^{i(k+\varepsilon)x} - e^{ikx})$$

and the normalization condition (III-151a) is

$$\frac{2}{\varepsilon} A^2 \int_0^\infty dx \frac{1 - \cos \varepsilon x}{x^2} = 1.$$

This gives

$$A^2 \pi = 1.$$

**Appendix E**
**Associated Laguerre polynomials**

## (1) Associated Laguerre polynomials

On substituting (III-140)

$$R_{n,l}(\rho) = e^{-1/2\rho}\rho^l W(\rho) \tag{E-1}$$

into (III-139), one obtains the equation

$$\rho\frac{d^2W}{d\rho^2} + [2(l+1) - \rho]\frac{dW}{d\rho} + (n-l-1)W = 0. \tag{E-2}$$

Consider the Laguerre polynomial $L_k(\rho)$ that satisfies the equation

$$\rho\frac{d^2L_k}{d\rho^2} + (1-\rho)\frac{dL_k}{d\rho} + kL_k = 0. \tag{E-3}$$

On differentiating this equation $j$ times, one gets

$$\rho\frac{d^2}{d\rho^2}\left(\frac{d^jL_k}{d\rho^j}\right) + (j+1-\rho)\frac{d}{d\rho}\left(\frac{d^jL_k}{d\rho^j}\right) + (k-j)\left(\frac{d^jL_k}{d\rho^j}\right) = 0. \tag{E-4}$$

Comparison of (III-143) and (III-141) shows that

$$W(\rho) = \frac{d^{2l+1}L_{n+l}(\rho)}{d\rho^{2l+1}} \tag{E-5a}$$

$$\equiv L_{n+l}^{2l+1}(\rho). \tag{E-5b}$$

$L_k^j(\rho)$ is called the *associated Laguerre polynomial.*

$L_k^j(\rho)$ can be obtained from the generating function

$$\phi(\rho,t) = (-1)^j\frac{\exp\left(-\dfrac{\rho t}{1-\rho}\right)}{(1-t)^{j+1}}t^j \tag{E-6}$$

$$= \sum_{k=j}^{\infty}\frac{1}{k!}L_k^j(\rho)t^k, \tag{E-7}$$

which gives $L_k^j$ as a polynomial[x] of $\rho$ of degree $k-j$,

---

[x] In the mathematical literature, the associated Laguerre polynomial $L_k^j$ (written as $\mathscr{L}_k^j(\rho)$ to avoid confusion with the $L_k^j$ defined by (E-7, E-8) above) is defined differently,

$$\mathscr{L}_k^j(\rho) = \sum_{v=0}^{k}(-1)^v\frac{(k+j)!}{(k-v)!(j+v)!v!}\rho^v \tag{E-9}$$

and is related to $L_k^j(\rho)$ as

$$\mathscr{L}_k^j(\rho) = \frac{(-1)^j}{(k+j)!}L_{k+j}^j(\rho). \tag{E-10}$$

It satisfied the recurrence relation

$$\rho\mathscr{L}_k^j = -(k+1)\mathscr{L}_{k+1}^j + (2k+j+1)\mathscr{L}_k^j - (k+j)\mathscr{L}_{k-1}^j. \tag{E-11}$$

$$L_k^j(\rho) = \sum_{v=0}^{k-j} (-1)^{v+j} \frac{(k!)^2}{(k-j-v)!(j+v)!\,v!} \rho^v. \tag{E-8}$$

$L_k^j(\rho)$ satisfies the recurrence relations[y]

$$L_{k-1}^{j-1} = \frac{1}{k} L_k^j - L_{k-1}^j, \tag{E-12a}$$

$$\rho L_k^j = -\frac{k-j+1}{k+1} L_{k+1}^j + (2k-j+1)L_k^j - k^2 L_{k-1}^j. \tag{E-12}$$

From (E-6) and (E-7), one obtains

$$\int_0^\infty e^{-\rho} \rho^{B+1} \phi(\rho,t)\phi(\rho,s)\,d\rho = \frac{(ts)^B}{(1-t)^{B+1}(1-s)^{B+1}} \int_0^\infty \rho^{B+1}$$

$$\times \exp\left[ -\left(1 + \frac{t}{1-t} + \frac{s}{1-s}\right)\rho\right] d\rho. \tag{E-13}$$

$$= \sum_{n,m=B} \frac{1}{n!\,m!} t^n s^m \int_0^\infty e^{-\rho} \rho^{B+1} L_n^B(\rho) L_m^B(\rho)\,d\rho. \tag{E-14}$$

$$(\text{E-13}) = \frac{(ts)^B (B+1)!(1-t)(1-s)}{(1-ts)^{B+2}} = (1 - t - s + ts) \sum_{k=0}^\infty \frac{(B+k+1)!}{k!} (ts)^{B+k}.$$

On comparing coefficients of equal powers of $t^n s^m$, one obtains, for normalization purpose,

$$\int_0^\infty e^{-\rho} \rho^{B+1} L_n^B(\rho) L_n^B(\rho)\,d\rho = \frac{(n!)^3 (2n - B - 1)}{(n-B)!}. \tag{E-15}$$

For hydrogen-like atoms, $\rho = (2Z/na)r$, $a = \hbar^2/\mu e^2$, $r$ in cgs units,

$$\int_0^\infty e^{-\rho} \rho^{2l+2} L_{n+l}^{2l+1}(\rho) L_{n+l}^{2l+1}(\rho)\,d\rho = \frac{2n[(n+l)!]^3}{(n-l-1)}. \tag{E-15a}$$

The associated Laguerre polynomials are related to the confluent hypergeometric functions

$$F(a,b,x) = 1 + \frac{a}{b \cdot 1} x + \frac{a(a+1)}{b(b+1)2!} x^2 + \cdots \tag{E-16}$$

$$L_{n+l}^{2l+1}(\rho) = \frac{[(n+l)!]^2}{(n-l-1)!(2l+1)!} F(l+1-n, 2l+2, \rho) \tag{E-17}$$

---

[y] For (E-12a), see Gradshteyn & Ryzhik, *Tables of Integrals, Series and Products* (Addison & Wesley, Reading, MA., 1970) §8, 971, 5. See the change of notations (D-10) above.

so that the normalized $R_{n,l}(\rho)$ is (see (III-141))

$$R_{n,l}(r) = \left[ \frac{(n+l)!}{2n(n-l-1)!\,[(2l+1)!]^2} \left( \frac{2Z}{na} \right)^3 \right]^{1/2} e^{-Zr/na}$$

$$\times \left( \frac{2Z}{na}r \right) F\left( l+1-n, 2l+2, \frac{2Z}{na}r \right), \qquad \text{(E-18)}$$

$$\int_0^\infty [R_{nl}(r)]^2 r^2\, dr = 1.$$

### (2) Integrals containing $L_m^c(a\rho)L_n^c(b\rho)$

The appearance of $(2Z/na)r$ as the argument in $L_{n+l}^{2l+1}(\rho)$ renders very difficult the evaluation of integrals involving two electrons of different quantum number $n$.

Consider the integral

$$\int_0^\infty e^{-\lambda\rho}\rho^{c+d}L_m^c(a\rho)L_n^{c'}(b\rho)\,d\rho,$$

where $c \neq c'$, $m \neq n$, $a \neq b$, and $\lambda$ may be complex. For $c \neq c'$, it is possible to raise the lower value, say $c'$, by repeated use of the relation (E-12a) until the two Laguerre polynomials have the same $c$. Thus we shall have to consider integrals of the form

$$J_d = \int_0^\infty e^{-\lambda\rho}\rho^{c+d}L_m^c(a\rho)\,L_n^c(b\rho)\,d\rho. \qquad \text{(E-19)}$$

Let

$$\sigma = \frac{\lambda - a}{\lambda}, \qquad \tau = \frac{\lambda - b}{\lambda}, \qquad \mu = \frac{a + b - \lambda}{\lambda}, \qquad \xi = \frac{\sigma\tau}{\mu}. \qquad \text{(E-20)}$$

After rather long calculations, one obtains the following results.[z]

$$J_0 = \frac{m!\,n!}{\lambda^{c+1}}\mu^{n-c}\sigma^{m-n}\sum_{s=0}^{n-c} \frac{(m+s)!}{(n-c-s)!\,(m-n+s)!\,s!}\xi^s, \qquad \text{(E-21)}$$

$$J_1 = \frac{m!\,n!}{\lambda^{c+2}}\mu^{n-c}\sigma^{m-n}\left\{ \sum_{s=0}^{n-c} \frac{(m+1+s)!}{(n-c-s)!\,(m-n+s)!\,s!}\xi^s \right.$$

---

[z] These formulas are given by T. Y. Wu, *Annual Reports, Acad. Sinica*, Taipei, (1972). Much more complicated expressions have been obtained by G. Elwert, *Z. Naturforschung* **10A**, 361 (1955); H. Kallmann & M. Päsler, *Z. Phys.* **128**, 178 (1950), using the method of Laplace transformation.

$$+ \frac{1}{\mu} \sum_{s=0}^{n-c-1} \frac{(m+s)!}{(n-c-1-s)!(m-n+s)!s!} \xi^s$$

$$- \frac{1}{\sigma} \sum_{s=0}^{n-c} \frac{(m+s)!}{(n-c-s)!(m-n-1+s)!s!} \xi^s$$

$$- \frac{\sigma}{\mu} \sum_{s=0}^{n-c-1} \frac{(m+1+s)!}{(n-c-1-s)!(m-n+1+s)!s!} \xi^s \Big\},$$

(E-22)

$$J_2 = \frac{m!\,n!}{\lambda^{c+3}} \mu^{n-c} \sigma^{m-n} \Big\{ \sum_{s=0}^{n-c} \frac{(m+2+s)!}{(n-c-s)!(m-n+s)!s!} \xi^s$$

$$+ \frac{4}{\mu} \sum_{s=0}^{n-c+1} \frac{(m+1+s)!}{(n-c-1-s)!(m-n+s)!s!} \xi^s$$

$$+ \frac{1}{\mu^2} \sum_{s=0}^{n-c-1} \frac{(m+s)!}{(n-c-2-s)!(m-n+s)!s!} \xi^s$$

$$- \frac{2}{\sigma} \sum_{s=0}^{n-c} \frac{(m+1+s)!}{(n-c-s)!(m-n-1+s)!s!} \xi^s$$

$$- \frac{2\sigma}{\mu} \sum_{s=0}^{n-c-1} \frac{(m+2+s)}{(n-c-1-s)!(m-n+1+s)!s!} \xi^s \Big\}.$$

(E-23)

In the above expressions, $m \geq n$. If $m < n$, one has simply to interchange $m$ and $n$, and $a$, $b$ at the same time (thereby interchanging $\sigma$ and $\tau$ also).

For discrete state problems, usually

$$\lambda = \frac{a+b}{2}$$

so that

$$\sigma = \frac{b-a}{b+a} = -\tau, \qquad \mu = 1, \qquad \xi = -\left(\frac{b-a}{b+a}\right)^2. \tag{E-24}$$

For scattering problems, $\lambda$ is complex.

### (3) Matrix elements for dipole moments

Matrix elements such as

$$\langle n, l | r | n', l-1 \rangle \equiv \int_0^\infty R_{n,l}\left(\frac{2Z}{na}r\right) r R_{n',l-1}\left(\frac{2Z}{n'a}r\right) r^2\, dr \tag{E-25}$$

are difficult, but have been found.* The result is

---

* W. Gordon, *Annalen d. Physik* **2**, 1031 (1929); cf. H. A. Bethe and E. E. Salpeter, *Quantum Mechanics of One- and Two-Electron Atoms* (Springer-Verlag, Heidelberg, 1955).

$$(\text{E-25}) = a\frac{(-1)^{n'-l}}{4(2l-1)!}\left[\frac{(n+l)!(n'+l-1)!}{(n-l-1)!(n'-l)!}\right]^{1/2}\frac{(4nn')^{l+1}(n-n')^{n+n'-2l-2}}{(n+n')^{n+n'}}$$

$$\times\left\{F\left(-n_r,-n'_r;2l;-\frac{4nn'}{(n-n')^2}\right)-\left(\frac{n-n'}{n+n'}\right)^2\right.\tag{E-26}$$

$$\left.\times F\left(-n_r-2,-n'_r;2l;-\frac{4nn'}{(n-n')^2}\right)\right\},$$

where

$$n_r = n - l - 1, \qquad n'_r = n' - l, \qquad a = \frac{\hbar^2}{\mu e^2},$$

$$\tag{E-27}$$

$$F(\alpha,\beta;\gamma;x) = 1 + \frac{\alpha\cdot\beta}{\gamma\cdot1}x + \frac{\alpha(\alpha+1)\beta(\beta+1)}{\gamma(\gamma+1)2!}x^2 + \cdots$$

For $n' = n$, the formula (E-26) is not valid, and has to be calculated separately.

$$\langle n,l|r|n,l-1\rangle = \frac{3}{2Z}an\sqrt{n^2-l^2}.\tag{E-28}$$

Integrals of the type

$$\left\langle n,l\left|\left(\frac{r}{a}\right)^k\right|n,l\right\rangle = \int_0^\infty\left(\frac{r}{a}\right)^k\left[R_{n,l}\left(\frac{2Z}{na}r\right)\right]^2 r^2\,dr\tag{E-29}$$

have been calculated, for $k$ from $-6$, $-5$, to 4.[†]

## Exercises

1. Find the eigenvalues and eigenfunctions of a two-dimensional hydrogen-like atom.

2. Show that the normalization procedure (III-151) for continuum wave function

$$\lim_{\varepsilon\to0}\left[\frac{1}{2\varepsilon}\int_0^\infty dx\rho(x)\left|\int_{\lambda-\varepsilon}^{\lambda+\varepsilon}\psi(x,\lambda')\,d\lambda'\right|^2\right] = 1$$

is equivalent to the following procedure

$$\lim_{\varepsilon\to0}\frac{1}{2\varepsilon}\int_0^\infty dx\rho(x)\psi(x,\lambda)\int_{\lambda-\varepsilon}^{\lambda+\varepsilon}\psi(x,\lambda')\,d\lambda' = 1.$$

Illustrate this with $\psi(x,\lambda) = \cos(\lambda x)$.

[†] Cf. H. A. Bethe and E. E. Salpeter, *loc. cit*; J. H. Van Vleck, *Proc. Roy. Soc.* London **A143** 679 (1934).

3. Discuss the quantum virial theorem

$$2\overline{E_{\text{kin}}} = \overline{\mathbf{r} \cdot \nabla V}$$

for (a) the harmonic oscillator, and (b) the Kepler problem.

4. Show that for hydrogen atom,

$$\langle n, l = n - 1 | r | n, l = n - 1 \rangle$$

is equal to the radius of the $n, l = n - 1$ orbit in the **Bohr** theory.

Show that for $n, l = n - 1$, the mean-square-deviation

$$(\Delta r)^2 = \overline{(r - \bar{r})^2} = \overline{r^2} - \bar{r}^2$$

is the smallest of all $l$ states for the same $n$.

5. Calculate the matrix element

$$\left\langle n, l \left| \left(\frac{r}{a}\right)^k \right| n, l \right\rangle = \int_0^\infty \left(\frac{r}{a}\right)^k \left[ R_{nl}\left(\frac{2Z}{na}r\right) \right]^2 r^2 \, dr$$

by the following method:

(a) $k = -1$, use the virial theorem;

(b) $k = -2$, write (III-129) in the form

$$(H - E)\Psi_{n,l} = 0, \qquad \Psi_{n,l}(r) = rR_{n,l}(r),$$

$$H = -\frac{d^2}{dr^2} - \frac{2Z}{r} + \frac{l(l+1)}{r^2},$$

$$E = -\frac{Z^2}{(n_r + l + 1)^2},$$

and regard $H$, $E$, $\Psi_{n,l}$ as continuous functions of $l$, and differentiate the Schrödinger equation;

(c) $k = -3$, differentiate the above Schrödinger equation with respect to $r$. Answer:

$$\left\langle n, l \left| \left(\frac{r}{a}\right)^{-1} \right| n, l \right\rangle = \frac{Z}{n^2},$$

$$\left\langle n, l \left| \left(\frac{r}{a}\right)^{-2} \right| n, l \right\rangle = \frac{Z^2}{n^3(l + \frac{1}{2})},$$

$$\left\langle n, l \left| \left(\frac{r}{a}\right)^{-3} \right| n, l \right\rangle = \frac{Z^2}{n^3(l + 1)(l + \frac{1}{2})l}.$$

# Chapter 4

# Probability Postulate and Uncertainty Principle

In the two preceding chapters, we have presented the mathematical aspects of matrix and wave mechanics, but have not taken up the physical interpretation of the wave $\psi$—the de Broglie wave for which Schrödinger set out to find an equation. The original interpretation of Schrödinger, that $e\psi^*\psi$ represents the charge density of an extended electron, was soon discarded since for a many-particle system, the $\psi$ is in configuration space, and a charge spread over configuration space has no clear physical meaning.

In June 1926, Born *proposed* a probability interpretation for $\psi^*\psi$—a probability whose nature is basically different from that in classical physics. It found acceptance from most quantum physicists.[a] This probability postulate has since become one of the basic postulates of quantum mechanics.

By mid 1926, there was already matrix mechanics, wave mechanics, and the probability interpretation.[b] Dirac developed the transformation theory in 1926 (December 1926, see Chap. 2, Sec. 3) which, one may say, is an extremely important step in the formulation of Quantum Mechanics as a complete system. In the transformation theory, the general equation (for a unitary transformation $U$)

$$\sum_{q''} F(q, q'') U(q'', a) = \sum_{q''} U(q, a'') F(a'', a) \tag{IV-1}$$

states the transformation of a matrix $F$ between a $q$- and an $a$-representation. If $F$ is the Hamiltonian, the $a$-representation is the $H$-representation itself, and $q$ the coordinate representation; then (IV-1) becomes

$$H\left(q, \frac{\hbar}{i}\frac{\partial}{\partial q}\right) U(q, E) = EU(q, E), \tag{IV-2}$$

which is the Schrödinger equation.[c] If in (IV-1) $F = Q = a$ function of $q$, then

---

[a] Ironically, with the exception of some who contributed so much to the founding of the quantum theory, namely, Einstein, Schrödinger, Planck, von Laue.

[b] Matrix mechanics was "completed" by the Born-Heisenberg-Jordan paper in November, 1925 (see Chap. 2). Wave mechanics, together with Schrödinger's observation of its equivalence with matrix mechanics, was "completed" in June, 1926 (see Chap. 3). Born's probability interpretation was proposed in June, 1926 (see the present chapter, Sec. 1 below).

[c] For (IV-1), see Chap. 2, (II-67). For (IV-2), see (A-6) at the end of Sec. 3, Ex. 2, Chap. 2. For (IV-3), see (II-68a).

$Q$ is diagonal in the $q$-representation with elements $Q', Q'', \ldots$ and (IV-1) leads to

$$Q(a', a') = \int U^*(q', a')Q'U(q', a')\,dq'. \qquad \text{(IV-3)}$$

With the probability interpretation of Born for

$$U^*(q, a)U(q, a) = \psi_a^*(q)\psi_a(q), \qquad \text{(IV-4)}$$

(IV-3) states that the diagonal elements of a physical quantity $Q$ in an $a$-representation are related to the diagonal elements of $Q$ in the $q$-, and therefore its own, representation, through the probability $|U(q', a)|^2\,dq' = |\psi_a(q')|^2\,dq'$ that, when $a$ is prescribed, $Q$ has eigenvalues between $Q'$ and $Q' + dQ'$. This suggests the "postulate" (in place of the postulate (IV-4)) that the "expectation value" of $Q$, when the Hamiltonian $H$ is known to have the eigenvalue $E$ (i.e., when the system is known to be in an eigenstate of $H$ of energy $E$), is

$$Q(E, E) = \int U^*(q, E)Q(q)U(q, E)\,dq, \qquad \text{(IV-5a)}$$

or

$$\langle E|Q|E\rangle = \int \psi_E^*(q)Q(q)\psi_E(q)\,dq. \qquad \text{(IV-5b)}$$

With this postulate of "expectation value", since two non-commuting matrices (or, operators) cannot be simultaneously diagonal in any representation, it follows that in the $E$-representation, the expectation value of $Q$ is *not* an eigenvalue of $Q$.

In his December 1926 paper, Dirac thus drew the conclusion that it is not possible to give (exact) numerical values for two conjugate variables (such as $q$ and $p$). This is in fact a qualitative statement of what Heisenberg formulated a few months later (March 1927) as the uncertainty principle.

Heisenberg analyzed, in a quantitative manner, the nature of the kinematical and mechanical concepts, such as the space coordinate $x$ and its conjugate momentum $p_x$. The uncertainty principle, not understandable on classical physics, has become a basic principle in quantum mechanics; it is an expression, or a consequence, of the Einstein-de Broglie relations, or/and of the commutation relation $pq - qp = (\hbar/i)1$, which are of course also "not understandable" on classical physics.

We shall treat the probability postulate and the uncertainty principle in historical order. To facilitate our presentation, we employ the slightly more modern technique of treating the scattering problem—integral Schrödinger equation—by which Born came to the idea of probability.

## 1. Schrödinger equation in integral form

### (1) Particle in a central field $V(r)$

The Schrödinger equation

$$\left[ -\frac{\hbar^2}{2m}\nabla^2 + V(r) \right]\psi(\mathbf{r}) = E\psi(\mathbf{r}) \tag{IV-6}$$

can be written in the form

$$(\nabla^2 + k^2)\psi = X(\mathbf{r}), \tag{IV-7}$$

with

$$E = \frac{\hbar^2 k^2}{2m}, \qquad X(r) = \frac{2m}{\hbar^2} V(r)\psi(\mathbf{r}). \tag{IV-8}$$

If we regard (IV-7) as an inhomogeneous equation, then, according to the theory of differential equation, the general solution is

$\psi(r) =$ complementary function $+$ a particular integral of (IV-7).

The complementary function is a solution of the homogeneous equation

$$(\nabla^2 + k^2)\phi(\mathbf{r}) = 0, \tag{IV-9}$$

namely,

$$\phi_k(\mathbf{r}) = e^{i\mathbf{k}\cdot\mathbf{r}}. \tag{IV-10}$$

A particular integral $\Psi(r)$ of (IV-7) can be obtained with the aid of the method of Green's functions.[d] Let

$$(\nabla^2 + k^2)G(\mathbf{r},\mathbf{r}') = -\delta(\mathbf{r} - \mathbf{r}'), \tag{IV-11}$$

where $\delta(\mathbf{r} - \mathbf{r}') = \delta(x - x')\delta(y - y')\delta(z - z')$ is the Dirac $\delta$-function in 3-dimensional space,

$$\iiint_v \delta(\mathbf{r} - \mathbf{r}')\,d^3r = 1 \qquad \text{for } \mathbf{r}' \text{ inside } v,$$

and $\tag{IV-12}$

$$\iiint f(r)\delta(\mathbf{r} - \mathbf{r}')\,d^3r = f(r').$$

Noting that

---

[d] The essence of the method of Green's functions is that by properly choosing the form of the Green's function $G(\mathbf{r},\mathbf{r}')$, one can obtain a particular solution of (IV-7) to satisfy a prescribed boundary (or asymptotic) condition on $\psi(r)$.

$$\iiint e^{i\boldsymbol{\kappa}\cdot(\mathbf{r}-\mathbf{r}')}\,d^3\kappa = (2\pi)^3\delta(\mathbf{r}-\mathbf{r}'), \tag{IV-13}$$

one sees that

$$\Psi(\mathbf{r}) = \phi_k(\mathbf{r}) - \iiint G(\mathbf{r},\mathbf{r}')X(\mathbf{r}')\,d^3r' \tag{IV-14}$$

satisfies (IV-7), or

$$\Psi(\mathbf{r}) = e^{i\mathbf{k}\cdot\mathbf{r}} - \iiint G(\mathbf{r},\mathbf{r}')\frac{2m}{\hbar^2}V(\mathbf{r}')\Psi(\mathbf{r}')\,d^3r' \tag{IV-15}$$

satisfies (IV-6). Since the function $\Psi(\mathbf{r})$ appears in the integral in (IV-15), this equation is an integral equation for $\Psi(\mathbf{r})$. It is equivalent to the Schrödinger equation (IV-6), but is particularly suitable for dealing with scattering problems by a proper choice of the Green's function for the desired asymptotic conditions.

Now the function

$$G(\mathbf{r},\mathbf{r}') = -\frac{1}{(2\pi)^3}\iiint \frac{1}{k^2-\kappa^2}\phi_\kappa(\mathbf{r})\phi_\kappa^*(\mathbf{r}')\,d^3\kappa \tag{IV-16}$$

can be seen to satisfy equation (IV-11). If we define the function

$$G^+(\mathbf{r},\mathbf{r}') = -\frac{1}{(2\pi)^3}\lim_{\varepsilon\to 0}\iiint \frac{e^{i\boldsymbol{\kappa}\cdot\boldsymbol{\rho}}}{k^2-\kappa^2+i\varepsilon}\,d^3\kappa, \qquad \varepsilon > 0 \tag{IV-17}$$

where

$$\boldsymbol{\rho} = \mathbf{r} - \mathbf{r}', \tag{IV-18}$$

it is seen that $G^+(\mathbf{r},\mathbf{r}')$ also satisfies (IV-11). We shall apply (IV-15), (IV-17) to the problem of the scattering of a particle by a central field.

### (2) Scattering of a particle by a potential field

Consider a particle of mass $m$, with initial momentum $\mathbf{p} = \hbar\mathbf{k}$ along the $z$-axis, scattered by a central field $V(r)$ centered at $r = 0$. The Schrödinger equation is (IV-15). We seek a solution that has the following asymptotic behavior

$$\Psi(\mathbf{r}) \to e^{ikz} + \frac{1}{r}e^{ikr}f(\vartheta), \tag{IV-19}$$

i.e., a plane wave, with $\mathbf{k} = (1/\hbar)\mathbf{p}$, along the direction of the $z$-axis, plus an elastically scattered wave with the same $|\mathbf{k}|$ and an amplitude $f(\vartheta)$ depending on the scattering angle $\vartheta$ but axially symmetric about the direction of the original $\mathbf{p}$, i.e., the $z$-axis. We shall show that this asymptotic condition is met by choosing $G^+(\mathbf{r}, \mathbf{r}')$ in (IV-17).

Consider the integral in (IV-17). In integrating over the directions of $\kappa$, take $\rho$ as the polar axis and obtain for the integral the expression

$$\frac{2\pi}{i\rho} \int_0^\infty \frac{e^{i\kappa\rho} - e^{-i\kappa\rho}}{k^2 - \kappa^2 + i\varepsilon} \kappa \, d\kappa. \tag{IV-20}$$

The integrand is an even function of $\kappa$, so that one may replace the lower limit of integration by $-\infty$ and divide the resulting integral by 2. For $\varepsilon \ll k^2$, (IV-20) may be replaced by

$$\frac{\pi}{i\rho} \int_{-\infty}^\infty \frac{e^{i\kappa\rho} - e^{-i\kappa\rho}}{\left(k + \dfrac{i\varepsilon}{2k} + \kappa\right)\left(k + \dfrac{i\varepsilon}{2k} - \kappa\right)} \kappa \, d\kappa. \tag{IV-21}$$

For the first integral, we take a closed path of integration as follows: along the real axis from $-\infty$ to $\infty$, then along a semi-circle, of infinite radius, in the upper half plane, in the counter clockwise sense, back to $\kappa = -\infty$. By Cauchy's theorem, the first integral is, since the pole is at $\kappa = k + (i\varepsilon/2k)$, in the limit $\varepsilon \to 0$,

$$-\frac{\pi}{i\rho} \frac{2\pi i}{2k} k e^{ik\rho}. \tag{IV-22}$$

For the second integral in (IV-21), the path of integration is along the real axis from $-\infty$ to $\infty$, a semi-circle of infinite radius in the lower half plane in the clockwise sense back to $\kappa = -\infty$. The pole is now at $\kappa = -(k + (i\varepsilon/2k))$, and the integral has the same value as (IV-22) but with the opposite sign. Hence $G(\mathbf{r}, \mathbf{r}')$ in (IV-17) is

$$G^+(\mathbf{r}, \mathbf{r}') = \frac{1}{4\pi|\mathbf{r} - \mathbf{r}'|} e^{ik|\mathbf{r} - \mathbf{r}'|}. \tag{IV-23}$$

Equation (IV-15) is then

$$\Psi^+(\mathbf{r}) = \phi_k(\mathbf{r}) - \frac{1}{4\pi} \iiint \frac{e^{ik|\mathbf{r} - \mathbf{r}'|}}{|\mathbf{r} - \mathbf{r}'|} \left(\frac{2m}{\hbar^2} V(r')\right) \Psi^+(\mathbf{r}') \, d^3r'. \tag{IV-24}$$

The asymptotic form of $\Psi^+(\mathbf{r})$ is obtained by noting

$$|\mathbf{r} - \mathbf{r}'| \cong r - r' \cos \Theta,$$

where $\Theta$ is the angle between $\mathbf{r}'$ and $\mathbf{r}$, and

$$k|\mathbf{r} - \mathbf{r}'| = kr - kr' \cos \Theta$$
$$= kr - \mathbf{k}' \cdot \mathbf{r}', \tag{IV-25}$$

where $\mathbf{k}'$ is the wave vector $\mathbf{k}' = |\mathbf{k}|(\mathbf{r}/|r|)$. Hence, from (IV-15),

$$\Psi^+(\mathbf{r}) = \phi_k(\mathbf{r}) - \frac{1}{4\pi}\frac{e^{ikr}}{r}\iiint e^{-i\mathbf{k}'\cdot\mathbf{r}'}\left[\frac{2m}{\hbar^2}V(r')\right]\Psi^+(\mathbf{r}')\,d^3r', \quad \text{(IV-26)}$$

which represents an outgoing scattered wave from the scattering center $r = 0$. The scattered amplitude $f(\vartheta)$ is

$$f(\vartheta) = -\frac{1}{4\pi}\iiint e^{-i\mathbf{k}'\cdot\mathbf{r}'}\frac{2m}{\hbar^2}V(r')\Psi^+(\mathbf{r}')\,d^3r'. \quad \text{(IV-27)}$$

If instead of (IV-27) we choose

$$G^-(\mathbf{r},\mathbf{r}') = -\frac{1}{(2\pi)^3}\lim_{\varepsilon\to 0}\iiint \frac{e^{i\mathbf{\kappa}\cdot\mathbf{\rho}}}{k^2-\kappa^2-i\varepsilon}\,d^3\kappa, \qquad \varepsilon > 0, \quad \text{(IV-28)}$$

entirely similar calculations as above will lead to, instead of (IV-26),

$$\Psi^-(\mathbf{r}) = \phi_k(\mathbf{r}) - \frac{e^{-ikr}}{r}\iiint e^{i\mathbf{k}'\cdot\mathbf{r}'}\left[\frac{2m}{\hbar^2}V(r')\right]\Psi^-(\mathbf{r}')\,d^3r', \quad \text{(IV-28a)}$$

which is an *incoming* (converging) spherical wave instead of an outgoing scattered wave. In most cases we shall only be interested in $\Psi^+(\mathbf{r})$.

From (IV-19), it is seen that the scattered wave intensity in the solid angle $\sin\vartheta\,d\vartheta\,d\phi$ is proportional to

$$|f(\vartheta)|^2\sin\vartheta\,d\vartheta\,d\phi. \quad \text{(IV-29)}$$

As $f(\vartheta)$ has the dimension of length, this is defined as the differential cross-section of scattering [e]

---

[e] From the time-dependent Schrödinger equation, one obtains the equation

$$\frac{\partial}{\partial t}(\Psi^*\Psi) + \frac{\hbar}{2im}\nabla(\Psi^*\nabla\Psi - \Psi\nabla\Psi^*) = 0,$$

from which one obtains, on integrating over a large sphere of radius $R$, the normal flux (the number of particles per unit time) within the solid angle $d\Omega$

$$F\,d\Omega = \frac{\hbar}{2im}\left(\Psi_{sc}^*\frac{\partial}{\partial r}\Psi_{sc} - \Psi_{sc}\frac{\partial}{\partial r}\Psi_{sc}^*\right)R^2\,d\Omega,$$

where

$$\Psi_{sc} \to \frac{1}{r}e^{ikr}f(\vartheta), \qquad r = R.$$

Thus

$$F\,d\Omega = \frac{\hbar k}{m}|f(\vartheta)|^2\,d\Omega.$$

The differential cross-section $d\sigma$ is $F\,d\Omega$ divided by the flux of incident particles (per unit area per second), i.e., by the velocity $v = \hbar k/m$, since $e^{i\mathbf{k}\cdot\mathbf{z}}$ is normalized to one particle per unit volume.

$$d\sigma = |f(\vartheta)|^2 \sin \vartheta \, d\vartheta \, d\phi. \tag{IV-29a}$$

The total cross-section $\sigma$ is

$$\sigma = 2\pi \int |f(\vartheta)|^2 \sin \vartheta \, d\vartheta. \tag{IV-29b}$$

Equation (IV-26) for $\Psi^+(\mathbf{r})$ is *not* a solution, since $\Psi^+(\mathbf{r}')$ in the integrand is the $\Psi^+(\mathbf{r})$ to be found. To obtain an approximate solution, one makes the assumption that the scattering field $V(r)$ is "weak" so that the scattered wave in (IV-27) is "small" compared with the incident wave $\phi_k(\mathbf{r}) = e^{i\mathbf{k}\cdot\mathbf{r}}$. In this case, one replaces $\Psi^+(\mathbf{r}')$ in $f(\vartheta)$ in (IV-27) by $e^{i\mathbf{k}\cdot\mathbf{r}}$ so that

$$f(\vartheta) = -\frac{m}{2\pi\hbar^2} \iiint e^{-i\mathbf{k}'\cdot\mathbf{r}'} V(\mathbf{r}') e^{i\mathbf{k}\cdot\mathbf{r}'} \, d^3 r', \tag{IV-30}$$

which can be evaluated. This approximation is due to Born, and the name *Born approximation* has been extended to other situations, almost beyond recognition.

The result in (IV-29b), (IV-30) can be obtained in other manners, for example, by the method of time-dependent perturbation theory. See Chap. 7, Sec. 2, (VII-96).

We shall express the integral equation (IV-14), (IV-15) in symbolic form by introducing the integral operator

$$\frac{1}{E - H_0 \pm i\varepsilon}, \quad \text{or} \quad \frac{1}{E - H_0}$$

defined by

$$\frac{1}{E - H_0} \phi_\kappa(\mathbf{r}) = \frac{1}{E - \kappa^2} \phi_\kappa(\mathbf{r}), \tag{IV-31}$$

where $\phi_\kappa$ is an eigenfunction of the operator $H_0$, i.e.,

$$H_0 \phi_\kappa = \kappa^2 \phi_\kappa,$$

and

$$\frac{1}{E - H_0} |\chi(\mathbf{r})\rangle = \frac{1}{E - H_0} \iiint d\kappa \iiint |\phi_\kappa(\mathbf{r})\rangle \langle \phi_\kappa(\mathbf{r}')|\chi(\mathbf{r}')\rangle \, d\mathbf{r}', \tag{IV-32}$$

where $\chi(\mathbf{r})$ is an arbitrary function, and $|\phi\rangle\langle\phi|$ is a projection operator, and $\iiint |\phi_\kappa(\mathbf{r})\rangle \langle \phi_\kappa(\mathbf{r}')| \, d\kappa = \delta_{\mathbf{r},\mathbf{r}'}$ is the "closure relation" of the complete set of the eigenfunctions $\phi_\kappa$ of $H_0$. Thus

$$\frac{1}{E - H_0 \pm i\varepsilon} \chi(\mathbf{r}) = \frac{1}{(2\pi)^3} \lim_{\varepsilon \to 0} \iiint d\mathbf{\kappa} \frac{1}{E - \kappa^2 \pm i\varepsilon} \iiint e^{i\mathbf{\kappa}\cdot\mathbf{r} - i\mathbf{\kappa}\cdot\mathbf{r}'} \chi(\mathbf{r}') \, d\mathbf{r}',$$

$$(IV\text{-}32a)$$

and equation (IV-15) now takes the form

$$\Psi^+(\mathbf{r}) = \phi_k(\mathbf{r}) + \frac{1}{E - H_0 + i\varepsilon} \left( \frac{2m}{\hbar^2} V(r) \right) \Psi^+(\mathbf{r}). \tag{IV-33}$$

## 2. Probability interpretation

As a result of the collision problem described in the preceding section, Born made the by now obvious and yet at that time (June, 1926) new and profoundly important step that the $|f(\vartheta)|^2$ in (II-23) is to be interpreted as the probability that an incoming electron represented by $e^{ikz}$ is scattered in the direction of $\mathbf{k}'$. It is perhaps not quite possible to pinpoint the way this idea was conceived by Born[f]; we shall simply take his own words in the June, 1926 article that, if one regards the results of the scattering theory as to signify a corpuscular electron, then there is only one (probability) interpretation possible.

The probability idea was soon more fully developed (July, 1926), and the intrinsic nature of this probability brought out, as distinct from the probability concept familiar in classical physics. Consider a single atom, and assume that the eigenstates $u_n$ of the Hamiltonian form a complete orthonormal set,

$$Hu_n = E_n u_n. \tag{IV-34}$$

Let $\psi$, an arbitrary (normalized) state, be expanded in this set

$$\psi = \sum_n \int C_n u_n, \tag{IV-35}$$

where

$$C_n = \int \psi u_n^* \, dq. \tag{IV-36}$$

The normalization of $\psi$ requires that

$$\sum_n \int |C_n|^2 = 1. \tag{IV-37}$$

---

[f] In 1955, in his Nobel Prize lecture, Born reminisced that he owed his ideas to the influence of Einstein's theory of the photon in which the electric field serves as a ghost field guiding or determining the motion of the photon, and of the strong evidence of electrons as particles from the work of his colleague, J. Franck, at Göttingen. But the suggestion of $|\psi(\vartheta)|^2$ and not $\psi$ itself, and the intrinsically probabilistic nature basically different from the probability concepts in the classical kinetic theory of gases, are revolutionally new.

The interpretation of (II-35) is then that $|C_n|^2$ is the probability that system in the state $\psi$ is in the eigenstate $u_n$ of the Hamiltonian $H$. If one calculates the matrix element of $H$, one obtains

$$\int \psi^* H \psi \, dq = \sum_n \int |C_n|^2 E_n. \tag{IV-38}$$

This states that in a general state $\psi$, the *expectation value*[g] of the energy is not a definite, single $E_j$, but all the $E_j$, each with a probability $|C_n|^2$, the $C_n$ being determined by (IV-36). Of course, as the result of an actual measurement of $H$, the system must be found to have one definite energy, say $E_m$. But the theory cannot predict beforehand which $E_j$ will come out; it can only predict the probabilities $|C_n|^2$ of their being found.

This situation is basically different from the classical theory in which the "probability" concept is used in the following sense. Take the Maxwell velocity distribution of the velocities of the molecules of a gas. There the probability concept is introduced because of our unwillingness (and practical inability) to follow the motion of a specific molecule in a gas made up of $10^{23}$ molecules; but according to *classical* dynamics, it is not *in principle* impossible to know the precise velocity of a molecule. In quantum mechanics, even when dealing with one single atom, one's knowledge is limited to a probability distribution $C_n$, although the *probabilities $C_n$* themselves are predicted with definiteness according to (IV-36).[h]

### 3. Uncertainty principle

In Sec. 1 above, we briefly indicated how, from the point of view of the transformation theory, Dirac (December, 1926) pointed out the impossibility of the simultaneous exact knowledge of two conjugate variables. In a perhaps still more formal transformation theory, Jordan also arrived at the same result. But the first strong indication of something fundamentally "wrong" in classical physics was the wave-particle dilemma expressed by the Einstein-de Broglie relations

---

[g] On the basis of the probability postulate (IV-5b) $\int \psi^* Q \psi \, dq$ is the "expectation value" of the physical quantity $Q$ when the system is in the state $\psi$. This *postulate* is to mean that, when a measurement (observation) of $Q$ is made on the system when it is in a state $\psi$, the result of the measurement is given by $\int \psi^* Q \psi \, dq$. When a system is known to be in an eigenstate $u_n$ of $H$ and if a measurement of $H$ is made, then the above postulate gives the eigenvalue $E_n$ uniquely,

$$\int u_n^* H u_n \, dq = E_n. \tag{IV-39}$$

[h] In the course of time, the change of $\psi$ is completely determined by the time-dependent Schrödinger equation, and therefore the probabilities $|C_n(t)|^2$ change in a deterministic way with time. But only the $\psi$ and the probabilities are causal; the expectation value $\int \psi^* Q \psi \, dq$ remains to be governed by a probability distribution.

                                                    

$$E = h\nu, \qquad p = \frac{h}{\lambda}, \tag{IV-40}$$

which, by ascribing classically mutually exclusive attributes to the same entities, brought to the fore the difficulties of classical physics at the level of the basic concepts. Then came the commutation relation

$$pq - qp = \frac{\hbar}{i} 1, \tag{IV-41}$$

which is clearly fundamentally non-classical in nature but which has been found from so many different points of view to have played a basic role in all formulations of quantum mechanics. That the Einstein-de Broglie relations and the commutation relation are not independent of each other was made clear by the transformation theory.[1] Thus even though Dirac and Jordan (December, 1926) had already arrived at a qualitative form of the uncertainty principle, it remained for Heisenberg to make an *anschaulich* analysis of the consequences of the "physical" Einstein-de Broglie relations on the nature of our knowledge of the kinematical and mechanical concepts such as the path and momentum of a particle. In March, 1927, his paper clearly brought out the basic difference between classical and quantum mechanics.

A little later, H. Weyl (1928) obtained the uncertainty relation from the commutation relation (IV-41) by means of the Schwarz inequality. Thus one may regard Heisenberg's original discovery as being based on the "physical" arguments represented by the Einstein-de Broglie relations, whereas Weyl's deduction is based on the mathematical implication of the commutation relation. One may say that the commutation relation is a mathematical expression of the Einstein-de Broglie relations. Some physicists like to take the uncertainty principle as a basic principle in quantum mechanics.

We shall follow the historical order of development below.

Heisenberg started with a view of physical (kinematical) concepts such as position and velocity (momentum) which is somewhat similar to Einstein's view concerning space and time in the special theory of relativity. We may recall that Einstein defined the concept of time for an observer in each inertial system by laying down an experimental procedure for its measurement, thereby clarifying the physical meaning of $t$ and $t'$ in Lorentz's transformation and rendering absolute time unphysical. In the same spirit Heisenberg examined the meaning of the concept of simultaneous position $x$ and its conjugate momentum $p_x$—the knowledge of which had been taken for granted from time immemorial and in Newton's Second Law of motion. Heisenberg's criteria were the Einstein-de Broglie relations.

---

[1] For example, see Chap. 3, (III-62).

Consider next the following thought-experiment of measuring the position of a particle by means of a microscope. Now the accuracy of the position is limited by diffraction effects which cause the image to blur. The resolving power, namely, the smallest separation $\Delta x$ between two points whose images are resolved, is given by

$$\Delta x = \frac{\lambda}{\sin \alpha}, \tag{IV-42}$$

where $\lambda$ is the wavelength of the illuminating light, and $2\alpha$ is the aperture angle of the objective lens of the microscope. Thus to achieve high accuracy in the measurement of the position (along the $x$-axis), one may use $\gamma$-rays (in the thought-experiment).

Now it is known from the Compton effect that when light is scattered by the particle into the objective of the microscope, the particle receives a recoil. But as the scattered photon may enter the objective in any direction within the aperature $2\alpha$, the corresponding recoiled particle may have a recoil velocity component over a corresponding range, as determined by the energy-momentum conservation laws in this Compton scattering. It can be seen (Chap. 1, Sec. 2) that corresponding to the angular spread $2\alpha$ of the scattered photon, the spread of the $x$-component of the recoil momentum is

$$\Delta p_x = \frac{h}{\lambda}\sin \alpha. \tag{IV-43}$$

Thus, from (IV-42), (IV-43), one has

$$\Delta x \Delta p_x \simeq h. \tag{IV-44}$$

Consider another experiment in which a beam of particles is incident perpendicularly on a screen with a slit of width $\Delta x$. Thus the indeterminacy of the position of the particle in the $x$-direction is $\Delta x$. Now from the wave point of view, a wave of wavelength $\lambda$ passing through a slit of width $\Delta x$ will undergo diffraction, resulting in a deflection of angle $\vartheta$ given by

$$\sin \vartheta \simeq \frac{\lambda}{\Delta x}. \tag{IV-45}$$

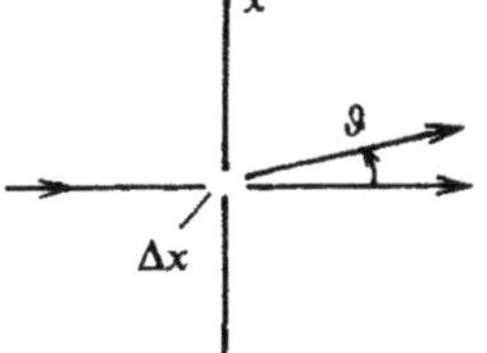


Fig. IV.1.

According to the de Broglie relation $p = h/\lambda$, the $x$-component of the momentum is $p \sin \vartheta = (h/\lambda) \sin \vartheta$. Due to diffraction the particle may be deflection either by $\vartheta$, or $-\vartheta$, so that the uncertainty in the $x$-component of the beam (after passing through the slit) is

$$\Delta p_x = \frac{h}{\lambda} \sin \vartheta. \tag{IV-46}$$

From (IV-45) and (IV-46), one has again

$$\Delta x \Delta p_x \simeq h. \tag{IV-47}$$

It is important to note that, while the interactions between the object of a measurement and the measuring devices cause finite but nonvanishing disturbances, they would not have been relevant in our present discussions if corrections for them could be made. This in principle can be done on classical physics. The point is that, in quantum mechanics corrections for the disturbances arising from the interactions *cannot* be made to an accuracy greater than that allowed by the constant $h$. To repeat, the effect of a measuring process disturbing the object of measurement, by itself, will not lead to the uncertainty relation; it is the limitation, set by the constant $h$ through the Einstein-de Broglie relations, on the accuracy of making the corrections that is in point. In classical physics, $h = 0$ and *in principle*, it is possible to make exact corrections and there is no "uncertainty relation".

During the 5th Solvay Congress in October, 1927, Einstein, who did not agree, philosophically, with the basic views of quantum mechanics, tried to exceed the limitations imposed by the uncertainty relation by thinking up thought-experiments, one after the other, but Bohr was able to show in each case the validity of the uncertainty relation by a consistent application of the Einstein-de Broglie relations. Einstein had to agree with the logical consistency of the system of quantum mechanics, although he remained unhappy with its basic philosophy. We shall come back to Einstein's view in Chap. 5, Sec. 7.

The above discussions are based on the particle-wave duality as expressed by the Einstein-de Broglie relations (IV-40). They are "anschaulich" (clear), but the results are qualitative. Heisenberg considered the representation of a particle by a wave packet, whose "intensity" has a distribution of a Gaussian curve about the point $x = 0$,

$$f(x) = A \exp\left( -\frac{x^2}{2\Delta^2} \right), \tag{IV-48}$$

where $A$ is a constant and $\Delta$ is the standard deviation,

$$\Delta^2 \equiv (\Delta x)^2 = \overline{|f|^2} - |\bar{f}|^2. \tag{IV-49}$$

The transformation from the $x$- to the $p_x$-representation is given by the unitary matrix $U(x, p_x)$ which is given by the (Schrödinger) equation

$$\frac{\hbar}{i}\frac{\partial}{\partial x} U(x, p_x) = p_x U(x, p_x), \tag{IV-50}$$

whose solution is

$$U(x, p) = e^{ip_x x/\hbar}, \tag{IV-51}$$

so that in the $p_x$-representation,

$$\sqrt{g(p)} = B \int_{-\infty}^{\infty} \sqrt{f(x)}\, e^{ip_x x/\hbar}\, dx. \tag{IV-52}$$

But this is just a Fourier transform, and one obtains

$$\sqrt{g(p)} = C \exp\left(-\left(\frac{\Delta}{\hbar}\right)^2 p_x^2\right) \tag{IV-53}$$

and

$$g(p) = C^2 \exp\left(-\frac{1}{2}\left(\frac{2\Delta}{\hbar}\right)^2 p_x^2\right), \tag{IV-54}$$

so that the standard deviation in $p$ is given by

$$(\Delta p_x)^2 = \left(\frac{\hbar}{2\Delta}\right)^2. \tag{IV-55}$$

Thus

$$\Delta x \Delta p_x = \frac{\hbar}{2} = \frac{h}{4\pi}. \tag{IV-56}$$

This equality sign holds only for Gaussian wave packets for which $\Delta x \Delta p_x$ is minimum. In general,

$$\Delta x \Delta p_x \geq \frac{h}{4\pi}. \tag{IV-57}$$

It was shown by H. Weyl that the uncertainty principle is a direct consequence of the commutation relation. The proof makes use of an inequality due to Schwarz.

For any two functions $u$ and $v$, there is the inequality (H. A. Schwarz, 1885)

$$\int u^* u\, d\tau \int v^* v\, d\tau \geq \frac{1}{4}\left|\int (u^* v + v^* u)\, d\tau\right|^2. \tag{IV-58}$$

The proof of this inequality is as follows: Let $\lambda$ be a real number, and form the expression

$$\int (\lambda u^* + v^*)(\lambda u + v)\, d\tau = \lambda^2 \int u^* u\, d\tau + \lambda \int (u^* v + v^* u)\, d\tau + \int v^* v\, d\tau.$$

The left-hand side is real and positive. Hence the quadratic form in $\lambda$ on the right cannot have real roots for $\lambda$. The condition for this is the inequality (IV-58).

Now let $\psi$ be any (wave) function, and let the expectation values of the conjugate variables $P$ and $Q$ in the state $\psi$ be, according to (IV-5b),

$$\bar{p} = \int \psi^* P \psi\, d\tau, \qquad \bar{q} = \int \psi^* Q \psi\, d\tau, \qquad \text{(IV-59)}$$

and let the standard deviations be

$$\Delta p = P - \bar{p}, \qquad \Delta q = Q - \bar{q}. \qquad \text{(IV-60)}$$

Let

$$u = (P - \bar{p})\psi, \qquad v = i(Q - \bar{q})\psi. \qquad \text{(IV-61)}$$

Equation (IV-58) becomes

$$\int (P - \bar{p})^* \psi^* (P - \bar{p})\psi\, d\tau \int (Q - \bar{q})^* \psi^* (Q - \bar{q})\psi\, d\tau$$

$$\geq \frac{1}{4}\left[ i \int (P - \bar{p})^* \psi^* (Q - \bar{q})\psi\, d\tau - i \int (Q - \bar{q})^* \psi (P - \bar{p})\psi\, d\tau \right]^2.$$

As $P - \bar{p}$, $Q - \bar{q}$ are hermitian (self-adjoint), and $\bar{p}$, $\bar{q}$ are numbers, we have

$$\int \psi^* (P - \bar{p})^2 \psi\, d\tau \int \psi^* (Q - \bar{q})^2 \psi\, d\tau \geq -\frac{1}{4}\left[ \int \psi^* (PQ - QP)\psi\, d\tau \right]^2.$$

Introducing the standard deviations as in (IV-48), and the commutation relation

$$PQ - QP = \frac{\hbar}{i} 1,$$

we have

$$(\Delta p)^2 (\Delta q)^2 \geq \frac{1}{4}\hbar^2$$

or

$$\Delta p \Delta q \geq \frac{\hbar}{2}, \qquad \text{(IV-61)}$$

which is the uncertainty relation.

The above results are summarized as follows: In Quantum Mechanics, physical quantities (called *observables* by Dirac) are represented by operators (which may have the special forms of matrices, differential operators, or of even more general nature such as Dirac's $q$-numbers, Born and Wiener's operators), and these operators obey the commutation relation. This is a basic postulate. Next the postulate is made that the values expected from the observation (measurement) of these *observables* on a system in a *state* $\psi$ are given by

$$\langle Q \rangle = \int \psi^* Q \psi \, d\tau. \qquad \text{(IV-62)}$$

From these postulates, it follows that two canonically conjugate variables such as $x$ and $p_x$ cannot be known to an accuracy greater than the limit set by the relation

$$\Delta p \Delta q \geq \frac{h}{4\pi}. \qquad \text{(IV-61)}$$

This is the Uncertainty Principle.[j]

This uncertainty principle, we must re-emphasize, is not a statement of limitations of accuracies in practical experimental measurements, but is a statement of the limitations, in basic principle, of the classical meaning of the concepts of a coordinate $x$ and its conjugate momentum $p_x$. On the basis of the uncertainty principle, it is not so much that $p_x$ and $x$ cannot be measured simultaneously to accuracies exceeding that principle; rather it is not even possible simultaneously to define the $p_x$, $x$ in the classical sense. As a consequence, the definition of the dynamical state of a system by the coordinates and their conjugate momenta at a given instant of time in classical dynamics has no exact meaning, and so have the equations of motion in classical dynamics, and their solutions—the trajectory of motion in time.

Thus in quantum mechanics, one gives up the classical dynamical description of the motion of a particle by deterministic equations of motion; for one does not have exact simultaneous knowledge of $x$ and $p_x$ at any given time so that it is not even possible to define an initial condition $x = x^0$, $p_x = p_x^0$ at $t = t_0$ for the integration of the second-order equations of motion. One either chooses to know the coordinates, or the momenta, but not both. The time variations of a system is then described by the usual Schrödinger equation in the former case, and its momentum representation in the latter.

---

[j] Note that Quantum Mechanics is a *physical* theory; according to it, the results of the mathematical formalism are to describe the outcome of actual measurements! In this sense the uncertainty relation is a *principle*.

**Exercises**

1. Take the one-dimensional Fourier transform (see (III-48))

$$\Psi(x,t) = \frac{1}{\sqrt{2\pi}} \int dk\, A(k) e^{i(kx-\omega t)}$$

and assume

$$A(k) = a \exp\left\{-\frac{(k-k_0)^2}{4s^2}\right\}, \qquad a = \text{const.}$$

Show that (for short time $t$)

$$\Psi(x,t) = \sqrt{2}\, sa\, e^{i(k_0 x - \omega_0 t)} \exp\{-s^2(x - v_g t)^2\},$$

where

$$\omega = \omega(k), \qquad v_g = \frac{\partial \omega}{\partial k}.$$

The above result is classical physics. If we now introduce the de Broglie relation

$$p = \frac{h}{\lambda} = \hbar k,$$

show that the width of $|A(p)|^2$ and the width of $|\Psi(x,t)|^2$ have the uncertainty relation,

$$\Delta p \Delta x = \frac{\hbar}{2}.$$

2. With the aid of (III-44), show that for a free particle

$$\frac{\partial \rho}{\partial t} + \text{div}\, \mathbf{J} = 0,$$

where

$$\rho = \Psi^*(\mathbf{r},t)\Psi(\mathbf{r},t), \qquad \mathbf{J} = \frac{\hbar}{2mi}(\Psi^*\nabla\Psi - (\nabla\Psi^*)\Psi).$$

Show that

$$\frac{\displaystyle\int J\, dx}{\displaystyle\int \Psi^*\Psi\, dx} \qquad \text{(one-dimensional case)}$$

is equal to

$$v = \frac{\displaystyle\int \frac{d\omega}{dk}|A(k)|^2\,dk}{\displaystyle\int |A(k)|^2\,dk},$$

where $A(k)$ is given in Problem 1, above. $v$ is the average value of the group velocity of the wave packet. Hint: Use Plancherel's theorem

$$\int \Psi^*(x,t)\Psi(x,t)\,dx = \int A^*(k,t)A(k,t)\,dk,$$

$$A(k,t) = A(k)e^{-i\omega t}.$$

3. From (III-44), the Schrödinger equation

$$-\frac{\hbar}{i}\frac{\partial \Psi}{\partial t} = \left(-\frac{\hbar^2}{2m}\nabla^2 + V\right)\Psi,$$

and the postulate that the expectation value of a quantity $Q$ is

$$\bar{Q} = \int \Psi^*Q\Psi\,dq,$$

show that

$$\frac{d}{dt}\bar{p} = -\int \Psi^*\nabla V\Psi\,dq,$$

$$\text{or}\quad \frac{d}{dt}\bar{p} = -\overline{\nabla V}.$$

This is known as Ehrenfest's theorem.

4. Show that the expectation value of the kinetic energy satisfies the quantum virial relation

$$2\bar{E}_{\text{kin}} = \frac{1}{m}\overline{p^2} = \overline{\mathbf{r}\cdot\nabla V}.$$

## References

M. Born, *Z. Phys.* **37**, 863 (June, 1926); *ibid.* **38**, 803 (July, 1926). In these two papers on the quantum mechanics of collision processes, Born introduced the probability interpretation of $\psi$. In the first paper, $\psi$ is given the probability meaning; this is corrected to $|\psi|^2$ in a "note added in proof". In the second paper, the probability idea is fully developed.

M. Born, *ibid.* **40**, 167–192 (Oct., 1926); M. Born and V. Fock, *ibid.* **51**, 165 (1928).

Born considered the variation of the state $\psi$ with time according to the time-dependent Schrödinger equation, and introduced the transition probability.

P. A. M. Dirac, *Proc. Roy. Soc.* London **A 112**, 661 (1926); **A 114**, 243 (1927), treat the variation of $\psi$ with time by perturbation theory also.

P. A. M. Dirac, *ibid* **A 113**, 621 (Dec., 1926). This paper deals with transformation theory.

W. Heisenberg, *Z. Phys.* **43**, 172–198 (March, 1927). This makes reference to the papers of Dirac, *loc. cit.*, and of P. Jordan, *Z. Phys.* **40**, 809 (1927) on the transformation theory.

Chapter 5

# Quantum Mechanics: General Theory

In Chaps. 2 and 3, we described matrix mechanics and wave mechanics in their conception and their mathematical aspects; in Chap. 4, we described the probability interpretation and the uncertainty principle and we began to see the growth of an integral system of Quantum Mechanics. In 1927, Bohr proposed his philosophical view of quantum mechanics under the name of a *complementarity principle*. With these developments, quantum mechanics matured into a coherent structure having a unified mathematical method and a self-consistent physical interpretation and philosophical point of view. The physical interpretation and philosophical view are known as the Copenhagen philosophy. The strong proponents were Bohr, Heisenberg, Born, Pauli, Dirac, Rosenfeld, but there were again prominent opponents. In the present chapter we shall present the Copenhagen view, the basic principles of quantum mechanics from a postulational point of view, and the mathematical notations of Dirac, which integrated the matrix and wave mechanics concepts and terminology.

## 1.  Complementarity principle

(i) We have firstly the particle-wave dilemma summarized by the Einstein-de Broglie relations

$$E = hv, \qquad p = \frac{h}{\lambda}. \qquad (V\text{-}1)$$

Electromagnetic *waves* exhibit, under certain conditions of observation, the "typically wave" attributes such as wavelengths, frequencies, interference, diffraction and polarization, but under other conditions, such as the photoelectric effect and the Compton effect, the "typically particle" attributes of energy and momentum. Electrons, neutrons and atoms behave under most conditions as particles, but they exhibit, under other conditions of observation, the typically wave property of diffraction. We have seen that the de Broglie relation $p = h/\lambda$ leads to the relations

$$x, \qquad p_x = \frac{\hbar}{i}\frac{\partial}{\partial x}, \qquad \text{etc.} \qquad (V\text{-}2)$$

which in turn lead to the commutation relation

$$p_x x - x p_x = \frac{\hbar}{i} 1, \qquad\qquad\text{(V-3)}$$

and we have seen in Chap. 4, Sec. 3, that either the Einstein-de Broglie relations or the commutation relation leads to the uncertainty relation

$$\Delta x \Delta p_x \geq \frac{h}{4\pi}. \qquad\qquad\text{(V-4)}$$

All the results (V-1), (V-2), (V-3), (V-4) are related together, and are all fundamentally non-classical in nature since the constant $h$ appears.

The Copenhagen view of this situation is as follows: The particle-wave dilemma arises from our concepts of "particles" and "waves" themselves; they have been formed from our experience with macroscopic phenomena. The Einstein-de Broglie relations must then be interpreted as stating that these concepts so originated are not adequate in dealing with the atomic domain, that one or the other of the two concepts is to be used depending on the particular phenomenon we are dealing with.[a] The *wave* and the *particle* attributes are what *we* have formed of them on the basis of our macroscopic experience. This process of formation of our "classical" concepts from one domain of "macroscopic" experience brings with it certain limitations on these concepts themselves, limitations which are brought out when these concepts are employed in another (atomic, or "microscopic") domain. The particle-wave duality is taken to indicate some limitations of our "classical" concepts; the "particle" and "wave" concepts are not to be regarded as contradictory, but rather as complementary, to each other.

The question may then be raised: Isn't it much better, to avoid confusion, to employ some more adequate sets of concepts in place of "particle" and "wave"? In the philosophy of the Copenhagen School, the view is that the "classical" concepts, i.e., those we have formed and acquired from our daily macroscopic experience, are the only concepts through which we can think and converse in, and without them, there even won't be physics. But with these "classical" concepts, it is necessary to take note of their complementary roles and to remedy their limitations by means of the Einstein-de Broglie relations and, in more mathematically precise form, the commutation relation.

---

[a] Thus when we are dealing with the propagation of electromagnetic waves in space, the *wave* concept is adequate; but when we are dealing with the interaction of electromagnetic "waves" with an atom or an electron, then the *photon* concept is the adequate one. Similarly, when we are describing an electron in J. J. Thomson's experiment for the measurement of $e/m$, the *particle* concept is adequate, but when we are dealing with the motion of an electron through a thin foil of matter, then the *wave* concept is the adequate one.

Note that all the above considerations arise only when we are concerned with actions (arising from interactions involved in our making an observation on a system) small enough compared with $h = 6.6 \times 10^{-27}$ erg. sec., i.e., in the atomic domain. As a microgram falling under gravity for 1 sec. has an action of 0.5 erg. sec; the quantum effect is insignificant and classical physics is adequate.

(ii) Consider the relationship between the complementarity idea and the uncertainty principle. The derivation of the relation (V-4) from the commutation relation (V-3) depends on the postulate of the expectation value

$$\langle P \rangle = \int \psi^* P \psi \, d\tau \tag{V-5}$$

(as a probability postulate), namely, the "expected value" of a measurement of a physical quantity (variable) on a system in the *state* represented by $\psi$ is given by (V-5). Thus the uncertainty relation so derived involves the theory of postulating (V-5) to represent the results of actual experiments. It is therefore more *anschaulich* to see the physical origin of the uncertainties in the thought-experiments (such as the $\gamma$-ray microscope and the diffraction slits) of Heisenberg. We discussed this in Chap. 4, Sec. 3 (following (IV-47)). There we saw that it was the unavoidable uncertainties, with limits set by $h$, in correcting for the disturbances caused by the interactions between the observed and the observing, that led to the uncertainty relation.

Thus the complementarity of the particle-wave concepts appears in the $p_x$, $x$ variables.

Bohr extended his ideas of complementarity as follows. Consider the motion of a particle represented by a wave packet. From (V-4)

$$\Delta x \Delta p_x \geq \frac{h}{4\pi},$$

where $\Delta x$ represents the width of the packet, and $\Delta p_x$ the range of $h/\lambda$ in the packet. From this, we have

$$\Delta t \Delta E > \frac{h}{4\pi}, \tag{V-6}$$

where[b]

---

[b] The second relation can be obtained simply from the relativistic expressions

$$E^2 = c^2 p^2 + m_0^2 c^4, \qquad p = \frac{m_0 v}{\sqrt{1 - \left(\dfrac{v}{c}\right)^2}} \quad \text{or} \quad p = E\left(\frac{v}{c^2}\right).$$

See Chap. 3, equation above (III-11).

$$\Delta t = \frac{\Delta x}{V_g}, \qquad \Delta E = v_g \Delta p_x,$$

$v_g$ is the group velocity of the wave packet and is equal to the particle velocity. Thus in the wave point of view, shortening the wave packet, thereby shortening the time the packet passes over a point $x$, causes a loss of accurate knowledge about the momentum and energy of the particle. Consequently, the space-time description and the momentum-energy description are complementary, in the sense that an exact space-time description precludes any momentum-energy law and *vice versa*.

The following table summarizes Bohr's ideas of the complementarity concepts

|  |  |  |
|---|---|---|
| Particle | Wave |  |
| $E, p$ | $v, \lambda$ | (V-7) |
| momentum-energy laws | space-time description |  |

Bohr even extended his complementarity ideas to fields other than physics, such as psychology.[c] There the ideas are less clearly defined and we shall refer to Bohr's *Atomic Theory and Description of Nature* (see References below).


## References

The complementarity concept was first proposed by Bohr at a meeting of the International Congress of Physics held in September, 1927, at Como, Italy. The substance of the lecture has been published by him in *Nature* **121**, 580 (1928); in *Atomic Theory and Description of Nature* (Cambridge Univ. Press, Cambridge, 1934).
W. Pauli Jr., article in *Handbuch der Physik* Bd. 24., 2nd ed. (Springer-Verlag, Berlin, 1933).


## 2.  Quantum mechanics: mathematical preliminaries

In the last three chapters we presented a brief story of the birth and developments of matrix mechanics and wave mechanics which by 1927 matured into a unified system now known by the name Quantum Mechanics.

---

[c] It seems that what Bohr meant by complementary concepts were two mutually exclusive descriptions as given in (V-7), while Pauli regarded as complementary, not two modes of descriptions such as *particle* and *wave*, but two classical concepts, such as coordinate $x$ and its conjugate momentum $p_x$, such that any experimental measurement of one precludes any experimental measurement of the other. It seems that Pauli's definition of complementarity is a restatement of the uncertainty relation, while Bohr's goes back a little deeper into the philosophy of the nature of our concepts. We shall not go into these and other views regarding "complementarity".

The purpose of giving these separate and chronological accounts is to present the reader with a clear picture of how and when the various important ideas evolved, in order to better understand and appreciate quantum mechanics. We are now ready to present quantum mechanics in postulational form. Before doing so, we shall recapitulate a few milestones:

| | |
|---|---|
| 1923, Sept.–1924, Nov. | Louis de Broglie proposed the idea of matter waves. |
| 1924, June | S. N. Bose proposed a new statistics for photons, immediately generalized by Einstein to molecules. |
| 1925, July | W. Heisenberg proposed his new ideas. |
| Sept. | M. Born & P. Jordan developed matrix mechanics based on Heisenberg's idea. |
| Nov. | M. Born, W. Heisenberg, P. Jordan "completed" matrix mechanics. |
| Nov. | P. A. M. Dirac developed his operator algebra and introduced the method of Quantum Poisson brackets. |
| 1926, Jan–June | E. Schrödinger developed wave mechanics. |
| Mar. | Schrödinger recognized the equivalence of matrix and wave mechanics. |
| Mar. | E. Fermi introduced a new statistics. |
| June | Born proposed the probability interpretation. |
| Dec. | P. A. M. Dirac and P. Jordan, developed transformation theory. |
| 1927, March | W. Heisenberg proposed the uncertainty principle. |
| Sept. | N. Bohr put forward the complementarity principle. |

In attempting a postulational presentation of quantum mechanics, we may divide the basic postulates into three categories, namely, those having to do with basic changes in the nature of physical concepts from classical physics (loosely represented as the complementarity principle), the probability postulate, and the Schrödinger time-dependent equation itself. We shall begin with the definitions of mathematical concepts and notations. In this we shall follow Dirac's *Principles of Quantum Mechanics*.[d]

## (1) Vector space, vectors

In ordinary three-dimensional space, we may choose any set of three orthogonal axes $X$, $Y$, $Z$, and unit vectors $\mathbf{i}$, $\mathbf{j}$, $\mathbf{k}$ along them, so that any vector $\mathbf{A}$ has components $A_x$, $A_y$, $A_z$ given by

---

[d] The following is an introduction to the substance of the first five chapters of Dirac's book in elementary form. The mature student cannot do better than study Dirac's work.

$$A_x = (\mathbf{A} \cdot \mathbf{i}), \qquad A_y = (\mathbf{A} \cdot \mathbf{j}), \qquad A_z = (\mathbf{A} \cdot \mathbf{k}). \tag{V-8}$$

$(A_x, A_y, A_z)$ may be called the coordinates of the vector $A$.

There are infinitely many sets of 3-axes which are related to $X$, $Y$, $Z$ by a rotation

$$\begin{pmatrix} X' \\ Y' \\ Z' \end{pmatrix} = \begin{pmatrix} a_{11} & a_{12} & a_{13} \\ a_{21} & a_{22} & a_{23} \\ a_{31} & a_{32} & a_{33} \end{pmatrix} \begin{pmatrix} X \\ Y \\ Z \end{pmatrix} \tag{V-9a}$$

with

$$\sum_k a_{ik} a_{jk} = \delta_{ij}. \tag{V-9b}$$

Vectors may be added; thus

$$\mathbf{B} = c_1 \mathbf{A}_1 + c_2 \mathbf{A}_2, \tag{V-10}$$

where $c_1$, $c_2$ are constants.

Vectors $\mathbf{A}_1$, $\mathbf{A}_2$, $\mathbf{A}_3$ are said to be linearly independent if no constants $c_1$, $c_2$, $c_3$ exist for which

$$\sum_{i=1}^{3} c_i A_i = 0. \tag{V-11}$$

The scalar product of two vectors is defined by

$$(\mathbf{A} \cdot \mathbf{B}) = \sum_i^3 A_i B_i \tag{V-12}$$

and the norm of $\mathbf{A}$ is

$$A^2 = (\mathbf{A} \cdot \mathbf{A}), \tag{V-13}$$

which is assumed to be finite.

Two vectors are orthogonal if

$$(\mathbf{A} \cdot \mathbf{B}) = 0. \tag{V-14}$$

Any arbitrary vector $\mathbf{A}$ can be expanded in the 3 basic unit vectors $\mathbf{i}, \mathbf{j}, \mathbf{k}$ of the set of $X$, $Y$, $Z$ axes, as shown in (V-8),

$$\mathbf{A} = A_x \mathbf{i} + A_y \mathbf{j} + A_z \mathbf{k}. \tag{V-15a}$$

But one may choose any other (orthogonal) set $X'$, $Y'$, $Z'$, and similarly, expand $\mathbf{A}$ in the 3 basic unit vectors $\mathbf{i}', \mathbf{j}', \mathbf{k}'$ of this set

$$\mathbf{A} = A'_x \mathbf{i}' + A'_y \mathbf{j}' + A'_z \mathbf{k}'. \tag{V-15b}$$

One may apply on $\mathbf{A}$ an operator $Q$ such that it transforms $\mathbf{A}$ into another vector $\mathbf{B}$

$$QA = B. \tag{V-16}$$

In the simplest case, $\mathbf{Q}$ may be a rotation

$$B_i = \sum_j^3 Q_{ij} A_j, \qquad \sum_k^3 Q^{ik} Q^{jk} = \delta_{ij},$$

or $Q$ may be a tensor (without the orthogonality relation). The operator $Q$ is linear if

$$Q(c_1 \mathbf{A}_1 + c_2 \mathbf{A}_2) = c_1 Q\mathbf{A}_1 + c_2 Q\mathbf{A}_2. \tag{V-17}$$

All the above are elementary.

Now we introduce the concept of a vector space of infinitely many dimensions, for which one may choose any set of infinitely many basic vectors which are orthogonal (and therefore independent). We denote such a set by $|k\rangle$, called *kets*, $k$ an index which may be discrete, or continuous, or both. But one may choose another set $|\alpha\rangle$. Two kets may be added (superposed) to form another ket,

$$c_1|k_1\rangle + c_2|k_2\rangle = |a\rangle. \tag{V-18}$$

An arbitrary ket may be expanded in any complete set of basic kets,

$$|F\rangle = \sum_k \int c_k |k\rangle \qquad \left( \sum \int \text{ includes } \int dk \right),$$

or
$$\tag{V-19}$$

$$|F\rangle = \sum \int d_\alpha |\alpha\rangle \qquad \left( \sum \int \text{ includes } \int d\alpha \right).$$

Now we introduce, corresponding to the space which is "spanned" by the complete set of basic kets $|k\rangle$, another space, called the dual space to that of $|k\rangle$, with a set of basic vectors denoted by $\langle k|$ called *bras*, in such a way that $\langle k_1|$ corresponds to $|k_1\rangle$. The $\langle k|$ and $|k\rangle$ are in *different* spaces and cannot be added (superposed) together. $\langle k|$ and $|k\rangle$ are "conjugate imaginary" (to distinguish from "complex conjugate") of each other. We assume that $\langle k|$ and $|k\rangle$ are so related that if $\alpha$ is a complex number, then $\alpha|k\rangle$ is a new ket, and the bra corresponding to $\alpha|k\rangle$ is $\alpha^*\langle k|$, where $\alpha^*$ is the complex conjugate of $\alpha$.

We define the product of two kets $|k\rangle$, $|j\rangle$ as the scalar product of $|k\rangle$ with the bra $\langle j|$ corresponding to $|j\rangle$ by the symbol,

$$\langle j|k\rangle = \text{a complex number.}$$

We assume (define) that

$$\langle k|j\rangle = \langle j|k\rangle^*  \qquad\text{(V-20)}$$

so that

$$\langle k|k\rangle = \text{real number,} \qquad\text{(V-21)}$$

i.e., the scalar product of $|k\rangle$ with itself is a real number. This is consistent with what was said about the bra of $\alpha|k\rangle$ is $\alpha^*\langle k|$, for

$$\alpha^*\langle k|\alpha|k\rangle = \alpha^*\alpha\langle k|k\rangle = \text{real.}$$

For finite norms $\langle k|k\rangle$, we may normalize $|k\rangle$,

$$\langle k|k\rangle = 1. \qquad\text{(V-21a)}$$

A set of basic ket $|k\rangle$ is complete and orthonormal if

$$\langle j|k\rangle = \delta_{jk} \qquad\text{(V-22)}$$

$$= \delta(j-k) \text{ for continuous } j,\, k,$$

and the expansion (V-19) is valid. In that case

$$c_k = \langle k|F\rangle,$$

or $\qquad\qquad\qquad\qquad\qquad\qquad\qquad\qquad\qquad\qquad\qquad\qquad$ (V-23)

$$d_\alpha = \langle \alpha|F\rangle.$$

The basic set $|k\rangle$ (or $|\alpha\rangle$) is said to form a representation; $\langle k|F\rangle$ are the "coordinates" of the vector $|F\rangle$ in the $|k\rangle$-representation, and are also called the *representatives* of $|F\rangle$ in the $|k\rangle$-representation.

Any complete set of basic kets forms a representation.

One can transform from one representation to another (from one basic set $|k\rangle$ to another basic set $|\alpha\rangle$) by a linear transformation (somewhat similar to the transformation (V-9).)


*(2) Linear operators, hermitian operators*

An operator $Q$ transforms one ket $|a\rangle$ (any ket) into another ket $|b\rangle$

$$Q|a\rangle = |b\rangle. \qquad\text{(V-24)}$$

$Q$ is said to be linear if

$$Q(c_1|a\rangle + c_2|b\rangle) = c_1 Q|a\rangle + c_2 Q|b\rangle. \qquad\text{(V-24a)}$$

An operator $Q$ operating on a bra is defined by

$$(\langle b|Q)|a\rangle = \langle b|(Q|a\rangle) \tag{V-25}$$

$$= \langle b|Q|a\rangle.$$

The *adjoint* $Q^\dagger$ of $Q$ is an operator such that if $Q|a\rangle = |c\rangle$, $Q^\dagger$ operating (to the left) on the bra $\langle a|$ leads to the bra $\langle c|$, i.e.,

$$Q|a\rangle = |c\rangle, \qquad \langle c| = \langle a|Q^\dagger. \tag{V-26}$$

Take the scalar product of $|c\rangle$ with an arbitrary $|b\rangle$. Then by (V-20), we have

$$\langle b|c\rangle = \langle c|b\rangle^*, \tag{V-27}$$

i.e., from (V-26),

$$\langle b|Q|a\rangle = \langle a|Q^\dagger|b\rangle^*. \tag{V-28}$$

If for $Q$, we take the adjoint $Q^\dagger$, and for $Q^\dagger$ we take the adjoint $(Q^\dagger)^\dagger$ $(\equiv Q^{\dagger\dagger})$ of $Q^\dagger$, then (V-28) becomes

$$\langle b|Q^\dagger|a\rangle = \langle a|Q^{\dagger\dagger}|b\rangle^*, \tag{V-29}$$

or, on taking the complex conjugate,

$$\langle b|Q^\dagger|a\rangle^* = \langle a|Q^{\dagger\dagger}|b\rangle.$$

From this and (V-28), we have

$$\langle b|(Q^{\dagger\dagger} - Q)|a\rangle = 0.$$

Since $|a\rangle$ and $|b\rangle$ are arbitrary, we have

$$Q^{\dagger\dagger} = Q, \tag{V-30}$$

i.e., the adjoint of the adjoint of a linear operator is the operator itself.

A self-adjoint (or, hermitian) operator is one whose adjoint is the operator itself, i.e.,

$$Q^\dagger = Q. \tag{V-31}$$

The following theorems follow from this definition:

**Theorem I.**   If $Q$ is a hermitian operator, then

$$\langle a|Q|b\rangle = \langle b|Q|a\rangle^*. \tag{V-32}$$

**Theorem II.**   If $P$, $Q$ are linear operators, then

$$(PQ)^\dagger = Q^\dagger P^\dagger. \tag{V-33}$$

**Definitions:**   If $Q$ is a (linear) operator and $|k\rangle$ are kets such that

$$Q|k\rangle = q_k|k\rangle, \qquad q_k = \text{a number,}$$

i.e., $Q|k\rangle$ leads to the same ket with a multiplicative constant (a complex number), then $|k\rangle$ is said to be an *eigenket* of $Q$, and $q_k$ an *eigenvalue* of $Q$, corresponding to the eigenket $|k\rangle$. We use the notation

$$Q|q_k\rangle = q_k|q_k\rangle. \tag{V-34}$$

**Theorem III.** If $Q$ is a hermitian operator, the eigenvalues are real, and the eigenkets are orthogonal.

**Proof:** Let $q_1, q_2$ be two eigenvalues of $Q$, and $|q_1\rangle, |q_2\rangle$ their corresponding eigenkets

$$Q|q_1\rangle = q_1|q_1\rangle, \qquad Q|q_2\rangle = q_2|q_2\rangle.$$

From these,

$$\langle q_2|Q|q_1\rangle = q_1\langle q_2|q_1\rangle,$$

$$\langle q_1|Q|q_2\rangle = q_2\langle q_1|q_2\rangle.$$

By (V-32), the last equation becomes

$$\langle q_2|Q|q_1\rangle = q_2^*\langle q_2|q_1\rangle.$$

$$\therefore (q_1 - q_2^*)\langle q_2|q_1\rangle = 0.$$

Setting $q_2 = q_1$, if $\langle q_1|q_1\rangle \neq 0$, then

$$q_1 = q_1^*. \tag*{QED}$$

If $q_1 - q_2 \neq 0$ (i.e., for a non-degenerate system), then

$$\langle q_2|q_1\rangle = 0. \tag*{QED}$$

**Theorem IV.** The eigenkets of a hermitian operator form a complete (orthonormal) set.

For self-adjoint differential operators (Sturm-Liouville problem), there is a similar theorem (Chap. 3, Sec. 4). For Hilbert space, one may take this theorem as an assumption.

**Theorem V.** If $A$, $B$ are two hermitian operators, then

$$AB - BA = 0, \tag{V-35}$$

if and only if there exist a set of kets which are the common eigenkets of $A$ and $B$.

**Proof:** If $|a, b\rangle$ is a common eigenket, i.e.,

$$A|a, b\rangle = a|a, b\rangle, \qquad B|a, b\rangle = b|a, b\rangle,$$

then

$$AB|a, b\rangle = Ab|a, b\rangle = ab|a, b\rangle$$

and

$$BA|a, b\rangle = Ba|a, b\rangle = ba|a, b\rangle$$

and this is consistent with (V-35). But if $AB - BA \neq 0$, there is contradiction and no common set of eigenkets exists. This theorem is similar to Theorem 18 for matrices (Chap. 2, Sec. 2).

## (3) Transformation of representations, unitary operators

Let $A$, $B$ be two hermitian operators, $|a\rangle$, $|b\rangle$ their respective complete set of eigenkets,

$$A|a_m\rangle = a_m|a_m\rangle, \qquad B|b_n\rangle = b_n|b_n\rangle. \tag{V-36}$$

Any $|a_m\rangle$ can be expanded in the set $|b_n\rangle$,

$$|a_m\rangle = \sum_k \int |b_k\rangle \langle b_k|a_m\rangle \tag{V-37a}$$

$$\equiv \sum_k \int |b_k\rangle U_{km}, \tag{V-37b}$$

which defines a matrix $U_{km}$. Similarly

$$|b_n\rangle = \sum_j \int |a_j\rangle \langle a_j|b_n\rangle \tag{V-38a}$$

$$= \sum_j \int |a_j\rangle \langle b_n|a_j\rangle^* \qquad \text{(by (V-20))}$$

$$= \sum_j \int |a_j\rangle U_{nj}^*$$

$$= \sum_j |a_j\rangle \tilde{U}_{jn}^*. \tag{V-38b}$$

Here

$$U_{km} = \langle b_k|a_m\rangle, \qquad \tilde{U}_{jn}^* = \langle a_j|b_n\rangle. \tag{V-39}$$

But $\tilde{U}^* = U^\dagger$, the adjoint of $U$,

$$\tilde{U}^* = U^\dagger,$$

hence (V-37b) and (V-38b) can be written

$$|a_m\rangle = \sum_{j,k} \int |a_j\rangle U_{jk}^\dagger U_{km}, \tag{V-40}$$

i.e.,

$$\sum_k \int U_{jk}^\dagger U_{km} = \delta_{jm}. \tag{V-41a}$$

Thus

$$U^\dagger = U^{-1}, \tag{V-41b}$$

$$\therefore\ U \text{ is unitary.}$$

Going back to (V-37b), (V-38b), we have

$$|a_m\rangle = \sum_k \int |b_k\rangle U_{km}, \qquad |b_n\rangle = \sum_j \int |a_j\rangle U_{jk}^\dagger, \tag{V-42}$$

$$\sum_k \int U_{jk}^\dagger U_{km} = \sum_k \int \langle a_j|b_k\rangle \langle b_k|a_m\rangle = \delta_{jm}, \tag{V-43}$$

$$\sum_k \int |\langle a_j|b_k\rangle|^2 = 1. \tag{V-44a}$$

Similarly[e]

$$\sum_j \int |\langle b_k|a_j\rangle|^2 = 1. \tag{V-44b}$$

From (V-37a), (V-43), it is seen that

$$\sum_k \int |b_k\rangle \langle b_k| = 1. \tag{V-45}$$

This relation is sometimes called the *closure relation*; it has its origin in the completeness of the set of kets $|b_k\rangle$.

$|b_k\rangle \langle b_k|$ is called a *projection operator*; when it acts on $|a_m\rangle$, $|b_k\rangle \langle b_k|a_m\rangle$ "projects" $|a_m\rangle$ onto $|b_k\rangle$.

*(4) Non-commutative hermitian operators satisfying*

$$PQ - QP = \frac{\hbar}{i}. \tag{V-46}$$

If we start with the $Q$-representation, i.e., with the eigenkets $|q_k\rangle$ of $Q$ as the basic kets

$$Q|q_k\rangle = q_k|q_k\rangle, \tag{V-47}$$

---

[e] (V-44a), (V-44b) are the analogues of the relations (V-9),

$$\cos^2(X'X) + \cos^2(X'Y) + \cos^2(X'Z) \equiv 1, \text{ etc.}$$

we wish to find the operator $P$ that satisfies (V-46) and to find the unitary transformation that transforms from the $Q$-representation (V-47) to the $P$-representation in which the basic kets are the eigenkets of $P$

$$P|p_n\rangle = p_n|p_n\rangle. \tag{V-48}$$

The eigenkets $|q_k\rangle$ of $Q$ for various eigenvalues $q_k$ may be regarded as functions of $q$, and we shall introduce the notation

$$|\psi(q)\rangle \qquad \text{for} \qquad |q_k\rangle. \tag{V-49}$$

We shall introduce an operator

$$\frac{1}{i}\frac{d}{dq} \tag{V-50}$$

and we shall show that it is hermitian. Let us assume that it is hermitian. By (V-25), the scalar product of $\langle\phi|(1/i)(\partial/\partial q)$ with $|\psi\rangle$ is

$$\left(\langle\phi|\frac{1}{i}\frac{d}{dq}\right)|\psi\rangle = \langle\phi|\left(\frac{1}{i}\frac{d}{dq}|\psi\rangle\right)$$

$$= \langle\phi|\frac{1}{i}\frac{d}{dq}|\psi\rangle. \tag{V-51}$$

The bra corresponding to $(1/i)(d/dq)|\psi\rangle$ (i.e., the conjugate imaginary to $(1/i)(\partial/\partial q)|\psi\rangle$) is, by (V-26),

$$\langle\psi|\left(\frac{1}{i}\frac{d}{dq}\right)^{\dagger} = \langle\psi|\frac{1}{i}\frac{d}{dq}, \qquad \text{by hypothesis.}$$

The scalar product of $|\psi\rangle$ with $\langle\phi|((1/i)(\partial/\partial q))$ is

$$\left(\langle\psi|\frac{1}{i}\frac{d}{dq}\right)|\phi\rangle^* = \langle\phi|\frac{1}{i}\frac{d}{dq}|\psi\rangle \quad \text{by} \quad \text{(V-32).} \tag{V-52}$$

This agrees with (V-51). Hence $(1/i)(d/dq)$ is hermitian. Let us apply the operator $(1/i)(\partial/\partial q)$ on $q|\psi(q)\rangle$

$$\frac{1}{i}\frac{d}{dq}(q|\psi(q)\rangle) = \frac{1}{i}\left[q\frac{d}{dq}|\psi(q)\rangle + |\psi(q)\rangle\right]$$

$$= \frac{1}{i}\left\{\left(q\frac{d}{dq}+1\right)\right\}|\psi(q)\rangle,$$

or as an operator equation,

$$\frac{1}{i}\left(\frac{d}{dq}q - q\frac{d}{dq}\right) = \frac{1}{i}. \tag{V-53}$$

Comparison with the commutation relation (V-46) shows that the hermitian operator

$$\frac{\hbar}{i}\frac{d}{dq}$$

satisfies the same relation as $P$. Hence we may take

$$P = \frac{\hbar}{i}\frac{d}{dq}. \tag{V-54}$$

This states that when one starts with the $q$-representation, the operator $P$ satisfying (V-46) is given by (V-54).

If we operate on $|\psi(q)\rangle$ by $(\hbar/i)(d/dq)$, we define

$$\frac{\hbar}{i}\frac{d}{dq}|\psi(q)\rangle = \frac{\hbar}{i}\left|\frac{d\psi}{dq}\right\rangle. \tag{V-55}$$

Let us take the representative in the $q$-representation of both side[f]

$$\langle q'|\frac{\hbar}{i}\frac{d}{dq}|\psi(q)\rangle = \left\langle q'\left|\frac{\hbar}{i}\frac{d\psi}{dq}\right.\right\rangle$$

$$= \frac{\hbar}{i}\frac{d}{dq'}\langle q'|\psi\rangle. \tag{V-56}$$

Since $|\psi\rangle$ in (V-56) is arbitrary, we have

$$\langle q'|\frac{\hbar}{i}\frac{d}{dq} = \frac{\hbar}{i}\frac{d}{dq'}\langle q'|. \tag{V-58}$$

From (V-51), using the closure relation (V-45) and the notation (V-57), we have

$$\int \langle\phi|\frac{\hbar}{i}\frac{d}{dq}|q'\rangle \, dq' \, \psi(q') = \int \langle\phi|q'\rangle \, dq' \frac{\hbar}{i}\frac{d}{dq'}\psi(q')$$

$$= -\int \frac{\hbar}{i}\frac{d}{dq'}\langle\phi|q'\rangle \, dq' \, \psi(q')$$

$$= -\int \langle\frac{\hbar}{i}\frac{d\phi}{dq}\Big|q'\rangle \, dq' \, \psi(q'); \tag{V-59}$$

[f] $\langle q'|\psi\rangle$, $\langle q'|d\psi/dq\rangle$ are, respectively, the representatives of $|\psi\rangle$, $|d\psi/dq\rangle$ in the $q$-representation. For convenience, we write

$$\psi(q') \equiv \langle q'|\psi\rangle, \qquad \frac{d\psi(q')}{dq'} \equiv \langle q'\left|\frac{d\psi}{dq}\right\rangle$$

$$= \frac{d}{dq'}\langle q'|\psi\rangle. \tag{V-57}$$

if the functions $\langle\phi|q'\rangle$, $\psi(q')$ vanish on the boundary,

$$\langle\phi|\frac{\hbar}{i}\frac{d}{dq} = -\langle\frac{\hbar}{i}\frac{d\phi}{dq}\bigg|. \tag{V-60}$$

By a procedure similar to that from (V-55) to (V-58), we obtain

$$\frac{\hbar}{i}\frac{d}{dq}|q'\rangle = -\frac{\hbar}{i}\frac{d}{dq'}|q'\rangle. \tag{V-61}$$

With $P = (\hbar/i)(d/dq)$ in (V-54), we can write (V-58), (V-61) in the form (using $p$ to represent $P$)

$$p|q'\rangle = -\frac{\hbar}{i}\frac{d}{dq'}|q'\rangle,$$

$$\langle q'|p = \frac{\hbar}{i}\frac{d}{dq'}\langle q'|. \tag{V-62}$$

Thus the operator $p$ on the ket $|q'\rangle$ is[8]

$$p = -\frac{\hbar}{i}\frac{d}{dq'}. \tag{V-63}$$

If we start with the $p$-representation (V-48), then, taking the representatives in the $q$-representation, we have

$$\langle q'|p|p'\rangle = \langle q'|p'|p'\rangle$$

$$= p'\langle q'|p'\rangle.$$

Using (V-62),

$$\frac{\hbar}{i}\frac{d}{dq'}\langle q'|p'\rangle = p'\langle q'|p'\rangle. \tag{V-64}$$

This is the familiar form when the notation (V-57) is used, namely, $\langle q'|p'\rangle$ is the representative of $|p\rangle$ in the $q$-representation, and $\langle q'|p'\rangle$ is the Schrödinger wave function $\psi(q)$. In fact (V-64) is the Schrödinger equation for the momentum $p$.[h]

From (V-39), it is seen that $\langle q'|p'\rangle$ is the transformation matrix element from the $p$-representation to the $q$-representation. (V-64) is the differential equation for $\langle q'|p'\rangle$. The solution is

$$\langle q'|p'\rangle = C(p')\exp(ip'q'/\hbar). \tag{V-65}$$

---

[8] Note that in (V-62), (V-63), $p = -(\hbar/i)(\partial/\partial q)$ when it operates on the ket $|q\rangle$. When $p$ operates on the representatives $\langle q|A\rangle$, $p$ is $(\hbar/i)(\partial/\partial q)$, as shown in (V-64).
[h] See Chap. 2, Sec. 3, Probs. 1 and 2.

From (V-37a) and (V-44a), we have

$$\langle p'|p''\rangle = \int_{-\infty}^{\infty} \langle p'|q'\rangle \, dq' \langle q'|p''\rangle$$

$$= C^*(p')C(p'') \int_{-\infty}^{\infty} \exp[-i(p'-p'')q'/\hbar] \, dq'$$

$$= |C(p')|^2 2\pi\hbar\delta(p'-p''). \tag{V-66}$$

From (V-27),

$$\langle p'|q'\rangle = C^*(p')\exp(-ip'q'/\hbar). \tag{V-67}$$

But had one started with the $p$-representation, an entirely similar procedure as above (with $p$, $q$ interchanged and $i$ changed into $-i$) would have led, instead of (V-54), (V-65), to

$$q = -\frac{\hbar}{i}\frac{\partial}{\partial p} \tag{V-68}$$

and

$$\langle p'|q'\rangle = C^*(q')\exp(-ip'q'/\hbar). \tag{V-69}$$

Comparison between (V-67) and (V-69) shows that $C^*(q') = C^*(p')$ must be independent of $q'$ or $p'$ and $C = $ constant. From the normalization condition (V-66), one has

$$|C| = \frac{1}{\sqrt{h}},$$

so that finally

$$\langle q'|p'\rangle = \frac{1}{\sqrt{h}}e^{ip'q'/\hbar}. \tag{V-70}$$

For an arbitrary ket, say $|X\rangle$, the relations between its representatives $\langle q'|X\rangle$, $\langle p'|X\rangle$ in the $q$- and $p$-representation are given by (V-37a),

$$\langle q'|X\rangle = \int \langle q'|p'\rangle \, dq' \langle p'|X\rangle$$

$$= \frac{1}{\sqrt{h}} \int_{-\infty}^{\infty} e^{ip'q'/\hbar} \, dq' \langle p'|X\rangle, \tag{V-71a}$$

$$\langle p'|X\rangle = \frac{1}{\sqrt{h}} \int_{-\infty}^{\infty} e^{-ip'q'/\hbar}\, dp' \langle q'|X\rangle. \tag{V-71b}$$

Thus $\langle q'|X\rangle$ and $\langle p'|X\rangle$ are Fourier transforms of each other.[i]

### (5) Unitary transformations[j]

Let $a$ be any linear operator; $a'$, $|a'\rangle$ its eigenvalues and eigenkets:

$$a|a'\rangle = a'|a'\rangle.$$

Let $S$ be an arbitrary linear operator whose inverse $S^{-1}$ exists,

$$S^{-1}S = 1.$$

Let $a$ be transformed by $S$ according to

$$A = SaS^{-1}. \tag{V-72}$$

Then one has

$$AS|a'\rangle = SaS^{-1}S|a'\rangle = Sa|a'\rangle$$
$$= a'S|a'\rangle, \tag{V-73}$$

i.e., $A$ has the eigenvalues $a'$ with the corresponding eigenkets $S|a'\rangle$. Thus

**Theorem:** The eigenvalues of any linear operator are invariant under transformations by arbitrary linear operators.

**Theorem:** Any linear operator $a$ can be expressed as the sum of two linear operators

$$a = b + ic, \tag{V-74}$$

where $b$, $c$ are hermitian operators.

Let $b$ be a hermitian operator, and let $U$ be a linear operator such that

$$B = UbU^{-1} \tag{V-75}$$

is also hermitian. From this, one has

$$BU = Ub, \tag{V-75a}$$

and taking the adjoint of both side, with (V-33), one has

$$U^\dagger B = bU^\dagger. \tag{V-76}$$

---

[i] In (IV-50), (IV-51), (IV-52), we now see that they are, respectively, (V-64), (V-70) and (V-71a). We have also seen them in Chap. 2, Sec. 3, Exercise 1, Eqns. (A-2)–(A-7).
[j] Most of the results of this section have already been known in matrix form (see Chap. 2, Sec. 2.).

(V-75a) gives

$$U^\dagger BU = U^\dagger Ub,$$

(V-76) gives

$$U^\dagger BU = bU^\dagger U.$$

Hence

$$U^\dagger Ub - bU^\dagger U = 0. \tag{V-77}$$

Thus $U^\dagger U$ commutes with any hermitian operator, and by the theorem of (V-74), with any linear operator. Hence $U^\dagger U$ is a number $c$ times the unit operator. Since, by (V-33),

$$(U^\dagger U)^\dagger = U^\dagger U,$$

$U^\dagger U$ is hermitian. Hence $c$ is a real number. Since for any ket $|A\rangle$,

$$\langle A|A \rangle = \text{a positive real number},$$

$$\langle A|U^\dagger U|A \rangle \text{ must be a real positive number.}$$

Hence we may choose $c = 1$ and

$$U^\dagger U = 1, \qquad \text{i.e., } U \text{ is unitary.}$$

**Theorem:**  A unitary transformation (V-75) leaves the hermitian character of an operator unchanged.

**Theorem:**  If $F$ is a hermitian operator,

$$U = e^{iF} \tag{V-78}$$

is a unitary operator.

   If $\varepsilon$ is an infinitesimal real number, and if in (V-78) $F$ is replaced by $\varepsilon F$, then

$$U = 1 + i\varepsilon F \tag{V-79}$$

defines an infinitesimal unitary transformation, for then (V-75) becomes

$$B = b + i\varepsilon(Fb - bF), \tag{V-80a}$$

where the hermiticity of $F$ ensures the hermiticity of $B$ since $i(Fb - bF)$ is hermitian.

   With the quantum Poisson bracket

$$[A, B] \equiv \frac{1}{i\hbar}(AB - BA),$$

(V-80a) can be expressed as

$$B = b + \varepsilon\hbar[b, F].\tag{V-80b}$$

## (6) Displacement operator

Here we are dealing with opertions involving the translational spatial displacements.

Let $|k\rangle$ be an arbitrary ket, and $|k\Delta\rangle$ the ket after a (parallel, spatial) displacement $\Delta$. Let $D$ be the operator

$$D|k\rangle = |k\Delta\rangle,\tag{V-81}$$

so that

$$\langle k\Delta| = \langle k|D^\dagger.$$

From

$$\langle k|D^\dagger D|k\rangle = \langle k\Delta|k\Delta\rangle = \langle k|k\rangle,$$

it follows that

$$D^\dagger D = 1,\tag{V-82}$$

i.e., $D$ is a unitary operator.

Let $\alpha$ be an arbitrary linear operator

$$\alpha|j\rangle = |k\rangle.$$

After the displacement $\Delta$, this relation is transformed into

$$\alpha_\Delta|j\Delta\rangle = |k\Delta\rangle.$$

Using (V-81), this becomes

$$\alpha_\Delta|j\Delta\rangle = |k\Delta\rangle = D|k\rangle = D\alpha|j\rangle = D\alpha D^{-1}|j\Delta\rangle,$$

i.e.,

$$\alpha_\Delta = D\alpha D^{-1}.\tag{V-83}$$

To find $D$, take an infinitesimal displacement $\delta x$, and define an operator $d_x$ by

$$d_x|k\rangle = \lim_{\delta x \to 0} \frac{|k\Delta\rangle - |k\rangle}{\delta x}$$

$$= \lim_{\delta x \to 0} \frac{D - 1}{\delta x}|k\rangle,\tag{V-84}$$

which can be written

$$D = 1 + \delta x\, d_x, \qquad \delta x \to 0,$$
$$D^\dagger = 1 + \delta x\, d_x^\dagger, \qquad \delta x \to 0. \tag{V-84a}$$

The condition $D^\dagger D = 1$ leads to $\delta x(d_x + d_x^\dagger) = 0$, i.e.,

$$d_x = -d_x^\dagger \tag{V-84b}$$

so that $d_x$ is a pure imaginary operator.

From (V-83), one has the infinitesimal operator

$$\alpha_\Delta = \alpha + \delta x(d_x \alpha - \alpha\, d_x), \tag{V-85a}$$

$$= \alpha + i\hbar \delta x[\alpha, d_x]. \tag{V-85b}$$

The meaning of $d_x$ can be obtained as follows. Suppose the operator $\alpha$ is the $x$ coordinate, and $D$ is the operator of displacing the measuring apparatus by $\delta x$. Hence the coordinate $x_\Delta = x - \delta x$, i.e., in (V-85a),

$$d_x x - x\, d_x = -1. \tag{V-86}$$

Comparing this with the commutation relation

$$px - xp = \frac{\hbar}{i},$$

one sees

$$d_x = -\frac{i}{\hbar} p \tag{V-87}$$

and for infinitesimal displacement,

$$D = 1 - \frac{i}{\hbar} p\delta x. \tag{V-88}$$

For finite displacement, from (V-78), (V-79), one may take

$$D_{x'} = \exp\left(-\frac{i}{\hbar} x'p\right). \tag{V-89}$$

Let $\psi(x)$ be the representative in the $x$-representation of an arbitrary ket. From (V-64),

$$p\psi(x) = \frac{\hbar}{i} \frac{d}{dx} \psi(x).$$

Hence[k]

---

[k] This is obtained by expanding (V-89) and using the Taylor theorem. Similarly for (V-91) below.

$$D_{x'}\psi(x) = \psi(x - x').  \tag{V-90}$$

For the $x$-representation,

$$x|x'\rangle = x'|x'\rangle,$$

one has from (V-62),

$$p|x\rangle = -\frac{\hbar}{i}\frac{d}{dx}|x\rangle,$$

and

$$D_{x'}|z\rangle = |x + x'\rangle.  \tag{V-91}$$

This means that $D_x, |x\rangle$ gives the eigenvalue $x + x'$, or[1]

$$x(D_{x'}|x\rangle) = (x + x')|x + x'\rangle  \tag{V-91a}$$

$$= (x + x')(D_{x'}|x\rangle).$$

From (V-89), or (V-90), one sees

$$(D_a)^2 = D_{2a}.  \tag{V-92}$$

Let $|k\rangle$ be the eigenket of $D_a$, and let $\phi_k(x)$ be the $x$-representative of $|k\rangle$. By (V-89)

$$D_a\psi_k(x) = e^{-ika}\psi_k(x), \qquad (k = \text{real number}).  \tag{V-93}$$

From (V-90), one has

$$\psi_k(x - a) = e^{-ika}\psi_k(x),  \tag{V-94}$$

which is a functional equation whose solution is

$$\psi_k(x) = e^{ikx}u_k(x),  \tag{V-95a}$$

where $u_k(x)$ is a periodic function of $x$, of period $a$

$$u_k(x) = u_k(x - a);  \tag{V-95b}$$

equations (V-95a, b) are characteristic of the electronic wave functions in a crystal lattice (or in metals). They come from the unitary operator $D_a$ for translational (spatial) displacements.

For rotational displacements, the unitary operator can be obtained in a similar manner. Let $\phi$ be the angle of rotation about the $z$-axis. The $z$-component of the angular momentum is

---

[1] This result is the same as that of Prob. 7, Chap. 2, Sec. 2.

$$L_z = \frac{\hbar}{i}\frac{\partial}{\partial\phi}.$$

The unitary operator for rotation through an angle $\alpha$ is (see (V-89))

$$D_\alpha = \exp\left(-\frac{i}{\hbar}\alpha L_z\right),$$

$$(D_\alpha)^2 = D_{2\alpha}, \qquad \text{(see (V-92))}$$

$$D_\alpha\psi(\phi) = \psi(\phi - \alpha), \qquad \text{(see (V-90))}$$

$$D_\alpha\psi_m(\phi) = e^{-im\alpha}\psi_m(\phi), \quad m = \text{integer}, \qquad \text{(see (V-93))}$$

$$\psi_m(r, \vartheta, \phi) = e^{im\phi}W_m(r, \vartheta, \phi), \qquad \text{(see (V-95a))}$$

$$W_m(r, \vartheta, \phi) = W_m(r, \vartheta, \phi - \alpha), \quad \alpha = \frac{2\pi}{n} \qquad \text{(see (V-95b))}.$$

### *(7) Time translation operator $U(t)$*

The nature of the problem is similar to that of spatial displacement of subsection (6) above. Let the ket $|k, t_0\rangle$ at time $t_0$ be translated to $|k, t\rangle$ at time $t$ (after $t_0$), and let the operator be $U$,

$$|k, t_0\rangle = U|k, t_0\rangle, \tag{V-96}$$

where $U(t - t_0)$ can be shown (see V-82, 83) to be unitary

$$U^\dagger U = 1.$$

Define the operator (see (V-84))

$$u_t \equiv \frac{d|k, t_0\rangle}{dt_0} = \lim_{t \to t_0}\frac{U - 1}{t - t_0}|k, t_0\rangle. \tag{V-97}$$

The $u_t$ can be shown to be a pure imaginary operator and by analogy with (V-86), (V-87) one suggests

$$i\hbar\frac{d}{dt_0} = \text{a hermitian operator.} \tag{V-98}$$

But the analogy with the spatial displacement ends here. We shall see that between the variable $t$ and the Hamiltonian $H$, there does *not* exist a relation analogous to (V-46), namely,

$$Ht - tH = -\frac{\hbar}{i}. \tag{V-99}$$

The argument for this statement will be given in the following section (Sec. 3) (V-105)–(V-108).

## Exercises

1. Show that the eigenfunctions of the displacement operator $d_x$, (V-87), are

$$e^{ikx}u_k(y, z).$$

Show that these functions are also the eigenfunctions of $p_x$ of eigenvalues $\hbar k$.

2. Let the unit cell of a metal crystal be represented by three vectors $\mathbf{a}_1$, $\mathbf{a}_2$, $\mathbf{a}_3$. Let the Hamiltonian for an electron in the crystal be

$$H = -\frac{\hbar^2}{2m}\nabla^2 + V(r),$$

$$V(r) = V(\mathbf{r} + n_1\mathbf{a}_1 + n_2\mathbf{a}_2 + n_3\mathbf{a}_3), \qquad n_1, n_2, n_3 = \text{integers},$$

$$= \text{a periodic potential.}$$

Show that the displacement operators $D_{a_1}$, $D_{a_2}$, $D_{a_3}$ (see (V-89)) commute with $H$. Show that

$$D_{a_j}\psi_k(\mathbf{r}) = \exp(-i\mathbf{k}\cdot\mathbf{a}_j)\psi_k(\mathbf{r}),$$

$$\psi_k(\mathbf{r}) = e^{i\mathbf{k}\cdot\mathbf{r}}u_k(\mathbf{r}).$$

3. For a central field, show that the rotation operator $D_\alpha$ (rotation about the $z$-axis through an angle $\alpha$) commutes with the Hamiltonian, and

$$D_\alpha V(r)\psi(r, \vartheta, \phi) = V(r)\psi(r, \vartheta, \phi - \alpha).$$

Show that for infinitesimal rotation $\alpha$, the operator $d_\alpha$

$$d_\alpha = -\frac{i}{\hbar}L_z$$

and from the commutation relation, show that the angular momentum components $L_x$, $L_y$, $L_z$ and $\mathbf{L}^2 = L_x^2 + L_y^2 + L_z^2$ commute with $H$.

4. An electron moves freely in one dimension in a metal crystal. The Hamiltonian

$$H = -\frac{\hbar^2}{2m}\frac{d^2}{dx^2}$$

has eigenfunctions

$$\phi(x) = \frac{1}{\sqrt{L}}e^{ikx}, \qquad k = \frac{2\pi n}{L}, \qquad n = \text{integer}.$$

Suppose $V(x)$ is a periodic perturbation

$$V(x) = V(x + a) = \sum V_{2\pi n/a} \exp\left(\frac{2\pi i n}{a} x\right).$$

Calculate, up to the second order, the effect of this perturbation. At what points $x$ will this calculation fail? Find the exact zeroth-order wave function at these points, and the energy spectrum of the system. Hint: The operator (V-89) commutes with the Hamiltonian.

## 3. Quantum mechanics in postulational form

We are now prepared to present the basic postulates of quantum mechanics.

### Complementarity postulates

*Postulate I.* The state of a physical system is represented by a vector (ket) in a linear vector space of infinitely many dimensions.

We shall make the following explanatory remarks:

(1) In view of the uncertainty relation, it is not possible to define, as in classical dynamics, the state of a system by specifying the coordinates and momenta of the particles at any given instant of time.

(2) A vector (ket) is to represent a state to allow the superposition of states. By hypothesis only the direction (not the *sense* such as the "arrow"), but not its length, defines a state. Thus multiplying a ket by any constant does not lead to a different state.

(3) The vector space of infinitely many dimensions is an abstract space, and not the configuration space nor the momentum space of the particles of a system.

(4) The changes of the state of a system are represented by changes in the ket. The Newtonian equations of motion are replaced by an equation for the change of the ket of a system.

*Postulate II.* All physical quantities ("observables") are represented by linear hermitian operators, which, acting on the state vector (ket), change the state of a system.

(1) Linearity is assumed for a linear theory in the state vectors to allow for superposition of state vectors.

(2) Hermitian operators are employed because hermitian operators have real eigenvalues. Real eigenvalues are needed on account of another basic postulate which connects the mathematical formalism (this and the preceding

postulate) with *physics*—the result of physical observations (see Probability Postulate in the following).

(3) The choice of the eigenvectors of an observable as the basis set of vectors immediately leads to the concepts of representations and transformation of representations. The requirement of invariance of the hermitian charactor of operators leads to unitary transformations.

(4) The transformation theory and the Probability Postulate (see IV below) build up Quantum Mechanics into a coherent system.

*Postulate III.* The hermitian operators for the coordinate $q$ and the conjugate momentum (in the sense of classical dynamics) obey the commutation relation

$$pq - qp = \frac{\hbar}{i} 1. \qquad \text{(V-99)}$$

(1) This relation, containing the Planck constant $h$, causes the system of quantum mechanics to have a fundamental break with classical physics.

(2) Together with the Probability Postulate (that connects the above mathematical formalism with physics), it leads to the uncertainty relation.

(3) In this sense, it may be regarded as the mathematical expression of the Einstein-de Broglie relations.

(4) It is a mathematical expression of the concept of complementarity.

(5) The commutation relation (V-99) may be taken to be a basic principle in quantum mechanics. The postulates I and II have been made for the construction of a mathematical system to cope with the non-classical commutation relation.

The above three postulates are essentially of a mathematical nature and have no real contact with the results of physical measurements. This contact is provided by the Probability Postulate.

*Postulate IV.* When a measurement is made of a physical quantity (represented by the hermitian operator $Q$) on a system in a state represented by a vector $|a\rangle$, the expectation value is given by

$$\langle Q \rangle = \langle a|Q|a \rangle, \qquad \text{(V-100)}$$

(1) This postulate contains the main physical contents of the theory; for it makes theoretical statements of the result of experimental measurements.

(2) If $|a\rangle$ is an eigenstate of $Q$, say $|a\rangle = |q_k\rangle$, then the expectation value of the measurement is

$$\langle Q \rangle = \langle q_k | Q | q_k \rangle = q_k \langle q_k | q_k \rangle$$

$$= q_k, \tag{V-101}$$

i.e., in this case, the measurement will definitely yield, with certainty, the eigenvalue $q_k$ of $Q$.

(3) If $|a\rangle$ in (V-100) is an arbitrary state, it can be expanded in the complete set of eigenkets of $Q$,

$$|a\rangle = \sum_k \int |q_k\rangle \langle q_k | a\rangle \tag{V-102}$$

and

$$\langle a| = \sum_k \int \langle a | q_k\rangle \langle q_k|$$

and

$$\langle Q \rangle = \sum_k \int |\langle a | q_k\rangle|^2 q_k. \tag{V-103}$$

This states that the result of measurement is not a definite one of the eigenvalues, not an in-between value of the $q_k$'s, but *one* of all the $q_k$'s each of which may come out with a probability $|\langle a|q_k\rangle|^2$. The probabilities themselves are definite enough, as they are given by (V-103) but the theory does not tell which *one* of the $q_k$'s will *actually* come out from the measurement.

(4) It is most important to note that the nature of this probability distribution

$$|\langle a|q_k\rangle|^2, \text{ for various } k,$$

is basically different from that in classical statistical physics. In the latter, probability is introduced, when dealing with an extremely large number of molecules, as a substitute for the exact knowledge which exits in principle. In quantum mechanics, even for one atom, a precise knowledge does *not* exist and probability is of an intrinsic nature.

For this reason, quantum mechanics differs from classical physics in a basic way.

(5) Then one may ask why such a probability concept is necessary. For one thing, since two non-commuting operators do not have simultaneous eigenkets, this probability postulate (employed in the derivation of the uncertainty relation) forms at least a *consistent* theory from the point of view of the *anschaulich* thought-experiments in Heisenberg's demonstration of the uncertainty relation on the Einstein-de Broglie relations.

(6) A much deeper and more difficult question is whether it is possible to assume the existence of "hidden variables" and the probability concept arises as the result of a sort of averaging over those hidden variables.[m] If this is the case, the probability will not be "intrinsic" but has the same classical meaning.

From the experimental point of view, when measurements are made, not on one atom, but on a large number of atoms, it is not possible to distinguish between the two kinds of probabilities. The question is a philosophical one—one concerning the nature of quantum mechanics. The concept of an intrinsic probability is repugnant to some philosophers of science; the most prominent ones have been Einstein, Planck and de Broglie. The question has been studied by von Neumann and the answer seems to be that an interpretation of probability in terms of hidden variables is inconsistent in the present system of quantum mechanics. We shall not go into this deep question, but refer to a review of critical studies by Belinfante.

The four postulates above have not included considerations regarding of the change of a system in time. For this, a separate postulate is furnished by the Schrödinger equation:

*Postulate V.* The change of the state of a physical system in time is governed by the equation

$$-\frac{\hbar}{i}\frac{\partial}{\partial t}|a,t\rangle = H|a,t\rangle,$$

or

$$\left(\frac{\hbar}{i}\frac{\partial}{\partial t} + H(q,p,t)\right)\Psi(q,t) = 0, \tag{V-104}$$

where $\Psi(q,t)$ is the representative $\langle q,t|a,t\rangle$ of $|a,t\rangle$ in the $q$-representation.

(1) This equation is an independent postulate, unlike the Schrödinger equation for momentum, (V-64)

$$\left(\frac{\hbar}{i}\frac{\partial}{\partial q} - p\right)\psi_p(q) = 0,$$

which can be derived from the commutation relation $pq - qp = \hbar/i$ on the transformation theory.

---

[m] In the classical kinetic theory of gases, the "hidden variables" are the coordinates and momenta of the individual molecules. With these variables, the dynamical theory is a deterministic one; but on averaging over, and therefore suppressing, these variables, the theory works with probable values.

(2) In classical dynamics, the time variable $t$ and the energy (Hamiltonian) $H$ formally behave as a pair of canonical conjugate variables $(t, -H)$. It might then be thought that, analogously to $(q, p)$, a commutation relation

$$Ht - tH = -\frac{\hbar}{i}\mathbf{1} \qquad\qquad \text{(V-105)}$$

exists, and if so, then one may have a time representation in which the hermitian operators are:

$$t, \qquad H = -\frac{\hbar}{i}\frac{\partial}{\partial t}, \qquad\qquad \text{(V-106)}$$

and an energy representation in which the hermitian operators are

$$H, \qquad t = \frac{\hbar}{i}\frac{\partial}{\partial E}. \qquad\qquad \text{(V-107)}$$

But if two hermitian operators $H$ and $t$ exist that satisfy the commutation relation (V-105), it can be proved that, if $E_a$, $\psi_a$ are the eigenvalue and eigenvector of $H$

$$H\psi_a = E_a\psi_a,$$

then

$$H(e^{i\varepsilon t/\hbar}\psi_a) = (E_a - \varepsilon)(e^{i\varepsilon t/\hbar}\psi_a), \qquad\qquad \text{(V-108)}$$

i.e., $E_a - \varepsilon$ is the eigenvalue of $H$, for arbitrary, real value of $\varepsilon$, with the eigenvector $e^{i\varepsilon t/\hbar}\psi_a$.[n] As $\varepsilon$ is real and arbitrary, it follows that $H$ has a continuous spectrum of eigenvalues extending from $-\infty$ to $+\infty$. But this is not true of systems in general; in fact negative infinite total energy is unphysical. Also systems exist whose energy spectrum is not continuous.

Thus no hermitian operator $t$ exists which satisfies a commutation relation like (V-105).[o] For this reason, it is not possible to derive the relations (V-106), (V-107) from a relation like (V-105).

The Schrödinger equation (V-104) must then be based on a different line of arguments.

In Sec. 2, (V-98), it is seen that

$$i\hbar\frac{d}{dt} = \text{a hermitian operator.}$$

<hr>

[n] See Prob. 7 at the end of Chap. 2, Sec. 2. Also (V-91a).
[o] This was pointed out by W. Pauli in his article in *Handbuch der Physik*, Bd. 24, 2nd ed. (Springer-Verlag, Heidelberg, 1933).

Hence it is reasonable to *postulate* (V-106)[p]

$$H = -\frac{\hbar}{i}\frac{\partial}{\partial t} \tag{V-109}$$

and hence the Schrödinger equation (V-104).

(3) The Schrödinger equation (V-104)

$$\left(\frac{\hbar}{i}\frac{\partial}{\partial t} + H\right)\Psi(q,t) = 0$$

is a first-order differential equation in $t$, but second-order with respect to $x$, $y$, $z$. Thus its form is not Lorentz covariant. Attempts to have a relativistic wave equation have been made.[q] These are the Klein-Gordan equation and the Dirac equation.

(4) An argument for an equation which is of the first order in $\partial/\partial t$ is as follows. From the normalization

$$\int \Psi^*(q,t)\Psi(q,t)\,dq = 1,$$

one has

$$\int\left(\Psi^*\frac{\partial\Psi}{\partial t} + \frac{\partial\Psi^*}{\partial t}\Psi\right)dq = 0. \tag{V-110}$$

To solve an initial value problem of an equation of first order in $\partial/\partial t$, one has to specify $\Psi$ at $t = t_0$. For a second order in $\partial/\partial t$, it is necessary to specify $\Psi$ and $\partial\Psi/\partial t$ at $t = t_0$, but (V-110) does not allow arbitrary $\Psi$ and $\partial\Psi/\partial t$.

(5) The Schrödinger equation (V-104) determines $\Psi(q,t)$ in the course of time completely deterministically, and hence the probability $|\Psi(q,t)|^2$, but only the *probability* distribution, and not the actual outcome of an experiment! One may say that quantum mechanics is a "causal theory of probabilities".

(6) There are "constants of motion" of the Schrödinger equation. For example, take a system with central symmetry, i.e., the Hamiltonian $H$ is invariant under parity operation (inversion with respect to a center, see Chap. 3, Sec. 4, (3), (III-115)–(III-120)) If at an instant $t_0$ the parity of the state of the system is *even*, the parity will remain *even* in time, if the system is not perturbed by external fields, such as electromognetic fields.

[p] This is the view of Dirac in *The Principles of Quantum Mechanics*, Sec. 27, 4th Ed. (Clarendon Press, Oxford, 1958).
[q] See references to Schrödinger, Klein, Gordon, de Broglie and Fock at the end of Sec. 3 of Chap. 3.

As another example, take an atom with two electrons, such as helium, whose Hamiltonian is symmetric with respect to the interchange of the two electrons. The space part (i.e., without the spins) of the wave function of the two electrons may be symmetric or antisymmetric with respect to this interchange, (although the total wave function including both the space and the spin parts must be antisymmetric, according to Pauli's principle, which constitutes another independent postulate). If at one instant the state is symmetric, this symmetry will remain unchanged in time when the state changes in accordance with the Schrödinger equation, provided the system is not perturbed by external fields affecting the electron spins.

Such "constants of motions" are essentially the consequence of the symmetry properties of the system.

(7) For systems whose Hamiltonian $H$ is not explicitly a function of time, then the assumption of the form

$$\Psi(q, t) = \psi_n(q) \exp\left(-\frac{iE_n t}{\hbar}\right) \tag{V-111}$$

leads to the time-independent Schrödinger equation

$$(H(q, p) - E_n)\psi_n(q) = 0. \tag{V-112}$$

We have seen a few examples of this equation in Chap. 3.

In subsection (6) above, we mentioned that for a system containing identical particles, there is the question of the symmetry of the state vector, or its representative $\psi(q_1, q_2, q_3, \ldots)$ in coordinate representation. This question arises as follows: Take a system containing two identical particles (for example, electrons), and let us denote them simply by the indices 1 and 2. The Schrödinger equation is

$$(H(1, 2) - E)\psi(1, 2) = 0.$$

Let $P$ be the operator interchanging the two identical particles. Hence in general

$$P^2\psi(1, 2) = \psi(1, 2),$$

so that the eigenvalue of $P^2$ is 1, and those of $P$ are $\pm 1$, and corresponding to them, the function $\psi(1, 2)$ is either symmetric with respect to $P$, i.e.,

$$P\psi_s(1, 2) = \psi_s(1, 2),$$

or antisymmetric, i.e.,

$$P\psi_a(1, 2) = -\psi_a(1, 2).$$

Both $\psi_s$, $\psi_a$ are solutions of the Schrödinger equation because

$$PH(1,2) = H(2,1) = H(1,2).$$

The question is: Does nature allow both kinds of solutions?

The answer to this question came, paradoxically, before the question was asked, i.e., before quantum mechanics was born. For the electrons in atoms, Pauli postulated, the exclusion principle early in 1925. The connection with the symmetry of the wave function was first given by Dirac (1926). This principle can now be stated as follows.

*Postulate VI.* For a system of identical particles of *half-odd-integral* spin (such as electrons, muons, protons, neutrons), the wave function is *antisymmetric* with respect to the interchange of two particles. For identical particles of *integral spin* (such as pions, deuterons, $\alpha$-particles), the $\psi$ is *symmetric* with respect to the interchange of two particles.

This principle (including the Pauli principle) really, so to speak, does not belong to the group of postulates that form the system of quantum mechanics. It is directly related to the statistics—the Fermi-Dirac and the Bose-Einstein statistics. The relationship among the symmetry, the spin and the statistics is regarded as a profound law of nature.

## 4. Representation; measurement

### (1) Representation

In Sec. 2, (2), we saw that two commuting operators have common eigenvectors. We shall now introduce the concept of a "complete set of commuting observables".

Let $\alpha$, $\beta$, $\gamma$, ... be mutually commuting observables and let $|s\rangle$ be their common eigenkets, i.e.,

$$\alpha|s\rangle = a|s\rangle, \qquad \beta|s\rangle = b|s\rangle, \ldots \tag{V-113}$$

Then any function of $\alpha$, $\beta$, $\gamma$, ... formed of sums and products, $f(\alpha,\beta,\gamma,\ldots)$, also has the same eigenkets

$$f(\alpha,\beta,\gamma,\ldots)|s\rangle = f(a,b,c,\ldots)|s\rangle. \tag{V-114}$$

If $\alpha$, $\beta$, $\gamma$, ... are mutually commuting observables, (and none is a function of the others) and if there does not exist any other observable that commutes with $\alpha$, $\beta$, $\gamma$, ..., then the set $\alpha$, $\beta$, $\gamma$, ... is said to form "a complete set of commuting observables".

If $|s\rangle$ are the eigenkets of a complete set of commuting observables, and if an arbitrary ket $|p\rangle$ can be expressed as

$$|p\rangle = f(\alpha, \beta, \gamma, \ldots)|s\rangle, \qquad\qquad\qquad \text{(V-115)}$$

then $f(\alpha, \beta, \gamma, \ldots)$ may be said to be the representation of $|p\rangle$.

It can be shown that this representation is unique; for if $f_1$ and $f_2$ are two representations

$$|p\rangle = f_1|s\rangle, \qquad |p\rangle = f_2|s\rangle,$$

then

$$(f_1 - f_2)|s\rangle = 0. \qquad\qquad\qquad \text{(V-116)}$$

If another arbitrary ket $|q\rangle$ is, like (V-115)

$$|q\rangle = g(\alpha, \beta, \gamma, \ldots)|s\rangle,$$

and since the set $\alpha$, $\beta$, $\gamma$, $\ldots$ is by hypothesis complete, $g$ and $f$ must commute, one has

$$(f_1 - f_2)|q\rangle = g(f_1 - f_2)|s\rangle$$
$$= 0, \qquad \text{by (V-116)}.$$

Since $|q\rangle$ is arbitrary, one has

$$f_1 = f_2.$$

The completeness of the set $\alpha$, $\beta$, $\gamma$, $\ldots$ is to be understood in the following sense. Let $W$ be an observable which commutes with each one of $\alpha$, $\beta$, $\gamma$, $\ldots$ and let

$$W|s\rangle = F(\alpha, \beta, \gamma, \ldots)|s\rangle. \qquad\qquad\qquad \text{(V-117)}$$

Take $W|p\rangle$. By (V-115),

$$W|p\rangle = Wf|s\rangle$$
$$= fW|s\rangle \qquad \text{since } Wf - fW = 0,$$
$$= fF|s\rangle \qquad \text{by (V-117)},$$
$$= Ff|s\rangle \qquad \text{since } Ff - fF = 0,$$
$$= F|p\rangle \qquad \text{by (V-115)}.$$

Since $|p\rangle$ is arbitrary, one has

$$W = F(\alpha, \beta, \gamma, \ldots), \qquad\qquad\qquad \text{(V-118)}$$

i.e., any observable that commutes with $\alpha$, $\beta$, $\gamma$, $\ldots$ must be a function of $\alpha$, $\beta$, $\gamma$, $\ldots$ Hence the set $\alpha$, $\beta$, $\gamma$, $\ldots$ is complete.

This concept of a complete set of commuting variables is not present in classical dynamics. For example, for a non-relativistic particle in a central field, the complete set of commuting variables consists of the total energy, the angular momentum, and its component along a certain axis. The eigenkets are $|n, l, m\rangle$ with

$$\langle r, \vartheta, \phi | n, l, m \rangle = R_{nl}(r)\Theta_{lm}(\vartheta)\Phi_m(\phi).$$

In the case of the simple harmonic oscillator, the complete set consists of one variable. In the Schrödinger-representation, the coordinate $x$ is the variable. One may also transform to the momentum-representation.

There is another *kind* of representation where the operators are not hermitian; for example, the Fock representation treated in Appendix B at the end of Chap. 3. The operators $b^\dagger$, $b$ are not self-adjoint (not hermitian). This representation is suitable for the quantization of *fields* for which the number of "quantized particles" is extremely huge and is not constant.


## *(2) Measurement*

The probability postulate IV, in Sec. 3 above, states that quantum mechanics predicts the result of a measurement of a physical quantity $Q$ on a system in the state represented by $|a\rangle$ to be[r]

$$\langle Q \rangle = \langle a | Q | a \rangle.$$

This postulate has been supplemented by the following assumptions:

(i) When a measurement of $Q$ on a system in state $|a\rangle$ is carried out, *one* of the eigenvalues of $Q$, say, $q_k$, is found. The system is then in the eigenstate $|q_k\rangle$ of $Q$. When a second measurement of $Q$ is made on the system, the result is now definitely $q_k$. One says that a measurement of an observable $A$ "forces" the system to go into *one* of the eigenstates $|a_k\rangle$ of $A$. The different eigenvalues $a_k$ will show up, with the probabilities $|\langle a_k | q \rangle|^2$.

(ii) Let $A$, $B$ be two commuting observables. They have common eigenstates $|a, b\rangle$. When a measurement of $A$ is made on a system in state $|q\rangle$, the result will be one of the eigenvalues, say $a_k$, of $A$, and the final state will be

$$A|a_1, b\rangle = a_1|a_1, b\rangle.$$

---

[r] In terms of the $q$-representatives of $|a\rangle$, namely

$$\langle q | a \rangle \equiv \psi_a(q),$$

one has

$$\langle Q \rangle = \int \psi_a^\dagger(q) Q \psi_a(q)\, dq.$$

When next a measurement of $B$ is made, the result is

$$B|a_1,b\rangle = b_1|a_1,b_1\rangle.$$

Another measurement of $A$ will certainly give $a_1$, and another measurement of $B$ will certainly give $b_1$. This will remain until the system is perturbed by the measurement of another observable that does not commute with $A$ and $B$. Successive measurements of commuting observables do not change the state, which is a common eigenstate, of the system.

(iii) If $A$ and $B$ do not commute, like the $p$, $q$ satisfying $pq - qp = \hbar/i$, as measurement of $q$, yields $q_k$ and leaves the system in state $|q_k\rangle$. A measurement of $p$ now will yield *one* of the eigenvalues, say $p_j$, of $p$, and leave the system in $|p_j\rangle$. A second measurement of $q$ now will yield, not $q_k$, but another *one* of the $q$'s, say $q_m$. In the former case, the probability of finding the value $p_j$ is $|\langle p_j|q_k\rangle|^2$; in the latter case, the probability of finding the value $q_m$ is $|\langle q_m|p_j\rangle|^2$. To summarize, a measurement of $p$ wipes out all previous knowledge of $q$, and *vice versa*. This is in agreement with the uncertainty principle.

(iv) The consequence of these "interpretations", or "postulates", concerning the theory of measurements in quantum mechanics is far-reaching. It means that quantum mechanics denies the possibility of simultaneous exact knowledge of both the coordinate and its conjugate momentum, so that it is not possible to make exact prediction of the state of motion as in classical dynamics. In fact the description of the motion of a particle by a continuous trajectory is not possible. One may replace the continuous curve by a series of dots which give the probable positions at small intervals apart. This is in fact the picture in Heisenberg's paper introducing the uncertainty relation.

## 5. Relation between Heisenberg's and Schrödinger's equations of motion

In Chap. 2, Sec. 4, we gave the equations of motion in matrix mechanics in the form

$$\dot{q} = [q, H], \qquad \dot{p} = [p, H], \tag{V-119}$$

where $[A, B]$ is the quantum Poisson brackets defined by

$$[A, B] = \frac{1}{i\hbar}(AB - BA). \tag{V-120}$$

From (V-119), it follows that any function $F$ formed of sums and products of $q$ and $p$ changes with time according to the equation

$$\dot{F} = [F, H]. \tag{V-121}$$

In Sec. 3, (V-104), we have the Schrödinger equation for the change of the state ket $|k, t\rangle$ with time,

$$-\frac{\hbar}{i}\frac{\partial}{\partial t}|k,t\rangle = H|k,t\rangle. \tag{V-122}$$

We shall now show that the two different equations, (V-121) giving the time rate of change of the operators (observables) ad (V-122) giving the time rate of change of the state (the ket, or its representative in some representation), actually differ from each other by a unitary transformation.

In Sec. 2, (V-96), we introduced the unitary operator for time translation

$$U(t-t_0)|k,t_0\rangle = |k,t\rangle. \tag{V-123}$$

On writing (V-122) in the form

$$i\hbar\frac{\partial}{\partial t}U|k,t_0\rangle = HU|k,t_0\rangle, \tag{V-124}$$

since $|k,t_0\rangle$ is arbitrary, one obtains the equation[s]

$$i\hbar\frac{\partial U}{\partial t} = HU. \tag{V-125}$$

From (V-123)

$$U^{-1}|k,t\rangle = |k,t_0\rangle, \tag{V-123a}$$

i.e., $U^{-1}$ reduces a changing ket $|k,t\rangle$ to a fixed ket $|k,t_0\rangle$ at time $t_0$.

Let $\alpha$ be an operator fixed in time operating on $|k,t\rangle$. Then by means of the unitary $U$, $\alpha$ is transformed into $\alpha_t$ which varies with time, such that

$$\langle k,t_0|\alpha_t|k,t_0\rangle = \langle k,t|\alpha|k,t\rangle. \tag{V-126}$$

From (V-123), the bra corresponding to $|k,t\rangle$ is $\langle k,t_0|U^{-1}$. Hence from (V-126), one obtains

$$\alpha_t = U^{-1}\alpha U, \tag{V-127}$$

or

$$U\alpha_t = \alpha U.$$

Differentiating with respect to $t$,

$$\frac{dU}{dt}\alpha_t + U\frac{d\alpha_t}{dt} = \alpha\frac{dU}{dt}$$

and using (V-125), one gets

---

[s] The solution of the Schrödinger equation by means of Eq. (V-125) for the unitary $U$ will be given in a later chapter.

$$\frac{d\alpha_t}{dt} = [\alpha_t, H_t],  \qquad \text{(V-128)}$$

where

$$H_t = U^{-1}HU, \qquad \text{(V-129)}$$

which transforms the time-fixed Hamiltonian $H$ into a varying $H_t$.

Letting $\alpha_t$ be $q$, $p$, one obtains from (V-128)

$$\dot{q} = [q, H] \qquad \dot{p} = [p, H],$$

which are Heisenberg's equations of motion (V-123).

Thus starting from the Schrödinger equation (V-122) in which the operators $H$, $q$, $p$ are all fixed in time but the kets are changing with time, one obtains, by means of a unitary transformation $U$ in (V-123), the Heisenberg equations of motion (V-119) in which the operators $H_t$, $q_t$, $p_t$ change with time[1] but the state kets $|k, t_0\rangle$ are fixed. We call these two views the *Schrödinger picture* and the *Heisenberg picture* respectively.

The complete equivalence of these two *pictures* can be affirmed by showing that they give the same expectation value for an operator $\alpha$. (After all, whether the operators are regarded as "moving" and the kets as "fixed", or *vice versa*, is not physically relevant; but the expectation values are.) This is in fact so, from (V-126). But, explicity, from (V-123) and (V-127), one has

$$|k, t\rangle = U|k, t_0\rangle,$$

$$\alpha_t = U^{-1}\alpha U. \qquad \text{(V-127)}$$

In the Heisenberg picture, the expectation value is

$$\langle k, t_0 | \alpha_t | k, t_0 \rangle = \langle k, t_0 | U^{-1}\alpha U | k, t_0 \rangle.$$

In the Schrödinger picture, the expectation value is

$$\langle k, t | \alpha | k, t \rangle = (\langle k, t_0 | U^{-1})\alpha(U | k, t_0 \rangle). \qquad \text{QED}$$

From (V-125), one obtains

$$U = \exp\left(-\frac{i}{\hbar}Ht\right) \qquad \text{(V-130)}$$

so that (V-123) gives

$$|k, t\rangle = \exp\left(-\frac{i}{\hbar}Ht\right)|k, 0\rangle, \qquad \text{(V-131)}$$

which is a formal solution of the Shcrödinger equation (V-122).

---

[1] $H_t$ actually is the same as $H$ and therefore does not change with time. See below.

(V-127) can be seen to be a solution of the Heisenberg equation (V-128), for

$$H \exp\left(\frac{i}{\hbar} H t\right) = \exp\left(\frac{i}{\hbar} H t\right) H$$

so that

$$H_t = H \tag{V-132}$$

and

$$\alpha_t = \exp\left(\frac{i}{\hbar} H t\right) \alpha \exp\left(-\frac{i}{\hbar} H t\right)$$

does satisfy (V-128).

Finally, if in the Schrödinger equation (V-122)

$$i\hbar \frac{\partial}{\partial t} |k, t\rangle = H |k, t\rangle,$$

we put

$$|k, t\rangle = \exp\left(-\frac{i}{\hbar} E t\right) |k, 0\rangle, \tag{V-133}$$

where $|k, 0\rangle$ is fixed in time, then one gets

$$(H - E)|k, 0\rangle = 0, \tag{V-134}$$

which is the time-independent Schrödinger equation. The eigenkets of $H$ are then the stationary states of $H$.


## 6.  Density matrix

In Chap. 4, Sec. 1, and in Chap. 5, Sec. 3, Postulate IV, we introduced the probability concept in quantum mechanics. We repeatedly emphasized that the probability concept here is of an intrinsic nature; it arises not from dealing with a large number of molecules as in the kinetic theory of gases, but even for a single atom.

But in actual situations, we do have systems consisting of a large number of atoms or molecules, and since they can be in different quantum states.[u] there does arise a statistical element of the classical kind. To treat such a system in quantum mechanics, von Neumann and Weyl introduced (1927) the concept of *density matrix*.

---

[u] Except for very special conditions when the various atoms or molecules are prepared in the same state, as for example, in a laser tube.

### (1) Pure and mixed state

To begin an introductory account, let us summarize what we have already treated before. Let us suppose that we know all the atoms in the system to be in the same state $|k\rangle$. Then the expectation value of a physical quantity $Q$ is

$$\langle Q \rangle_p \equiv \langle k|Q|k \rangle. \tag{V-135}$$

If $|n\rangle$ are the complete set of eigenkets of some observable $A$, one may express $|k\rangle$ in terms* of $|n\rangle$

$$|k\rangle = \sum_n C_n^{(p)}|n\rangle. \tag{V-136}$$

Then

$$\langle Q \rangle_p = \sum_{n,m} C_n^{(p)*} C_m^{(p)} \langle n|Q|m \rangle. \tag{V-137}$$

Note that here $\langle n|Q|m \rangle$ are independent of $|k\rangle$. If $|n\rangle$ are the eigenkets $|q_k\rangle$ of $Q$ itself, then

$$|k\rangle = \sum_j C_j|q_j\rangle, \qquad C_j = \langle q_j|k \rangle, \tag{V-138}$$

$$\langle Q \rangle_p = \sum_j |C_j|^2 q_j, \tag{V-139}$$

$$\sum_j |C_j|^2 = 1. \tag{V-139a}$$

$|C_j|^2$ is the probability that the state $|k\rangle$ is in the state $|q_j\rangle$.

Let us now consider a system of atoms which are in different states, and assume that we do not have sufficient information concerning their states. We can no longer represent by $|p\rangle$ the state of all the atoms, and must replace (V-137) by the average value

$$\langle Q \rangle_m = \sum_p W_p \langle p|Q|p \rangle. \tag{V-140}$$

Here $\langle p|Q|p \rangle$ is the expectation value $\langle Q \rangle_p$ of $Q$ when the atom is in a (pure) state $|p\rangle$ (such as $|k\rangle$ in (V-136)), and $W_p$ is the probability that the system is in the pure state $|p\rangle$, so that

$$0 \le W_p \le 1, \qquad \sum_p W_p = 1. \tag{V-141}$$

Note that the $W_p$ are basically different from the $|C_j|^2$ in (V-138, 139) which

---

* In (V-136), (V-137), the expansion may or may not have to cover the complete set of $|n\rangle$, depending on the nature of the state $|k\rangle$. The subscript $p$ and the superscript $(p)$ indicate that here we are concerned with only "pure" states. (See below.) We shall omit the $(p)$ after (V-137) for simplicity.

are "intrinsic", the $W_p$ arise from the large number of atoms in the system, the average in (V-140) being of the classical nature. The average $\langle Q \rangle_m$ is said to be an average over "mixed states", in contrast to the $\langle Q \rangle_p$ in (V-137), (V-139) which is an average over "pure states", a terminology due to Weyl.

### (2) Density operator and density matrix

We define the density operator $\rho$ by

$$\rho = \sum_p W_p |p\rangle \langle p|, \tag{V-142}$$

where the probabilities $W_p$ are those in (V-140). Expressing the ket $|p\rangle$ in terms of $|n\rangle$ as in (V-136), we have

$$\rho = \sum_p W_p \sum C_m^{(p)} C_n^{(p)*} |m\rangle \langle n|, \tag{V-143}$$

where

$$C_m = \langle m|p \rangle, \qquad C_n^* = \langle p|n \rangle.$$

The density matrix is defined as

$$\langle m|\rho|n \rangle_m = \sum_p W_p \langle m|p \rangle \langle p|n \rangle. \tag{V-144}$$

Thus the density matrix has $N^2$ complex numbers if $N$ states are involved in $|n\rangle$.

**Theorem:**   The density operator is a hermitian operator.

**Proof:**
$$\langle m|\rho|n \rangle^* = \sum_p W_p \langle m|p \rangle^* \langle p|n \rangle^*$$

$$= \sum_p W_p \langle n|p \rangle \langle p|m \rangle$$

$$= \langle n|\rho|m \rangle. \quad \text{QED} \tag{V-145}$$

This hermitian property reduces $\langle m|\rho|n \rangle$ to $N^2$ real numbers.

If the system is in a pure state, i.e.,

$$W_p = 1,$$

then

$$\rho_p = |p\rangle \langle p| = \sum_{n,m} C_n^* C_m |m\rangle \langle n| \tag{V-146}$$

and

$$\langle m|\rho_p|n \rangle^* = \langle n|\rho_p|m \rangle, \qquad \text{by (V-145)}.$$

*(3) Trace*

**Theorem:**
$$\langle Q \rangle_m = \mathrm{Tr}(\rho Q). \tag{V-147}$$

**Proof:**
$$
\begin{aligned}
\mathrm{Tr}(\rho Q) &= \mathrm{Tr} \sum_p W_p(|p\rangle\langle p|Q) \cdot \\
&= \mathrm{Tr} \sum_p W_p |p\rangle\langle p|p\rangle\langle p|Q \\
&= \mathrm{Tr} \sum W_p |p\rangle\langle p|Q|p\rangle\langle p| \\
&= \mathrm{Tr} \sum W_p \langle p|Q|p\rangle |p\rangle\langle p| \\
&= \sum_p W_p \langle p|Q|p\rangle \, \mathrm{Tr}(|p\rangle\langle p|) \\
&= \sum W_p \langle p|Q|p\rangle \langle p|p\rangle \\
&= \langle Q \rangle_m, \qquad \text{by (V-140).} \qquad\qquad \text{QED}
\end{aligned}
$$

For a system in a pure state, $W_p = 1$ and (V-147) still holds

$$\mathrm{Tr}(\rho_p Q) = \langle Q \rangle_p. \tag{V-147a}$$

*(4) Normalization*

If in (V-147) we set $Q = 1$, we get

$$\mathrm{Tr}\,\rho = 1. \tag{V-148}$$

Or, from (V-144),

$$
\begin{aligned}
\mathrm{Tr}\,\rho &= \sum_n \langle n|\rho|n\rangle \\
&= \sum_p W_p \sum_m |\langle m|p\rangle|^2 \\
&= \sum W_p, \qquad \text{using (V-139a),} \\
&= 1 \qquad\qquad \text{using (V-141).} \tag{V-149}
\end{aligned}
$$

The normalization reduces the number of real numbers in $\langle m|\rho|n\rangle$ from $N^2$ to $N^2 - 1$.

*(5) $\rho^2$ and the condition for a pure state*

From (V-142),

$$
\begin{aligned}
\rho^2 &= \left( \sum W_p |p\rangle\langle p| \right)\left( \sum W_{p'} |p'\rangle\langle p'| \right) \\
&= \sum_{p,p'} W_p W_{p'} |p\rangle\langle p|p'\rangle\langle p'| \\
&= \sum_p W_p^2 |p\rangle\langle p|. \tag{V-150}
\end{aligned}
$$

Thus $\rho^2$ is obviously different from $\rho$ for mixed state ($W_p \neq 1$). Now

$$
\begin{aligned}
\mathrm{Tr}\,\rho^2 &= \sum_p W_p^2 \sum_n \langle n|p\rangle \langle p|n\rangle \\
&= \sum_p W_p^2 \sum_n |C_n^{(p)}|^2 \\
&= \sum W_p^2, \qquad \text{using (V-139a),} \\
&\leq 1, \qquad\qquad \text{using (V-141).}
\end{aligned}
\qquad\text{(V-151)}
$$

For a system in a pure state, $W_p = 1$,

$$
\begin{aligned}
\rho_p^2 &= (|p\rangle\langle p|)(|p'\rangle\langle p'|) \\
&= |p\rangle\langle p| \\
&= \rho_p \qquad \text{(using V-146),}
\end{aligned}
\qquad\text{(V-152)}
$$

so that

$$
\begin{aligned}
\mathrm{Tr}\,\rho_p^2 &= \mathrm{Tr}\,\rho_p \\
&= 1 \qquad \text{(using V-148).}
\end{aligned}
\qquad\text{(V-153)}
$$

From (V-147a),

$$
\langle Q\rangle_p = \mathrm{Tr}(\rho_p Q),
$$

we have

$$
\langle k|Q|k\rangle = \sum_{m,n} \langle m|\rho_p|n\rangle \langle n|Q|m\rangle.
\qquad\text{(V-154)}
$$

If we choose a set of basic kets that diagonalize $\rho_p$, this relation can be satisfied by choosing

$$
\langle m|\rho_p|n\rangle = \delta_{mk}\delta_{nk},
\qquad\text{(V-155)}
$$

i.e., $\rho_p$ has the eigenvalue 1 and the rest equal to zero.

Any one of (V-152), (V-153), (V-155) is the necessary and sufficient condition for pure state.

### (6) Physical meaning of density matrix and mixed state

From the definition (V-144)

$$
\langle n|\rho|n\rangle_m = \sum_p W_p |\langle p|n\rangle|^2,
\qquad\text{(V-156)}
$$

it is seen that $\langle n|\rho|n\rangle_m$ is the probability that the system is in the state $|n\rangle$. This probability, with the $W_p$, is of the nature of the probability in an ensemble in classical statistical mechanics.

If we definitely know that all the atoms in the system are in the state $|p\rangle$, we say that the system is in a pure state, and $W_p = 1$ so that

$$\langle n|\rho|n\rangle_p = |\langle p|n\rangle|^2, \tag{V-157a}$$

which in the usual notation is

$$\langle n|\rho|n\rangle_p = |\psi_n^{(p)}|^2. \tag{V-157b}$$

To further clarify the meaning of mixed states and density matrix, consider a closed system $S$ in which $S_1(x)$ is a subsystem, $x$ stands for the variables in $S_1$; $x$, $\xi$ the variables of the whole system $S$. On account of the interactions between the $x$ and the $\xi$, the state function $\Psi(x, \xi)$ cannot be expressed in the form

$$\Psi(x, \xi) = \Phi(\xi)\phi(x). \tag{V-158}$$

If $\phi_n(x)$ are a complete set of eigenkets of "a complete set of commutable variables","  then $\Psi(x, \xi)$ can be expressed in the form

$$\Psi(x, \xi) = \sum_n \int \Phi_n(\xi)\phi_n(x). \tag{V-159}$$

If an observable $Q$ acts only on the variables $x$, the expectation value of $Q$ is

$$\langle Q\rangle = \iint \Psi^*(x, \xi)Q(x)\Psi(x, \xi)\,dx\,d\xi \tag{V-160a}$$

$$= \sum_{n,m} \int \sum \int \Phi_n^*(\xi)\Phi_m(\xi)\,d\xi \int \phi_n^*(x)Q\phi_m(x)\,dx. \tag{V-160b}$$

We define the density operator

$$\rho_{mn} \equiv \langle m|\rho|n\rangle = \int \Phi_n^*(\xi)\Phi_m(\xi)\,d\xi. \tag{V-161}$$

Then

$$\langle Q\rangle = \sum_{n,m} \langle m|\rho|n\rangle \langle n|Q|m\rangle$$

$$= \mathrm{Tr}(\rho Q) \qquad \text{by (V-145).} \tag{V-160c}$$

(V-161) expresses the meaning of the density matrix in the situation represented by (V-159).

We can express the density operator in another form. We rewrite (V-160a) in the following form

$$\langle Q\rangle = \int dx\, Q(x)\left[\int \Psi^*(x', \xi)\Psi(x, \xi)\,d\xi\right]_{x'=x}. \tag{V-162}$$

---

" In the sense of Sec. 4, (1), (V-114) above.

Now define $\rho(x, x')$ as

$$\rho(x, x') \equiv \int \Psi^*(x', \xi)\Psi(x, \xi)\, d\xi$$

$$= \sum_{n,m} \int \sum \int \Phi_n^*(\xi)\Phi_m(\xi)\, d\xi \,\, \phi_n^*(x')\phi_m(x)$$

$$= \sum_{n,m} \int \sum \langle m|\rho|n\rangle \phi_n^*(x')\phi_m(x) \qquad \text{by (V-161)}, \qquad \text{(V-163a)}$$

$$= \sum_{n,m} \rho_{mn}\phi_n^*(x')\phi_m(x) \qquad\qquad \text{by (V-161)}. \qquad \text{(V-163)}$$

$\rho(x, x')$ is the matrix element of $\rho$ in the $x$-representation. (V-162) can now be expressed as

$$\langle Q\rangle = \int dx\, Q(x)\rho(x, x')|_{x'=x}$$

$$= \int Q(x)\rho(x, x')\delta(x' - x)\, dx\, dx'. \qquad \text{(V-164)}$$

If $Q(x)\delta(x' - x)$, the matrix of $Q$ in the $x$-representation, is written in the form

$$Q(x)\delta(x' - x) = \langle x'|Q|x\rangle, \qquad \text{(V-165)}$$

then

$$\langle Q\rangle = \iint \rho(x, x')\langle x'|Q|x\rangle\, dx'\, dx$$

$$= \mathrm{Tr}(\rho Q), \qquad \text{(V-166)}$$

which is of the same form as (V-160c).

If there is no interaction between the variables $\xi$ in $S$ and $x$ in $S_1$, then the separable form (V-158) is possible,

$$\Psi(x, \xi) = \Phi_n(\xi)\phi_n(x) \qquad \text{(V-158)}$$

and the system is in a pure state. In this case, the $\langle m|\rho|n\rangle$ in (V-163a) can be reduced by (V-155); so that (V-163) becomes

$$\rho(x, x') = \phi_n^*(x)\phi_n(x'). \qquad \text{(V-167)}$$

*(7) Transformation property of $\rho$*

If $Q$ and $\rho$ are transformed by the same (unitary) matrix $U$

$$Q' = U^\dagger Q U, \qquad \rho' = U^\dagger \rho U, \qquad \text{(V-168)}$$

then from (V-147), $\langle Q\rangle_m$ can be shown to be invariant,[x] for

$$\langle Q'\rangle_m = \mathrm{Tr}(\rho'Q')$$

$$= \mathrm{Tr}(U^\dagger \rho Q U)$$

$$= \mathrm{Tr}(\rho Q), \quad \text{by Theorem 11, Chap. 2, Sec. 2.}$$

$$= \langle Q\rangle_m. \tag{V-169}$$

Conversely, from the invariance $\langle Q'\rangle_m = \langle Q\rangle_m$, one can obtain the transformation (V-168).

### (8) Liouville equation in quantum mechanics

Consider a time varying system so that we replace (V-136) by

$$|k, t\rangle = \sum C_n^{(p)}(t)|q_n\rangle, \tag{V-170}$$

or, in terms of the representatives in the $x$-representation,

$$\psi^{(p)}(t) = \sum C_n^{(p)}(t)\psi_n(x), \tag{V-170a}$$

where $(p)$ again denotes pure state.

On using the notation (V-161) for the matrix element in (V-144), (V-143), we have

$$\langle m|\rho|n\rangle \equiv \rho_{mn}(t) = \sum_p W_p C_n^{(p)*}(t) C_n^{(p)}(t). \tag{V-171}$$

From the Schrödinger equation

$$i\hbar\frac{\partial \Psi(p)}{\partial t} = H\Psi^{(p)}, \tag{V-172}$$

we get

$$i\hbar\frac{\partial C_m^{(p)}}{\partial t} = \sum \langle m|H|n\rangle C_n^{(p)}, \tag{V-173}$$

where

$$\langle m|H|n\rangle = \int \psi_m^* H\psi_n\, dx. \tag{V-174}$$

From (V-171) and (V-173),

$$\frac{\partial}{\partial t}\rho_{mn}(t) = \frac{1}{i\hbar}\sum W_p\left[ -\sum_k \langle n|H|k\rangle^* C_k^{(p)*} C_m^{(p)} + \sum_k C_n^{(p)*}\langle m|H|k\rangle C_k^{(p)} \right]$$

---

[x] The invariance holds also for any similarity transformation.

$$= \frac{1}{i\hbar}\sum_k \left[ -\langle k|H|n\rangle \rho_{mk} + \langle m|H|k\rangle \rho_{kn}\right]$$

$$= \frac{1}{i\hbar}(H\rho - \rho H)_{mn} \tag{V-175}$$

$$= [H,\rho], \tag{V-175a}$$

where $[H,\rho]$ is the quantum Poisson bracket. This equation has the same form as the Liouville equation in classical dynamics and statistical mechanics.[y] According to Heisenberg's equations of motion, (V-121),

$$\frac{d\rho}{dt} = [\rho, H] + \frac{\partial \rho}{\partial t}. \tag{V-177}$$

By (V-175a), we have

$$\frac{d\rho}{dt} = 0.$$

Thus (V-175a) is the quantum mechanical analogue of the classical Liouville equation.

It is seen from (V-175a) that the equation is invariant under time reversal.[z] This invariance has its origin in the (Wigner) time-reversal invariance of the Schrödinger equation.

If in (V-170a) the $\psi_n(x)$ are the eigenfunctions of the Hamiltonian $H$, then (V-175) simplifies to

$$\frac{\partial \rho_{mn}}{\partial t} = \frac{1}{i\hbar}(E_m - E_n)\rho_{mn}(t), \tag{V-178}$$

whose solution is

$$\rho_{mn}(t) = \rho_{mn}(0)\exp\left( -\frac{i(E_m - E_n)t}{\hbar}\right). \tag{V-179}$$

[y] In classical statistical mechanics, the Liouville equation

$$\frac{\partial \rho}{\partial t} = (H, \rho) \tag{V-176}$$

expresses the constancy of the density $\rho(q, p, t)$ for stationary ensembles.

[z] In quantum mechanics, the time-reversal operation $t \to -t$ is to be accompanied by a complex conjugation $i \to -i$. This is called the *Wigner time reversal*. Thus, upon the reversal $t \to -t$, the sign of the quantum Poisson bracket is to change. This leaves (V-175a) invariant. Also, under this Wigner time-reversal, the Schrodinger equation $(i\hbar(\partial/\partial t) - H)\Psi = 0$ is invariant. Instead of the conjugation $i \to -i$ of Wigner, one may take the alternative view that in classical dynamics, the reversal of time is accompanied by a reversal of the sign of the Poisson brackets, so that in quantum mechanics, one similarly changes the sign of all quantum Poisson brackets. See T. Y. Wu, *Am. J. Phys.* **26**, 568 (1958).

*(9) Density matrix and irreversibility of macroscopic processes*

The quantum Liouville equation (V-175) is invariant under time reversal as we have just shown. The question is: How do you propose to describe the irreversible macroscopic processes? This problem is in fact similar to the same problem in classical statistical mechanics and kinetic theory of gases. In the quantum theory, one can adapt the equation for the density matrix (V-175) to irreversible processes by introducing the so-called random-phase-approximation. We shall not go into this problem here.*

### References

J. von Neumann, *Göttingen Nachr.* **245**, 273 (1927); *Mathematical Foundations of Quantum Mechanics* (Princeton Univ. Press, Princeton, 1955).
H. Weyl, *Z. Phys.* **46**, 1 (1927).
P. A. M. Dirac, *Proc. Camb. Phil. Soc.* **25**, 62; **26**, 376 (1929); **27**, 240 (1930).
U. Fano, *Rev. Mod. Phys.* **29**, 74 (1957).

### 7. The Copenhagen school and Einstein's philosophy

We have seen in this and the preceding chapters that Quantum Mechanics is a system consisting of (1) a mathematical structure, (2) a physical interpretation, and (3) a philosophical attitude, as represented, respectively, by (1) the Postulates I, II, III, V, (2) the Probability Postulate IV, and (3) the complementarity concept. The probability postulate and the complementarity concept form what is known as the *Copenhagen philosophy*; it is specifically related to quantum mechanics, but is of broader import as it has to do with the philosophy concerning the *nature* of theories in physics (or, in science, in general). This was brought about by the many discussions between Einstein who held a different philosophy and Bohr who represented the Copenhagen School.

We have seen that Einstein introduced the quantum theory of radiation $[E = h\nu$ (1905) and later on, $p = h/\lambda$ (1916)$]$ on statistical fluctuation considerations, and the concept of transition probabilities (1916–7). But when the probability interpretation of quantum mechanics was proposed by Born and accepted (1926) into the new theory, when Heisenberg (1927) showed that the Einstein-de Broglie relations led to the uncertainty relation, and a little later Bohr introduced the complementarity point of view, Einstein could not accept these views. Einstein believed in a fundamental deterministic, causal Nature;

---

* For a discussion of the quantum Liouville equation, see, for example, T. Y. Wu and D. C. Riviere *Helv. Phys. Acta* **34**, 661 (1961).

he could not accept Nature as being intrinsically governed by probabilistic laws; he could not accept a physical theory which fundamentally denied the possibility of ever knowing simultaneously the position and the momentum of a particle.

For a time (around 1927), Einstein at first tried, by thought-experiments, to exceed the limit imposed by the uncertainty relations $\Delta x \Delta p \simeq h$, $\Delta t \Delta E \simeq h$, but later accepted the consistency of quantum mechanics, i.e., the inescapability of the uncertainty relations so long as one accepts the premise—the "complementarity postulates" having their origin in the Einstein-de Broglie relations, and the "probability postulate". But this acceptance of the internal consistency did not convert Einstein to the basic philosophy of the Copenhagen School.

In 1935, in a famous paper Einstein, B. Podolsky and N. Rosen expressed his philosophy toward a physical theory in clear form. We shall present a brief summary of the essential point of this famous paper.

The basic philosophy of Einstein, Podolsky and Rosen is that a physical theory—in the present case, quantum mechanics—must contain information of all physical realities, and "physical reality" they define as follows: *If, without in any way disturbing a system, we can predict with certainty the value of a physical quantity, then there exists an element of physical reality corresponding to this quantity.*

They held a theory to be incomplete if it did not furnish a proper account of physical quantities having an element of physical reality, and in this sense, they concluded that quantum mechanics did not give a complete description of all elements of physical reality.

They considered the following example: Let two particles 1 and 2 interact for a while and separate to a great distance apart so that they no longer interact and may be considered mutually independent and free. The wave *function is assumed to be*

$$\Psi(x_1, x_2) = \int_{-\infty}^{\infty} e^{ip(x_1 - x_2 + x_0)/\hbar} \, dp. \tag{V-180}$$

This state is one in which the sum of the momenta of the two particles has the eigenvalue 0, as seen from

$$(p_1 + p_2)\Psi(x_1, x_2) = \frac{\hbar}{i}\left(\frac{\partial}{\partial x_1} + \frac{\partial}{\partial x_2}\right)\Psi$$

$$= \int_{-\infty}^{\infty} (p - p)e^{ip(x_1 - x_2 + x_0)/\hbar} \, dp$$

$$= 0. \tag{V-181}$$

From this it follows that a measurement of $p_1$ on particle 1 will yield an exact value of the momentum $p_2$ of particle 2.

But $\Psi(x_1, x_2)$ in (V-180) can also be re-expressed in the form

$$\Psi(x_1, x_2) = 2\pi\delta(x_1 - x_2 + x_0)$$

$$= 2\pi \int \delta(x_1 - x)\delta(x - x_2 + x_0)\,dx. \qquad (V\text{-}182)$$

The operator $\eta \equiv x_2 - x_1$ acting on $\delta(x_1 - x_2 + x_0)$ gives

$$\eta\delta(x_0 - \eta) = x_0\delta(x_0 - \eta), \qquad (V\text{-}183)$$

i.e., $\Psi(x_1, x_2) = 2\pi\delta(x_1 - x_2 + x_0)$ is the eigenfunction of $x_2 - x_1$ for the exact eigenvalue $x_2 - x_1 = x_0$. Thus a measurement of $x_1$ of particle 1 will give the exact value for $x_2 = x_1 + x_0$.[†]

From these considerations, Einstein, Podolsky and Rosen concluded that, both $p_2$ and $x_2$ can be exactly predicted by measurements on particle 1 alone without perturbing the particle 2. This, according to the criterion above for "element of physical reality", means that both $p_2$ and $x_2$ are elements of physical reality. The philosophy of these authors towards the *nature* of a physical theory is that a complete theory should contain a complete description (or information) of all the elements of physical realities. Now in quantum mechanics, the possibility of the simultaneous knowledge of $p_2$ and $x_2$ is fundamentally denied. One must then say that either (1) a measurement of particle 1 always affects the other particle even though they are very far apart, or (2) the quantum mechanical theory does not give a complete description of all the physical realities. But it is unreasonable to believe that a measurement on one particle cannot be made without perturbing the other even when they are very far apart. These authors hence believed that one is forced to conclude that quantum mechanical description is not complete.

Let us see how this contention is answered according to the Copenhagen view. The answer, within the internally self-consistent system of quantum mechanics, is very simple indeed: When one measures $p_1$ obtaining a definite value $p_1^0$, say, one indeed deduces from (V-180) the exact value $-p_1^0$ for $p_2$. After this measurement, the wave function of the system "collapses" from $\Psi(x_1, x_2)$ in (V-180) to

$$e^{ip_1^0(x_1 - x_2 + x_0)/\hbar}$$

and in this state, $x_1$ is completely not known, and so is $x_2$. If immediately after

---

[†] The discussion above, (V-180)–(V-183), may be summarized by saying that, while $p_1$, $q_1$ and $p_2$, $q_2$ satisfy $p_1 q_1 - q_1 p_1 = \hbar/i$, $p_2 q_2 - q_2 p_2 = \hbar/i$, the operators $P = p_1 + p_2$, $Q = q_1 - q_2$ commute, i.e., $PQ - QP = 0$.

the measurement of $p_1$, one measures $x_1$ and obtains the value $x_1^0$, the wave function $e^{ip_1^0 x_1/\hbar}$ collapses to

$$\delta(x_1 - x_1^0) = \frac{1}{2\pi\hbar} \int_{-\infty}^{\infty} e^{ip_1(x_1 - x_1^0)/\hbar} \, dp_1$$

so that $p_1$ is completely unknown. When $p_1$ is unknown, so is $p_2$.

Of course, all these arguments are familiar, well-known in quantum mechanics, and in fact explicitly accepted by Einstein, Podolsky and Rosen. Therefore the situation can be summarized as follows:

Einstein, Podolsky and Rosen accepted the commutation relation and the consequent uncertainty relation. They showed, by an example, the possibility of predicting the exact values of $p_2$ and $x_2$ of a particle 2 on measurements of $p_1$ and $x_1$ on particle 1 widely separated from particle 2. Thus $p_2$ and $x_2$ are "elements of physical reality". It is their basic philosophy that a complete physical theory should contain a complete account of all elements of physical reality.

The answer according to the Copenhagen philosophy is simple and based on the self-consistency of quantum mechanics, namely, $p_2$ and $x_2$ are indeed predictable exactly, but *not* simultaneously, and a pair of *exact* but *not simultaneous* values of $p_2$, $x_2$ are not useful in the sense that they cannot be used for predicting the future development of the state of the system. The Copenhagen school insists that quantum mechanics, with the complementarity, uncertainty relations and the concept of intrinsic probability, is not only logically self-consistent, but is a *complete* theory; that it answers all the questions that one may meaningfully ask.

Thus the "controversy" between Einstein and the Copenhagen School is one of philosophical attitude. Einstein's ideal of a physical theory is one that is ultimately causal, deterministic as in classical physics, and gives a "complete description" of all elements of physical reality. The Copenhagen School believes that the present quantum mechanics does give a complete description of physical phenomena. Such a difference in basic philosophical attitude cannot be resolved by arguments, certainly not by arguments based only on the internal self-consistency of the present quantum mechanics.

It is true that up to the present, no alternative to the present quantum mechanics has been found, despite the dissatisfaction with the theory by many originators such as Einstein, Schrödinger, von Laue, and the many attempts to modify the interpretation of the theory, such as those of de Broglie, Bohm and others. But it is not in the spirit of science to ridicule these "objectors", just as it is not right to deny the possibility of a non-Euclidean geometry on arguments based on the logical consistency of Euclidean geometry.

## References


P. A. M. Dirac, *The Principles of Quantum Mechanics*, 4th ed. (Clarendon Press, Oxford, 1958).

A. Einstein, B. Podolsky and N. Rosen, *Phys. Rev.* **47**, 777 (1935).

N. Bohr, *Phys. Rev.* **48**, 696 (1935).

N. Bohr, *Dialectica* **2**, 312 (1948).

N. Bohr, in *Albert Einstein: Philosopher-Scientist*, P. Schilpp, ed. (Tudor, New York, 1949).

D. Bohm, *Phys. Rev.* **85**, 166, 180 (1952); **87**, 389; **88**, 1458 (1953) attempt a classical interpretation of quantum mechanics.

L. de Broglie, "Physique quantique restera-t-elle indeterministe" in *Nouvelles Perspectives en Microphysique* (Albin Michel, Paris, 1956).

L. de Broglie, *Nuovo Cimento*, Ser. X, Vol. 1 (1955).

L. de Broglie, *Une Textative d'Interpretation Causale et non Linéare de la Mecánique Ondutatoire* (Gauthier-Villars, Paris, 1956). This is on the double-solution theory.

J. P. Vigier, *Structure des Micro-Objects dans d'Interpretation Causale de la Théorie des Quante* (Gauthier-Villars, Paris, 1955).

J. von Neumann, *Mathematical Foundations of Quantum Mechanics* (Princeton Univ. Press, Princeton, 1955).

F. J. Belinfante, *A Survey of Hidden Variables Theories* (Pergamon Press, New York, 1973).

# Chapter 6

# Perturbation Theory: Stationary State Problem

For systems whose Hamiltonian is explicitly independent of time, the time-independent Schrödinger equation is

$$H\psi = E\psi. \tag{VI-1}$$

In Chap. 3, we gave the exact solution of this equation for a few problems, but, in general, this equation cannot be solved in terms of known functions in closed form. One method of obtaining approximate solutions is the perturbation theory given by Schrödinger. The theory is an adaptation of the method developed by Lord Rayleigh for the problem of vibrating strings (1878).

The theory is based on comparing the Hamiltonian $H$ of the given system with the $H_0$ of another system which has been solved already, i.e.,

$$H_0\psi_n^0 = E_n^0\psi_n^0 \tag{VI-2}$$

of which the eigenvalues $E_n^0$ and eigenfunctions $\psi_n^0$ are already known. The $\psi_n^0$ are assumed to form a complete set of normalized and orthogonal functions. It is assumed that the difference $H - H_0$ is "small". $H$ is then expressed in the form

$$H = H_0 + \lambda H^{(1)} + \lambda^2 H^{(2)} + \cdots, \tag{VI-3}$$

where $\lambda$ is a parameter satisfying

$$0 \le \lambda \le 1,$$

such that $\lambda = 0$ corresponds to the zeroth-order system (VI-2) and $\lambda = 1$ corresponds to the system (VI-1) under consideration. It is also assumed that the eigenvalues $E_n$ and eigenfunctions $\psi_n$ of $H$ can be expressed in the form

$$E_n = E_n^0 + \lambda E_n^{(1)} + \lambda^2 E_n^{(2)} + \cdots, \tag{VI-4}$$

$$\psi_n = \psi_n^0 + \lambda\psi_n^{(1)} + \lambda^2\psi_n^{(2)} + \cdots, \tag{VI-5}$$

where $E_n^{(1)}$, $E_n^{(2)}$, ..., $\psi_n^{(1)}$, $\psi_n^{(2)}$ ... are the first and second-order changes due to $H^{(1)}$, $H^{(2)}$, ... to be found.

On substituting (VI-3, 4, 5) into (VI-1) and equating the coefficients of various powers of $\lambda$ to zero, which is arbitrary between $\lambda = 0$ and $\lambda = 1$, one obtains:

$$\lambda^0: \qquad (H^0 - E_n^0)\psi_n^0 = 0, \tag{VI-2}$$

$$\lambda^1: \qquad (H^0 - E_n^0)\psi_n^{(1)} + (H^{(1)} - E_n^{(1)})\psi_n^0 = 0, \tag{VI-6}$$

$$\lambda^2: \qquad (H^0 - E_n^0)\psi_n^{(2)} + (H^{(1)} - E_n^{(1)})\psi_n^{(1)} + (H^{(2)} - E_n^{(2)})\psi_n^0 = 0, \tag{VI-7}$$

$$\cdot \qquad \cdot \qquad \cdot$$
$$\cdot \qquad \cdot \qquad \cdot$$

One proceeds to calculate $E_n^{(1)}$, $\psi_n^{(1)}$, $E_n^{(2)}$, $\psi_n^{(2)}$, ... in succession.

### 1. Perturbation theory for non-degenerate systems

If the zeroth-order system $H^0$ is non-degenerate, i.e., the eigenvalue $E_n^0$ is different for different eigenstates $\psi_n^0$, one proceeds as follows. On expanding

$$\psi_n^{(1)} = \sum \int A_{ni}\psi_i^0, \tag{VI-8}$$

$$(H^{(1)} - E_n^{(1)})\psi_n^0 = \sum \int B_{ni}\psi_i^0, \tag{VI-9}$$

one obtains, by using (VI-6),

$$B_{nn} = 0, \tag{VI-10a}$$

$$A_{ni} = \frac{B_{ni}}{E_n^0 - E_i^0}, \qquad n \neq i, \tag{VI-10b}$$

$$B_{ni} = \int \psi_i^{0*} H^{(1)} \psi_n^0 \, dq \equiv \langle i|H^{(1)}|n\rangle, \tag{VI-10c}$$

$$E_n^{(1)} = \int \psi_n^{0*} H^{(1)} \psi_n^0 \, dq \equiv \langle n|H^{(1)}|n\rangle. \tag{VI-10d}$$

Thus, up to the second order $\lambda^2$, one finally obtains

$$E_n = E_n^0 + \langle n|H^{(1)}|n\rangle + \sum_i' \frac{|\langle n|H^{(1)}|i\rangle|^2}{E_n^0 - E_i^0} + \langle n|H^{(2)}|n\rangle, \tag{VI-11}$$

$$\psi_n = \psi_n^0 + \sum_i' \frac{\langle i|H^{(1)}|n\rangle}{E_n^0 - E_i^0}\psi_i^0 + \sum_{m,i}' \frac{\langle i|H^{(1)}|m\rangle\langle m|H^{(1)}|n\rangle}{(E_n^0 - E_m^0)(E_n^0 - E_i^0)}\psi_i^0$$

$$+ \sum_i' \frac{\langle i|H^{(1)}|n\rangle}{E_n^0 - E_i^0}\psi_i^0, \tag{VI-12}$$

where the prime in $\sum'$ signifies the exclusion of $i = n$. By the hypothesis of non-degeneracy, $E_n \neq E_i$ for $n \neq i$.

A few examples are as follows:

*(1) Anharmonic oscillator*

$$H = H^0 + H^{(1)} + H^{(2)},$$

$$H^0 = \frac{1}{2\mu}p^2 + \frac{1}{2}k_1 x^2,$$

$$H^{(1)} = \frac{1}{6}k_2 x^3,$$

$$H^{(2)} = \frac{1}{24}k_3 x^4,$$

$$E_n^{(1)} = 0$$

$$E_n^{(2)} = \left(\frac{k_2}{6}\right)^2 \left\{\frac{|\langle n|x^3|n-3\rangle|^2}{-3\hbar\omega} + \frac{|\langle n|x^3|n-1\rangle|^2}{-\hbar\omega}\right.$$

$$\left. + \frac{|\langle n|x^3|n+1\rangle|^2}{\hbar\omega} + \frac{|\langle n|x^3|n+3\rangle|^2}{3\hbar\omega}\right\} + \frac{k_3}{24}\langle n|x^4|n\rangle. \quad \text{(VI-13)}$$

The matrix elements of $x^3$, $x^4$ are already known from Chap. 3, Appendix A, (A-10a). The result is exactly the same as that obtained in the matrix mechanics method of Chap. 2, (II-140).


*(2) Stark effect in alkali metal atoms*

If one treats such atoms as Li, Na, K, ... as a one-electron system (the valence electron moving in a central but non-Coulomb field), then the result on the angular part of $\psi^0(r)$ of Chap. 3, Sec. 4 (3) can be applied,

$$\psi_n^0(r, \vartheta, \phi) = R_{nl}(r)\Theta_{lm}(\vartheta)\Phi_m(\phi). \quad \text{(VI-14)}$$

For an atom (such as Li, Na, K, ...) in an external static (uniform) electric field $\varepsilon$ in the $z$-direction, the perturbation is

$$H^{(1)} = e(\boldsymbol{\varepsilon} \cdot \mathbf{r}) = e\varepsilon r \cos\vartheta. \quad \text{(VI-15)}$$

The nonvanishing matrix elements of $H^{(1)}$ are between states of opposite parities.[a] Thus

---

[a] From (III-122), the non-vanishing elements are

$$\langle n', l+1, m|z|n, l, m\rangle = -\sqrt{\frac{(l-m+1)(l+m+1)}{(2l+1)(2l+3)}} \;\langle n'|r|n\rangle, \quad \text{(VI-16)}$$

$$\langle n'|r|n\rangle = \int_0^\infty R_{n',l+1} r R_{n,l} r^2\, dr.$$

$$E_n^{(1)} = 0, \tag{VI-17a}$$

$$E_n^{(2)} = (e\varepsilon)^2 \sum_k{}' \frac{|\langle n|z|k\rangle|^2}{E_n^0 - E_k^0}, \tag{VI-17b}$$

$$\langle n|z|k\rangle = \int \psi_n^{0*} z \psi_k^0 \, dr, \tag{VI-17c}$$

where the summation is over states $\psi_k^0$ different from $\psi_n^0$, in fact over $\psi_k^0$ of opposite parity to $\psi_n^0$. Thus for a central non-Coulomb field, there is only the so-called quadratic Stark effect, in contradistinction to the Coulomb field (hydrogenic atoms) where there is a linear Stark effect. See Sec. 2 in the following.

The wave function is, up to first order in $e\varepsilon$,

$$\psi_n(r) = \psi_n^0 + e\varepsilon \sum{}' \frac{\langle k|z|n\rangle}{E_n^0 - E_k^0} \psi_k^0, \tag{VI-18}$$

which shows that a state $\psi_n^0$ is now mixed up with states $\psi_k^0$ of the opposite parity.[b]

The matrix element of the electric moment of the perturbed atom is, up to $O(e\varepsilon)$ and in terms of the matrix elements of the unperturbed atom (VI-17c),

$$\int \psi_m^* er \psi_n \, dr = \langle m|er|n\rangle + e\varepsilon \sum_k{}' \frac{\langle m|z|k\rangle}{E_m^0 - E_k^0} \langle k|er|n\rangle$$

$$+ e\varepsilon \sum_k{}' \langle m|er|k\rangle \frac{\langle k|z|n\rangle}{E_n^0 - E_k^0}. \tag{VI-19}$$

The first term is different from zero only when the two states $\psi_m^0$, $\psi_n^0$ have different parities and angular momentum quantum numbers $l'$ and $l$ differing by 1 (see (III-122)). The second and third terms do not vanish only between states $\psi_m^0$, $\psi_n^0$ of the same parity, and $l' - l = 0, \pm 2$, and give the matrix elements of the induced electric dipole moment due to the external electric field.

These results are relevant when we come to the selection rules for radiation in Chap. 8, Sec. 2(1).

### (3) Sommerfeld's relativistic correction

In Chap. 1, Sec. 5, we referred to Sommerfeld's correction to Bohr's theory of the hydrogen atom for the relativistic variation of the mass of the electron

---

[b] In spectroscopic notation (Chap. 1, Sec. 9), an $S$ state $\psi_n^0(S)$ in the perturbed atom has some components of $P$ states $\psi_k^0(P)$ mixed up with it; a $P$ state $\psi_n^0(P)$ has $S$ and $D$ states $\psi_k^0(S)$, $\psi_k^0(D)$ mixed up with it, etc.

with velocity in elliptic orbits. We shall recalculate this effect according to quantum mechanics.

The Hamiltonian of a hydrogenic atom is taken to be

$$H = m_0 c^2 \left( \frac{1}{\sqrt{1 - \beta^2}} - 1 \right) + V(r) \tag{VI-20}$$

$$m_0 = \text{rest mass}, \qquad \beta = \frac{v}{c}, \qquad V(r) = -\frac{Ze^2}{r}.$$

Let $p(p_x, p_y, p_z)$ be the momentum

$$p_x = m_0 v_x \frac{1}{\sqrt{1 - \beta^2}},$$

so that

$$\frac{1}{\sqrt{1 - \beta^2}} = \sqrt{1 + \frac{1}{m_0^2 c^2} p^2}, \tag{VI-21}$$

and

$$H = \frac{1}{2m_0} p^2 - \frac{1}{2m_0} p^2 \left( \frac{p}{2m_0 c} \right)^2 + \cdots + V(r). \tag{VI-22}$$

In the hydrogen atom, $(p/2m_0 c)^2$ is approximately $(v/2c)^2 \ll 1$, so that the second term in $H$ can be treated by the perturbation theory method. The zeroth-order system is given by

$$\left[ \frac{p^2}{2m_0} + V(r) \right] \psi^0 = E^0 \psi^0$$

so that

$$\frac{p^2}{2m_0} \left( \frac{p^2}{2m_0} \psi \right) = \frac{p^2}{2m_0} (E - V)\psi$$

$$= (E - V)^2 \psi^c. \tag{VI-23}$$

The Schrödinger equation of the perturbed system is

$$-\frac{\hbar^2}{2m_0} \nabla^2 \psi + \left[ V - E - \frac{1}{2m_0 c^2} (V - E)^2 \right] \psi = 0. \tag{VI-24}$$

---

$^c$ $(p^2/2m_0)(E - V)\psi$ is not equal to $(E - V)(p^2/2m_0)\psi$. But (VI-23) is correct. See Sucher and Foley, *Phys. Rev.* **95**, 966 (1954), and Wu and Tauber, *ibid*, **106**, 1767 (1955). See also reference at the end of this chapter.

Separating variables

$$\psi(r) = \frac{1}{r} R(r)\Theta(\cos \vartheta)\Phi(\phi),$$

one gets

$$\frac{d^2 R}{dr^2} + \left[ \frac{2m_0}{\hbar^2}(E - V) + \frac{1}{\hbar^2 c^2}(E - V)^2 - \frac{l(l+1)}{r^2} \right] R = 0$$

or, with $\alpha = (e^2/\hbar c) \simeq (1/137)$,

$$\frac{d^2 R}{dr^2} + \left\{ \frac{E^2}{\hbar^2 c^2} + \frac{2m_0 E}{\hbar^2} + \frac{2m_0 Z e^2}{\hbar^2 r}\left(1 + \frac{E}{m_0 c^2}\right) + \frac{1}{r^2}[Z^2 \alpha^2 - l(l+1)] \right\} R$$

$$= 0. \tag{VI-25}$$

This equation is of the form

$$\frac{d^2 R}{dr^2} + \left( A + \frac{B}{r} + \frac{C}{r^2} \right) R = 0, \tag{VI-25a}$$

where $A$, $B$, $C$ are constants. Using the same procedure as that for solving (III-129), one sets in (VI-25)

$$R(r) = \exp(-\sqrt{-A}\,r)r^\gamma \sum_k a_k r^k \tag{VI-26}$$

and obtains the indicial equation

$$\gamma(\gamma - 1) = l(l+1) - Z^2 \alpha^2;$$

one root is

$$\gamma = \sqrt{(l + \tfrac{1}{2})^2 - Z^2 \alpha^2} + \tfrac{1}{2}. \tag{VI-27}$$

The recurrence relation for the $a_k$ is

$$a_{k+1} = \frac{2\sqrt{-A}(\gamma + k) - B}{(\gamma + k)(\gamma + k + 1) + C} a_k. \tag{VI-28}$$

Setting $a_{k+1} = a_{k+2} \cdots = 0$ to reduce the infinite series in (VI-26) to a polynomial, i.e.,

$$2\sqrt{-A}(\gamma + k) - B = 0,$$

one obtains

$$\frac{E}{m_0 c^2} + 1 = \left\{ 1 + \frac{Z^2 \alpha^2}{(k + \tfrac{1}{2} + \sqrt{(l + \tfrac{1}{2})^2 - Z^2 \alpha^2})^2} \right\}^{-1/2}. \tag{VI-29}$$

Let

$$n = k + l + 1 \tag{VI-30}$$

and expand (VI-29) in powers of $Z^2 \alpha^2$, one gets

$$E = -\frac{Z^2 Rhc}{n^2} \left\{ 1 + \frac{Z^2 \alpha}{n^2} \left( \frac{n}{l + \frac{1}{2}} - \frac{3}{4} \right) + \cdots \right\}, \tag{VI-31}$$

$$R = \frac{2\pi^2 m_0 e^4}{ch^3}.$$

The first term in (VI-31) is the Bohr term; the second term (in $Z^2 \alpha^2$) is the quantum mechanical value of the Sommerfeld classical dynamical version given in (I-63). As Sommerfeld's formula (I-63) seems to be in fairly good agreement with the observed fine structure of the $H_\alpha$ line ($n = 3 \rightarrow n = 2$), this new value (VI-31) might seem to have destroyed that agreement.

But if one remembers that there is the spin-orbit interaction correction to be included, one gets from (I-101)

$$\Delta E_s = \frac{RhcZ^4 \alpha^2}{n^4} \begin{cases} \dfrac{n}{(l+1)(2l+1)}, & l = 0, 1, \ldots, \\[2ex] -\dfrac{n}{l(2l+1)}, & l = 1, 2, \ldots, \end{cases} \tag{VI-32}$$

and

$$\Delta E_{\text{rel}} + \Delta E_s = -\frac{RhcZ^4 \alpha^2}{n^4} \begin{cases} \dfrac{n}{l+1} - \dfrac{3}{4}, & l = 0, 1, 2, \ldots, \\[2ex] \dfrac{n}{l} - \dfrac{3}{4}, & l = 1, 2, \ldots, \end{cases} \tag{VI-33}$$

which gives the same numerical value for the energy correction as Sommerfeld's formula (I-63)

$$\Delta E_{\text{Somm.}} = -\frac{RhcZ^4 \alpha^2}{n^4} \left( \frac{n}{k} - \frac{3}{4} \right), \tag{VI-34}$$

$$k = 1, 2, \ldots, n.$$

In Chap. 8, Sec. 1, (2), we shall treat the spin-orbit interaction $\Delta E_s$, and in Chap. 8, Sec. 3, we shall discuss the fine structure of the hydrogenic atom spectral lines as given by (VI-33) and (VI-34).

## 2. Perturbation theory: degenerate systems

For systems $H^0$ such that certain eigenstates are degenerate, i.e., the $\alpha$ independent eigenfunctions

$$\psi_{n_1}^0, \psi_{n_2}^0, \ldots, \psi_{n_\alpha}^0 \tag{VI-35}$$

all correspond to the same eigenvalue $E_n^0$, the formulas (VI-10b), (VI-11, 12) do not hold. In such cases, one proceeds as follows.

Assume that the $\psi_{n_1}^0, \ldots, \psi_{n_\alpha}^0$ are orthogonal and normalized (or, orthogonalized by the Schmidt process; see Chap. 2, Sec. 2, Theorem 14, proof), i.e.,

$$(\psi_{n_i}^0, \psi_{n_j}^0) = \delta_{n_i, n_j}, \tag{VI-36}$$

$$(\psi_{n_i}^0, H^0 \psi_{n_j}^0) = E_n^0 \delta_{n_i, n_j} \text{ for all } n_i. \tag{VI-37}$$

Let $A$ be a $(\alpha \times \alpha)$ unitary matrix

$$AA^\dagger = 1, \qquad \left( \sum_{n_k} A_{n_i, n_k} A_{n_j, n_k}^* = \delta_{n_i, n_j} \right) \tag{VI-38}$$

and transform $\psi_{n_i}^0$ into $\Phi_{n_i}^0$

$$\Phi_{n_i}^0 = \sum_{n_k} A_{n_i, n_k} \psi_{n_k}^0, \tag{VI-39}$$

which can be readily seen to remain orthonormal,

$$(\Phi_{n_i}^0, \Phi_{n_j}^0) = \delta_{n_i, n_j}. \tag{VI-40}$$

With (V-38), it can readily be seen that

$$(\phi_{n_i}^0, H^0 \phi_{n_j}^0) = E_n^0 \delta_{n_i, n_j}; \tag{VI-41}$$

we introduce the notation

$$\begin{aligned}
\mathcal{H}_{n_i, n_j}^{(1)} &\equiv (\Phi_{n_i}^0, H^{(1)} \Phi_{n_j}^0), \\
H_{n_i, n_j}^{(1)} &\equiv (\psi_{n_i}^0, H^{(1)} \psi_{n_j}^0).
\end{aligned} \tag{VI-42}$$

We transform $H^{(1)}$ by $A$ into a diagonal matrix $\mathcal{H}^{(1)}$

$$A^\dagger H^{(1)} A = \mathcal{H}^{(1)},$$

or

$$H^{(1)} A = A \mathcal{H}^{(1)}, \tag{VI-43}$$

or

$$\sum_{n_k} H_{n_i, n_k}^{(1)} A_{n_k, n_j} = \sum_{n_k} A_{n_i, n_k} \mathcal{H}_{n_k, n_j}^{(1)}$$

$$= A_{n_i, n_j} \mathcal{H}_{n_j, n_j}^{(1)},$$

or

$$(H_{n_1 n_1}^{(1)} - \mathcal{H}_{n_1, n_j}^{(1)}) A_{n_1 n_j} + H_{n_1 n_2}^{(1)} A_{n_2 n_j} + H_{n_1 n_3}^{(1)} A_{n_3 n_j} + \cdots = 0,$$

$$H_{n_2 n_1}^{(1)} A_{n_1 n_j} + (H_{n_2 n_2}^{(1)} - \mathcal{H}_{n_j, n_j}^{(1)}) A_{n_2 n_j} + H_{n_2 n_3}^{(1)} A_{n_3 n_j} + \cdots = 0,$$

$$H_{n_3 n_1}^{(1)} - A_{n_1 n_j} + H_{n_3 n_2}^{(1)} A_{n_2 n_j} + (H_{n_3 n_3}^{(1)} - \mathcal{H}_{n_j, n_j}^{(1)}) A_{n_3 n_j} + \cdots = 0,$$

The condition for non-identically vanishing $A_{n_i, n_j}$ is the determinental equation

$$\det \| H^{(1)}_{n_k, n} - \mathcal{H}^{(1)}_{n_j, n_j} \delta_{n_k, n} \| = 0. \tag{VI-44}$$

There are $\alpha$ roots for $\mathcal{H}^{(1)}_{n_j, n_j}$, which may be all distinct, but there may still be repeated roots. In the former case, one says that the original degeneracy has been completely removed by the perturbation $H^{(1)}$; in the latter case, only partially removed. Let us assume that the $\alpha$ roots, i.e., eigenvalues, of $H^{(1)}$ are all distinct. Then the transformation

$$A^\dagger (H^0 + H^{(1)}) A = \mathcal{H}^0 + \mathcal{H}^{(1)}$$

transforms $(H^0 + H^{(1)})_{n_i, n_j}$ into a diagonal matrix $(\mathcal{H}^0 + \mathcal{H}^{(1)})_{n_i, n_j}$.

Let states outside of the $\psi^0_{n_1} \cdots \psi^0_n$ be denoted by $\psi^0_m$, i.e., $E^0_{n_1} - E^0_m \neq 0$. Then,

$$E_{n_i} = E^0_n + \mathcal{H}^{(1)}_{n_i, n_i} + \sum_m{}' \frac{|\mathcal{H}^{(1)}_{n_i, m}|^2}{E^0_{n_i} - E^0_m} + \mathcal{H}^{(2)}_{n_i, n_i} + \cdots, \tag{VI-45a}$$

$$E_m = E^0_m + H^{(1)}_{m, m} + \sum_k{}' \frac{|H^{(1)}_{m, k}|^2}{E^0_m - E^0_k} + H^{(2)}_{m, m} + \sum_{n_i}{}' \frac{|\mathcal{H}^{(1)}_{m, n_i}|^2}{E^0_m - E^0_{n_i}} + \cdots, \tag{VI-45b}$$

$$\phi_{n_i} = \phi^0_{n_i} + \sum_m{}' \frac{\mathcal{H}^{(1)}_{m, n_i}}{E^0_{n_i} - E^0_m} \psi^0_m + \cdots, \tag{VI-45c}$$

$$\psi_m = \psi^0_m + \sum_k{}' \frac{H^{(1)}_{k, m}}{E^0_m - E^0_k} \psi^0_k + \sum_{n_i}^{n_d} \frac{\mathcal{H}^{(1)}_{n_i, m}}{E^0_m - E^0_{n_i}} \phi^0_{n_i}, \tag{VI-45d}$$

where

$$\mathcal{H}^{(1)}_{n_i, m} = (\phi^0_{n_i}, H^{(1)} \psi^0_m), \text{ etc.}$$

To summarize, we see that by the transformation (VI-43), $\mathcal{H}^{(1)}_{n_i, n_j}$ no longer appear with the denominator $E^0_{n_i} - E^0_{n_j} = 0$.

We shall give a few examples.

### (1) Stark effect in (non-relativistic) hydrogenic atoms

The Hamiltonian of a hydrogenic atom in a uniform static field $\varepsilon$ in the $z$-direction is[d]

$$H = \frac{1}{2\mu} p^2 - \frac{Ze^2}{r} + H^{(1)},$$

$$H^{(1)} = -e\varepsilon z. \tag{VI-46}$$

---

[d] We do not include the electron spin here, or, we consider $\varepsilon$ to be strong so that the fine structure of energy levels due to spin-orbit interaction can be ignored and the zeroth-order system (free atom) is degenerate.

The zeroth-order eigenvalues and eigenfunctions are

$$E^0_{n,l,m} = E^0_n \text{ independent of } l, m,$$

$$\psi^0_{n,l,m} = R_{nl}(r)\, Y_{l,m}(\vartheta, \phi).$$

For a given $n$, $l$ can take on the values $0, 1, \ldots, n-1$, and for any $l$, $m$ takes on $-l, -l+1, \ldots, l-1, 2l+1$ values, so that the state $n$ is $n^2$-fold degenerate.[e]
Consider the $n = 2$ level, which is 4-fold degenerate, with wave functions

$$R_{20}Y_{00}, \qquad R_{21}Y_{1,-1}, \qquad R_{21}Y_{1,0}, \qquad R_{21}Y_{1,1}. \tag{VI-47}$$

The non-vanishing matrix elements of $e\varepsilon z$ are those given by (VI-16) and by (E-28) of Chap. 3, Sec. 5, Appendix E. Thus

$$\langle 2,1,0|z|2,0,0\rangle = -\frac{3}{Z}a, \qquad a = \frac{\hbar^2}{m_0 e^2} = \text{Bohr radius}, \tag{VI-48}$$

$$\langle 2,0,0|z|2,1,0\rangle = -\frac{3}{Z}a.$$

The determinantal equation (VI-44) is

$$
\begin{array}{cc}
 & \begin{array}{cccc} l & 0 & 1 & 1 & 1 \\ m & 0 & 1 & 0 & -1 \end{array}
\end{array}
$$

$$
\begin{array}{cc}
\begin{array}{cc} l & m \\ 0 & 0 \\ 1 & 1 \\ 1 & 0 \\ 1 & -1 \end{array}
&
\begin{vmatrix}
-\mathcal{H}^{(1)} & 0 & -\dfrac{3}{Z}a & 0 \\[2mm]
0 & -\mathcal{H}^{(1)} & 0 & 0 \\[2mm]
-\dfrac{3}{Z}a & 0 & -\mathcal{H}^{(1)} & 0 \\[2mm]
0 & 0 & 0 & -\mathcal{H}^{(1)}
\end{vmatrix}
= 0.
\end{array}
\tag{VI-49}
$$

The roots and the eigenfunctions $\phi^0$ are

$$
E^{(1)} = \begin{cases} -\dfrac{3}{Z}ae, \\[3mm] 0, \\[3mm] 0, \\[3mm] +\dfrac{3}{Z}ae, \end{cases}
\qquad
\phi^0 = \begin{cases} \phi^0_1 = \dfrac{1}{\sqrt{2}}(R_{20}Y_{00} + R_{21}Y_{10}), \\[3mm] \phi^0_3 = R_{21}Y_{11}, \\[3mm] \phi^0_4 = R_{21}Y_{1,-1}, \\[3mm] \phi^0_2 = \dfrac{1}{\sqrt{2}}(R_{20}Y_{00} - R_{21}Y_{10}). \end{cases}
\tag{VI-50}
$$

---

[e] With electron spin, the degeneracy is $2n^2$-fold. See Chap. 1, Sec. 11, (I-115).

The two eigenvalues $E^{(1)} = 0$ are still degenerate, on account of the rotational symmetry of the system around the direction of the electric field $\varepsilon$.

Similar calculations can be carried out for the other $n$ levels.

The zeroth-order Schrödinger equation for a hydrogen atom in a Coulomb field can be solved by the method of separation of variables in more than one set of coordinates. In Appendix F at the end of this section, we shall introduce parabolic coordinates which are particularly suitable for the linear Stark effect treated above, and for the scattering of an electron by a Coulomb field.

## Appendix F
### Parabolic coordinates: Stark effect

The parabolic coordinates $\xi$, $\eta$, $\phi$ are defined in terms of the rectangular coordinates by

$$x = \sqrt{\varepsilon\eta}\cos\phi, \qquad y = \sqrt{\xi\eta}\sin\phi, \qquad z = \tfrac{1}{2}(\xi - \eta),$$

or

$$\xi = r - z, \qquad \eta = r + z, \qquad \phi = \tan^{-1}\left(\frac{y}{x}\right), \tag{F-1}$$

so that

$$r = \tfrac{1}{2}(\xi + \eta), \tag{F-1a}$$

and

$$dx\,dy\,dz = \tfrac{1}{4}(\xi + \eta)\,d\xi\,d\eta\,d\phi. \tag{F-2}$$

The Schrödinger equation of a hydrogenic atom $(Z)$ can be shown to be

$$\frac{\partial}{\partial\xi}\left(\xi\frac{\partial\Psi}{\partial\xi}\right) + \frac{\partial}{\partial\eta}\left(\eta\frac{\partial\Psi}{\partial\eta}\right) + \frac{1}{4}\left(\frac{1}{\xi} + \frac{1}{\eta}\right)\frac{\partial^2\Psi}{\partial\phi^2} + \frac{\mu}{2\hbar^2}[E(\xi + \eta) + 2Ze^2]\Psi = 0. \tag{F-3}$$

This equation can be solved by

$$\Psi(\xi,\eta,\phi) = \frac{1}{\sqrt{2}}e^{im\phi}F(\xi)G(\eta), \tag{F-4}$$

leading to

$$\frac{d}{d\xi}\left(\xi\frac{dF}{d\xi}\right) + \frac{\mu}{2\hbar^2}\left[E\xi + 2Ze^2\beta_1 - \frac{m^2\hbar^2}{2\mu\xi}\right]F = 0, \tag{F-5}$$

$$\frac{d}{d\eta}\left(\eta\frac{dG}{d\eta}\right) + \frac{\mu}{2\hbar^2}\left[E\eta + 2Ze^2\beta_2 - \frac{m^2\hbar^2}{2\mu\eta}\right]G = 0, \qquad \text{(F-6)}$$

where

$$m = 0, \pm 1, \pm 2, \ldots \qquad \text{(F-7)}$$

$$\beta_1 + \beta_2 = 1. \qquad \text{(F-8)}$$

These two equations can be solved in terms of the associated Laguerre polynomials (see Chap. 3, Sec. 5, Appendix E),

$$F(u) = u^{1/2|m|}e^{-u/2}L_{k_1+|m|}^{|m|}(u), \qquad \text{(F-9a)}$$

$$G(v) = u^{1/2|m|}e^{-v/2}L_{k_2+m}^{|m|}(v), \qquad \text{(F-9b)}$$

where

$$u = \frac{\xi}{na}, \qquad v = \frac{\eta}{na}, \qquad a = \frac{\hbar^2}{\mu e^2} = \text{Bohr radius},$$

$$k_1 + k_2 + |m| + 1 = n,$$

$$n\beta_1 = k_1 + \tfrac{1}{2}|m| + \tfrac{1}{2}, \qquad \text{(F-10)}$$

$$n\beta_2 = k_2 + \tfrac{1}{2}|m| + \tfrac{1}{2},$$

$$k_1, k_2 \geq 0, \text{ integers.}$$

The eigenvalue $E_n^0$ is once again the Bohr formula

$$E_n^0 = -\frac{Z^2 Rhc}{n^2}, \qquad n = 1, 2, 3, \ldots \qquad \text{(F-11)}$$

The normalized eigenfunction is

$$\Psi_{n,k,m}^0(\xi, \eta, \phi) = \frac{1}{\sqrt{2\pi}}e^{im\phi}\sqrt{\frac{2k_1!k_2!}{a^3 n^4[(k_1 + |m|)!(k_2 + |m|)!]^3}} \times F(u)F(v). \qquad \text{(F-12)}$$

If the atom is in an electric field $\varepsilon$ in the $z$-direction, the perturbation potential is

$$H^{(1)} = -e\varepsilon z = -\tfrac{1}{2}e\varepsilon(\xi - \eta). \qquad \text{(F-13)}$$

This will lead to an additional term

$$\begin{aligned}
&\tfrac{1}{2}e\varepsilon\xi^2 \quad &&\text{inside } [\qquad] \text{ of (F-5),}\\
&-\tfrac{1}{2}e\varepsilon\eta^2 \quad &&\text{inside } [\qquad] \text{ of (F-6).}
\end{aligned} \qquad \text{(F-14)}$$

The non-vanishing matrix elements of $H^{(1)}$ are

$$\langle n, k_1, m | H^{(1)} | n, k_1', m' \rangle = -\frac{3}{2} n(k_1 - k_2) ae\delta_{k_1, k_1'} \delta_{m, m'}, \qquad \text{(F-15)}$$

so that $H^{(1)}$ is diagonal in $k_1$ and $m$,

$$E^{(1)}_{n, k_1, m} = \langle n, k_1, m | H^{(1)} | n, k_1, m \rangle$$

$$= -\frac{3}{2} n(k_1 - k_2) e\varepsilon a. \qquad \text{(F-15a)}$$

Hence the function $\Psi^0_{n, k, m}(\xi, \eta, \phi)$ is already the function $\Phi^0_m$ of (VI-39), (VI-45c). Comparison between (F-15a) and (VI-50) for $n = 2$ shows that[f]

$$\frac{1}{\sqrt{2}}(R_{20} Y_{00} + R_{21} Y_{10}) = \Phi^0_1 = \Psi^0_{2, 1, 0},$$

$$\begin{aligned} R_{21} Y_{1,1} &= \Phi^0_3 \\ R_{21} Y_{1,-1} &= \Phi^0_4 \end{aligned} = \begin{cases} \psi^0_{2, 0, 1}, \\ \Psi^0_{2, 0, 1}, \end{cases} \qquad \text{(F-16)}$$

$$\frac{1}{\sqrt{2}}(R_{20} Y_{00} - R_{21} Y_{10}) = \Phi^0_2 = \psi^0_{2, 0, 0}.$$

## Exercises

1. Consider the two-dimensional harmonic oscillator of Prob. 4 at the end of Chap. 3, Sec. 4. Calculate the energy of the oscillator when perturbed by

$$H^{(1)} = k_1 \hbar\omega\rho^4$$

and the degeneracy of state $n$.

2. Consider the three-dimensional isotropic harmonic oscillator of Prob. 5 at the end of Chap. 3, Sec. 4. Show that the energy when perturbed by

$$H^{(1)} = k_1 \hbar\omega\rho^4$$

is given by

$$E^{(1)}_{n, l} = \frac{1}{2} k_1 \left[ 3\left(n + \frac{3}{2}\right)^2 - \left(l + \frac{1}{2}\right)^2 + 1 \right] \hbar\omega.$$

[f] The relation between the $\Phi^0_{n_i}$ (combinations of $R_{nl} Y_{lm}$ in spherical polar coordinates) and the $\Psi^0_{n, k_1, m}(\xi, \eta, \phi)$ can of course be obtained by means of the relations of $R_{nl}$; $\Psi^0_{n, k_1, m}$ and the associated Laguerre polynomials. That procedure is however lengthy. The method of comparing the roots of the determinent (VI-49) and the expression (F-15a) is shorter.

Show that this perturbation splits the $\frac{1}{2}(n + 1)(n + 2)$-fold degenerate level $E_n^0$ into

$$\tfrac{1}{2}(n + 2) \text{ levels for } n = \text{even integer,}$$

$$\tfrac{1}{2}(n + 1) \text{ levels for } n = \text{odd integer.}$$

3. Given the matrix of the Hamiltonian to be

$$\begin{vmatrix} H_{11}^0 + \lambda H_{11}^{(1)} & \lambda H_{12}^{(1)} \\ \lambda H_{21}^{(1)} & H_{22}^0 + \lambda H_{22}^{(1)} \end{vmatrix},$$

$$H_{12}^{(1)} = H_{21}^{(1)}.$$

(a) Obtain the exact eigenvalues $E_1$ and $E_2$.

(b) Calculate the approximate eigenvalues by the perturbation method, assuming

$$\frac{\lambda H_{12}^{(1)}}{H_{22}^0 - H_{11}^0} \ll 1.$$

(c) Compare these approximate results with the expansions of $E_1$, $E_2$ in powers of $\lambda$.

(d) Discuss the case where $H_{11}^0 = H_{22}^0$.

4. Given a Hamiltonian whose matrix is

$$H(\xi) = \begin{vmatrix} H_{11} & H_{12} \\ H_{21} & H_{22} \end{vmatrix},$$

where $H_{12}(\xi) = H_{21}(\xi)$, $\xi$ being a parameter determined by some external field. The system is such that for the two asymptotic limits of $\xi \ll 1$ and $\xi \gg 1$, the two eigenvalues $W_{11}$, $W_{22}$ are different from each other. Show that there exists no value $\xi_0$ of $\xi$ for which

$$W_{11}(\xi_0) = W_{22}(\xi_0),$$

i.e., there is no value of $\xi$ for which the system is accidentally degenerate. This result is known as the *non-crossing theorem*.

5. Show that, if in (VI-22) one treats the term

$$-\frac{1}{2m_0} p^2 \left( \frac{p}{2m_0 c} \right)^2$$

and its equivalent $-(E - V)^2$ in (VI-23) in a first-order perturbation calculation, the result is the same as that given by (VI-31).

6. Obtain the Stark effect energy corrections for the $n = 3$ state for hydrogenic atoms (corresponding to (VI-49)). Also obtain the wave functions of these levels in parabolic coordinates (similar to the relations in (F-16) in Appendix F).

## References

E. Schrödinger, *Annalen d. Physik* **80**, 437 (1926). This paper deals with perturbation theory for the time-dependent Schrödinger equation, Stark effect for hydrogen atom and parabolic coordinates.

E. U. Condon and G. H. Shortley, *The Theory of Atomic Spectra* (Cambridge Univ. Press, Cambridge, 1934). Chapter 5, Sec. 3 is on the relativistic correction. Note that a statement "$p^2$ commutes with $(E - V)$" on p. 114, would be incorrect; however, up to the second order correction, one has $(1/8m_0^2 c^2)p^4\psi = (1/2c^2)(E - V)^2\psi$.

Chapter 7

# Time-Dependent Systems

Up to now, we have treated stationary state problems. For time-dependent systems, in nonrelativistic quantum mechanics, the basic theory is the Schrödinger equation

$$\frac{\hbar}{i}\frac{\partial \Psi}{\partial t} + H\Psi = 0. \tag{VII-1}$$

In his fourth communication on wave mechanics (June, 1926), Schrödinger developed a perturbation theory for this time-dependent equation. This was followed by the work of Born and Dirac; we shall consider these methods in the following.

## 1. Theory of the Raman effect and inverse Stark effect

In 1928, the Indian physicist C. V. Raman discovered that when a substance, a liquid of carbon tetrachloride $CCl_4$ for example, is irradiated by a monochromatic light of frequency $v$, the spectrum of the scattered radiation shows, in addition to the frequency $v$, spectral lines of frequencies $v - v_k$ where the differences $v_k$ are characteristic of the substance. It is soon found that $v - v_k$ correspond to the vibrational frequencies of the molecule of the substance. In fact the presence of such frequencies $v - v_k$ had already been implied in the work of Kramers and Heisenberg (1925) on the quantum theory of dispersion and suggested even earlier by Smekal (1923).[a] From the time span between 1925 and 1928, it appears that, the communication between the theoretical and the experimental physicists was not very close.

Let the electric field of a monochromatic beam of radiation of frequency $v$ be

$$\varepsilon = \varepsilon_0 e^{-2\pi i v t} + \varepsilon_0^* e^{2\pi i v t}. \tag{VII-2}$$

Let the electric moment of the molecule of the irradiated substance be **M**. The Hamiltonian of the system $H$ is

$$H = H^0 + H^{(1)}$$

$$= H^0 - (\mathbf{M} \cdot \varepsilon). \tag{VII-3}$$

---

[a] See Chap. 1, Sec. 8.

For the unperturbed system, the Schrödinger equation is

$$\left(H^0 - i\hbar\frac{\partial}{\partial t}\right)\Psi^0 = 0,$$

$$\left(H^0 - i\hbar\frac{\partial}{\partial t}\right)^* \Psi^{0*} = 0, \tag{VII-4a}$$

and

$$\Psi_n^0 = \psi_n^0 \exp(-iE_n t/\hbar), \tag{VII-4b}$$

$$(H^0 - E_n)\psi_n^0 = 0. \tag{VII-4c}$$

For the perturbed system, the Schrödinger equation is

$$\left(H^0 - i\hbar\frac{\partial}{\partial t}\right)\Psi = (\mathbf{M}\cdot\boldsymbol{\varepsilon})\Psi. \tag{VII-5}$$

Let

$$\Psi_n = \Psi_n^0 + \Psi_n^{(1)}, \tag{VII-6}$$

where

$$|\Psi_n^{(1)}| \ll |\Psi_n^0|. \tag{VII-6b}$$

If one neglects $(\mathbf{M}\cdot\boldsymbol{\varepsilon})\Psi_n^{(1)}$ compared with $(\mathbf{M}\cdot\boldsymbol{\varepsilon})\Psi_n^0$, (VII-5) leads to

$$\left(H^0 - i\hbar\frac{\partial}{\partial t}\right)\Psi_n^{(1)} = (\mathbf{M}\cdot\boldsymbol{\varepsilon})\Psi_n^0, \tag{VII-7}$$

$$\left(H^0 - i\hbar\frac{\partial}{\partial t}\right)^* \Psi_n^{(1)*} = (\mathbf{M}\cdot\boldsymbol{\varepsilon})\Psi_n^{0*}.$$

Now on making the *Ansatz*

$$\begin{aligned}
\Psi_n^{(1)} &= \psi_n^{(+)}\exp[-i(E_n + h\nu)t/\hbar], \\
&\quad + \psi_n^{(-)}\exp[-i(E_n - h\nu)t/\hbar],
\end{aligned} \tag{VII-8}$$

one obtains

$$[H^0 - (E_n + h\nu)]\psi_n^{(+)} = (\boldsymbol{\varepsilon}_0\cdot\mathbf{M})\psi_n^0, \tag{VII-9a}$$

$$[H^0 - (E_n - h\nu)]\psi_n^{(-)} = (\boldsymbol{\varepsilon}_0^*\cdot\mathbf{M})\psi_n^0, \tag{VII-9b}$$

and two other equations (VII-9c, 9d) in which the $\psi_n^{(+)}$, $\psi_n^{(-)}$, $\psi_0$, $\varepsilon$, $\varepsilon_0^*$ above are replaced by their complex conjugates. On expanding

$$\psi_n^{(+)} = \sum_r A_{nr}\psi_r^0, \tag{VII-10a}$$

$$(\boldsymbol{\varepsilon}_0 \cdot \mathbf{M})\psi_n^0 = \sum_r \langle r|\boldsymbol{\varepsilon}_0 \cdot \mathbf{M}|n\rangle \psi_r^0, \tag{VII-10b}$$

where

$$\langle r|\boldsymbol{\varepsilon}_0 \cdot \mathbf{M}|n\rangle = \int \psi_r^{0*}(\boldsymbol{\varepsilon}_0 \cdot \mathbf{M})\psi_n^0 \, d\tau, \quad \text{etc.,} \tag{VII-10c}$$

and putting these into (VII-9a, b, c, d), one obtains

$$\psi_n^{(+)} = \sum_r \frac{\langle r|\boldsymbol{\varepsilon}_0 \cdot \mathbf{M}|n\rangle}{E_r - (E_n + h\nu)}\psi_r^0, \qquad \psi_n^{(-)} = \sum_r \frac{\langle r|\boldsymbol{\varepsilon}_0^* \cdot \mathbf{M}|n\rangle}{E_r - (E_n - h\nu)}\psi_r^0, \tag{VII-11a}$$

$$\psi_n^{(+)} = \sum_r \frac{\langle n|\boldsymbol{\varepsilon}_0^* \cdot \mathbf{M}|r\rangle}{E_r - (E_n + h\nu)}\psi_r^{0*}, \qquad \psi_n^{(-)} = \sum_r \frac{\langle n|\boldsymbol{\varepsilon}_0 \cdot \mathbf{M}|r\rangle}{E_r - (E_n - h\nu)}\psi_r^{0*}. \tag{VII-11b}$$

The matrix elements of the $x$-component of the electric moment of the system are given by

$$\int \Psi_k^* M_x \Psi_n \, d\tau$$

$$= \int (\Psi_k^0 + \Psi_k^{(1)})^* M_x (\Psi_n^0 + \Psi_n^{(1)}) \, d\tau$$

$$= \langle k|M_x|n\rangle e^{2\pi i \nu_{kn} t}$$

$$+ \frac{1}{h}\sum_r \left\{ \frac{\langle k|M_x|r\rangle \langle r|\boldsymbol{\varepsilon}_0 \cdot \mathbf{M}|n\rangle}{\nu_{rn} - \nu} + \frac{\langle k|\boldsymbol{\varepsilon}_0 \cdot \mathbf{M}|r\rangle \langle r|M_x|n\rangle}{\nu_{rn} + \nu} \right\} e^{-2\pi i(\nu + \nu_{nk})t}$$

$$+ \frac{1}{h}\sum_r \left\{ \frac{\langle k|M_x|r\rangle \langle r|\boldsymbol{\varepsilon}_0^* \cdot \mathbf{M}|n\rangle}{\nu_{rn} + \nu} + \frac{\langle k|\boldsymbol{\varepsilon}_n^* \cdot \mathbf{M}|r\rangle \langle r|M_x|n\rangle}{\nu_{rn} - \nu} \right\} e^{2\pi i(\nu - \nu_{nk})t}$$

$$+ \text{terms proportional to } \varepsilon_0 \varepsilon_0^*. \tag{VII-12}$$

The interpretation of the above result is as follows: The first term on the right-hand side, $\langle k|M_x|n\rangle e^{-2\pi i \nu_{nk} t}$, is the matrix element of the unperturbed system with the frequency $\nu_{nk} = (E_n - E_k)/h$; the second and the third terms are the matrix elements of the induced electric moment with frequency $\nu + \nu_{nk}$ and $\nu_{nk} - \nu$. The terms in $\varepsilon_0 \varepsilon_0^*$ will be regarded as small and neglected here.

Now according to O. Klein[b], in the spirit of the correspondence principle, the electric moment

$$\langle i|M|j\rangle e^{-2\pi i \nu_{ij} t} \tag{VII-13}$$

emits no radiation if $\nu_{ij} < 0$, and is equal to

[b] O. Klein, *Z. Phys.* **41**, 407 (1927).

$$\langle i|M|j\rangle e^{-2\pi i v_{ij}t} + \langle i|M|j\rangle^* e^{2\pi i v_{ij}t}$$

if $v_{ij} > 0$. Thus the first terms of (VII-12) leads to spontaneous emission of radiation of frequency $v_{nk}$ if $E_n > E_k$. The second term represents the radiation (due to the induced dipole moment) of frequency $v + v_{kn}$ for $v + v_{kn} > 0$. Here we distinguish three cases:

(i) $E_n - E_k > 0$. The "scattered" radiation has a frequency $v + v_{nk}$ higher than that of the exciting radiation, leading to an anti-Stokes Raman line.

(ii) $E_k - E_n > 0$. The scattered radiation has frequency $v - v_{kn} < v$, leading to a Stokes Raman line.

(iii) $E_k - E_n = 0$. In this case, if $\psi_k^0$ and $\psi_n^0$ are one and the *same* state $n$, then the second term of (VII-12) is

$$\sum_{\sigma}^{x,y,z} \frac{1}{\hbar}\varepsilon_{0\sigma} \sum_n \left\{ \frac{\langle k|M_x|r\rangle \langle r|M_\sigma|n\rangle}{v_{rn} - v} + \frac{\langle k|M_\sigma|r\rangle \langle r|M_x|n\rangle}{v_{rn} + v} \right\} e^{-2\pi i v t},$$

$$(\text{VII-14})$$

which gives rise to the *coherent* part of the Rayleigh scattering. If $\psi_k^0$ and $\psi_n^0$ are two degenerate states with $E_n = E_k$, then in (VII-14), the two independent phase factors in $\psi_n^0 \exp(-2\pi i[(E_n t/h) - \delta_n])$, $\psi_n^0 \exp(-2\pi i[(E_k t/h) - \delta_k])$ will render the scattering incoherent.

The above three cases are shown in the following figure.

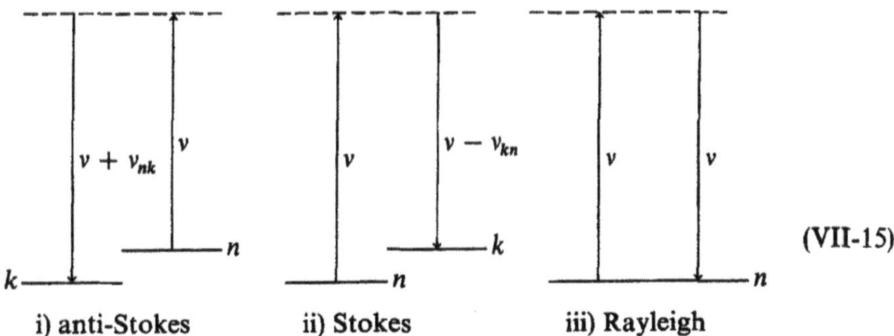


$$(\text{VII-15})$$

The third term in (VII-12) represents the emission of radiation of frequency $v_{nk} - v$ if $v_{nk} - v > 0$. In this case we must have $E_n > E_k$, as shown in the following figure.



$$(\text{VII-16})$$

From (VII-15), it is seen that, on account of the Boltzmann factor for the distribution of the molecules in the two states $E_n$ and $E_k$, the Stokes Raman lines have greater intensity than the anti-Stokes lines.

The induced moment (VII-14) for coherent Rayleigh scattering can be put into the form

$$\langle n|M|n\rangle^{(1)} = \sum_{\sigma=x,y,z} (\alpha_{x\sigma}^{(0)})_{nn}\varepsilon_\sigma, \qquad \varepsilon_\sigma = \varepsilon_{0\sigma}e^{-2\pi i\nu t}, \qquad \text{(VII-17)}$$

where $\alpha_{xy}^{(0)}$ is the polarizability of the system,

$$(\alpha_{x\sigma}^{(0)})_{nn} = \frac{1}{h}\sum_r \left\{ \frac{\langle n|M_x|r\rangle\langle r|M_\sigma|n\rangle}{\nu_{rn}-\nu} + \frac{\langle n|M_\sigma|r\rangle\langle r|M_x|n\rangle}{\nu_{rn}+\nu} \right\}, \quad \text{(VII-18)}$$

which is a property of the unperturbed system. It is a tensor since $\alpha_{xy}^{(0)}$ transforms as the product of two vector components $M_x M_y$. Here only the diagonal element $(n, n)$ of the ground state $n$ is involved for coherent Rayleigh scattering.

Similarly for the Raman scattering, the induced moment is

$$\langle k|M|n\rangle^{(1)} = \sum_\sigma (\alpha_{x\sigma}^{(0)})_{kn}\,\varepsilon_\sigma, \qquad \varepsilon_\sigma = \varepsilon_{0\sigma}e^{-2\pi i(\nu-\nu_{kn})t}. \qquad \text{(VII-19)}$$

Here the polarizability tensor matrix element $(\alpha_{x\sigma}^{(0)})_{kn}$ is associated with the states $k$ and $n$ (for example, two vibrational states of a molecule).

From (VII-12), it is seen that for Raman scattering involving a transition $n \to k$ in (VII-15) to appear, it is necessary that there exists at least one state $\psi_r^0$ for which $\langle k|M_\sigma|r\rangle$ and $\langle r|M_\sigma|n\rangle$ do not vanish together. This is very important. For example, if the system has central symmetry, this condition for the appearance of Raman scattering is that the states $k$ and $n$ be of the *same* parity. In this case the "intermediate" states $r$ have opposite parity to that of $k$ and $n$.

Thus the symmetry of the (change of the) polarizability (in the transition $n \to k$) of the system determines the selection rule of Raman scattering.

A special case of the general theory as represented in (VII-12) is when the frequency of the exciting radiation $\nu$ approaches the limit zero. In this case, the electric field $\varepsilon$ becomes a static field. Equation (VII-12) then gives the $(k, n)$ matrix element of the induced electric moment associated with the frequency $\nu_{kn}$ ($E_k - E_n > 0$ for emission, $E_k - E_n < 0$ for absorption). This induced moment gives rise to the emission or absorption lines $\nu_{kn} > 0$ when the system (molecule, say) is in a static electric field.

This case is particularly interesting if the system has (central) symmetry such that the dipole transition $k \leftrightarrow n$ for the unperturbed system is "forbidden", i.e., $\langle k|M|n\rangle = 0$, as for example when states $k, n$ have the same (parity) symmetry. Thus the forbidden dipole transitions $^2S - n^2S$, $^2S - n^2D$ in the alkali metal atoms are observed in the absorption spectrum of the atomic vapor in an electric field. The $m^2P$ states are then the "intermediate" or "perturbing" states $r$ in the expression (VII-12). This is known as the inverse Stark effect.[c] We have now seen here that the general theory of the Raman effect includes the inverse Stark effect as a special case.

## 2. Perturbation theory of transitions of Dirac[d]

### (1) Method of variation of constants

Suppose the Schrödinger equation of (unperturbed) system

$$ih\frac{\partial \Psi}{\partial t} = H^0 \Psi^0(q, t) \tag{VII-20}$$

has already been solved, i.e., the eigenvalues and eigenfunctions are known

$$H^0 \psi_n(q) = E_n^0 \psi_n(q), \tag{VII-21}$$

$$\Psi_n^0(q, t) = e^{-iE_n t/\hbar} \psi_n(q). \tag{VII-22}$$

Then an arbitrary solution of (VII-20) can be expressed as

$$\Psi^0(q, t) = \sum_m a_m^0 \psi_m(q) \exp\left(-i\frac{E_m t}{\hbar}\right). \tag{VII-23}$$

If at time $t = 0$,

$$\Psi^0(q, 0) = \text{a given function } f(q),$$

then

$$a_m^0 = \int \psi_m^*(q) f(q)\, dq, \tag{VII-24}$$

and the initial-value problem of (VII-20) is solved.

Let the system be perturbed by a potential $H_1$ and we wish to study the effect of this perturbation. Let the Schrödinger equation of the perturbed system be

---

[c] The inverse, quadratic Stark effect in Rb and Cs atoms was first observed and measured by Y. T. Yao, *Z. Phys.* **77**, 307 (1932).

The possibility of observing "forbidden" or "inactive" vibrational transitions of molecules in infrared absorption in strong electric field was suggested by E. U. Condon, *Phys. Rev.* **41**, 759 (1932). The connection between the general theory of the Raman effect and the suggestion of Condon was pointed out by T. Y. Wu, *Vibrational Spectra and Structure of Polyatomic Molecules* (National Univ., Peking Press 1939), p. 64; 2nd ed. (Ediv. Bros., Ann Arbor, Mich., 1946).

[d] P. A. M. Dirac, *Proc. Roy. Soc.* London **A112**, 66 (1926); **A114**, 243 (1927).

$$ i\hbar\frac{\partial \Psi}{\partial t} = (H^0 + \lambda H_1)\Psi(q, t), \qquad\qquad \text{(VII-25)} $$

where $\lambda$ is an arbitrary parameter

$$ 0 \le \lambda \le 1. $$

We shall assume that $H_1$ is small compared with $H^0$

$$ |\langle \psi_m | H_1 | \psi_m \rangle| \ll |\langle \psi_m | H_0 | \psi_m \rangle| \qquad\qquad \text{(VII-26)} $$

and assume that a solution $\Psi(q, t)$ of (VII-25) can be expanded in the form

$$ \Psi(q, t) = \sum a_n(t)\psi_n(q)\exp\left(-i\frac{E_n}{\hbar}t\right), \qquad\qquad \text{(VII-27)} $$

where the coefficients $a_n(t)$ are no longer constants as in (VII-23, 24), but are functions of time. From (VII-27) and (VII-25), one obtains a system of equations for $a_n(t)$,

$$ \frac{da_n}{dt} = i\hbar \sum a_m(t)\langle n|H_1|m\rangle \exp\left\{\frac{i}{\hbar}(E_n - E_m)t\right\}, \qquad \text{(VII-28)} $$

where

$$ \langle n|H_1|m\rangle = \int \psi_n^* H_1 \psi_m \, dq. \qquad\qquad \text{(VII-29)} $$

We assume further that

$$ a_n(t) = a_n^0 + \lambda a_n^{(1)}(t) + \lambda^2 a_n^{(2)}(t) + \cdots, \qquad\qquad \text{(VII-30)} $$

where $a_n^0$ is the constant in (VII-23) determined by the initial condition (VII-24).

We now solve the initial value problem (VII-25) with the initial condition

$$ \text{at } t = 0, \qquad a_n^0 = 1, \qquad a_m^0 = 0 \qquad \text{for } m \ne n. \qquad \text{(VII-31)} $$

Substituting (VII-30) in (VII-28) and equating the coefficients of each power of $\lambda$ to zero, we get

$$ \lambda^0: \qquad\qquad \frac{da_k^0}{dt} = 0, $$

$$ \lambda^1: \qquad\qquad \frac{da_k^{(1)}}{dt} = i\hbar \sum_m \langle k|H_1|m\rangle a_m^0 \exp\left\{\frac{i}{\hbar}(E_k - E_m)t\right\}. \qquad \text{(VII-32)} $$

From the normalization condition, we get

$$\int \Psi^{0*}\Psi^0\, dq = a_n^{0*} a_n^0 = 1,$$

$$\int \Psi^*\Psi\, dq = \sum a_m^*(t) a_m(t) = 1. \tag{VII-33}$$

$|a_m(t)|^2$ give the probabilities the various states $\psi_m(q)$ appear at time $t$ in (VII-27), after $H_1$ has been switched on at $t = 0$.

Consider the case of a periodic perturbation

$$H_1 = V(q)[e^{i\omega t} + e^{-i\omega t}]. \tag{VII-34}$$

On integrating (VII-32), one obtains

$$a_k^{(1)} = -\langle k|V|n\rangle \left\{ \frac{\exp\left[\dfrac{i}{\hbar}(E_k - E_n + \hbar\omega)t\right] - 1}{E_k - E_n + \hbar\omega} \right.$$

$$\left. + \frac{\exp\left[\dfrac{i}{\hbar}(E_k - E_n - \hbar\omega)t\right] - 1}{E_k - E_n - \hbar\omega} \right\}. \tag{VII-35}$$

Let $\psi_n$ be the ground state so that $E_k - E_n > 0$, and let $h\nu \simeq E_k - E_n$ (near resonance absorption). In this case the first term can be neglected compared with the second, so that transition probability in the time interval $t$ is

$$|a_k^{(1)}(t)|^2 = 4|\langle k|V|n\rangle|^2 \frac{\sin^2 x}{x^2}\left(\frac{t}{2\hbar}\right)^2, \tag{VII-36}$$

where

$$x = \left(\frac{E_k - E_n - \hbar\omega}{2\hbar}\right) t. \tag{VII-36a}$$

The function $\sin^2 x/x^2$ has a very sharp maximum at $x = 0$ so that the probability of the transition $n \to k$ is large when $\hbar\omega \simeq E_k - E_n$.

There are two circumstances for the resonance condition

$$\hbar\omega = E_k - E_n. \tag{VII-37}$$

(i) $H_1$ has a sharp frequency $\omega$, but the state $E_k$ either is in the continuum or has a width $\Delta E_k$.

Let $\rho(E_k)\, dE_k$ be the number of states $\psi_k$ lying between $E_k$ and $E_k + dE_k$. Thus the probability of the system going from $\psi_n$ to any one of the states near $E_k$ in time $t$ is

$$w(t) = \int |a_k^{(1)}(t)|^2 \rho(E_k)\, dE_k. \tag{VII-38}$$

If, within $\Delta E_k$, $\rho(E_k)$ and $|\langle k|V|n\rangle|^2$ are very slowly varying functions of $E_k$, then, from (VII-36),

$$w(t) = |\langle k|V|n\rangle|^2 \rho(E_k) \int_{-\infty}^{\infty} \left(\frac{2t}{\hbar}\right) \frac{\sin^2 x}{x^2}\, dx, \tag{VII-39}$$

$$= \frac{2\pi}{\hbar} |\langle k|V|n\rangle|^2 \rho(E_k) t. \tag{VII-39a}$$

This is the probability for the transition $n \to k$ during the time $t$. The probability per unit time is

$$P = \frac{w(t)}{t} = \frac{2\pi}{\hbar} |\langle k|V|n\rangle|^2 \rho(E_k). \tag{VII-40}$$

(ii) $E_k$ and $E_n$ are sharp, but $H_1$ has a frequency spread near resonance (VII-37).

Let $\sigma(\nu)\, d\nu$ be the number of states of $H_1$ between $\nu$ and $\nu + d\nu$. The probability $w(t)$ in (VII-38) is, in this case, following similar arguments as above,

$$w(t) = \int |a_k^{(1)}(t)|^2 \sigma(\nu)\, d\nu$$

$$= |\langle k|V|n\rangle|^2 \sigma(\nu) \int_{-\infty}^{\infty} \left(\frac{t}{\pi\hbar^2}\right) \frac{\sin^2 x}{x^2}\, dx$$

$$= \frac{1}{\hbar^2} |\langle k|V|n\rangle|^2 \sigma(\nu) t \tag{VII-41}$$

and the probability per unit time is

$$P = \frac{w(t)}{t} = \frac{1}{\hbar^2} |\langle k|V|n\rangle|^2 \sigma(\nu). \tag{VII-41a}$$

We shall make the following remarks concerning the above theory:

(i) When the perturbation $H_1$ is time-dependent, the treatment above can be applied.

(ii) The time $t$ in (VII-39a) and (VII-41) must not be too long so that $w(t)$ is still $\ll 1$; but must not be too short; for example, it must be long compared with the "periodic time" of the electron orbital motion, say.

(iii) The time-dependent theory can still be used when $H_1$ is not time-dependent as for example in the scattering of a particle by a potential field $V$. In such cases, it is necessary to introduce some process of switching on and off the interaction $V$ at certain instants of time. This switching may be done abruptly as follows:

$$H = H_0, \qquad -\infty < t < 0,$$

$$H = H_0 + V, \qquad 0 \le t \le \tau,$$

$$H = H_0, \qquad \tau < t,$$

and the problem treated in the manner described above. Or, the switching may be carried out in an adiabatic manner, which is more conveniently treated by means of the integral form of the Schrödinger equation. Since the switching process is equivalent to replacing the static $V$ by $Vf(t)$, the solution of (VII-25) in general depends on $f(t)$. Hence it is necessary that the physical result be independent of the parameters appearing in the function $f(t)$. The theory has been developed for this question, but we shall refer the reader to the literature.[e]

We shall treat a few problems to illustrate the method of this section.

## (2) Einstein's $B_n^m$, $A_n^m$ coefficients

In Chap. 1, Sec. 6, we defined $B_n^m u(v)$, where $u(v)\,dv$ is the radiation energy density in the frequency range $v$ and $v + dv$, as the transition per unit time of a system in the radiation field. According to (VII-36), we have

$$B_n^m u(v) = \frac{1}{t} |a_m^{(1)}(t)|^2; \qquad\qquad \text{(VII-42)}$$

we shall now calculate $B_n^m$.

In classical electrodynamics, for a particle of mass $\mu$ and charge $e$ in an electromagnetic field $(A, \phi)$, the Hamiltonian is[f]

---

[e] Cf. T. Y. Wu and T. Ohmura, *The Quantum Theory of Scattering* (Prentice-Hall, N.J., 1962) Secs. R, 5, 6.

[f] The equation of motion in the theory of Lorentz is

$$\ddot{\mathbf{r}} = e\mathbf{E} + \frac{e}{c}[\dot{\mathbf{r}} \times \mathbf{B}],$$

which is the Lagrange equation of the Lagrangian $L$

$$L = \frac{1}{2}\mu\dot{\mathbf{r}}^2 - e\phi + \frac{e}{c}(\mathbf{A} \cdot \dot{\mathbf{r}}).$$

The generalized momentum $p$ is

$$p_x = \frac{\partial L}{\partial \dot{x}} = \mu\dot{x} + \frac{e}{c}A_x, \text{ etc.}$$

The Legendre transformation

$$H = \sum_x^z p_x \dot{x} - L$$

gives (VII-43).

$$H = \frac{1}{2\mu}\left(p - \frac{e}{c}A\right)^2 + e\phi. \tag{VII-43}$$

The Schrödinger equation is[g]

$$i\hbar\frac{\partial\Psi}{\partial t} = \frac{1}{2\mu}\left[\mathbf{p}^2 - \frac{2e}{c}(\mathbf{A}\cdot\mathbf{p}) + 2\mu e\phi\right]\Psi \tag{VII-44a}$$

or

$$i\hbar\frac{\partial\Psi}{\partial t} = \left(-\frac{\hbar^2}{2\mu}\nabla^2 + e\phi + H_1\right)\Psi, \tag{VII-44b}$$

$$H_1 = -\frac{e}{\mu c}(\mathbf{A}\cdot\mathbf{p}) = i\frac{e\hbar}{\mu c}(\mathbf{A}\cdot\nabla). \tag{VII-44c}$$

Let the vector potential $\mathbf{A}(r, t)$ be a plane wave

$$\mathbf{A} = \mathbf{A}_0 e^{i(\mathbf{k}\cdot\mathbf{r}-\omega t)} + \mathbf{A}_0^* e^{-i(\mathbf{k}\cdot\mathbf{r}-\omega t)}, \tag{VII-45}$$

where $A_0$ is a constant (vector). The $\langle m|V|n\rangle$ in (VII-42) is now

$$-\frac{e}{\mu c}\langle m|\mathbf{A}\cdot\mathbf{p}|n\rangle = i\frac{e\hbar}{\mu c}\int \psi_m^*(\mathbf{A}_x\cdot\nabla)\psi_n\,dr, \tag{VII-46}$$

where $\psi_m$, $\psi_n$ are the eigenfunctions of the unperturbed system. One expands the plane wave

$$e^{i\mathbf{k}\cdot\mathbf{r}} = e^{ikr\cos\alpha}$$

$$= \sqrt{\frac{\pi}{2kr}}\sum_{l=0}^{\infty}(2l + 1)i^l J_{l+(1/2)}(kr)P_l(\cos\alpha). \tag{VII-47}$$

If the wave length $\lambda = 2\pi/k$ of the wave is long compared with dimension of the system (i.e., the size $r$ of the region in which the wave functions $\psi_m$, $\psi_n$ are appreciable),

$$\frac{2\pi r}{\lambda} = kr \ll 1,$$

---

[g] From (VII-43), there appear $(\mathbf{p}\cdot\mathbf{A}) + (\mathbf{A}\cdot\mathbf{p})$. Now

$$(\mathbf{p}\cdot\mathbf{A})\Psi = (\mathbf{A}\cdot\mathbf{p})\Psi + \frac{\hbar}{i}\operatorname{div}\mathbf{A}\Psi.$$

From the Lorentz relation

$$\operatorname{div}\mathbf{A} + \frac{1}{c}\frac{\partial\phi}{\partial t} = 0$$

and for a static field $\phi$ (such as that in atoms or molecules ), $\operatorname{div}\mathbf{A} = 0$. The term $(e^2/c^2)\mathbf{A}^2$ is small (contributing to the diamagnetic susceptibility) and is neglected here.

one may expand $\sqrt{\pi/2kr}\,J_{l+(1/2)}(kr)$ and keep only the lowest power term which is proportional to $(kr)^l$. Then

$$e^{ikr\cos\alpha} = 1 + 3ikrP_1(\cos\alpha) - 5(kr)^2 P_2(\cos\alpha) + \cdots \qquad \text{(VII-48)}$$

On taking only the first term—the so-called dipole approximation[h]—(VII-46) becomes

$$\langle m|V|n\rangle = i\frac{e\hbar}{\mu c}\int \psi_m^*(\mathbf{A}_0\cdot\nabla)\psi_n\,dr$$

$$= -i\frac{e\hbar}{\mu c}\frac{\mu}{\hbar^2}(E_m - E_n)\langle m|\mathbf{A}_0\cdot\mathbf{r}|n\rangle. \qquad \text{(VII-49)}$$

In evaluating the integral $\langle m|\mathbf{A}_0\cdot\mathbf{r}|n\rangle$, we take vector $\mathbf{A}_0$ as the polar axis so that $\mathbf{A}_0\cdot\mathbf{r} = A_0 r\cos\vartheta$. Using (VII-49) in (VII-36), and from (VII-42), one obtains

$$u(v_{mn})B_n^m = \frac{1}{t}\left(\frac{2\pi e v_{mn}}{\hbar c}\right)^2 |A_0|^2\langle m|r\cos\vartheta|n\rangle\frac{\sin^2 x}{x^2}t^2. \qquad \text{(VII-51)}$$

The radiation field energy density is

$$u(v)\,dv = \frac{1}{4\pi}\left|\frac{1}{c}\frac{\partial A}{\partial t}\right|^2$$

$$= \frac{2\pi v^2}{c^2}|A_0|^2.$$

From (VII-36a),

$$t\,dv = \frac{t}{2\pi}d\omega = \frac{1}{\pi}dx.$$

[h] From the Schrödinger equations

$$\nabla^2\psi_m^\dagger + \frac{2\mu}{\hbar^2}(E_m - e\phi)\psi_m^* = 0,$$

$$\nabla^2\psi_n + \frac{2\mu}{\hbar^2}(E_n - e\phi)\psi_n = 0,$$

one obtains

$$-\frac{2\mu}{\hbar^2}(E_m - E_n)\int \psi_m^\dagger x\psi_n\,dr = -\int (\psi_m^* x\nabla^2\psi_n - \psi_n x\nabla^2\psi_m^*)\,dr$$

$$= -\int (\psi_m^*\nabla^2(x\psi_n) - x\psi_n\nabla^2\psi_m^*)\,dr$$

$$= 2\int \psi_m^*\frac{\partial}{\partial x}\psi_n\,dr,$$

or

$$i\mu\omega_{mn}r_{mn} = p_{mn}. \qquad \text{(VII-50)}$$

Hence, on integrating the right-hand side of (VII-51) over $\mathbf{x}$ as in (VII-41),

$$B_n^m = \frac{2\pi}{\hbar^2}|\langle m|er\cos\vartheta|n\rangle|^2. \tag{VII-52}$$

As $er\cos\vartheta$ is the component of the electric moment $er$ along the direction of $\mathbf{A}_0$, and as the direction of $er$ is random relative to $\mathbf{A}_0$, one has

$$|\langle m|r\cos\vartheta|n\rangle|^2 = \tfrac{1}{3}|\langle m|\mathbf{r}|n\rangle|^2.$$

Hence

$$B_n^m = \frac{2\pi}{3\hbar^2}|\langle m|er|n\rangle|^2 \tag{VII-53}$$

and from Chap. 1, Sec. 6,

$$A_n^m = \frac{64\pi^4 v_{mn}^3}{3hc^3}|\langle m|er|n\rangle|^2. \tag{VII-54}$$

Note that perturbation theory allows $B_n^m$ and $B_n^m$ to be calculated as above, but not the $A_n^m$ coefficient directly except through the relation $A_n^m = (8\pi h v^3/c^3)B_n^m$ from Einstein's theory. The $A_n^m$ coefficients have been obtained by Dirac on his theory of radiation field (1927).

### (3) Theory of dispersion

In Chap. 1, Sec. 8, we saw the translation by Ladenberg and Kramers, in the spirit of the correspondence principle, of the classical theory of dispersion into the quantum theory. The induced dipole moment is given by (I-77Q) (summed over all $k$)

$$M = \frac{e^2 E}{\mu}\sum_k \frac{f_{kn}}{\omega_{kn}^2 - \omega^2}, \tag{VII-55}$$

where $\mu$ the mass of electron, and $E$ is the electric field $E = E_0 e^{i\omega t}$.

Let the Schrödinger equation of the unperturbed system (an atom, say) be

$$(H_0 - E_n)\psi_n(r) = 0. \tag{VII-56}$$

The perturbing field $H_1$ is given by (VII-44c) and (VII-45)

$$H_1 = -\frac{e}{\mu c}(\mathbf{A}\cdot\mathbf{p}) = i\frac{e\hbar}{\mu c}(\mathbf{A}\cdot\nabla), \tag{VII-44c}$$

$$\mathbf{A} = \mathbf{A}_0 e^{i(\mathbf{k}\cdot\mathbf{r}-\omega t)} + \mathbf{A}_0^* e^{-i(\mathbf{k}\cdot\mathbf{r}-\omega t)}, \tag{VII-45}$$

and for the perturbed system,

$$ i\hbar \frac{\partial \Psi}{\partial t} = (H_0 + H_1)\Psi, \qquad \text{(VII-57)} $$

$$ \Psi(r, t) = \sum a_n(t)\psi_n(r)\exp\left(-i\frac{E_n}{\hbar}t\right), \qquad \text{(VII-27)} $$

$$ \frac{da_k}{dt} = i\hbar \sum_m \langle k|H_1|m\rangle a_m(t)\exp(i\omega_{km}t). \qquad \text{(VII-28)} $$

For the vector potential **A**, we again take the dipole approximation as in (VII-48, 49), i.e.,

$$ e^{i\mathbf{k}\cdot\mathbf{r}} \cong 1, $$

and replace **A** by

$$ A_x = \frac{cE_0}{2\omega i}(e^{-i\omega t} - e^{i\omega t}) \qquad \text{(VII-58)} $$

so that the radiation field is an electric field $E_x$ along the $x$-direction

$$ E_x = -\frac{1}{c}\frac{\partial A_x}{\partial t} = E_0 \cos\omega t, \qquad \text{(VII-59)} $$

and the perturbation $H_1$ is replaced by

$$ H_1 = -exE_x = -exE_0 \cos\omega t. \qquad \text{(VII-60)} $$

On integrating (VII-28) above with the initial condition that at $t = 0$

$$ a_n = 1, \qquad a_m = 0 \quad \text{for} \quad m \neq n, \qquad \text{(VII-61)} $$

one obtains

$$ a_k(t) = \frac{eE_0}{2\hbar}x_{kn}\left\{\frac{e^{i(\omega_{kn}+\omega)t} - 1}{\omega_{kn} + \omega} + \frac{e^{i(\omega_{kn}-\omega)t} - 1}{\omega_{kn} - \omega}\right\}. \qquad \text{(VII-62)} $$

Then from (VII-27)

$$ \Psi(r, t) = \psi_n\exp(-iE_nt/\hbar) + \sum_{k\neq n} a_k(t)\exp(-iE_kt/\hbar)\psi_k. \qquad \text{(VII-63)} $$

The expectation value of the electric moment is

$$ (\Psi, ex\Psi) = ex_{nn} + e\sum_{k\neq n}(a_k^*(t)e^{-i\omega_{nk}t}x_{kn} + a_k(t)e^{i\omega_{nk}t}x_{nk}). \qquad \text{(VII-64)} $$

Substituting (VII-62) into this, one obtains

$$ (\Psi, ex\Psi) = ex_{nn} + \frac{e^2E_0}{\hbar}\sum_{k\neq n}\frac{2\omega_{kn}}{\omega_{kn}^2 - \omega^2}|x_{kn}|^2(\cos\omega t - \cos\omega_{nk}t). \qquad \text{(VII-65)} $$

Comparison with the expression (VII-55) leads to the quantum mechanical expression for the oscillator strength $f_{kn}$

$$f_{kn} = 2\frac{\mu}{\hbar}\omega_{kn}x_{kn}x_{nk}$$

$$= \frac{\mu}{\hbar}(\omega_{kn}x_{kn}x_{nk} - \omega_{nk}x_{nk}x_{kn}). \qquad \text{(VII-66)}$$

By means of (VII-50) we obtain

$$p_{kn} = i\mu\omega_{kn}x_{kn},$$

$$f_{kn} = \frac{i}{\hbar}(p_{nk}x_{kn} - x_{nk}p_{kn}). \qquad \text{(VII-67)}$$

It follows that

$$\sum_k f_{kn} = \frac{i}{\hbar}(px - xp)_{nn}$$

$$= 1, \qquad \text{(VII-68)}$$

by the commutation relation. This is the Thomas-Kuhn sum rule (I-84a) found also by Kramers and Heisenberg (1925).

From (VII-66), it is seen that

$$f_{kn} > 0 \qquad \text{for} \qquad \hbar\omega_{kn} = E_k - E_n > 0, \qquad \text{(VII-69a)}$$

$$f_{kn} < 0 \qquad \text{for} \qquad \hbar\omega_{kn} = E_k - E_n < 0. \qquad \text{(VII-69b)}$$

If $E_n$ is an excited state, $f_{kn}$ can be negative, and this leads to the so-called negative absorption observed by Ladenburg and Kopfermann (1928).[1]

The following table gives the $f_{n3}$ values for the $3s\,^2S - np\,^2p$ transitions of the Na atom.

| $3S - np$ | (Å) | $f_{n3}$ (VII-66) | $f_{n3}$ Observed (relative) | |
|---|---|---|---|---|
| $n = 3$ | 5893 | 0.975 | 1 | |
| 4 | 3309 | 0.014,4 | 0.014,4 | (VII-70) |
| 5 | 2853 | 0.002,41 | 0.002,11 | |
| 6 | 2680 | 0.000,98 | 0.000,65 | |

[1] See Chap. 1, Sec. 8, following (I-82).

### (4) Rearrangement collisions

Consider the following problem: Two (composite) particles $A$ and $B$ (such as atoms or molecules) initially are far apart, so that the Hamiltonian of the system is

$$H_0(A + B) = H_0(A) + H_0(B), \tag{VII-71}$$

and the eigenvalues and eigenfunctions are given by

$$[H_0(A + B) - E_n^0(A + B)]u_n = 0. \tag{VII-72}$$

When $A$, $B$ approach each other, they interact, with a potential $V_{AB}$. The Schrödinger equation of the system is

$$\left[ i\hbar \frac{\partial}{\partial t} - H_0(A + B) \right] \Psi = V_{AB}\Psi. \tag{VII-73}$$

After the collision, $A$ and $B$ may separate again into $A$ and $B$ (in a so-called "direct scattering")

(i) $$A + B \to A + B, \tag{VII-74}$$

or, they may separate as two other particles $C$ and $D$ (in a so-called "rearrangement collision")

(ii) $$A + B \to C + D. \tag{VII-75}$$

In case (i), the final state may be a state $u_m$ of $H_0(A + B)$, and the transition probability $n \to m$ can be calculated by the method of this section. Thus, from (VII-27), we now have

$$\Psi = \sum_m a_m(t)u_m \exp(-iE_m^0(A + B)t/\hbar) \tag{VII-76}$$

and the initial condition is

$$a_n(0) = 1, \qquad a_m(0) = 0 \qquad \text{for } m \neq n. \tag{VII-77}$$

We shall assume that the interaction $V_{AB}$ is "small" in the following sense:

$$\left| \int d\tau u_m^* V_{AB}[\Psi - u_n e^{-iE_n^0 t/\hbar}] \right| \ll \left| \int d\tau u_m^* V_{AB} u_n e^{-iE_n^0 t/\hbar} \right|, \tag{VII-78}$$

where $E_n^0 \equiv E_n^0(A + B)$ of (VII-72). The transition probability of the system $A + B$ from state $n$ to state $m$ is, from (VII-40),

$$P_{A+B} = \frac{2\pi}{\hbar} |\langle u_m | V_{AB} | u_n \rangle|^2 \rho(E_m^0), \tag{VII-79}$$

where $\rho(E_m^0)$ is the density of states at $E_m^0(A + B)$.

For the rearrangement collision (VII-75), let the Hamiltonian of the system when $C$ and $D$ are very far apart be

$$H_0(C + D) = H_0(C) + H_0(D), \tag{VII-80}$$

and the eigenvalues and eigenfunctions be

$$[H_0(C + D) - E_n^0(C + D)]v_n = 0. \tag{VII-81}$$

When $C$ and $D$ are close to each other, i.e., before they go off after the collision, let their interaction be $V_{CD}$. The Schrödinger equation is

$$\left[i\hbar\frac{\partial}{\partial t} - H_0(C + D)\right]\Phi = V_{CD}\Phi. \tag{VII-82}$$

Similarly to (VII-76), let us expand $\Phi$ in terms of the $v_m$ of $H_0(C + D)$ in (VII-81),

$$\Phi = \sum_m b_m(t)v_m \exp(-iE_m^0(C + D)t/\hbar). \tag{VII-83}$$

Now $\Phi$ in (VII-83) and $\Psi$ in (VII-73) both describe the same system $A + B$, or $C + D$; they differ only in using two sets of eigenfunctions $v_m$, $u_n$ for expansion. Hence we substitute $\Phi$ of (VII-83) into the left-hand side of (VII-82), and $\Psi$ of (VII-73) for $\Phi$ on the right-hand side of (VII-82). We make the assumption (approximation) that

$$\left|\int d\tau v_m^* V_{CD}[\Psi - u_n \exp(-iE_n^0(A + B)t/\hbar)]\right|$$

$$\ll \left|\int d\tau v_m^* V_{CD} u_n \exp(-iE_n^0(A + B)t/\hbar)\right|. \tag{VII-84}$$

By a procedure similar to that leading to (VII-79), one obtains the probability of the rearrangement transition $A + B \rightarrow C + D$

$$P_{C+D} = \frac{2\pi}{\hbar}|\langle v_m| V_{CD}|u_n\rangle|^2 \rho(E_m^0(C + D)). \tag{VII-85}$$

Consider now the two expressions for the Hamiltonian of the system

$$H = H_0(A + B) + V_{AB}, \tag{VII-86a}$$

$$H = H_0(C + D) + V_{CD}. \tag{VII-86b}$$

In calculating $P_{C+D}$ for the rearrangement transition, it would seem to appear ambiguous whether, in (VII-85), $V_{CD}$ should be used in association with $v_m$ for the *final* state of process $A + B \rightarrow C + D$, or $V_{AB}$ should be used in association with $u_n$ for the *initial* state of the same process. There is, however, no ambiguity; from the hermitian nature of the Hamiltonian in (VII-86a, b), it can be shown that[j]

$$(V_{CD}v_m, u_n) = (v_m, V_{AB}u_n).\qquad\text{(VII-87)}$$

Hence $P_{C+D}$ in (VII-85) is in fact equal to

$$P_{C+D} = \frac{2\pi}{\hbar}|\langle v_m|V_{AB}|u_n\rangle|^2\rho(E_m^0(C + D)).\qquad\text{(VII-88)}$$

We must make an important remark concerning the probabilities for direct and rearrangement collisions respectively. From (VII-76) and (VII-83), we have:

for direct scattering, a total probability of 1,

$$\sum_m |a_m(t)|^2 = 1,\qquad\text{(VII-89)}$$

for rearrangement collision, a total probability of 1,

$$\sum_m |b_m(t)|^2 = 1.\qquad\text{(VII-90)}$$

These results imply no contradiction in the sense that direct scattering would seem to have excluded some rearrangement collisions, or vice versa. It is just that $H_0(A + B)$ and $H_0(C + D)$ do not commute and the eigenfunctions $u_n$, $v_n$ form two distinct complete sets so that a state $u_n$ of $H_0(A + B)$ is a superposition of the states $v_m$ of $H_0(C + D)$, and vice versa. The Schrödinger equation

$$i\hbar\frac{\partial\Psi}{\partial t} = H\Psi\qquad\text{(VII-91)}$$

together with a given initial condition completely determines the $\Psi$ of the system at subsequent times. This can be analyzed with respect to the eigenstates of either $H_0(A + B)$, or $H_0(C + D)$, leading to $P_{A+B}$, or $P_{C+D}$, respectively. This is a basic feature of quantum mechanics.

### *(5) Transition probability and cross section*

As another application of the theory of Subsection (1), let us consider the scattering of a particle by a static field $V(r)$. The unperturbed system is a free particle so that the eigenstates are plane waves

---

[j] The proof depends simply on the hermiticity of the various operators in (VII-86a, b), and because of conservation of energy,

$$E_n^0(A + B) = E_n^0(C + D).$$

$$(H_0 - E)\psi(r) = 0, \qquad E_k = \frac{\hbar^2 k^2}{2m}, \tag{VII-92}$$

$$\psi(r) = \frac{1}{\sqrt{\Omega}} e^{i\mathbf{k}\cdot\mathbf{r}}, \tag{VII-93}$$

where $\psi(r)$ is normalized to one particle in the volume $\Omega = L^3$. We inquire about the probability per second that a particle originally in the state

$$\psi_n(r) = \frac{1}{\sqrt{\Omega}} e^{i\mathbf{k}\cdot\mathbf{r}}$$

makes a transition to a state

$$\psi_m(r) = \frac{1}{\sqrt{\Omega}} e^{i\mathbf{k}'\cdot\mathbf{r}}, \qquad |\mathbf{k}'| = |\mathbf{k}|.$$

The density $\rho(E_k)$, i.e., the number of states per unit energy range, can be obtained as follows. The number of states in $dk_x\, dk_y\, dk_z$, where $k = 2\pi/\lambda$, is

$$\frac{L^3}{(2\pi)^3} dk_x\, dk_y\, dk_z = \frac{\Omega}{(2\pi)^3} k^2\, dk \sin\vartheta\, d\vartheta\, d\phi,$$

where $(\vartheta, \phi)$ gives the direction of the momentum $\mathbf{p}' = \hbar\mathbf{k}'$ of the scattered particle. $\rho(E_k)\, dE_k$ is then

$$\rho(E_k)\, dE_k = \frac{\Omega}{(2\pi\hbar)^3} p^2\, dp\, d\cos\vartheta\, d\phi$$

$$= \frac{\Omega}{(2\pi\hbar)^3} mp\, dE_k\, d\cos\vartheta\, d\phi. \tag{VII-94}$$

The transition probability per unit time is, from (VII-40),

$$P = \frac{2\pi}{\hbar} \frac{mp}{(2\pi\hbar)^3 \Omega} \left| \int e^{-i\mathbf{k}'\cdot\mathbf{r}} V(r) e^{i\mathbf{k}\cdot\mathbf{r}}\, dr \right|^2 d\cos\vartheta\, d\phi. \tag{VII-95}$$

This transition involves the incidence of $v/L = p/mL$ particles per second. We define a differential cross-section $d\sigma$ of the scattering by

$$d\sigma = \frac{P(\text{no. of scattering in solid angle } d\cos\vartheta\, d\phi \text{ per sec.})}{\text{no. of particles crossing unit area per sec.}}$$

$$= \frac{P}{\dfrac{v}{L\cdot L^3}} = \frac{P}{v}\Omega.$$

Hence

$$d\sigma = \left(\frac{m}{2\pi\hbar^2}\right)^2 \left|\int e^{i\mathbf{k}'\cdot\mathbf{r}} V(r) e^{i\mathbf{k}\cdot\mathbf{r}} \, dr\right|^2 d\cos\vartheta \, d\phi. \tag{VII-96}$$

This result is identical with the first Born approximation obtained by the stationary state method in Chap. 4, Sec. 1 (2), (IV-31), (IV-33).

## 3. Unitary (time-translation) operator

A more general method of treating the transitions of a system due to a perturbation $V$ is due to Dirac. The method consists in introducing a unitary operator $U(t, t_0)$ that effects the transformation from a state $\Psi(r_0, t_0)$ at a space-time $r_0$, $t_0$ to a state $\Psi(r, t)$ at $r$, $t$ in accordance with the Schrödinger equation

$$i\hbar\frac{\partial\Psi}{\partial t} = (H_0 + V)\Psi, \tag{VII-97}$$

where $H_0$ is time-independent and has stationary states, and $V$ may or may not depend explicitly on time.

We first make a unitary transformation of $V$ into $\mathscr{H}(t)$

$$\mathscr{H}(t) = \exp(iH_0 t/\hbar) V \exp(-iH_0 t/\hbar), \tag{VII-98}$$

$$\Psi'(t) = \exp(iH_0 t/\hbar)\Psi(t). \tag{VII-99}$$

Equation (VII-97) then becomes

$$i\hbar\frac{\partial\Psi'}{\partial t} = \mathscr{H}\Psi'. \tag{VII-100}$$

This equation contains only the "perturbation" or "interaction" $V$ through $\mathscr{H}$, and has come to be referred to as Schrödinger's equation in the "interaction picture".

Let this equation be expressed in terms of the kets $|\alpha, t\rangle$ instead of $\Psi'(t)$ which are the representatives of $|\alpha, t\rangle$ in the coordinate (i.e., Schrödinger) representation,

$$i\hbar\frac{\partial}{\partial t}|\alpha, t\rangle = \mathscr{H}(t)|\alpha, t\rangle. \tag{VII-101}$$

Let $U(t, t_0)$ be a unitary (linear) operator

$$U^\dagger(t, t_0) U(t, t_0) = 1 \tag{VII-102}$$

such that, for any $\alpha$, the following relation holds

$$|\alpha, t\rangle = U(t, t_0)|\alpha, t_0\rangle. \tag{VII-103}$$

From (VII-102) and (VII-103), the following properties follow:

$$U(t, t) = 1, \qquad \text{(VII-104a)}$$

$$U(t, t_1)U(t_1, t_0) = U(t, t_0), \qquad \text{(VII-104b)}$$

$$U^\dagger(t, t_0) = U(t_0, t). \qquad \text{(VII-104c)}$$

Using (VII-103) in (VII-100), one obtains the equation

$$i\hbar \frac{\partial U(t, t_0)}{\partial t} = \mathscr{H}(t)U(t, t_0). \qquad \text{(VII-105)}$$

This equation, and the boundary condition (VII-104a), can be replaced by the integral equation

$$U(t, t_0) = 1 - \frac{i}{\hbar} \int_{t_0}^{t} dt' \, \mathscr{H}(t')U(t', t_0). \qquad \text{(VII-106)}$$

The operator $U(t, t_0)$ can be represented by the unitary (projection) operator

$$U(t, t_0) = \int |\alpha'', t\rangle \, d\alpha'' \langle \alpha'', t_0|, \qquad \text{(VII-107)}$$

for this satisfies (VII-103),

$$U(t, t_0)|\alpha, t_0\rangle = \int |\alpha'', t\rangle \, d\alpha'' \langle \alpha'', t_0|\alpha, t_0\rangle$$

$$= \int |\alpha'', t\rangle \, d\alpha'' \delta(\alpha'' - \alpha)$$

$$= |\alpha, t\rangle, \qquad \text{QED}$$

and it also satisfies the equation (VII-106),

$$-\frac{i}{\hbar} \int_{t_0}^{t} \mathscr{H}(t')U(t', t_0) \, dt' = \int_{t_0}^{t} \frac{\partial}{\partial t'} U(t', t_0) \, dt' \qquad \text{by (VII-105)}$$

$$= \int_{t_0}^{t} dt' \frac{\partial}{\partial t'} \int |\alpha'', t'\rangle \, d\alpha'' \langle \alpha'', t_0|$$

$$= \int \{|\alpha'', t\rangle - |\alpha'', t_0\rangle\} \, d\alpha'' \langle \alpha'', t_0|$$

$$= U(t, t_0) - 1. \qquad \text{QED}$$

To obtain the transition probabilities, let us form the matrix element of $U(t, t_0)$

$$\langle \alpha', t_0|U(t, t_0)|\alpha'', t_0\rangle = \langle \alpha', t_0|\alpha'', t\rangle. \qquad \text{(VII-108)}$$

The right-hand side is the probability amplitude of the transition from the state $|\alpha', t_0\rangle$ at time $t_0$ to the state $|\alpha'', t\rangle$ at time $t$. It is seen that the transition probability $|\langle \alpha', t_0 | \alpha'', t\rangle|^2$ is independent of the time order of $t_0$ and $t$.

The total transition probability from $|\alpha', t_0\rangle$ to all states $|\alpha'', t\rangle$ at time $t$ is unity, as

$$\sum \int |\langle \alpha', t_0 | \alpha'', t\rangle|^2 \, d\alpha'' = \sum \int \langle \alpha', t_0 | \alpha'', t\rangle \, d\alpha'' \langle \alpha'', t | \alpha', t_0\rangle$$

$$= \langle \alpha', t_0 | \alpha', t_0\rangle$$

$$= 1. \tag{VII-109}$$

Now, let us specify the $|\alpha, t\rangle$ above to be the eigenstates of $H_0$ of (VII-97),

$$i\hbar \frac{\partial}{\partial t} |E_n, t\rangle = H_0 |E_n, t\rangle, \tag{VII-110}$$

where $E_n$ are the eigenvalues of $H_0$. We shall assume that $V$ in (VII-97) is small. Then the integral equation (VII-106) can be solved by iteration. Let $t_0 = 0$ and let $U(t) \equiv U(t, 0)$. Then

$$U(t) = 1 - \frac{i}{\hbar} \int_0^t dt' \, \mathscr{H}(t') U(t')$$

$$= 1 + U_1(t) + U_2(t) + U_3(t) + \cdots, \tag{VII-111}$$

where

$$U_1(t) = -\frac{i}{\hbar} \int_0^t \mathscr{H}(t') \, dt', \tag{VII-112}$$

$$U_n(t) = \left(-\frac{i}{\hbar}\right)^n \int_0^t dt_1 \, \mathscr{H}(t_1) \int_0^{t_1} dt_2 \, \mathscr{H}(t_2) \cdots \int_0^{t_{n-1}} dt_n \, \mathscr{H}(t_n).$$

For the transition $|E_0, 0\rangle \to |E_n, t\rangle$, the transition probability amplitude is, to the first order in $\mathscr{H}$, from (VII-108) and (VII-98),

$$\langle E_0, 0 | U_1(t) | E_n, 0\rangle = -\frac{i}{\hbar} \int_0^t \langle E_0, 0 | V | E_n, 0\rangle \exp(i(E_0 - E_n)t'/\hbar) \, dt'. \tag{VII-113}$$

If $V$ is independent of $t$, the transition probability is

$$|\langle E_0, 0 | U_1(t) | E_n, 0\rangle|^2 = |\langle E_0 | V | E_n\rangle|^2 \frac{\sin^2 x}{x^2} \left(\frac{t}{\hbar}\right)^2, \tag{VII-114}$$

$$x = \frac{(E_n - E_0)t}{2\hbar}.$$

This result is identical with (VII-36) obtained by Dirac's method of variation of constants.

If $\langle E_0|V|E_n\rangle$ vanishes (for symmetry reasons) or is very small, it is then necessary to go to the second order in $\mathcal{H}$. Thus

$$\langle E_n|U(t)|E_0\rangle = -\frac{1}{\hbar^2}\int_0^t dt_1 \int_0^{t_1} dt_2 \sum_m \langle E_n|V|E_m\rangle\langle E_m|V|E_0\rangle$$

$$\times \exp[i(E_n - E_m)t_1/\hbar]\exp[i(E_m - E_0)t_2/\hbar]$$

$$= \sum \frac{\langle E_n|V|E_m\rangle\langle E_m|V|E_0\rangle}{E_m - E_0}$$

$$\times \left\{\frac{\exp[i(E_n - E_0)t/\hbar] - 1}{E_n - E_0} - \frac{\exp[i(E_n - E_m)t/\hbar] - 1}{E_n - E_m}\right\}.$$

$$\text{(VII-115)}$$

The probability, up to the second order in $\mathcal{H}$, is, from (VII-113) and (VII-115),

$$|\langle E_n|U_1(t)|E_0\rangle + \langle E_n|U_2(t)|E_0\rangle|^2. \qquad \text{(VII-116)}$$

It is necessary to add the amplitudes before squaring, for there is an interference effect—a basic consequence of the probability postulate of quantum mechanics.

From this point on, the calculation follows the same procedure of Sec. 2, (1), (VII-36) ff.


## 4. Method of integral equation and Green's function

In Chap. 4, Sec. 1, (1), we expressed the time-independent Schrödinger equation in the form of an integral equation. One practical advantage of the integral form is that by suitably choosing the Green's function, the asymptotic condition (for scattering problems) can be embodied into the equation.

In the present section, we shall show that the time-dependent Schrödinger equation can also be put in the form of an integral equation, and again by suitably choosing its Green's function, the equation becomes specially convenient for treating scattering problems.

To be specific, let us consider the following problem: the scattering of a particle in a potential field.[k]

Let $H_0$ be the Hamiltonian of the free particle, and $V$ the scattering potential. The Schrödinger equations are

[k] In Chap. 4, Sec. 1, we treated the same problem, but from a different point of view. There we represented the incident and scattered particle by a steady stream in a steady state and the time-independent Schrödinger equation was used. Here the state of the system (the incident and scattered particle) changes with time and the time-dependent Schrödinger equation is needed. Of course the results obtained with both methods and from two different points of view are identical.

$$(H_0 - E_k)e^{i\mathbf{k}\cdot\mathbf{r}} = 0, \tag{VII-117}$$

$$E_k = \frac{1}{2m}\hbar^2 k^2 = \hbar\omega_k,$$

$$\left(i\hbar\frac{\partial}{\partial t} - H_0 - V\right)\Psi(r,t) = 0. \tag{VII-118}$$

### (1) The unperturbed system $H_0$

The Green's function $G_0(\mathbf{r},t)$ is the solution of the equation

$$\left(i\hbar\frac{\partial}{\partial t} - H_0\right)G_0(r,t) = \delta(\mathbf{r}-\mathbf{r}_1)\delta(t-t_1), \tag{VII-119}$$

$$G_0(\mathbf{r}-\mathbf{r}_1,t-t_1) = \frac{1}{(2\pi)^4}\int\int_\Gamma \frac{\exp[i\mathbf{k}\cdot(\mathbf{r}-\mathbf{r}_1)-i\omega(t-t_1)]}{\hbar\omega - E_k}\,d\mathbf{k}\,d\omega, \tag{VII-120}$$

where $\Gamma$ denotes the path of integration in the complex $\omega$-plane. The path $\Gamma$ is chosen to satisfy a causality condition that will be made clear presently,

$$G_0(\mathbf{r}-\mathbf{r}_1,t-t_1) = 0 \qquad \text{for} \qquad t-t_1 < 0. \tag{VII-121}$$

This condition is met by choosing, for $t-t_1 < 0$, $\Gamma$ to be the contour from $-\infty$ to $+\infty$ along the real axis and a large semi-circle in the upper $\omega$-plane in the counter-clockwise sense back to $\omega = -\infty$, and for $t-t_1 > 0$, $\Gamma$ to be from $-\infty$ to $\infty$ and a large semi-circle in the lower $\omega$-plane in the clockwise sense back to $\omega = -\infty$. This defines $G_0(\mathbf{r}-\mathbf{r}_1,t-t_1)$ to be

$$G_0(\mathbf{r}-\mathbf{r}_1,t-t_1)$$

$$= \begin{cases} -\dfrac{i}{(2\pi)^3\hbar}\displaystyle\int \exp\{i\mathbf{k}\cdot(\mathbf{r}-\mathbf{r}_1)-i\omega_k(t-t_1)\}\,d\mathbf{k}, & \text{for } t-t_1 > 0, \\[2ex] 0, & \text{for } t-t_1 < 0, \end{cases} \tag{VII-122}$$

$$= \begin{cases} \left[\dfrac{m}{2\pi i\hbar(t-t_1)}\right]^{3/2}\exp\left\{\dfrac{im(\mathbf{r}-\mathbf{r}_1)^2}{2\hbar(t-t_1)}\right\}, & t-t_1 > 0, \\[2ex] 0, & t-t_1 < 0. \end{cases} \tag{VII-122a}$$

This $G_0(\mathbf{r}-\mathbf{r}_1,t-t_1)$ may be regarded as the representative, in the coordinate representation, of the operator $G_0(t-t_1)$ defined by

$$\left(i\hbar\frac{\partial}{\partial t} - H_0\right)G_0(t-t_1) = \delta(t-t_1). \tag{VII-123}$$

To show this, we have

$$
\begin{aligned}
G_0(t - t_1) &= \frac{1}{2\pi} \lim_{\varepsilon \to 0} \int_\Gamma \frac{e^{-i\omega(t-t_1)}}{\hbar\omega - E_k + i\varepsilon} \, d\omega, \qquad \varepsilon > 0 \\[2mm]
&= -\frac{i}{2\pi} \lim_{\varepsilon \to 0} \int_0^\infty d\xi \int_\Gamma \exp\{i\xi(\hbar\omega - E_k + i\varepsilon) - i\omega(t - t_1)\} \, d\omega \\[2mm]
&= -i \int_0^\infty d\xi \exp(-i\xi H_0)\delta[\xi\hbar - (t - t_1)] \\[2mm]
&= \begin{cases} -\dfrac{i}{\hbar}\exp\{-iH_0(t - t_1)/\hbar\}, & t - t_1 > 0, \\[3mm] 0, & t - t_1 < 0. \end{cases}
\end{aligned}
\qquad \text{(VII-124)}
$$

The $\langle r|G|r_1\rangle$ matrix element of $G_0(t - t_1)$ is

$$
\begin{aligned}
\langle r|G_0(t - t_1)|r_1\rangle &= -\frac{i}{\hbar} \int \langle r|E_k\rangle \exp\{-iE_k(t - t_1)/\hbar\} \langle E_k|r_1\rangle \, d\mathbf{k} \\[2mm]
&= -\frac{i}{(2\pi)^3\hbar} \int e^{i\mathbf{k}\cdot\mathbf{r}} \exp\{-i\omega_k(t - t_1)\} e^{-i\mathbf{k}\cdot\mathbf{r}_1} \, d\mathbf{k},
\end{aligned}
$$

$$\text{(VII-125)}$$

which is (VII-122).

In the form (VII-124) or (VII-125), we have found a meaning for $G_0(\mathbf{r} - \mathbf{r}_1, t - t_1)$ as follows.

Let $|\alpha, t\rangle$ be a ket of

$$
\left(i\hbar\frac{\partial}{\partial t} - H_0\right)|\alpha, t\rangle = 0, \qquad\qquad \text{(VII-126)}
$$

whose formal solution is

$$
|\alpha, t\rangle = \exp\{-iH_0(t - t_1)/\hbar\}|\alpha, t_1\rangle. \qquad\qquad \text{(VII-127)}
$$

Taking the representative in the coordinate representation, we have

$$
\langle r|\alpha, t\rangle = \int \langle r|\exp\{-iH_0(t - t_1)/\hbar\}|r_1\rangle \, dr_1 \langle r_1|\alpha, t_1\rangle. \quad \text{(VII-128)}
$$

This states that

$$
\langle r|\exp\{-iH_0(t - t_1)/\hbar\}|r_1\rangle \qquad\qquad \text{(VII-129a)}
$$

is the transition probability amplitude from a state $\langle r_1|\alpha, t_1\rangle$ at time $t_1$ to go, in accordance with the Schrödinger equation (VII-128), to the state $\langle r|\alpha, t\rangle$ at time $t$ ($t > t_1$). By (VII-124), this is equal to

$$ih\langle r|G_0(t - t_1)|r_1\rangle, \qquad\qquad \text{(VII-129b)}$$

and by (VII-125) and (VII-122), this is equal to

$$i\hbar G_0(\mathbf{r} - \mathbf{r}_1, t - t_1). \qquad\qquad \text{(VII-129c)}$$

Let

$$i\hbar G_0(2, 1) \equiv i\hbar G_0(\mathbf{r}_2 - \mathbf{r}_1, t_2 - t_1). \qquad\qquad \text{(VII-130)}$$

Then $i\hbar G_0(2, 1)$ is the probability amptitude, of a system with Hamiltonian $H_0$, of a transition from the state $\langle r_1|\alpha, t_1\rangle = \Psi_\alpha(\mathbf{r}_1, t_1)$ to the state $\langle r_2|\alpha, t_2\rangle = \Psi_\alpha(\mathbf{r}_2, t_2)$. This meaning of the Green's function $G_0(\mathbf{r}_2 - \mathbf{r}_1, t_2 - t_1)$ is very important. The reason for imposing the causality condition (VII-121) is now clear.

The Schrödinger equation (VII-126) and its integral form (VII-128) can be expressed in the form

$$\Psi(\mathbf{r}_2, t_2) = i\hbar \int G_0(\mathbf{r}_2 - \mathbf{r}_1, t_2 - t_1)\Psi(\mathbf{r}_1, t_1)\, d\mathbf{r}_1. \qquad \text{(VII-131)}$$

### (2) The perturbed system $H = H_0 + V$

The Schrödinger equation is (VII-118)

$$\left(i\hbar\frac{\partial}{\partial t} - H_0\right)|\alpha, t\rangle = V|\alpha, t\rangle, \qquad\qquad \text{(VII-132)}$$

where the perturbation $V$ may be a function of $t$. It can be shown that this equation is satisfied by

$$|\alpha, t\rangle = |H_0, t\rangle + \int_{-\infty}^{t} G_0(t - t')V(t')|\alpha, t'\rangle\, dt', \qquad \text{(VII-133)}$$

$$\left(i\hbar\frac{\partial}{\partial t} - H_0\right)|H_0, t\rangle = 0,$$

as can be seen from (VII-123) and (VII-126). This can be expressed in a more familiar form by taking the representatives in the coordinate representation, and with $\Psi_0 \equiv \langle r|H_0, t\rangle$, etc.,

$$\Psi(\mathbf{r}, t) = \Psi_0(\mathbf{r}, t) + \int_{-\infty}^{t} dt' \int d\mathbf{r}' G_0(\mathbf{r} - \mathbf{r}', t - t')V(\mathbf{r}', t')\Psi(\mathbf{r}', t'), \quad \text{(VII-133a)}$$

which is the Schrödinger equation (VII-132) in integral form.[1]

---

[1] This may be compared with the corresponding situation for the time-independent problem treated in Chap. 4, Sec. 1, (IV-6) and (IV-18).

Let $G(r, t)$ be the solution of

$$\left(i\hbar\frac{\partial}{\partial t} - H\right)G(\mathbf{r}, t) = \delta(\mathbf{r} - \mathbf{r}_1)\delta(t - t_1), \qquad \text{(VII-134)}$$

where $H = H_0 + V$.

The Schrödinger equation (VII-132) in the form

$$\left(i\hbar\frac{\partial}{\partial t} - H\right)\Psi(\mathbf{r}, t) = 0 \qquad \text{(VII-135)}$$

can then be expressed, analogously to (VII-126) and (VII-131), in the form

$$\Psi(\mathbf{r}_2 - \mathbf{r}_1, t_2 - t_1) = i\hbar\int G(\mathbf{r}_2 - \mathbf{r}_1, t_2 - t_1)\Psi(\mathbf{r}_1, t_1)\,d\mathbf{r}_1. \quad \text{(VII-136)}$$

Before we proceed to obtain $G(\mathbf{r}_2 - \mathbf{r}_1, t_2 - t_1)$ for $H = H_0 + V$ from the $G_0(\mathbf{r}_2 - \mathbf{r}_1, t_2 - t_1)$ for $H_0$, let us illustrate the method of Green's function by the problem of the scattering of a particle by a static potential field $V$, i.e., $H = H_0 + V$ is explicitly independent of time. Let the eigenfunction of $H_0$ be $\psi^0(r)$, and

$$\Psi_0(r, t) = e^{-iEt/\hbar}\psi^0(r),$$

where $E$ is the (constant) energy of the particle. Let

$$\Psi(r, t) = e^{-iEt/\hbar}\psi(r).$$

For the system $H = H_0 + V$, the equation for $\Psi(r, t)$ is (VII-133a). By using the method of integration of Chap. 4, Sec. 1, one obtains from (VII-133a)

$$\psi(r) = \psi_0(r) - \frac{1}{4\pi}\int\frac{e^{ik|\mathbf{r} - \mathbf{r}_1|}}{|r - r_1|}\left(\frac{2m}{\hbar^2}V(\mathbf{r}_1)\right)\psi(\mathbf{r}_1)\,d\mathbf{r}_1, \qquad \text{(VII-137)}$$

which is identical with (IV-27).


## 5.  Green's function as propagator

In Chap. 4, Sec. 1, we expressed the Green's function in the form of an integral operator for the time-independent Schrödinger equation[m]

---

[m] The equation (VII-138) in integral form, for the case of an outgoing wave, is (IV-18)

$$\psi^+(\mathbf{r}) = \psi^0(\mathbf{r}) - \int G_0(\mathbf{r}, \mathbf{r'})\left(\frac{2m}{\hbar^2}V(\mathbf{r'})\psi^+(\mathbf{r'})\right)d^3r',$$

where

$$(H_0 - E)\psi^0 = 0, \qquad \psi_k^0 = e^{i\mathbf{k}\cdot\mathbf{r}}, \qquad E_k = k^2,$$

$$(H_0 + V - E)\psi(r) = 0. \qquad \text{(VII-138)}$$

For the time-dependent Schrödinger equation (VII-126)

$$\left(i\hbar\frac{\partial}{\partial t} - H_0\right)\Psi^0(\mathbf{r}, t) = 0,$$

we have seen that the integral form is (VII-131)

$$\Psi^0(\mathbf{r}_2, t_2) = i\hbar \int G_0(\mathbf{r}_2 - \mathbf{r}_1, t_2 - t_1)\Psi^0(\mathbf{r}_1, t_1)\,d\mathbf{r}_1,$$

where the Green's function is an integral operator (VII-120)

$$G_0(\mathbf{r} - \mathbf{r}_1, t - t_1) = \frac{1}{(2\pi)^4}\lim_{\varepsilon\to 0}\int d\omega \int d^3k \frac{1}{\hbar\omega - E_k + i\varepsilon}e^{i(\mathbf{k}\cdot\mathbf{r} - \omega t) - i(\mathbf{k}\cdot\mathbf{r}_1 - \omega t_1)}$$

$$\equiv \frac{1}{\hbar\omega - H_0 + i\varepsilon}. \qquad \text{(VII-140)}$$

Similarly, for the equation (VII-132)

$$\left(i\hbar\frac{\partial}{\partial t} - (H_0 + V)\right)\Psi(\mathbf{r}, t) = 0,$$

the Green's function is the solution of (VII-134), and the integral form is (VII-136)

$$\Psi(\mathbf{r}_2, t_2) = i\hbar \int G(\mathbf{r}_2 - \mathbf{r}_1, t_2 - t_1)\Psi(\mathbf{r}_1, t_1)\,d\mathbf{r}_1.$$

$G(\mathbf{r} - \mathbf{r}_1, t - t_1)$ is an integral operator

$$G(\mathbf{r} - \mathbf{r}_1, t - t_1) = \frac{1}{\hbar\omega - H + i\varepsilon}. \qquad \text{(VII-141)}$$

To obtain a relation between $G_0$ and $G$, let us introduce the notations

---

and (IV-28),

$$\psi^+(\mathbf{r}) = \psi^0(\mathbf{r}) + \frac{1}{E - H_0 + i\varepsilon}\left(\frac{2m}{\hbar^2}V(\mathbf{r}')\right)\psi^+(\mathbf{r}),$$

where $H_0 = -\nabla^2$, and the integral operator is defined by

$$\frac{1}{E - H_0 + i\varepsilon}\chi(r) = \frac{1}{(2\pi)^3}\lim_{\varepsilon\to 0}\int d\kappa \frac{1}{k^2 - \kappa^2 + i\varepsilon}\int d\mathbf{r}'\,\psi^0_\kappa(\mathbf{r})\psi^0_\kappa(\mathbf{r}')\chi(\mathbf{r}').$$

Thus

$$-G_0(\mathbf{r}, \mathbf{r}') = \frac{1}{E - H_0 + i\varepsilon}. \qquad \text{(VII-139)}$$

$$G_0(\lambda) \equiv G_0(\mathbf{r} - \mathbf{r}_1, t - t_1) = \frac{1}{\left(i\hbar\dfrac{\partial}{\partial t} - H_0 + i\varepsilon\right)} \equiv \frac{1}{\lambda - H_0}, \qquad \text{(VII-142a)}$$

$$G(\lambda) \equiv G(\mathbf{r} - \mathbf{r}_1, t - t_1) = \frac{1}{\left(i\hbar\dfrac{\partial}{\partial t} - H + i\varepsilon\right)} \equiv \frac{1}{\lambda - H}; \qquad \text{(VII-142b)}$$

$\lambda - H_0$, $\lambda - H$ are differential operators. From these one has

$$(\lambda - H_0)\frac{1}{\lambda - H} = (\lambda - H + V)\frac{1}{\lambda - H}$$

$$= 1 + V\frac{1}{\lambda - H}$$

$$= 1 + VG$$

and

$$\frac{1}{\lambda - H_0}(\lambda - H_0)\frac{1}{\lambda - H} = G_0 + G_0 VG$$

$$G = G_0(1 + VG). \qquad \text{(VII-143a)}$$

Similarly, one obtains the following relations

$$G = (1 + GV)G_0, \qquad \text{(VII-143b)}$$

$$G_0 = G(1 - VG_0), \qquad \text{(VII-143c)}$$

$$G_0 = (1 - G_0 V)G. \qquad \text{(VII-143d)}$$

These forms are conveniently suitable for iteration if the perturbating potential $V$ is weak. Thus, from (VII-143a),

$$G = G_0 + G_0 VG_0 + G_0 VG_0 VG_0 + \cdots \qquad \text{(VII-144)}$$

The meaning of this series is as follows. On introducing the abbreviation

$$G_0(n + 1, n) \equiv G_0(\mathbf{r}_{n+1} - \mathbf{r}_n, t_{n+1} - t_n), \qquad \text{(VII-145)}$$

then

$$G_0 VG_0 VG_0 \ldots VG_0$$

$$= \int \cdots \int G_0(n + 2, n + 1) V(\mathbf{r}_{n+1}, t_{n+1}) \, d\mathbf{r}_{n+1} \, dt_{n+1}$$

$$\cdot G_0(n + 1, n) V(\mathbf{r}_n, t_n) \, d\mathbf{r}_n \, dt_n \cdots V(\mathbf{r}_2, t_2) \, d\mathbf{r}_2 \, dt_2 \, G(2, 1), \quad \text{(VII-146)}$$

where the $t_n$ obey the sequence

$$t_{n+2} > t_{n+1} > t_n > \cdots > t_2 > t_1, \qquad \text{(VII-147)}$$

and $G_0 V G_0 \ldots V G_0 = 0$ for $t_n$ not obeying this order. This is due to the causality condition (VII-121). The integral

$$\iint G_0(3,2) V(\mathbf{r}_2, t_2) G_0(2,1) \, d\mathbf{r}_2 \, dt_2 \qquad \text{(VII-148)}$$

is the transition probability amplitude for a *free* particle (since $G_0$ is the Green's function for $H_0$) propagating from $(\mathbf{r}_1, t_1)$, reaching $(\mathbf{r}_2, t_2)$ and being scattered by $V$, and then propagating as a free particle to $(\mathbf{r}_3, t_3)$. Similarly

$$\int \cdots \int G_0(4,3) V(\mathbf{r}_3, t_3) G_0(3,2) V(\mathbf{r}_2, t_2) G_0(2,1) \, d\mathbf{r}_3 \, dt_3 \, d\mathbf{r}_2 \, dt_2$$

is the probability amptitude for the transition for a free particle propagating from $(\mathbf{r}_1, t_1)$ to $(\mathbf{r}_2, t_2)$, being scattered by $V$ at $(\mathbf{r}_2, t_2)$, propagating again freely to $(\mathbf{r}_3, t_3)$, being scattered by $V$ at $(\mathbf{r}_3, t_3)$, and propagating freely again to $(\mathbf{r}_4, t_4)$. This gives the name "propagator" to the Green's function $G(\mathbf{r} - \mathbf{r}_1, t - t_1)$.

Accordingly, the transition probability amptitude from $\Psi_n(\mathbf{r}_1, t_1)$ to $\psi_m(\mathbf{r}_3, t_3)$ in the first order approximation (i.e., first order in $V$) is,

$$a_{mn}^{(1)} = i\hbar \int \cdots \int \Psi_m^{0*}(\mathbf{r}_3, t_3) G_0(\mathbf{r}_3 - \mathbf{r}_2, t_3 - t_2) \, d\mathbf{r}_3 \, V(\mathbf{r}_2)$$

$$\times \, G_0(\mathbf{r}_2 - \mathbf{r}_1, t_2 - t_1) \Psi_n^0(\mathbf{r}_1, t_1) \, d\mathbf{r}_2 \, dt_2 \, d\mathbf{r}_1. \qquad \text{(VII-149)}$$

Now, from (VII-120) or (VII-122a), $G_0(\mathbf{r} - \mathbf{r}', t - t')$ is hermitian,

$$G_0(\mathbf{r} - \mathbf{r}', t - t') = G_0^*(\mathbf{r}' - \mathbf{r}, t' - t). \qquad \text{(VII-150)}$$

Hence

$$\Psi_m^*(\mathbf{r}_2, t_2) = -i\hbar \int G_0^*(\mathbf{r}_2 - \mathbf{r}_3, t_2 - t_3) \Psi_m^*(\mathbf{r}_3, t_3) \, d\mathbf{r}_3,$$

$$\Psi_n(\mathbf{r}_2, t_2) = i\hbar \int G_0(\mathbf{r}_2 - \mathbf{r}_1, t_2 - t_1) \Psi_n(\mathbf{r}_1, t_1) \, d\mathbf{r}_1,$$

and (VII-149) becomes

$$a_{mn}^{(1)} = \frac{-1}{i\hbar} \iint \Psi_m^*(\mathbf{r}_2, t_2) V(\mathbf{r}_2) \Psi_n(\mathbf{r}_2, t_2) \, d\mathbf{r}_2 \, dt_2. \qquad \text{(VII-151)}$$

The probability of the transition $\psi_n^0 \to \psi_m^0$ in the time interval $t_3 - t_1$ is given by the formula (VII-38)

$$w(t) = \int |a_{mn}^{(1)}|^2 \rho(E_m)\, dE_m \tag{VII-152}$$

and the probability per unit time is (VII-40)

$$P = \frac{2\pi}{\hbar} |\langle m|V|n\rangle|^2 \rho(E_k). \tag{VII-153}$$

The application of the theory above to the problem of scattering by a static potential field merely illustrates the theory, which is of a much more general nature. See the works of Feynman for further discussions and applications.

### 6. Uncertainty relation for energy and time

In Chap. 4, (IV-61) we have deduced from the commutation relation

$$pq - qp = \frac{\hbar}{i} \tag{VII-154}$$

the uncertainty relation

$$\Delta p \Delta q \geq \frac{\hbar}{2} \tag{VII-155}$$

for the standard deviations $\Delta p$, $\Delta q$ of *simultaneous* measurement of $p$ and $q$. The emphasis is on the word "simultaneous".

In classical mechanics, $-H$ and $t$ formally have the property of being a pair of canonical conjugate variables. In the special theory of relativity,

$$(x, y, z, ict)$$

and $\tag{VII-156}$

$$\left(p_x, p_y, p_z, \frac{iE}{c}\right)$$

are 4-vectors under Lorentz transformations. From these considerations, it would seem that a relation like

$$Ht - tH = \frac{\hbar}{i} \tag{VII-157}$$

and a relation

$$\Delta E \Delta t \simeq \hbar \tag{VII-158}$$

would hold. But it has been shown that no hermitian operator $t$ exists[n] that

---

[n] See Chap. 5, Sec. 3, (V-105)–(V-109).

satisfies the commutation relation (VII-157), so that the method of (IV-58)–(IV-61) for deducing the relation $\Delta p \Delta q \geq \hbar/2$ from $pq - qp = \hbar/i$ does not obtain.

However, an energy and time uncertainty relation of the form (VII-158) does exist, although it has a very different origin and meaning from $\Delta p \Delta q \geq \hbar/2$. A theory for such a relation as $\Delta E \Delta t \sim h$ is as follows.

Let us consider a system (an atomic system, say) and we are measuring the energy states $E_m$, $E_n$ of the system by means of an apparatus. The exact procedure of the measuring process is not relevant; its essential element consists in coupling the system being measured with some measuring apparatus whose behavior obeys the macroscopic laws of classical physics—for example, the deflection of a needle.[o] If the coupling (interaction) between the system with the apparatus causes the system to make a transition from state $n$ to state $k$ $(E_n - E_k > 0)$ in the time $t$, during which the apparatus has received an amount of energy (in the form of a quantum of radiation $\hbar\omega$, say), the theory of the last section shows that the transition probability is given by (VII-36)

$$|a_k(t)|^2 = 4|\langle k|V|n\rangle|^2 \left(\frac{t}{2\hbar}\right)^2 \frac{\sin^2 x}{x^2}, \qquad \text{(VII-159)}$$

where

$$x = \left(\frac{E_k - E_n + \hbar\omega}{2\hbar}\right) t. \qquad \text{(VII-160)}$$

Now $\varepsilon \equiv \hbar\omega$, being measured by "classical" instruments, is in principle knowable exactly. The factor $\sin^2 x/x^2$ has a narrow maximum at $x = 0$, corresponding to exact resonance $E_k = E_n - \hbar\omega$, i.e., to exact knowledge of the energy change in the atom. The function $\sin^2 x/x^2$, however, has a "width", and this width means that the $|a_k(t)|^2$ has appreciable values for a range of values of $\Delta(E_k - E_n)$ and $\Delta t$ given by

$$\Delta(E_k - E_n)\Delta t \simeq \hbar. \qquad \text{(VII-161)}$$

The meaning of this relation is this: If the measurement of energy is made in a time interval $\Delta t$, the energy (change) is known only up to $\hbar/\Delta t$.

The relation (VII-161) has been obtained from the factor $\sin^2 x/x^2$ in (VII-159) and is therefore independent of the exact form and strength of the perturbation (interaction) $V$, i.e., is a general relation.

From this relation, one obtains an important application, namely, to the concept of the "width" of a state. Consider a state $k$ of a system perturbed by

[o] As already discussed in Sec. 3 of Chap. 4, it is the limitations in our knowledge of the inaccuracies arising from the interactions between the observed and the measuring instruments that are the origin of the uncertainty relation $\Delta p \Delta q \geq \hbar/2$. Now we are measuring the energy, instead of the momentum or coordinate, of, say, a composite system such as an atom.

a number of external causes, such as radiation field and collision with other particles, and let the total transition probabilities $P$ per second of $k$ to all other states $j$ of the system be

$$P = \sum_j P_{\text{rad.}} + \sum_j P_{\text{col.}} \cdot \qquad \text{(VII-162)}$$

The state $k$ then has a "lifetime" $T$

$$T = \frac{1}{P}, \qquad \text{(VII-163)}$$

and according to (VII-39), a width $\Delta E$

$$\Delta E_k \simeq \frac{\hbar}{T} = \hbar P, \qquad \text{(VII-164)}$$

i.e., the uncertainty in the energy of state $k$ corresponding to the interval $\Delta t = T$ during which the measurement of energy is to be made. This is equivalent to a width $\Delta E_k$ of the energy level.

## References


E. Schrödinger, *Annalen d. Physik* **81**, 109 (1926) deals with perturbation theory of the time-independent Schrödinger equation and Raman effect.

P. A. M. Dirac, *Proc. Roy. Soc.* London **A 112**, 661 (1926), **A 114**, 243 (1927) discuss the method of "variation of constants".

T. Y. Wu and T. Ohmura, *Quantum Theory of Scattering* (Prentice-Hall, N.J., 1962) Sec. O, 1, iii) Rearrangement scattering, Sec. O, 3, Green's function; propagator.

P. A. M. Dirac, *Principles of Quantum Mechanics*, 4th ed. (Clarendon Press, Oxford, 1958). Sections 27, 44 treat the unitary matrix ($U$ is denoted by $T$ in Dirac's book).

R. P. Feynman, *Rev. Mod. Phys.* **20**, 367 (1948), *Phys. Rev.* **76**, 749 (1949) treat the Green's function method.

$$\text{Chapter 8}$$

# The Hydrogen Atom

The quantum mechanics of atoms is not only historically important because its many initial successes helped its rapid development and establishment in the late 1920's, but is important because it has led to many concepts and methods that have played a basic role in other fields, such as the structure of molecules, metals, and even atomic nuclei. In the present and the following chapter, we shall treat a few topics on the one-electron and many-electron atoms.

In Chap. 3, Sec. 5, we studied the hydrogen atom on the (non-relativistic) Schrödinger equation without the electron spin. We shall in the present chapter still use the Schrödinger equation but with the electron spin added to it, "from the outside" so to speak.

## 1. Electron spin

### (1) Operators and eigenvectors

In Chap. 1, Sec. 8, we saw the introduction of the electron spin by Uhlenbeck and Goudsmit in 1925. When the electron spin is empirically and formally added to Schrödinger's wave mechanics, many problems in atomic spectroscopy can be satisfactorily treated. Electron spin became an integral part of quantum mechanics after the relativistic wave equation of the electron of Dirac (1928) was introduced.

Empirically, *before* the Dirac theory, Pauli introduced for the spin angular momentum components three hermitian matrices (operators)

$$\sigma_x = \begin{pmatrix} 0 & 1 \\ 1 & 0 \end{pmatrix}, \qquad \sigma_y = \begin{pmatrix} 0 & -i \\ i & 0 \end{pmatrix}, \qquad \sigma_z = \begin{pmatrix} 1 & 0 \\ 0 & -1 \end{pmatrix}, \qquad \text{(VIII-1)}$$

which have the following properties

$$\sigma_x\sigma_y = i\sigma_z, \qquad \sigma_y\sigma_z = i\sigma_x, \qquad \sigma_z\sigma_x = i\sigma_y, \qquad \text{(VIII-2a)}$$

$$\sigma_i\sigma_j + \sigma_j\sigma_i = 2\delta_{ij}, \qquad \text{(VIII-2b)}$$

$$\sigma_x^2 = \sigma_y^2 = \sigma_z^2 = \begin{pmatrix} 1 & 0 \\ 0 & 1 \end{pmatrix}. \qquad \text{(VIII-2c)}$$

The spin angular momenta operators are defined as

$$s_x = \frac{1}{2}\sigma_x, \qquad s_y = \frac{1}{2}\sigma_y, \qquad s_z = \frac{1}{2}\sigma_z \tag{VIII-3}$$

and

$$s_x^2 + s_y^2 + s_z^2 = \frac{3}{4}\begin{pmatrix} 1 & 0 \\ 0 & 1 \end{pmatrix}, \tag{VIII-4}$$

$$\frac{1}{2}(s_x + is_y) = \begin{pmatrix} 0 & 1 \\ 0 & 0 \end{pmatrix}, \qquad \frac{1}{2}(s_x - is_y) = \begin{pmatrix} 0 & 0 \\ 1 & 0 \end{pmatrix}. \tag{VIII-5}$$

$s_x, s_y, s_z$ satisfy the relations (II-103)

$$[s_x\hbar, s_y\hbar] = s_z\hbar, \qquad \text{etc.,} \tag{VIII-6}$$

where the quantum Poisson brackets are

$$[A, B] = \frac{1}{i\hbar}(AB - BA).$$

The eigenvalues of $s_z\hbar$ and $s^2\hbar^2$ are $m_s = \pm\frac{1}{2}\hbar$ and $\frac{1}{2}(1 + \frac{1}{2})\hbar^2$ respectively, and the eigenvectors can be expressed in the form

$$s_z\hbar\begin{pmatrix} 1 \\ 0 \end{pmatrix} = \frac{1}{2}\hbar\begin{pmatrix} 1 \\ 0 \end{pmatrix}, \qquad s_z\hbar\begin{pmatrix} 0 \\ 1 \end{pmatrix} = -\frac{1}{2}\hbar\begin{pmatrix} 0 \\ 1 \end{pmatrix}, \tag{VIII-7a}$$

$$s^2\hbar^2\begin{pmatrix} 1 \\ 0 \end{pmatrix} = \frac{3}{4}\hbar^2\begin{pmatrix} 1 \\ 0 \end{pmatrix} \tag{VIII-7b}$$

$$= \frac{1}{2}\left(1 + \frac{1}{2}\right)\hbar^2\begin{pmatrix} 1 \\ 0 \end{pmatrix}.$$

This shows that $s_z$ and $s^2$ have the common eigenvectors $\begin{pmatrix} 1 \\ 0 \end{pmatrix}$, $\begin{pmatrix} 0 \\ 1 \end{pmatrix}$.

In the literature, various alternative notations for the eigenvectors (or, eigenfunctions) are used. Thus

$$\begin{pmatrix} 1 \\ 0 \end{pmatrix} \equiv \chi_{1/2} \equiv \alpha, \qquad \begin{pmatrix} 0 \\ 1 \end{pmatrix} \equiv \chi_{-1/2} \equiv \beta, \tag{VIII-8}$$

so that (VIII-7a, 7b) become

$$s_z\alpha = \frac{1}{2}\alpha, \qquad s_z\beta = -\frac{1}{2}\beta, \qquad s^2\begin{Bmatrix} \alpha \\ \beta \end{Bmatrix} = \frac{1}{2}\left(1 + \frac{1}{2}\right)\begin{Bmatrix} \alpha \\ \beta \end{Bmatrix},$$

$$\frac{1}{2}(s_x + is_y)\begin{Bmatrix} \alpha \\ \beta \end{Bmatrix} = \begin{Bmatrix} 0 \\ \alpha \end{Bmatrix},$$

$$\frac{1}{2}(s_x - is_y)\begin{Bmatrix} \alpha \\ \beta \end{Bmatrix} = \begin{Bmatrix} \beta \\ 0 \end{Bmatrix}. \tag{VIII-9}$$

The eigenvectors $\binom{1}{0}$, $\binom{0}{1}$ are orthogonal and normalized,

$$(1 \quad 0)\binom{1}{0} = 1, \qquad (0 \quad 1)\binom{0}{1} = 1, \qquad (1 \quad 0)\binom{0}{1} = 0, \quad \text{(VIII-10)}$$

or

$$\alpha^\dagger \alpha = 1, \qquad \beta^\dagger \beta = 1, \qquad \alpha^\dagger \beta = \beta^\dagger \alpha = 0,$$

$$\chi_{1/2}^\dagger \chi_{1/2} = 1, \qquad \chi_{-1/2}^\dagger \chi_{-1/2} = 1, \qquad \chi_{1/2}^\dagger \chi_{-1/2} = \chi_{-1/2}^\dagger \chi_{1/2} = 0. \quad \text{(VIII-10a)}$$

*(2) Spin-orbit interactions*

In Chap. 1, Sec. 10, we obtained from classical electromagnetic theory the expression (I-98) for the spin-orbit interaction energy[a]

$$H_{\text{s.o.}} = 2\mu_B^2 \frac{Z}{r^3} (\mathbf{l} \cdot \mathbf{s}), \tag{VIII-11}$$

$$\mu_B = \frac{e\hbar}{2mc} = \text{Bohr magneton},$$

$l\hbar$, $s\hbar$ = the orbital and spin angular momentum, respectively.

This interaction $H_{\text{s.o.}}$ is to be treated as a perturbation. As the hydrogenic atom in a state $\psi_{n,l,m_l}(r, \vartheta, \phi)$ is degenerate in $l$ and $m_l$ and is independent of $m_s$, one must use the perturbation theory for degenerate systems. One may start with any representation, calculate the matrix $H_{\text{s.o.}}$ and transform to a representation in which $H_{\text{s.o.}}$ becomes diagonal, thereby obtaining the eigenvalues of $H_{\text{s.o.}}$.

Let us start with the so-called $(m_l, m_s)$-representation in which a ket is

$$|n, l, m_l, m_s\rangle, \tag{VIII-12a}$$

or a wave function

$$\psi_{n,l,m_l} \chi_{m_s} = R_{n,l}(r) Y_{l,m_l}(\vartheta, \phi) \chi_{m_s}. \tag{VIII-12b}$$

Then

$$\langle n, l, m_l, m_s | H_{\text{s.o.}} | n', l', m_l', m_s' \rangle$$

$$= 2Z\mu_B^2 \left\langle n, l \left| \frac{1}{r^3} \right| n', l' \right\rangle \langle l, m_l, m_s | l_x s_x + l_y s_y + l_z s_z | l', m_l', m_s' \rangle. \tag{VIII-13}$$

The non-vanishing matrix elements of $l_x$, $l_y$, $l_z$ are given by (III-125, 126),

---

[a] This expression includes the empirical value $g_s = 2$ for the gyromagnetic ratio for the spin magnetic moment, and the Thomas correction factor $\frac{1}{2}$. See Chap. 1, Sec. 10. Both the value $g_s = 2$ (instead of the value 1 of classical electromagnetic theory) and the Thomas factor $\frac{1}{2}$ come out automatically from Dirac's theory.

$$\langle l, m_l | l_x | l, m_l - 1 \rangle = i \langle l, m_l | l_y | l, m_l - 1 \rangle$$

$$= \frac{1}{2}\sqrt{(l - m_l + 1)(l + m_l)},$$

$$\langle l, m_l | l_x | l, m_l + 1 \rangle = i \langle l, m_l | l_y | l, m_l + 1 \rangle$$

$$= \frac{1}{2}\sqrt{(l - m_l)(l + m_l + 1)},$$

$$\langle l, m_l | l_z | l, m_l \rangle = m_l$$

and those of $s_x$, $s_y$, $s_z$ are, from (VIII-7, 9),

$$\langle m_s | s_x | m_s - 1 \rangle = i \langle m_s | s_y | m_s - 1 \rangle = \frac{1}{2},$$

$$\langle m_s | s_x | m_s + 1 \rangle = -i \langle m_s | s_y | m_s + 1 \rangle = \frac{1}{2},$$

$$\langle m_s | s_z | m_s \rangle = m_s.$$

From these, one obtains

$$\langle l, m_l, m_s | \mathbf{l} \cdot \mathbf{s} | l, m_l - 1, m_s + 1 \rangle = -\frac{1}{2}\sqrt{(l - m_l + 1)(l + m_l)}, \qquad \text{(VIII-14a)}$$

$$\langle l, m_l, m_s | \mathbf{l} \cdot \mathbf{s} | l, m_l + 1, m_s - 1 \rangle = -\frac{1}{2}\sqrt{(l - m_l)(l + m_l + 1)}, \qquad \text{(VIII-14b)}$$

$$\langle l, m_l, m_s | \mathbf{l} \cdot \mathbf{s} | l, m_l, m_s \rangle = m_l m_s. \qquad \text{(VIII-14c)}$$

From (VIII-14a, b, c), it is seen that

$$(\mathbf{l} \cdot \mathbf{s}) \text{ is diagonal with respect to } m_l + m_s \equiv m. \qquad \text{(VIII-15)}$$

This is not accidental, but is a consequence of the commutation relation

$$(\mathbf{l} \cdot \mathbf{s})(l_z + s_z) - (l_z + s_z)(\mathbf{l} \cdot \mathbf{s}) = 0, \qquad \text{(VIII-16)}$$

which follows from the relations (II-103) and (VIII-6). From (II-104), one has the commutation

$$\mathbf{l}^2(\mathbf{l} \cdot \mathbf{s}) - (\mathbf{l} \cdot \mathbf{s})\mathbf{l}^2 = 0, \qquad \text{(VIII-17)}$$

and from (VIII-4)

$$\mathbf{s}^2(\mathbf{l} \cdot \mathbf{s}) - (\mathbf{l} \cdot \mathbf{s})\mathbf{s}^2 = 0. \qquad \text{(VIII-18)}$$

If we define $\mathbf{j}$ as the vector sum of $\mathbf{l}$ and $\mathbf{s}$,

$$\mathbf{j} = \mathbf{l} + \mathbf{s} \qquad \text{(VIII-19a)}$$

or

$$\mathbf{j}^2 = \mathbf{l}^2 + \mathbf{s}^2 + 2(\mathbf{l}\cdot\mathbf{s}), \tag{VIII-19b}$$

$$j_z = l_z + s_z, \tag{VIII-19c}$$

then it follows from (VIII-16, 17, 18) that

$$\mathbf{j}^2(\mathbf{l}\cdot\mathbf{s}) - (\mathbf{l}\cdot\mathbf{s})\mathbf{j}^2 = 0, \tag{VIII-20}$$

i.e., there exists a representation in which the four matrices

$$\mathbf{j}^2, \mathbf{l}^2, \mathbf{s}^2, j_z \text{ are all diagonal,} \tag{VIII-21}$$

and they have a common eigenvector, namely

$$|l,j,m\rangle. \tag{VIII-22}$$

The eigenvalues of $\mathbf{l}^2, \mathbf{j}^2, j_z$ are given by

$$\mathbf{l}^2|l,j,m\rangle = l(l+1)|l,j,m\rangle, \tag{VIII-22a}$$

$$\mathbf{j}^2|l,j,m\rangle = j(j+1)|l,j,m\rangle, \tag{VIII-22b}$$

$$j_z|l,j,m\rangle = m|l,j,m\rangle, \tag{VIII-22c}$$

where

$$l - \frac{1}{2} \le j \le l + \frac{1}{2}, \tag{VIII-23a}$$

$$-j \le m \le j. \tag{VIII-23b}$$

This representation is called the $(j,m)$-representation.[b]

Going back to (VIII-13), one sees that $H_{\text{s.o.}}$ is diagonal in $l$, but is not in $n$. The non-diagonal matrix elements of $H_{\text{s.o.}}$ will show their effect in second-order perturbation. This second-order effect is usually small and can be neglected.[c]

One constructs the matrix of $(\mathbf{l}\cdot\mathbf{s})$, for a given $l$, in the $(m_l, m_s)$-representation of (VIII-12a).

As an example, take the case $l = 2$, $-2 \le m_l \le 2$, $-\frac{1}{2} \le m_s \le \frac{1}{2}$, and $-\frac{5}{2} \le m \le \frac{5}{2}$. The matrix

$$\langle m_l, m_s|\mathbf{l}\cdot\mathbf{s}|m_l', m_s'\rangle$$

---

[b] Note that $j_z = l_z + s_z$ commutes with $\mathbf{j}^2$, but $l_z$, $s_z$ separately do not. Hence $j^2$ and $j_z$ are simultaneously diagonal, and $j$ and $m \ (= m_l + m_s)$ are "good" quantum numbers, but $m_l$ and $m_s$ separately are not.

[c] Except for certain situations such as the anomalous intensity ratio (deviation from the value 2) of the two components of the Cs principal series doublets. See E. Fermi, *Z. Phys.* **59**, 680 (1929).

is as follows

|       |        |       |       | $m$   | $\frac{5}{2}$ | $\frac{3}{2}$ | | $\frac{1}{2}$ | | $-\frac{1}{2}$ | | $-\frac{3}{2}$ | | $-\frac{5}{2}$ |
|-------|--------|-------|-------|-------|---|---|---|---|---|---|---|---|---|---|
|       |        |       |       | $m_l$ | 2 | 2 | 1 | 1 | 0 | 0 | $-1$ | $-1$ | $-2$ | $-2$ |
| $m$   | $m_l$  | $m_s$ | $m_s$ |       | $\frac{1}{2}$ | $-\frac{1}{2}$ | $\frac{1}{2}$ | $-\frac{1}{2}$ | $\frac{1}{2}$ | $-\frac{1}{2}$ | $\frac{1}{2}$ | $-\frac{1}{2}$ | $\frac{1}{2}$ | $-\frac{1}{2}$ |
| $\frac{5}{2}$ | 2 | $\frac{1}{2}$ | | | $\odot$ | | | | | | | | | |
| $\frac{3}{2}$ | 2 | $-\frac{1}{2}$ | | | | $\odot$ | $*$ | | | | | | | |
|               | 1 | $\frac{1}{2}$  | | | | $\triangle$ | $\odot$ | | | | | | | |
| $\frac{1}{2}$ | 1 | $-\frac{1}{2}$ | | | | | | $\odot$ | $*$ | | | | | |
|               | 0 | $\frac{1}{2}$  | | | | | | $\triangle$ | $\odot$ | | | | | |
| $-\frac{1}{2}$ | 0 | $-\frac{1}{2}$ | | | | | | | | $\odot$ | $*$ | | | |
|                | $-1$ | $\frac{1}{2}$ | | | | | | | | $\triangle$ | $\odot$ | | | |
| $-\frac{3}{2}$ | $-1$ | $-\frac{1}{2}$ | | | | | | | | | | $\odot$ | $*$ | |
|                | $-2$ | $\frac{1}{2}$  | | | | | | | | | | $\triangle$ | $\odot$ | |
| $\frac{5}{2}$ | $-2$ | $-\frac{1}{2}$ | | | | | | | | | | | | $\odot$ |

$$\odot : m_l' = m_l, \qquad m_s' = m_s,$$

$$* : m_l' = m - \frac{1}{2}, \qquad m_l = m + \frac{1}{2},$$

$$\triangle : m_l' = m + \frac{1}{2}, \qquad m_l = m - \frac{1}{2}.$$

$$\text{(VIII-24)}$$

Let

$$\xi_{n,l} \equiv 2Z\mu_B^2 \left\langle n, l \left| \frac{1}{r^3} \right| n, l \right\rangle. \tag{VIII-25a}$$

This matrix element for hydrogenic atom is known from Problem 5 at the end of Chap. 3,

$$\xi_{n,l} = 2Z\mu_B^2 \frac{Z^3}{n^3 l(l + \frac{1}{2})(l + 1)a^3}. \tag{VIII-25b}$$

The determinantal equation (for a given $n$, $l$)

$$\det \| \langle n, l, m_l, m_s | H_{\text{s.o.}} | n, l, m_l', m_s' \rangle - E\delta_{m_l, m_l'}\delta_{m_s, m_s'} \| = 0, \tag{VIII-26}$$

as seen from (VIII-23), has two linear roots

$$E = \frac{l}{2}\xi_{n,l}, \tag{VIII-27a}$$

and $2l$ quadratic factors

$$
\begin{vmatrix}
\langle m_l, m_s | \mathbf{l}\cdot\mathbf{s} | m_l, m_s \rangle - \dfrac{E}{\xi_{n,l}} & \langle m_l, m_s | \mathbf{l}\cdot\mathbf{s} | m_l - 1, m_s + 1 \rangle \\[2ex]
\langle m_l - 1, m_s + 1, | \mathbf{l}\cdot\mathbf{s} | m_l, m_s \rangle & \langle m_l - 1, m_s + 1 | \mathbf{l}\cdot\mathbf{s} | m_l - 1, m_s + 1 \rangle - \dfrac{E}{\xi_{n,l}}
\end{vmatrix} = 0
$$

$$\text{(VIII-27b)}$$

or, from (VIII-14a, b, c),

$$
\begin{vmatrix}
-\tfrac{1}{2}(m + \tfrac{1}{2}) - E/\xi_{n,l} & -\tfrac{1}{2}\sqrt{(l - m + \tfrac{1}{2})(l + m + \tfrac{1}{2})} \\[2ex]
-\tfrac{1}{2}\sqrt{(l - m + \tfrac{1}{2})(l + m + \tfrac{1}{2})} & \tfrac{1}{2}(m - \tfrac{1}{2}) - E/\xi_{n,l}
\end{vmatrix} = 0,
$$

$$\text{(VIII-27c)}$$

whose roots are

$$
E = \frac{l}{2}\xi_{n,l}, \qquad -\frac{l+1}{2}\xi_{n,l}. \tag{VIII-27d}
$$

On combining (VIII-27a) and (VIII-27d), one obtains for (VIII-25) the eigenvalues, for a given $n$, $l$,

$$
E = \begin{cases}
\dfrac{l}{2}\xi_{n,l}, & \overset{l\ s}{\uparrow\uparrow} \quad (2l + 2)\text{-fold degenerate,} \\[3ex]
-\dfrac{l+1}{2}\xi_{n,l}, & \uparrow\downarrow \quad 2l\text{-fold degenerate.}
\end{cases} \tag{VIII-28}
$$

### (3) Transformation between $(m_l, m_s)$- and $(j, m)$-representations

From (VIII-27a, b, c, d), we have obtained the eigenvalues of $(\mathbf{l}\cdot\mathbf{s})$

$$
\mathbf{l}\cdot\mathbf{s} = \begin{cases}
\dfrac{l}{2}, \\[3ex]
-\dfrac{l+1}{2},
\end{cases} \tag{VIII-29}
$$

and from (VIII-22a, 22b), (VIII-19b), (VIII-7b), we see the values of $j$ corresponding to these two values of $\mathbf{l}\cdot\mathbf{s}$ are

$$
j = \begin{cases}
l + \dfrac{1}{2} & \text{for } \mathbf{l}\cdot\mathbf{s} = \dfrac{l}{2} \qquad \overset{l\ s}{\uparrow\uparrow}, \\[3ex]
l - \dfrac{1}{2} & \text{for } \mathbf{l}\cdot\mathbf{s} = -\dfrac{l+1}{2} \quad \uparrow\downarrow.
\end{cases} \tag{VIII-30}
$$

The common eigenvectors of $\mathbf{j}^2$ and $\mathbf{l}\cdot\mathbf{s}$, denoted by $|l, j, m\rangle$ in (VIII-22), are given in terms of $|l, m_l, m_s\rangle$, from (VIII-27c), by:

for $j = l + \frac{1}{2}$,

$$|l,j,m\rangle = \sqrt{\frac{l+m+\frac{1}{2}}{2l+1}}\,|l,m-\tfrac{1}{2},\tfrac{1}{2}\rangle - \sqrt{\frac{l-m+\frac{1}{2}}{2l+1}}\,|l,m+\tfrac{1}{2},-\tfrac{1}{2}\rangle; \quad \text{(VIII-31a)}$$

for $j = l - \frac{1}{2}$,

$$|l,j,m\rangle = \sqrt{\frac{l-m+\frac{1}{2}}{2l+1}}\,|l,m-\tfrac{1}{2},\tfrac{1}{2}\rangle + \sqrt{\frac{l+m+\frac{1}{2}}{2l+1}}\,|l,m+\tfrac{1}{2},-\tfrac{1}{2}\rangle. \quad \text{(VIII-31b)}$$

The inverse transformation is, for $|l, m_l, m_s\rangle$ in terms of $|l, j, m\rangle$,

for $m_l = m - \frac{1}{2}$, $m_s = \frac{1}{2}$,

$$|l,m_l,m_s\rangle = \sqrt{\frac{l+m+\frac{1}{2}}{2l+1}}\,|l,l+\tfrac{1}{2},m\rangle + \sqrt{\frac{l-m+\frac{1}{2}}{2l+1}}\,|l,l-\tfrac{1}{2},m\rangle; \quad \text{(VIII-32a)}$$

for $m_l = m + \frac{1}{2}$, $m_s = -\frac{1}{2}$,

$$|l,m_l,m_s\rangle = -\sqrt{\frac{l-m+\frac{1}{2}}{2l+1}}\,|l,l+\tfrac{1}{2},m\rangle + \sqrt{\frac{l+m+\frac{1}{2}}{2l+1}}\,|l,l-\tfrac{1}{2},m\rangle.$$

$$\text{(VIII-32b)}$$

In the $|l, j, m\rangle$ representation $\mathbf{j}^2$, $\mathbf{l}^2$, $\mathbf{s}^2$, $(\mathbf{l}\cdot\mathbf{s})$, $j_z$ are all diagonal. Thus,

$$\langle j,m|\mathbf{l}\cdot\mathbf{s}|j',m'\rangle = \tfrac{1}{2}\langle j,m|\mathbf{j}^2 - \mathbf{l}^2 - \mathbf{s}^2|j',m'\rangle$$

$$= \tfrac{1}{2}[j(j+1) - l(l+1) - s(s+1)]\delta_{jj'}\delta_{mm'},$$

$$\langle j,\mathbf{m}|\mathbf{l}\cdot\mathbf{s}|j,m\rangle = \begin{cases} \dfrac{l}{2} & \text{for } j = l + \dfrac{1}{2}, \\[2ex] -\dfrac{l+1}{2} & \text{for } j = l - \dfrac{1}{2}. \end{cases} \quad \text{(VIII-33)}$$

## 2.  Selection rules

For the absorption and emission of electric dipole radiation, the transition probabilities are given by the Einstein coefficients (VII-53, 54)

$$A_n^k = \frac{\pi}{3h}\left(\frac{4\pi\nu_{kn}}{c}\right)^3 |\langle k|e\mathbf{r}|n\rangle|^2,$$

$$B_n^k = \frac{(2\pi)^3}{3h^2}|\langle k|e\mathbf{r}|n\rangle|^2.$$

For magnetic dipole radiation transitions, the electric moment $e\mathbf{r}$ is to be replaced by the magnetic dipole moment

$$\mathbf{M} = -\frac{e}{2mc}(\mathbf{l} + g\mathbf{s})\hbar, \qquad g = 2 \tag{VIII-34}$$

$$= -(\mathbf{l} + g\mathbf{s})\mu_B.$$

The selection rules are determined by the matrix elements $\langle k|e\mathbf{r}|n\rangle$ and $\langle k|(\mathbf{l} + g\mathbf{s})|n\rangle$.

For electric dipole transitions, the selection rules are:

(i)  The Laporte rule is

$$\begin{Bmatrix} \text{even} \\ \text{odd} \end{Bmatrix} \text{parity} \leftrightarrow \begin{Bmatrix} \text{odd} \\ \text{even} \end{Bmatrix} \text{parity}, \tag{VIII-35}$$

which is a completely general rule for *any* system possessing central symmetry. For one electron in a central field, states with $\begin{Bmatrix} \text{even} \\ \text{odd} \end{Bmatrix}$ value of $l$ have $\begin{Bmatrix} \text{even} \\ \text{odd} \end{Bmatrix}$ parity, and the general parity rule has the explicit form

$$\Delta l = \pm 1. \tag{VIII-36}$$

See (III-122).

(ii)  For $j$, $m$, the selection rules are[d]

$$\Delta m = 0, \pm 1, \tag{VIII-37}$$

$$\Delta j = 0, \pm 1, \text{ excluding } j = 0 \leftrightarrow j' = 0. \tag{VIII-38}$$

From (II-101), one obtains for the orbital angular momenta $L_x$, $L_y$, $L_z$,

$$L_z z - z L_z = 0, \tag{VIII-39a}$$

$$L_z(x + iy) - (x + iy)L_z = (x + iy)\hbar, \tag{VIII-39b}$$

$$(x + iy)(L_x + iL_y) - (L_x + iL_y)(x + iy) = 0. \tag{VIII-39c}$$

Since the spin operators $s_x$, $s_y$, $s_z$ and the $x$, $y$, $z$, $L_x$, $L_y$, $L_z$ operate on different coordinates, we have,

$$J_z = L_z + s_z, \tag{VIII-40}$$

$$J_z z - z J_z = 0, \tag{VIII-41a}$$

$$J_z(x + iy) - (x + iy)J_z = (x + iy)\hbar, \tag{VIII-41b}$$

$$J_z(x - iy) - (x - iy)J_z = -(x - iy)\hbar, \tag{VIII-41c}$$

$$(x + iy)(J_x + iJ_y) - (J_x + iJ_y)(x + iy) = 0. \tag{VIII-41d}$$

---

[d](VIII-37, 38) can be proved by the method of the theory of groups. The following direct proof is due to Dirac. (Capital letters are used for angular momentum operators.)

In the $(j, m)$-representation $\mathbf{J}^2$, $J_z$, $H_{s.o.}$ are all diagonal, and

$$\mathbf{J}^2|j, m\rangle = j(j + 1)\hbar^2|j, m\rangle,$$

$$J_z|j, m\rangle = m\hbar|j, m\rangle.$$

From (VIII-41a, 41b, 41c), one obtains, respectively,

$$(m' - m'')\langle m'|z|m''\rangle = 0, \tag{VIII-42a}$$

$$(m' - m'' - 1)\langle m'|x + iy|m''\rangle = 0, \tag{VIII-42b}$$

$$(m' - m'' + 1)\langle m'|x - iy|m''\rangle = 0. \tag{VIII-42c}$$

For $\langle m'|z|m''\rangle$ not to vanish, one must have

$$\Delta m = 0.$$

For $\langle m'|x \pm iy|m''\rangle$ not to vanish, one must have

$$\Delta m = \pm 1.$$

Thus the selection rules are

$$\Delta m = 0, \pm 1. \tag{VIII-43}$$

From

$$\mathbf{J}^2(J_x + iJ_y) - (J_x + iJ_y)\mathbf{J}^2 = 0,$$

it follows that $J_x + iJ_y$ is diagonal with respect to $j$.

The only non-vanishing matrix elements of

$$J_x + iJ_y = L_x + iL_y + s_x + is_y,$$

are

$$\langle j', m'|J_x + iJ_y|j', m' - 1\rangle. \tag{VIII-44a}$$

From (VIII-42b), the non-vanishing matrix elements of $x + iy$ are

$$\langle j', m'|x + iy|j'', m' - 1\rangle. \tag{VIII-44b}$$

From (VIII-41d), one has

$$\langle j', m'|x + iy|j'', m' - 1\rangle\langle j'', m' - 1|J_x + iJ_y|j'', m' - 2\rangle$$

$$= \langle j', m'|J_x + iJ_y|j', m' - 1\rangle\langle j', m' - 1|x + iy|j'', m - 2\rangle. \tag{VIII-45}$$

For the right-hand side to be different from zero, one must have

$$-j' + 1 \le m' \le j',$$

$$-j' + 1 \le m' \le j'' + 2. \tag{VIII-46}$$

For the left-hand side to be different from zero, one must have

$$-j'' + 1 \leq m' \leq j',$$
$$-j'' + 2 \leq m' \leq j'' + 1. \tag{VIII-47}$$

To satisfy these requirements, if $j' = j''$, one must have

$$-j' + 2 \leq m' \leq j'; \tag{VIII-48a}$$

and if $j'' = j' + 1$, one must have

$$-j' + 1 \leq m' \leq j'; \tag{VIII-48b}$$

and if $j'' = j' - 1$, one must have

$$-j' + 3 \leq m' \leq j'. \tag{VIII-48c}$$

But if $j'' = j' + 2$, then (VIII-47) leads to the impossible

$$-j' - 1 \leq m' \leq j'. \tag{VIII-48d}$$

Hence

$$\Delta j = j'' - j' = 0, \pm 1. \tag{VIII-49}$$

The case $j'' = j' = 0$ must be excluded, as seen from (VIII-48a), $2 \leq m \leq 0$.

For magnetic dipole transitions, the selection rules are:

(i) The parity rule is

$$\begin{Bmatrix} \text{even} \\ \text{odd} \end{Bmatrix} \text{parity} \leftrightarrow \begin{Bmatrix} \text{even} \\ \text{odd} \end{Bmatrix} \text{parity}, \tag{VIII-50}$$

which is opposite to that for electric dipole transition.

(ii) For $l$ and $m$,

$$\Delta l = 0, \pm 1,$$
$$\Delta m = 0, \pm 1. \tag{VIII-51}$$

(See (III-125, 126)[e] for $l_x$, $l_y$, $l_z$ and (VIII-9) for $s_x$, $s_y$, $s_z$.)

## 3. Fine structure

In Chap. 6, Sec. 1 (3), we already gave the theory of the fine structure of hydrogenic atoms on the Schrödinger theory, with $\Delta E_{rel}$ arising from the relativistic correction, and $\Delta E_s$ arising from the spin-orbit interaction.

---

[e] Since in $M$, (VIII-34), $l$ and $s$ are additive, one has (i) for $\Delta m_s = 0$, $\Delta m_l = 0, \pm 1$, and (ii) for $\Delta m_l = 0$, $\Delta m_s = 0, \pm 1$. In general, $\Delta m = \Delta(m + m_s) = 0, \pm 1$.

From (VIII-25b) and (VIII-28), we have now

$$\Delta E_s = \frac{RhcZ^4\alpha^2}{n^3 l(l+\frac{1}{2})(l+1)} \begin{cases} \dfrac{l}{2}, & l = 0, 1, 2, \\[2mm] -\dfrac{l+1}{2}, & l = 1, 2, 3, \end{cases} \qquad \text{(VIII-52)}$$

which is (VI-32), given in (I-101) in advance.

The sum of $\Delta E_{rel}$ and $\Delta E_s$ is

$$\Delta E_{rel} + \Delta E_s = -\frac{RhcZ^4\alpha^2}{n^4} \begin{cases} \dfrac{n}{l+1} - \dfrac{3}{4}, & l = 0, 1, 2, \\[2mm] \dfrac{n}{l} - \dfrac{3}{4}, & l = 1, 2, 3. \end{cases} \qquad \text{(VIII-53)}$$

This is to be compared with the Sommerfeld formula on classical theory

$$E_{\text{Somm}} = -\frac{RhcZ^4\alpha^2}{n^4}\left(\frac{n}{k} - \frac{3}{4}\right), \qquad k = 1, 2, \ldots \qquad \text{(VIII-54)}$$

where $k$ is the azimuthal quantum number and is replaced by $l + 1$ in quantum mechanics, It is seen that (VIII-53), (VIII-54) give the same numerical values for the energy changes, except that a level

$$k = 1 \qquad \text{in Sommerfeld's theory} \qquad \text{(VIII-54a)}$$

is now a degenerate pair of levels

$$\begin{cases} l = 0 & \text{for} \quad j = l + \frac{1}{2}, \\ l = 1 & \text{for} \quad j = l - \frac{1}{2}. \end{cases} \qquad \text{(VIII-52a)}$$

The selection rules

$$\Delta k = \pm 1 \qquad \text{in Sommerfeld's theory,} \qquad \text{(VIII-55)}$$

and (VIII-36, 38)

$$\Delta l = \pm 1, \qquad \text{in quantum mechanics,}$$
$$\Delta j = 0, \pm 1, \qquad\qquad\qquad\qquad\quad \text{(VIII-56)}$$

give rise to different transitions in the two theories, as shown below.

Consider the $n = 2, n = 3$ levels. The following table gives $\Delta E_{rel}$ of (VI-31), $\Delta E_s$ of (VIII-52), the sum $\Delta E_{rel} + \Delta E_s$ of (VIII-53), and Sommerfeld's $\Delta E_{\text{Somm}}$ of (VIII-54).

In the Fig. VIII-1, we show the fine structure of the energy levels $n = 2$ and $n = 3$, and the dipole transitions according to the selection rules (VIII-56), of the "$H_\alpha$ line" of a hydrogenic atom ($Z$).

| | Relativistic $\Delta E_{\text{rel}}$ (VI-31) | Spin-orbit $\Delta E_{\text{s}}$ (VIII-52) | $\Delta E_{\text{rel}} + \Delta E_{\text{s}}$ (VIII-53) | Sommerfeld $\Delta E_{\text{rel}}$ (VIII-54) | |
|---|---|---|---|---|---|
| | $l$ | $j$ | | $k$ | |
| | $2 \quad -\left(\dfrac{3}{5/2} - \dfrac{3}{4}\right)$ | $\dfrac{5}{2} \quad +\dfrac{1}{5}$ | $-\left(\dfrac{3}{3} - \dfrac{3}{4}\right)$ | $3 \quad -\left(\dfrac{3}{3} - \dfrac{3}{4}\right)$ | |
| $n = 3$ | | $\dfrac{3}{2} \quad -\dfrac{3}{10}$ | $-\left(\dfrac{3}{2} - \dfrac{3}{4}\right)$ | $2 \quad -\left(\dfrac{3}{2} - \dfrac{3}{4}\right)$ | $\dfrac{RhcZ^4\alpha^2}{3^4}$ |
| | $1 \quad -\left(\dfrac{3}{3/2} - \dfrac{3}{4}\right)$ | $\dfrac{3}{2} \quad +\dfrac{1}{2}$ | $-\left(\dfrac{3}{2} - \dfrac{3}{4}\right)$ | | |
| | | $\dfrac{1}{2} \quad -1$ | $-\left(\dfrac{3}{1} - \dfrac{3}{4}\right)$ | | |
| | $0 \quad -\left(\dfrac{3}{1/2} - \dfrac{3}{4}\right)$ | $\dfrac{1}{2} \quad +3$ | $-\left(\dfrac{3}{1} - \dfrac{3}{4}\right)$ | $1 \quad -\left(\dfrac{3}{1} - \dfrac{3}{4}\right)$ | (VIII-57) |
| | $1 \quad -\left(\dfrac{2}{3/2} - \dfrac{3}{4}\right)$ | $\dfrac{3}{2} \quad +\dfrac{1}{3}$ | $-\left(\dfrac{2}{3} - \dfrac{3}{4}\right)$ | $2 \quad -\left(\dfrac{2}{2} - \dfrac{3}{4}\right)$ | |
| $n = 2$ | | $\dfrac{1}{2} \quad -\dfrac{2}{3}$ | $-\left(\dfrac{2}{1} - \dfrac{3}{4}\right)$ | | $\dfrac{RhcZ^4\alpha^2}{2^4}$ |
| | $0 \quad -\left(\dfrac{2}{1/2} - \dfrac{3}{4}\right)$ | $\dfrac{1}{2} \quad +2$ | $-\left(\dfrac{2}{1} - \dfrac{3}{4}\right)$ | $1 \quad -\left(\dfrac{2}{1} - \dfrac{3}{4}\right)$ | (VIII-58) |

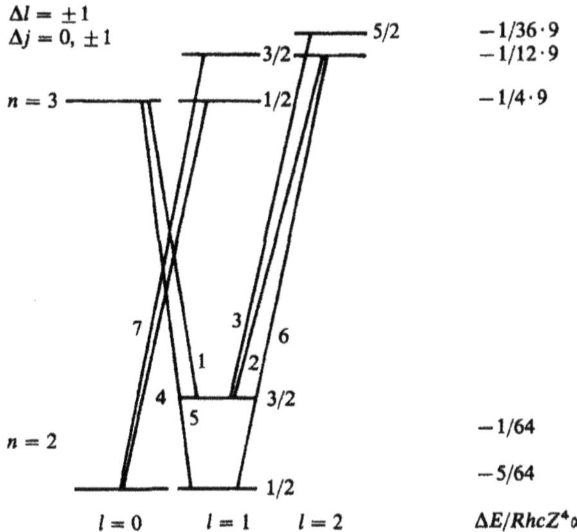


Fig. VIII.1. Dipole transitions in the $H_\alpha$ line of a hydrogenic atom of nuclear charge $Ze$. $\Delta l = \pm 1, \Delta j = 0, \pm 1$.

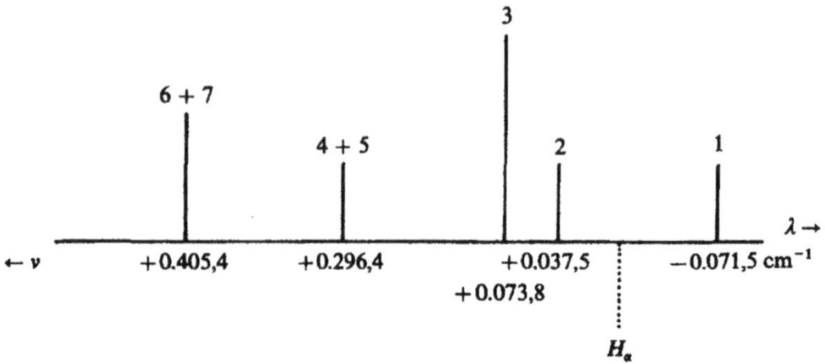

Fig. VIII.2. Fine structure of $H_\alpha$ line of $H$.

In the Fig. VIII-2, the fine structure components of the $H_\alpha$ line are shown for hydrogen ($Z = 1$), according to (VIII-53) of quantum mechanics and (VIII-57) and (VIII-58).

In Sommerfeld's old quantum theory, the energy levels are *numerically* the same as in Fig. VIII-1, but the quantum numbers are different, as shown in (VIII-54a) and (VIII-52a), and the selection rule (VIII-55) now leads only to the fine-structure components 1, 3, 6, 7 (i.e., the components 2, 4, 5 are absent).

Early spectroscopic measurements of Paschen (1916) on the $H_\alpha$ line of ionized helium support the existence of fine structure. In 1928, Dirac developed a relativistic wave equation of the electron which gave the following expression for the energy for a hydrogenic atom

$$\frac{E}{m_0 c^2} + 1 = \left\{ 1 + \frac{Z^2 \alpha^2}{[n - j - \frac{1}{2} + \sqrt{(j + \frac{1}{2})^2 - Z^2 \alpha^2}]^2} \right\}^{-1/2} . \quad \text{(VIII-53a)}$$

Note that (VIII-53a) is the same as the Sommerfeld formula (VI-29) if one replaces $k$ in (VI-29) by $n - l - 1$, and then $l$ by $j$. On expanding this in powers of $(Z\alpha)^2$ and setting $j = l \pm \frac{1}{2}$ in the first two terms, one obtains exactly the result (VIII-53)!

More accurate interferometric measurements seem to favor the quantum mechanical result over the Sommerfeld theory, although there seem to be some small discrepancy between the theoretical structure (Fig. VIII-2) and the observed intensity distribution.

The problem of the fine structure of the hydrogenic levels does not stop here.

In 1947, at Columbia University, using the technique of atomic beams perfected there, W. E. Lamb, Jr., and his student Retherford studied the microwave transitions $2\,^2S_{1/2} \leftrightarrow 2\,^2P_{1/2}$, (i.e., $n = 2$, $l = 0$ to $n = 2$, $l = 1$) of hydrogen in a magnetic field. It was found that the $2\,^2S_{1/2}$ level ($n = 2$, $l = 0$, $j = \frac{1}{2}$) was shifted upward by about $0.0353$ cm$^{-1}$ (relative to $2\,^2P_{1/2}$). This, known as the Lamb shift, is extremely important in stimulating and supporting the development and results of the quantum electrodynamics of Schwinger,

Feynman and Tomonaga (in 1946). Briefly, the Lamb shift is the difference (interaction between electron and electromagnetic field in vacuum polarized by the Coulomb field of the nucleus)—(interaction between electron and the electromagnetic field in vacuum). The probability density $|\psi(r)|^2$ at $r = 0$ (at the nucleus) is different from zero only for S states ($l = 0$) so that $^2S_{1/2}$ is shifted while $^2P_{1/2}$ is hardly shifted at all.[f]

As a result of this Lamb shift of $2\,^2S_{1/2}$ the component lines "4" and "7" are shifted by about $0.035 \text{ cm}^{-1}$ toward the long wavelength side, so that they are separated from "5" and "6" respectively. The resulting fine structure pattern seems to be in better agreement with the unresolved intensity pattern of the $H_\alpha$ line in interferometric measurements.

### 4. Zeeman effect

The Hamiltonian of a hydrogenic atom in a magnetic field $\mathbf{B} = \text{curl}\,\mathbf{A}$ is given by (VII-44b)

$$H = H_0 + H_1,$$

$$H_0 = \frac{1}{2\mu}(P_x^2 + P_y^2 + P_z^2) + V(r), \tag{VIII-59}$$

$$H_1 = \frac{e}{\mu c}(\mathbf{A}\cdot\mathbf{p}),$$

where $\mu$ is mass and $-e$ the charge of the electron. We choose the vector potential $\mathbf{A}$

$$A_x = -\frac{1}{2}yB, \qquad A_y = \frac{1}{2}xB, \qquad A_z = 0 \tag{VIII-60}$$

to represent a magnetic field $\mathbf{B}$ along the $z$-direction, and

$$H_1 = \frac{e}{2\mu c}(xp_y - yp_x)B,$$

$$= \mu_B B l_z, \qquad \mu_B = \frac{e\hbar}{2\mu c} = \text{Bohr magneton}$$

and $l_z\hbar$ is the angular momentum about the direction of $\mathbf{B}$.

---

[f] Another important result of the quantum electrodynamics of Schwinger-Feynman-Tomonaga is that the gyromagnetic ratio $g$ in

$$\frac{\text{spin magnetic moment}}{\text{spin angular momentum}} = g\frac{e}{2m_0 c}$$

is not 2 as in Dirac's relativistic theory, but is

$$g = 2\left(1 + \frac{1}{2\pi}\alpha + \cdots\right), \qquad \alpha = \frac{e^2}{\hbar c} = \frac{1}{137}. \tag{VIII-58}$$

This correction has been spectroscopically verified by P. Kusch (1947).

When the electron spin is included, we have

$$H_1 = \mu_B B(l_z + g s_z), \qquad g = 2. \tag{VIII-61}$$

From (VIII-16), it is known that $l_z + s_z = j_z$ commutes with $\mathbf{j}^2$, but $s_z$ does not. Hence $\mathbf{j}^2$ does not commute with $H_1 = \mu_B B(j_z + s_z)$, i.e.,

$$Hj_z - j_z H = 0,$$

but

$$H\mathbf{j}^2 - \mathbf{j}^2 H \neq 0. \tag{VIII-62}$$

Thus in a magnetic field, only $m$ (the eigenvalue of $j_z$) is a good quantum number, and $j$ is not, i.e., in $H$-representation $\mathbf{j}^2$ is not diagonal.

On the other hand, $H = H_0 + H_1$ is diagonal in the $(m_l, m_s)$-representation, as

$$\langle m_l, m_s | H_1 | m_l', m_s' \rangle = \mu_B B \langle m_l, m_s | l_z + 2 s_z | m_l', m_s' \rangle$$

$$= (m_l + 2 m_s) \mu_B B \delta_{m_l, m_l'} \delta_{m_s, m_s'}$$

$$= (m + m_s) \mu_B B \delta_{m_l, m_l'} \delta_{m_s, m_s'}. \tag{VIII-63}$$

As

$$-(l + 1) \leq m_l + 2 m_s \leq l + 1, \tag{VIII-64}$$

$H_1$ has $2l + 3$ distinct eigenvalues, of which 4 are distinct and $(2l - 1)$ doubly degenerate. The total number of eigenvalues is $2(2l - 1) + 4 = 2(2l + 1)$.


*(1) Strong magnetic field, Paschen-Bach effect*

When the electron spin-orbit interaction $H_{s.o.}$ is small compared with $H_1 = \mu_B B(l_z + 2 s_z)$, one can start with the $(m_l, m_s)$-representation in which $H_1$ is diagonal (VIII-63), and $H_{s.o.}$ is treated by the perturbation method. To the first order in $H_{s.o.}$, one calculates the diagonal element of $H_{s.o.}$ in (VIII-11),

$$\langle n, l, m_l, m_s | H_{s.o.} | n, l, m_l, m_s \rangle = m_l m_s \tag{VIII-65}$$

(see (VIII-14c) and (VIII-25a, b)). Hence the energy change due to $H_1 + H_{s.o.}$ is

$$\Delta E = \Delta E_1 + \Delta E_{s.o.}$$

$$= (m + m_s) \mu_B B + m_l m_s \xi_{n, l}, \qquad \mu_B B \gg \xi_{n, l}. \tag{VIII-66}$$

If we neglect the term $\xi_{n, l}$, the energy level is split into a number of equidistant components, and the pattern of spectral lines is similar to the normal Zeeman effect. This is known as the **Paschen-Back** effect. See Chap. 1 (I-95b).

### (2) Weak magnetic field

If $H_1$ is very small compared with[s] $H_{\text{s.o.}}$, it is then convenient to start with the $(j, m)$-representation in which $H_{\text{s.o.}}$ is diagonal, and to calculate $H_1$ by the perturbation method.

$$\langle n, l, j, m | H_1 | n, l, j, m \rangle = \mu_B B \langle j, m | l_z + 2s_z | j, m \rangle$$

$$= \mu_B B \langle l \pm \tfrac{1}{2}, m | l_z + 2s_z | l \pm \tfrac{1}{2}, m \rangle. \quad \text{(VIII-67)}$$

The $|j, m\rangle$ can be expressed in terms of $|m_l, m_s\rangle$ by the transformation (VIII-31a, b), and in the $|m_l, m_s\rangle$ representation,

$$(l_z + 2s_z)|m_l = m - \tfrac{1}{2}, m_s = \tfrac{1}{2}\rangle = (m + \tfrac{1}{2})|m - \tfrac{1}{2}, \tfrac{1}{2}\rangle,$$

$$(l_z + 2s_z)|m_l = m + \tfrac{1}{2}, m_s = -\tfrac{1}{2}\rangle = (m - \tfrac{1}{2})|m + \tfrac{1}{2}, -\tfrac{1}{2}\rangle.$$

One obtains for (VIII-67),

$$j = l \pm \frac{1}{2}, \qquad \Delta E_1 = \left( \frac{2l + 1 \pm 1}{2l + 1} \right) m \mu_B B,$$

$$= m \begin{Bmatrix} g_1 \\ g_2 \end{Bmatrix} \mu_B B, \qquad \text{(VIII-68)}$$

$$g_1 = \frac{2l + 2}{2l + 1} \quad \text{for} \quad j = l + \frac{1}{2}, \qquad g_2 = \frac{2l}{2l + 1} \quad \text{for} \quad j = l - \frac{1}{2}, \quad \text{(VIII-69)}$$

$$g_1 + g_2 = 2.$$

The energy change due to $H_{\text{s.o.}} + H_1$ is, for $\xi_{n,l} \gg \mu_B B$,

$$\Delta E = \begin{cases} \dfrac{l}{2} \xi_{n,l} + m g_1 \mu_B B, \\[2mm] -\dfrac{l + 1}{2} \xi_{n,l} + m g_2 \mu_B B. \end{cases} \qquad \text{(VIII-70)}$$

This gives the anomalous Zeeman effect; $g_1, g_2$ are the Landé $g$ factors for this special case $(s = \tfrac{1}{2}, j = l \pm \tfrac{1}{2})$. See Chap. 1, (I-95a), for the $^2S_{1/2}$, $^2P_{1/2}$ levels of alkali atoms.

### (3) Arbitrary magnetic field

For $\xi_{n,l} = \mu_B B$, say, it is necessary to diagonalize $H_{\text{s.o.}} + H_1$. In this case, one may start with any representation, say, the $(m_l, m_s)$, and form the deter-

---

[s] This case is realized in the alkali atom, regarded as a one-electron system. For hydrogen, the fine structure due to $H_{\text{s.o.}}$ and the relativistic correction is so small that for ordinary **B**, the system may be regarded as degenerate.

minantal equation

$$\det \| \langle m_l, m_s | (\mathbf{l} \cdot \mathbf{s}) \xi_{n,l} + (l_z + 2s_z) \mu_B B | m_l', m_s' \rangle - E \delta_{m_l m_l'} \delta_{m_s m_s'} \| = 0. \quad \text{(VIII-71)}$$

In a calculation analogous to (VIII-24)–(VIII-28), one finds two linear factors giving

$$\Delta E_1 = (m - \tfrac{1}{2}) \tfrac{1}{2} \xi_{n,l} + (m + \tfrac{1}{2}) \mu_B B, \qquad (m = l + \tfrac{1}{2}),$$

$$\Delta E_2 = -(m + \tfrac{1}{2}) \tfrac{1}{2} \xi_{n,l} + (m - \tfrac{1}{2}) \mu_B B, \qquad (m = l + \tfrac{1}{2}),$$

$$\text{(VIII-72)}$$

and $2l$ quadratic factors

$$\begin{vmatrix} -\tfrac{1}{2}(m + \tfrac{1}{2})\xi + (m - \tfrac{1}{2})\mu_B B - \Delta E & \tfrac{1}{2}\sqrt{(l - m + \tfrac{1}{2})(l + m + \tfrac{1}{2})} \\ \tfrac{1}{2}\sqrt{(l - m + \tfrac{1}{2})(l + m + \tfrac{1}{2})} & \tfrac{1}{2}(m - \tfrac{1}{2})\xi + (m + \tfrac{1}{2})\mu_B B - \Delta E \end{vmatrix} = 0.$$

$$\text{(VIII-73)}$$

It is seen that these roots (VIII-72), (VIII-73) do lead to the two limiting cases $\mu_B B \gg \xi_{n,l}$, $\xi_{n,l} \gg \mu_B B$ in (VIII-66), (VIII-70) respectively.

The case for weak $B$ in (VIII-70) has been illustrated for the $^2S_{1/2}$, $^2P_{1/2}$ levels in alkali atom (one-electron) in Chap. 1. Sec. 9, (I-95a), and the transition from weak $B$ to strong $B$ illustrated in Fig I-1.

### (4) Non-crossing theorem

In (VIII-72), it is seen that the energy of the level $(m_l = l, m_s = \tfrac{1}{2})$ in a magnetic field is directly proportional to $B$. Similarly, the level $(m_l = -l, m_s = -\tfrac{1}{2})$ is also directly proportional to $B$. In Fig. (I-1), the "transition curves" ($E$ vs $B$) from weak to strong $B$ in such cases (linear factor in the determinental equation (VIII-73)) are straight lines.

In (VIII-73), the two roots for the energy change for a given value of $m = m_l + m_s$ are given by a quadratic equation. The "transition curves" are illustrated in (I-1) by a pair of dashed lines for $m = \tfrac{1}{2}$, and a pair of dotted lines for $m = -\tfrac{1}{2}$.

It can be shown that the two roots of (VIII-73) are not equal to each other for any value of $B$ (or, any value of the parameter $\varepsilon = \mu_B B / \xi_{n,l}$).

The proof of this statement is as follows. Let the matrix of (VIII-73) be transformed by $U$ into a diagonal from $W$

$$W = U^{-1} H U$$

$$= \begin{vmatrix} W_{11}(\varepsilon) & 0 \\ 0 & W_{22}(\varepsilon) \end{vmatrix}. \quad \text{(VIII-74)}$$

Let us assume that for a certain value $\varepsilon_0$ of the parameter, the two roots are equal, $W_{11}(\varepsilon_0) = W_{22}(\varepsilon_0)$, so that the matrix $W$ at $\varepsilon_0$ becomes

$$W(\varepsilon_0) = \begin{vmatrix} W_{11}(\varepsilon_0) & 0 \\ 0 & W_{11}(\varepsilon_0) \end{vmatrix} = W_{11}(\varepsilon_0)\begin{vmatrix} 1 & 0 \\ 0 & 1 \end{vmatrix}.$$

This matrix will remain a diagonal matrix in any representation (i.e., upon any transformation $U^{-1}W(\varepsilon_0)U$). But this is not true in the $(m_l, m_s)$-representation, say. Hence $W_{11}(\varepsilon_0) = W_{22}(\varepsilon_0)$ cannot be true.

This theorem of "non-crossing" of states is not confined to the problem of this subsection, but is of a general nature applicable to many other problems, such as the energy states belonging to the same symmetry class in a system consisting of two atoms, or the limiting situations being a "united atom" and a pair of "separated atoms".

For another proof of this theorem, see Problem 4 at the end of Chap. 6.

### 5. Scattering of an electron by the hydrogen atom

#### (1) Born approximation

Let **r** be the coordinate of the electron in the scattering atom, **R** that of the incoming and scattered electron. The Hamiltonian of the system is

$$H = H_0(r) + H_0(R) + V(r, R),$$

$$H_0(r) = -\frac{\hbar^2}{2\mu}\nabla_r^2 - \frac{e^2}{r}, \qquad H_0(R) = -\frac{\hbar^2}{2\mu}\nabla_R^2, \qquad \text{(VIII-75)}$$

$$V(r, R) = -e^2\left(\frac{1}{R} + \frac{1}{|\mathbf{R} - \mathbf{r}|}\right).$$

Let

$$(H_0(r) - E_n)|n\rangle = 0, \qquad \text{(VIII-76)}$$

$$\left(H_0(R) - \frac{\hbar^2 k^2}{2\mu}\right)\Big|k\Big\rangle = 0. \qquad \text{(VIII-77)}$$

The representatives of these kets are

$$\psi_n(r) = \text{hydrogenic wave functions}, \qquad \text{(VIII-76a)}$$

$$\frac{1}{\sqrt{\Omega}}e^{i\mathbf{k}\cdot\mathbf{R}}, \quad \text{normalized to one electron in volume } \Omega. \quad \text{(VIII-77a)}$$

In the following, we shall neglect the effect of electron exchange arising from the indistinguishability of the two electrons and the Pauli principle. We shall treat $V(r, R)$ by the perturbation theory method, to the first order, Born approximation.

From (VII-96), (or (IV-31, 33)), we have the differential cross-section

$$d\sigma = \left(\frac{\mu}{2\pi\hbar^2}\right)^2 \left| \langle n, k| V(r, R)|n', k'\rangle \right|^2 d\cos\vartheta\, d\phi, \qquad \text{(VIII-78)}$$

where $n' = n$, $|k'| = |k|$ corresponds to elastic scattering, and $n' \neq n$, $k'^2 < k^2$ corresponds to inelastic scattering (the excitation of the atom from state $n$ to state $n'$).

From energy conservation,

$$E_{n'} + \frac{1}{2\mu}\hbar^2 k'^2 = E_n + \frac{1}{2\mu}\hbar^2 k^2. \qquad \text{(VIII-79)}$$

Let

$$Q^2 = k^2 + k'^2 - 2kk'\cos\vartheta, \qquad \text{(VIII-80)}$$

$$\mathbf{Q} = \mathbf{k}' - \mathbf{k},$$

i.e., $\mathbf{Q}\hbar$ is the change in the momentum of the electron in the scattering process. Now[h]

$$e^2 \left\langle n, k \left| \frac{1}{R} - \frac{1}{|\mathbf{R} - \mathbf{r}|} \right| n', k' \right\rangle = -\frac{4\pi e^2}{\Omega Q^2}\langle n|1 - e^{i\mathbf{Q}\cdot\mathbf{r}}|n'\rangle. \qquad \text{(VIII-81)}$$

The evaluation of the integral

$$\langle n|e^{i\mathbf{Q}\cdot\mathbf{r}}|n'\rangle \qquad \text{(VIII-82)}$$

in spherical polar coordinates is lengthy in general, but is considerably simpler in parabolic coordinates.[i]

The above treatment involves the use of plane waves $e^{i\mathbf{k}\cdot\mathbf{r}}$, $e^{i\mathbf{k}'\cdot\mathbf{r}}$ for the incident and the scattered electron. Such an approximation is referred to as the Born approximation. Its accuracy is not good for low electron energies ($<100$ electron volts), but improves at high energies.

---

[h] 
$$\left\langle k \left| \frac{1}{R} \right| k' \right\rangle = \frac{1}{\Omega}\iiint \frac{1}{R} e^{i\mathbf{Q}\cdot\mathbf{R}} d^3R = \frac{2\pi}{\Omega}\iint R\, dR\, e^{iQR\cos\alpha}\, d\cos\alpha$$

$$= \frac{4\pi}{\Omega Q}\int_0^\infty \sin QR\, dR = \frac{4\pi}{\Omega Q^2}.$$

$$\left\langle k \left| \frac{1}{|\mathbf{R} - \mathbf{r}|} \right| k' \right\rangle = \frac{1}{\Omega}\iiint \frac{1}{|\mathbf{R} - \mathbf{r}|} e^{i\mathbf{Q}\cdot\mathbf{R}} d^3R = \frac{e^{i\mathbf{Q}\cdot\mathbf{r}}}{\Omega}\iiint \frac{1}{|\mathbf{R} - \mathbf{r}|} e^{i\mathbf{Q}\cdot(\mathbf{R}-\mathbf{r})} d^3R$$

$$= \frac{4\pi}{\Omega Q^2} e^{i\mathbf{Q}\cdot\mathbf{r}}.$$

[i] In parabolic coordinates (Chap. 6, Appendix F),

$$\mathbf{Q}\cdot\mathbf{r} = Qz = \frac{1}{2}Q(\xi - \eta),$$

so that the integral in (VIII-82) is reduced to the product of two integrals in $\xi$ and $\eta$. These integrals can be evaluated by the method of Chap. 3, Appendix E (E-21, 22, 23).

The effect of the symmetry of the wave function with respect to the interchange of the two electrons should in principle be included in the theory. The reader is referred to the literature on this subject.

## Exercises

1. Let $\chi$ be the spin wave function, $\sigma_i$ the Pauli matrices. Show that the Schrödinger equation for $\chi$ is

$$i\hbar \frac{\partial \chi}{\partial t} = -\mu_B (\boldsymbol{\sigma} \cdot \mathbf{B}),$$

**B** being a magnetic field. Let

$$\chi = a(t)\begin{pmatrix} 1 \\ 0 \end{pmatrix} + b(t)\begin{pmatrix} 0 \\ 1 \end{pmatrix}$$

and $\omega = eB/2mc$, the Larmor frequency. Show that

$$a(t) = a_0 e^{-i\omega t}, \qquad b(t) = b_0 e^{i\omega t},$$

$$a_0 = e^{i\delta}\cos\frac{\vartheta}{2}, \qquad b_0 = e^{i\delta}\sin\frac{\vartheta}{2},$$

$\delta$, $\vartheta$ being constants. Show that

$$(\chi, \sigma_z \chi) = \cos\vartheta,$$

$$(\chi, \sigma_x \chi) = \sin\vartheta\cos(2\omega t + \delta - \gamma),$$

$$(\chi, \sigma_y \chi) = \sin\vartheta\sin(2\omega t + \delta - \gamma), \qquad \gamma = \text{const.}$$

Show that

$$\boldsymbol{\sigma}(t) = \exp(iHt/\hbar)\boldsymbol{\sigma}\exp(-iHt/\hbar),$$

$$\frac{d\boldsymbol{\sigma}(t)}{dt} = \frac{i}{\hbar}(H\boldsymbol{\sigma} - \boldsymbol{\sigma}H),$$

$$\frac{d\boldsymbol{\sigma}}{dt} = 2\boldsymbol{\omega}_B \times \boldsymbol{\sigma},$$

$$\frac{d\mathbf{s}}{dt} = \mu_B \boldsymbol{\sigma} \times \mathbf{B},$$

where $\omega_B = -(e/2mc)\mathbf{B}$.

2. The magnetic moment of the electrons in an atom is

$$\mathbf{M} = \frac{1}{2c}\sum_i e[\mathbf{r}_i \times \mathbf{v}_i].$$

On account of Larmor procession $\omega_L$

$$\mathbf{v}_i = \mathbf{u}_i + [\omega_L \times \mathbf{r}_i].$$

Show that, if the electron distribution is spherically symmetric, the component of $\mathbf{M}$ in the direction of an external magnetic field $\mathbf{B}$ is

$$M_z = M_z^0 - \frac{e^2 B}{6mc^2} \sum_i (\Psi, r_i^2 \Psi),$$

where $\Psi$ is the wave function of the electrons in the atom.

3. Obtain the Born approximation for the cross section for elastic scattering of an electron by a hydrogen atom in the ground state, using parabolic coordinates.

## References


P. A. M. Dirac, *Principles of Quantum Mechanics*, 4th ed. (Clarendon Press, Oxford, 1958). Section 40 deals with selection rules.

T. Y. Wu and T. Ohmura, *Quantum Theory of Scattering* (Prentice-Hall, N.J., 1962). Chapter L is on the scattering of electrons by the hydrogen atom.

W. E. Lamb, Jr. and R. C. Retherford, *Phys. Rev.* **72**, 241 (1947) discovered the "Lamb shift".

P. Kusch and H. M. Foley, *Phys. Rev.* **72**, 1256 (1947) discovered, from atomic beam measurements of spectral lines, the anomolous gyromagnetic ratio $g \simeq 2(1 + 2\pi\alpha)$.

J. von Neumann and E. P. Wigner, *Phys. Z.* **32**, 467 (1929), gave the non-crossing theorem for atomic energy levels of the same $J$ in many-electron configurations to be treated in Chap. 9, Sec. 3.

Chapter 9

# Two- and Many-Electron Atoms

When one deals with a system containing two or more than two electrons, one meets two new features not found with a one-electron system. One arises from the symmetry of any physical quantity with respect to the interchange of two electrons which are physically indistinguishable. The other is a mathematical one—the lack of an exact solution of the Schrödinger equation in terms of known functions in closed forms, even for the next-to-the-simplest of all atoms, namely helium.

In the present chapter we shall treat the two-electron (helium-like) and the many-electron atoms.

## 1. Symmetry property

Let the Hamiltonian of an $N$-electron system be denoted by

$$H(\mathbf{r}_1, \mathbf{r}_2, \ldots \mathbf{r}_N).$$

This Hamiltonian, and in fact any physical quantity of the atom, must be unchanged upon any permutation of the electrons since they are "identical", or "nondistinguishable".

Let $\Psi(1, 2, \ldots N; t)$ be the wave function of the system, and let $P_{12}$ be an operator that permutes electrons 1 and 2,

$$P_{12}\Psi(1, 2, 3, \ldots, N; t) = \Psi(2, 1, 3, \ldots, N; t). \tag{IX-1}$$

From this, it follows that

$$(P_{12})^2\Psi(1, 2, 3, \ldots, N; t) = \Psi(1, 2, 3, \ldots, N; t)$$

so that $(P_{12})^2$ has the eigenvalue 1, and

$$P_{12}\Psi = \pm\Psi. \tag{IX-2}$$

The function $\Psi$ is thus symmetric (antisymmetric) with respect to the permutation of any two electrons, corresponding to the eigenvalue $1$ $(-1)$ of the operator $P_{12}$.

It is a law of nature that for a system of electrons, only antisymmetric wave function exists.

From the Schrödinger equation

$$i\hbar \frac{\partial \Psi}{\partial t} = H \Psi(1, 2, \ldots, t),$$ (IX-3)

it follows that if at any instant $\Psi$ is antisymmetric in the above sense, it remains antisymmetric in the course of time, since the Hamiltonian $H$ is symmetric in all the electrons.

In $\Psi(1, 2, \ldots)$ above, the name index $i$, $i = 1, 2, \ldots$, denotes collectively the spatial and spin coordinates of electron $i$.

Consider the simple case of a two-electron system. If the interactions arising from the spins are very weak so that the Hamiltonian can be written in the form

$$H(1, 2) = H_0(\mathbf{r}_1, \mathbf{r}_2) + H_s,$$ (IX-4)

$$H_s \ll H_0,$$

the Schrödinger equation can be treated by the perturbation method so that

$$(H_0 - E)\psi(\mathbf{r}_1, \mathbf{r}_2) = 0$$ (IX-5)

and the wave function $\Psi(1, 2)$ approximated by

$$\Psi(1, 2) = \psi(\mathbf{r}_1, \mathbf{r}_2)\chi(s_1, s_2).$$ (IX-6)

By the same argument leading to (IX-2), the function $\psi(\mathbf{r}_1, \mathbf{r}_2)$ must be either symmetric $\psi_s$, or antisymmetric $\psi_a$. The requirement of antisymmetric $\Psi(1, 2)$ can now be met by two possibilities, namely

$$\psi_s(\mathbf{r}_1, \mathbf{r}_2)\chi_a(s_1, s_2),$$ (IX-7a)

or

$$\psi_a(\mathbf{r}_1, \mathbf{r}_2)\chi_s(s_1, s_2).$$ (IX-7b)

Let us assume further that the spins of the two electrons are very weakly coupled, i.e.,

$$H_s = H_{s_1} + H_{s_2},$$ (IX-8a)

so that

$$\chi(1, 2) = \chi(1)\chi(2).$$ (IX-8b)

As the basic kets of the spin of one electron are, according to (VIII-8),

$$\begin{pmatrix} 1 \\ 0 \end{pmatrix}, \begin{pmatrix} 0 \\ 1 \end{pmatrix}, \quad \text{or } \chi_{1/2}, \chi_{-1/2}, \quad \text{or } \alpha, \beta,$$ (IX-9)

it is possible to form explicit products of these basic kets corresponding to the symmetric and antisymmetric $\chi(1, 2)$,

$$\left.\begin{array}{c} \chi_1^s \\ \chi_0^s \\ \chi_{-1}^s \end{array}\right\} = \left\{\begin{array}{l} \alpha_1 \alpha_2, \\ \dfrac{1}{\sqrt{2}}(\alpha_1 \beta_2 + \beta_1 \alpha_2), \\ \beta_1 \beta_2, \end{array}\right. \qquad \text{(IX-10a)}$$

$$\chi_{-1}^a = \frac{1}{\sqrt{2}}(\alpha_1 \beta_2 - \beta_1 \alpha_2), \qquad \text{(IX-10b)}$$

and for two-electron systems, it is possible to have two distinct classes of states (IX-7a) and (IX-7b).

When there are $N \geq 3$ electrons, there are $2^N$ linearly independent combinations of the products $\chi_{m_s}(1)\chi_{m_s'}(2)\chi_{m_s''}(3)\ldots$, of which there are totally symmetric functions, but no totally antisymmetric ones. For example, for 3 electrons, one may form 8 orthogonal combinations

$$\chi_1 = \alpha_1 \alpha_2 \alpha_3,$$

$$\chi_2 = \beta_1 \beta_2 \beta_3,$$

$$\chi_3 = \frac{1}{\sqrt{3}}(\alpha_1 \beta_2 \alpha_3 + \beta_1 \alpha_2 \alpha_3 + \alpha_1 \alpha_2 \beta_3), \qquad \text{(IX-11a)}$$

$$\chi_4 = \frac{1}{\sqrt{3}}(\beta_1 \alpha_2 \beta_3 + \alpha_1 \beta_2 \beta_3 + \beta_1 \beta_2 \alpha_3),$$

which are totally symmetric, and

$$\chi_5 = \frac{1}{\sqrt{6}}(\alpha_1 \alpha_2 \beta_3 + \alpha_1 \beta_2 \alpha_3 - 2\beta_1 \alpha_2 \alpha_3),$$

$$\chi_6 = \frac{1}{\sqrt{6}}(\beta_1 \beta_2 \alpha_3 + \beta_1 \alpha_2 \beta_3 - 2\alpha_1 \beta_2 \beta_3), \qquad \text{(IX-11b)}$$

which are symmetric in 2, 3, and

$$\chi_7 = \frac{1}{\sqrt{2}}\alpha_1(\alpha_2 \beta_3 - \beta_2 \alpha_3),$$

$$\chi_8 = \frac{1}{\sqrt{2}}\beta_1(\beta_1 \alpha_3 - \alpha_2 \beta_3), \qquad \text{(IX-11c)}$$

which are antisymmetric in 2, 3. $\chi_5, \chi_6, \chi_7, \chi_8$ do not have definite symmetries in the pair 1, 2 or 1, 3. That it is not possible to form a totally antisymmetric spin wave function for 3 or more than 3 electrons is due to the fact that in $\chi_{m_s}$, $m_s$ has only 2 values.

## 2. Two-electron atom: energy

Consider a helium-like atom. The Hamiltonian is

$$H_0(\mathbf{r}_1, \mathbf{r}_2) = -\frac{\hbar^2}{2\mu}\nabla_1^2 - \frac{Ze^2}{r_1} - \frac{\hbar^2}{2\mu}\nabla_2^2 - \frac{Ze^2}{r_2} + \frac{e^2}{r_{12}}. \tag{IX-12}$$

Let

$$S_z = s_z(1) + s_z(2),$$

$$L_z = l_z(1) + l_z(2),$$

$$J_z = L_z + S_z, \tag{IX-13}$$

$$\mathbf{L} = \mathbf{l}_1 + \mathbf{l}_2, \qquad \mathbf{S} = \mathbf{s}_1 + \mathbf{s}_2,$$

$$\mathbf{J} = \mathbf{L} + \mathbf{S}.$$

On considerations similar to those of Chap. 8, Sec. 1, (1), it can be seen that

$$L_z, S_z, J_z, \mathbf{L}^2, \mathbf{S}^2, \mathbf{J}^2 \tag{IX-14}$$

all commute with one another so that there exists a representation in which all these operators are diagonal matrices. Also they all commute with the Hamiltonian $H_0$ in (IX-12). The eigenvalues are, respectively,[a]

$$M_L, M_s, M, L(L + 1), S(S + 1), J(J + 1). \tag{IX-15}$$

$H_0$ is diagonal with respect to these quantum numbers. The eigenvalues of $H_0$ have the following properties:

(i) $H_0$ is degenerate in $M_L$, $M_s$, $M$ (independent of $M_L$, $M_s$, $M$);
(ii) $H_0$ is degenerate in $J$, since $H_0$ is independent of the electron spins and is independent of how the $l_i$ and $s_i$ are coupled by spin-orbit interactions;
(iii) Although $H_0$ does not contain the electron spins, one must not conclude that the eigenvalues of $H_0$ are independent of $S$. The electron spins affect the symmetry of the eigenfunctions of $H_0$ through the requirement (IX-7a, b), and greatly affect the eigenvalues. This was first brought out by Heisenberg (1926). To see this, let $\phi(\mathbf{r}_1, \mathbf{r}_2)$ be an approximate solution of

---

[a] From (IX-10a, b), it is seen

$$S_z\chi_1^s = \chi_1^s, \qquad S_z\chi_0^s = 0, \qquad S_z\chi_{-1}^s = -\chi_{-1}^s,$$
$$S_z\chi_0^a = 0. \tag{IX-16}$$

i.e., the eigenvalues $M_s$ of $S_z$ are $1, 0, -1$ and $0$, etc.

$$H_0\psi(\mathbf{r}_1,\mathbf{r}_2) = E\psi(\mathbf{r}_1,\mathbf{r}_2). \qquad \text{(IX-17)}$$

Since $H_0$ is symmetric in $\mathbf{r}_1$ and $\mathbf{r}_2$, $\phi(\mathbf{r}_2,\mathbf{r}_1)$ is also an approximate solution. Let

$$\Delta \equiv \iint \phi^*(\mathbf{r}_1,\mathbf{r}_2)\phi(\mathbf{r}_2,\mathbf{r}_1)\,d\mathbf{r}_1\,d\mathbf{r}_2, \qquad \text{(IX-18)}$$

and form

$$\Phi^s = \frac{1}{\sqrt{2(1+\Delta)}}[\phi(\mathbf{r}_1,\mathbf{r}_2) + \phi(\mathbf{r}_2,\mathbf{r}_1)],$$

$$\Phi^a = \frac{1}{\sqrt{2(1-\Delta)}}[\phi(\mathbf{r}_1,\mathbf{r}_2) - \phi(\mathbf{r}_2,\mathbf{r}_1)], \qquad \text{(IX-19)}$$

and the two antisymmetric functions

$$\Psi_s = \Phi^s\chi^a, \qquad \Psi_a = \Phi^a\chi^s. \qquad \text{(IX-20)}$$

Then

$$\iint \Phi^{s*}H_0\Phi^s\,d\mathbf{r}_1\,d\mathbf{r}_2 - \iint \Phi^{a*}H_0\Phi^a\,d\mathbf{r}_1\,d\mathbf{r}_2$$

$$= \frac{2}{1-\Delta^2}\left[\iint \phi^*(\mathbf{r}_1,\mathbf{r}_2)H_0\phi(\mathbf{r}_2,\mathbf{r}_1)\,d\mathbf{r}_2\,d\mathbf{r}_2\right.$$

$$\left. - \Delta \iint \phi^*(\mathbf{r}_1,\mathbf{r}_2)H_0\phi(\mathbf{r}_1,\mathbf{r}_2)\,d\mathbf{r}_1\,d\mathbf{r}_2\right], \qquad \text{(IX-21)}$$

which is different from zero. In the following, we shall see that for reasonable $\phi(\mathbf{r}_1,\mathbf{r}_2)$, this difference is positive, showing that the energy of the state $\Psi_a$, antisymmetric in the spins (singlet state), is higher than that of the triplet state.

*(1) Perturbation theory*

The Schrödinger equation (IX-17) has so far not been solved exactly in terms of known functions in closed forms. The first approximate method that suggests itself is the perturbation theory that treats the Coulomb interaction between the two electrons as a perturbation,

$$H = H^0 + \frac{e^2}{r_{12}}, \qquad \text{(IX-22a)}$$

$$H^0 = -\frac{\hbar^2}{2\mu}\nabla_1^2 - \frac{Ze^2}{r_1} - \frac{\hbar^2}{2\mu}\nabla_2^2 - \frac{Ze^2}{r_2}. \qquad \text{(IX-22b)}$$

The eigenfunctions of $H^0$ are products of hydrogenic functions

$$\Psi^0(\mathbf{r}_1, \mathbf{r}_2) = \psi_{n_1}(\mathbf{r}_1)\psi_{n_2}(\mathbf{r}_2), \tag{IX-23a}$$

where

$$\psi_{n_1}(\mathbf{r}_1) = R_{n,l}(r_1)\, Y_{l_1, m_{l_1}}(\vartheta_1, \phi_1), \text{ etc.}, \tag{IX-23b}$$

and

$$H^0\Psi^0 = (E_{n_1} + E_{n_2})\Psi^0. \tag{IX-23c}$$

One forms the triplet and singlet state functions

$$\left.\begin{matrix} ^3\Psi^0 \\ ^1\Psi^0 \end{matrix}\right\} = \frac{1}{2}[\psi_{n_1}(\mathbf{r}_1)\psi_{n_2}(\mathbf{r}_2) \mp \psi_{n_1}(\mathbf{r}_2)\psi_{n_1}(\mathbf{r}_1)]\begin{cases} \chi^s_{m_s} \\ \chi^a_{m_s} \end{cases}. \tag{IX-24}$$

Then

$$\iint {}^1\Psi^{0*}H^1\Psi^0\, d\mathbf{r}_1\, d\mathbf{r}_2 = (E_{n_1} + E_{n_2}) + \left\langle n_1, n_2 \left| \frac{e^2}{r_{12}} \right| n_1, n_2 \right\rangle,$$

$$+ \left\langle n_1, n_2 \left| \frac{e^2}{r_{12}} \right| n_2, n_1 \right\rangle, \tag{IX-25a}$$

$$\iint {}^3\Psi^{0*}H^3\Psi^{0*}\, d\mathbf{r}_1\, d\mathbf{r}_2 = (E_{n_1} + E_{n_2}) + \left\langle n_1, n_2 \left| \frac{e^2}{r_{12}} \right| n_1, n_2 \right\rangle$$

$$- \left\langle n_1, n_2 \left| \frac{e^2}{r_{12}} \right| n_2, n_1 \right\rangle, \tag{IX-25b}$$

where

$$\left\langle n_1, n_2 \left| \frac{e^2}{r_{12}} \right| n_1, n_2 \right\rangle = \iint \psi^*_{n_1}(\mathbf{r}_1)\psi^*_{n_2}(\mathbf{r}_2)\frac{e^2}{r_{12}}\psi_{n_1}(\mathbf{r}_1)\psi_{n_2}(\mathbf{r}_2)\, d\mathbf{r}_1\, d\mathbf{r}_2, \tag{IX-26a}$$

$$\left\langle n_1, n_2 \left| \frac{e^2}{r_{12}} \right| n_2, n_1 \right\rangle = \iint \psi^*_{n_1}(\mathbf{r}_1)\psi^*_{n_2}(\mathbf{r}_2)\frac{e^2}{r_{12}}\psi_{n_1}(\mathbf{r}_2)\psi_{n_2}(\mathbf{r}_1)\, d\mathbf{r}_1\, d\mathbf{r}_2. \tag{IX-26b}$$

The integral $\langle n_1, n_2 |(e^2/r_{12})| n_1, n_2 \rangle$ is the Coulomb interaction between the two electrons, one in state $\psi_{n_1}$ and the other in state $\psi_{n_2}$. It is called the Coulomb energy (or, the "direct integral"). $\langle n_1, n_2 |(e^2/r_{12})| n_2, n_1 \rangle$ has no classical analogue, neither electron being in state $\psi_{n_1}$ or $\psi_{n_2}$. It is called the "exchange energy" (or, the "exchange integral"). It is positive, so that

$$E(\text{singlet state}) - E(\text{triplet state}) > 0. \tag{IX-27}$$

The integrals $\langle n_1, n_2 |(e^2/r_{12})| n_1, n_2 \rangle$, $\langle n_1, n_2 |(e^2/r_{12})| n_2, n_1 \rangle$ can be evaluated by means of the expansion

$$\frac{1}{|\mathbf{r}_1 - \mathbf{r}_2|} = \sum_{k=0}^{\infty} \sum_{-k}^{k} m\, \frac{2}{2k+1} \frac{r_<^k}{r_<^{k+1}} \times \Theta_{k,m}(x_1)\Theta_{k,m}(x_2)e^{im(\phi_1-\phi_2)}, \quad \text{(IX-28)}$$

$$x_1 = \cos\vartheta_1, \qquad x_2 = \cos\vartheta_2,$$

$\vartheta_1$, $\phi_1$, $\vartheta_2$, $\phi_2$ being the polar angles of $\mathbf{r}_1$ and $\mathbf{r}_2$, and $r_<$, $r_>$ the smaller and the larger one, respectively, of $r_1$ and $r_2$. $\Theta_{k,m}$ are the normalized associated Legendre polynomials (Chap. 3, Appendix C, (C-27)).

On inserting (IX-28) into $\langle n_1, n_2 | (1/r_{12}) | n_1, n_2 \rangle$, upon integrating over $\phi_1$ and $\phi_2$, one finds that, for the integral not to vanish, $m$ must be 0; writing for brevity

$$n = (n, l, m) \quad \text{for } n_1, l_1, m_{l_1},$$

$$n' = (n', l', m') \quad \text{for } n_2, l_2, m_{l_2},$$

one obtains

$$\left\langle n, n' \left| \frac{1}{r_{12}} \right| n, n' \right\rangle = \sum_{k=0} \frac{2}{2k+1} \int \Theta_{l,m}(x_1)\Theta_{lm}(x_1)\Theta_{k,0}(x_1)\,dx_1$$

$$\times \int \Theta_{l',m'}(x_2)\Theta_{l',m'}(x_2)\Theta_{k0}(x_2)\,dx_2$$

$$\times \iint [R_{n,l}(r_1)]^2 \frac{r_<^k}{r_>^{k+1}} [R_{n',l'}(r_2)]^2 r_1^2\, dr_1 r_2^2\, dr_2. \quad \text{(IX-29a)}$$

Similarly,

$$\left\langle n, n' \left| \frac{1}{r_{12}} \right| n', n \right\rangle = \sum_{k=0} \frac{2}{2k+1} \left[ \int \Theta_{l,m}(x_1)\Theta_{l',m'}(x_1)\Theta_{k,m-m'}(x_1)\,dx_1 \right]^2$$

$$\times \iint R_{n,l}(r_1)R_{n',l'}(r_1) \frac{r_<^k}{r_>^{k+1}} R_{n,l}(r_2)R_{n'l'}(r_2)r_1^2\, dr_1 r_2^2\, dr_2.$$

$$\text{(IX-29b)}$$

On introducing the notations

$$c^k(l, m; l', m') = \sqrt{\frac{2}{2k+1}} \int_{-1}^{1} \Theta_{l,m}\Theta_{l',m'}\Theta_{k,m-m'}\,dx, \qquad \text{(IX-30)}$$

$$a^k(l, m; l', m') = c^k(l, m; l, m)c^k(l', m'; l', m'), \qquad \text{(IX-31a)}$$

$$b^k(l, m; l', m') = |c^k(l, m; l', m')|^2, \qquad \text{(IX-31b)}$$

$$F^{(k)}(n, l; n', l') = e^2 \iint [R_{nl}(r_1)]^2 \frac{r_<^k}{r_>^{k+1}} [R_{n',l'}(r_2)]^2 r_1^2\, dr_1 r_2^2\, dr_2, \quad \text{(IX-32a)}$$

$$G^{(k)}(n, l; n', l') = e^2 \iint R_{n,l}(r_1) R_{n',l'}(r_1) \frac{r_<^k}{r_>^{k+1}} R_{n,l}(r_2) R_{n',l'}(r_2) r_1^2 \, dr_1 \, r_2^2 \, dr_2,$$

$$(\text{IX-32b})$$

one can express (IX-29a, b) as

$$\left\langle n, n' \left| \frac{e^2}{r_{12}} \right| n, n' \right\rangle = \sum_{k=0} a^k(l, m; l', m') F^{(k)}(n, l; n', l'), \qquad (\text{IX-33a})$$

$$\left\langle n, n' \left| \frac{e^2}{r_{12}} \right| n', n \right\rangle = \sum_{k=0} b^k(l, m; l', m') G^{(k)}(n, l; n', l'), \qquad (\text{IX-33b})$$

and

$$\left.\begin{array}{l} \text{Singlet } {}^1\text{E} \\ \text{Triplet } {}^3\text{E} \end{array}\right\} = E_n + E_{n'} + \left\langle n, n' \left| \frac{e^2}{r_{12}} \right| n, n' \right\rangle \pm \left\langle n, n' \left| \frac{e^2}{r_{12}} \right| n', n \right\rangle. \quad (\text{IX-34})$$

The integrals $F^{(k)}$, $G^{(k)}$ are

$$F^{(k)}(n, l; n', l') = e^2 \int_0^\infty dr_1 \, r_1^2 [R_{n,l}(r_1)]^2 \left\{ \frac{1}{r_1^{k+1}} \int_0^{r_1} r_2^{k+2} [R_{n',l'}(r_2)]^2 \, dr_2 \right.$$

$$\left. + r_1^k \int_{r_1}^\infty \frac{1}{r_2^{k-1}} [R_{n',l'}(r_2)]^2 \, dr_2 \right\}, \qquad (\text{IX-35a})$$

$$G^{(k)}(n, l; n', l') = 2e^2 \int_0^\infty dr_1 \, r_1^{k+2} R_{n,l}(r_1) R_{n',l'}(r_1)$$

$$\times \int_{r_1}^\infty \frac{1}{r_2^{k-1}} R_{n,l}(r_2) R_{n',l'}(r_2) \, dr_2. \qquad (\text{IX-35b})$$

As an example, take the ground state $1s^2 \ {}^1S$ of a helium-like atom.

$$\psi_{1s}(r) = \frac{1}{\sqrt{4\pi}} 2 \left( \frac{Z}{a} \right)^{3/2} \exp\left( -\frac{Zr}{a} \right), \ a = \frac{\hbar^2}{me^2}, \ \text{Bohr radius.} \quad (\text{IX-36})$$

For the singlet state, the spin wave function $\chi^a_{M_s} = \chi^a_0$,

$$\Psi^0(1, 2) = \psi_{1s}(r_1) \psi_{1s}(r_2) \chi^a_0.$$

One finds

$$E(1s^2) = E_{1s} + E_{1s} + F^{(0)}(1, 0; 1, 0)$$

$$= -2 \frac{Z^2 e^2}{2a} + \frac{5}{8} \frac{Ze^2}{a}$$

$$= -\frac{Ze^2}{a} \left( Z - \frac{5}{8} \right). \qquad (\text{IX-37})$$

The ionization potential is

$$E_{\text{ion}} = E(1s) - E(1s^2)$$

$$= -\frac{Z^2 e^2}{2a} - E(1s^2)$$

$$= \frac{Ze^2}{2a}\left(Z - \frac{5}{4}\right). \tag{IX-38}$$

The following table gives the results for $H^-$ (the negative ion of hydrogen), He, $Li^+$, $Be^{++}$, and the experimental (spectroscopic) results for comparison

| | $E(1s^2\ {}^1S)$ | | $E_{\text{ion}}$ (in units of $e^2/2a$) | | |
| Z | (IX-37) | Exp. | (IX-38) | Exp. | % diff. |
|---|---|---|---|---|---|
| 1 ($H^-$) | $-0.75$ | $-1.055$ | $-0.25$ | 0.055 | 550 |
| 2 (He) | $-5.50$ | $-5.81$ | 1.50 | 1.81 | 17 |
| 3 ($Li^+$) | $-14.25$ | $-14.51$ | 5.25 | 5.56 | 5.6 |
| 4 ($Be^{++}$) | $-27.00$ | $-27.305$ | 11.00 | 11.307 | 2.7 |

$$\tag{IX-39}$$

It is seen that the results obtained by the perturbation theory improve as $Z$ increases, as expected. But for $H^-$, the calculated $E_{\text{ion}}$ has the wrong sign. The reason is clear: the electron-electron interaction $e^2/r_{12}$ cannot be regarded as "small" compared with the electron-nucleus interaction $-e^2/r_1$, $-e^2/r_2$.

## (2) Ritz variational method

The basic principle is the variational form of the Schrödinger equation (see Chap. 3, Problem 3 at the end of Sec. 4). The modified form of the variational method due to Ritz is to replace the true eigenfunction $\psi$ in the integral

$$E = \int \Psi^* H \Psi \, d\tau \tag{IX-40}$$

by an approximate function $\phi(c_1, c_2, \ldots, c_n)$ containing some parameters $c_1$, $c_2, \ldots c_n$ and an approximate value $W$ of the true eigenvalue $E$ is obtained from

$$\frac{\partial E}{\partial c_i} = 0, \qquad i = 1, 2, \ldots, n. \tag{IX-41}$$

The accuracy of this method depends on the choice of the trial wave function $\phi$—its functional form and its adjustability by means of the variational parameters $c_i$. It can be proved that the approximate eigenvalue $W$ so obtained is always higher than the true eigenvalue $E$.[b]

As an example, take again the $1s^2\ ^1S$ state of the helium-like atoms. Here we take for $\phi$ the form of the hydrogenic function (IX-36), but with $Z$ replaced by an "effective charge" parameter $\alpha$ so that

$$\phi_{1s}(r) = \frac{1}{\sqrt{\pi}}\left(\frac{\alpha}{a}\right)^{3/2}\exp\left(-\frac{\alpha}{a}r\right),\qquad\text{(IX-45)}$$

so that (IX-37) is now

$$W = \left\{2(\alpha^2 - 2Z\alpha) + \frac{5}{4}\alpha\right\}\frac{e^2}{2a}.$$

$\alpha$ is given by $\partial W/\partial\alpha = 0$ to be

$$\alpha = Z - \frac{5}{16},$$

so that

$$W = -\left[\frac{e^2}{a}\left(Z - \frac{5}{16}\right)^2\right],\qquad\text{(IX-46)}$$

$$E_{\text{ion}} = \left[Z^2 - \frac{5}{4}Z + \frac{25}{128}\right]\frac{e^2}{2a}.\qquad\text{(IX-47)}$$

The results for $H^-$, $He$, $Le^+$, $Be^{++}$ are given in the following table.

---

[b] Let the true eigenvalues and eigenfunctions of $H$ be

$$(H - E_n)\psi_n = 0,\qquad\text{(IX-42)}$$

$E_0$ being the ground state. For the ground state, let the trial wave function $\phi$ be expanded in the complete set $\psi_n$

$$\phi = \sum\int a_n\psi_n,\qquad \sum\int|a_n|^2 = 1.\qquad\text{(IX-43)}$$

An approximate value $W_0$ of $E_0$ is

$$W_0 = \int\phi^* H\phi\,d\tau = \sum\int a_m^* a_n\int\psi_m^* H\psi_n\,d\tau$$

$$= \sum\int|a_n|^2 E_n,$$

so that

$$W_0 - E_0 = \sum\int|a_n|^2(E_n - E_0) > 0.\quad\text{QED}\qquad\text{(IX-44)}$$

| | $E(1s^2\ {}^1S)$ | | $E_{ion}$ (in units of $e^2/2a$) | | |
|---|---|---|---|---|---|
| Z | (IX-46) | Exp. | (IX-47) | Exp. | % diff. |
| 1 (H$^-$) | $-0.945$ | $-1.055$ | $-0.055$ | 0.055 | 200 |
| 2 (He) | $-5.695$ | $-5.81$ | 1.695 | 1.81 | 6.4 |
| 3 (Li$^+$) | $-14.445$ | $-14.51$ | 5.445 | 5.56 | 2.1 |
| 4 (Be$^{++}$) | $-27.195$ | $-27.305$ | 11.195 | 11.307 | 1.0 |

$$\text{(IX-48)}$$

It is seen that there is some improvement over the perturbation theory method.

### (3) Hartree-Fock method

The starting point is the variational method. The trial (variational) wave function for the singlet and triplet state is

$$\left.\begin{array}{r}{}^1\Psi\\{}^3\Psi\end{array}\right\} = \frac{1}{\sqrt{2}}\{\phi_a(\mathbf{r}_1)\phi_b(\mathbf{r}_2) \pm \phi_b(\mathbf{r}_1)\phi_a(\mathbf{r}_2)\}\left\{\begin{array}{l}\chi^a\\\chi^s\end{array}\right. \tag{IX-49}$$

Here unlike the Ritz method of the last subsection (2), the functions $\phi_a$, $\phi_b$ are not given any assumed functional form, but are completely flexible, in the following sense. On substituting (IX-49) into

$$\delta \iint \Psi^* H \Psi \, dr_1 \, dr_2 = 0, \tag{IX-50a}$$

subject to the auxiliary condition

$$\iint \Psi^* \Psi \, dr_1 \, dr_2 = 1, \tag{IX-50b}$$

and making independent variations $\delta\phi_a$, $\delta\phi_b$, one obtains

$$\{H_0(1) + E_{bb} + G_{bb}(\mathbf{r}_1) - E\}\phi_a(\mathbf{r}_1) = \mp\{E_{ba} + G_{ba}(\mathbf{r}_1)\}\phi_b(\mathbf{r}_1), \tag{IX-51a}$$

$$\{H_0(2) + E_{aa} + G_{aa}(\mathbf{r}_2) - E\}\phi_b(\mathbf{r}_2) = \mp\{E_{ab} + G_{ab}(\mathbf{r}_2)\}\phi_a(\mathbf{r}_2), \tag{IX-51b}$$

where

$$H_0(r) = -\frac{\hbar^2}{2m}\nabla^2 - \frac{Ze^2}{r},$$

$$E_{aa} = \int \phi_a^*(\mathbf{r})H_0(r)\phi_a(\mathbf{r})\,d\mathbf{r},$$

$$E_{ba} = \int \phi_b^*(\mathbf{r}) H_0(\mathbf{r}) \phi_a(\mathbf{r})\, d\mathbf{r}, \tag{IX-52}$$

$$G_{bb}(\mathbf{r}_1) = \int \frac{e^2}{r_{12}} \phi_b^*(\mathbf{r}_2) \phi_b(\mathbf{r}_2)\, d\mathbf{r}_2,$$

$$G_{ba}(\mathbf{r}_1) = \int \frac{e^2}{r_{12}} \phi_b^*(\mathbf{r}_2) \phi_a(\mathbf{r}_2)\, d\mathbf{r}_2, \qquad \text{etc.}$$

$G_{bb}(\mathbf{r}_1)$ is the potential energy of electron 1 at $r_1$ due to electron 2 in state $\phi_b$; $G_{ba}$ is the "exchange energy" of electron 1 at $r_1$ with electron 2 in "state $\phi_a$ and $\phi_b$"—a term without classical analogue; $E - E_{bb}$ is the eigenvalue of electron 1; etc.

The functions $\phi_a$, $\phi_b$ are now given by a pair of simultaneous differential-integral equations (IX-51a, b). This method is due to Hartree (1928) who did not include the symmetrized form (IX-49) but only the product $\phi_a(\mathbf{r}_1)\phi_b(\mathbf{r}_2)$, and therefore obtained (IX-51a, b) without the exchange terms on the right-hand sides. The inclusion of the exchange effect by using the form (IX-49) is due to Fock and Slater (1930).

The method is called the self-consistent field method in the following sense: Consider (IX-51a, b) without the exchange term on the right-hand sides. The wave function $\phi_a(1)$ for electron 1 is determined by $H_0(1)$ and the potential $G_{bb}(\mathbf{r}_1)$ due to electron 2, $G_{bb}(\mathbf{r}_1)$ being of course determined by $\phi_b(2)$, while $\phi_b(2)$ for electron 2 is determined by $H_0(2)$ and the potential $G_{aa}(\mathbf{r}_2)$ due to electron 1, $G_{aa}(\mathbf{r}_2)$ being of course determined by $\phi_a(1)$. The coupled equations for $\phi_a$ and $\phi_b$ determine $\phi_a$, $\phi_b$ in a "self-consistent" way for the effect of exchange.

The coupled differential-integral equations are usually solved by numerical methods and now by computers. Since the one-electron wave functions are not restricted by any assumed functional *forms*, the self-consistent field method gives better results than the modified Ritz variational method. In fact, subject to the limitation that the many-electron wave function is constructed from (properly symmetrized) products of one-electron wave functions, the self-consistent field method gives the best result. The use of one-electron wave function product, however, is in itself a severe limitation, since it does not take proper account of the correlation effect of the electrons. To make this point clear, consider the ground state $1s^2\,{}^1S$ of a two-electron atom. Any wave function of the form

$$\Psi(\mathbf{r}_1, \mathbf{r}_2) = \psi_a(\mathbf{r}_1)\psi_a(\mathbf{r}_2)$$

will give the same probability $|\Psi(\mathbf{r}_1, \mathbf{r}_2)|^2$

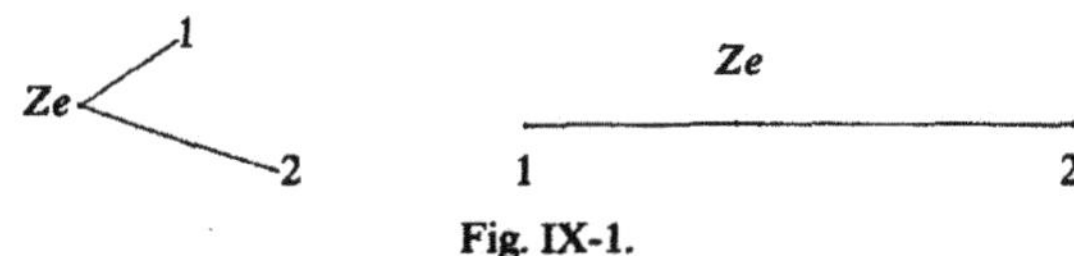

Fig. IX-1.

for the two configurations of the two electrons in the figure (in which $\mathbf{r}_1$ and $\mathbf{r}_2$ have the same distance from the nucleus in the two figures), while one should expect the configuration in which the two electrons are farther apart should have a greater probability on account of the repulsion between the two electrons.

We shall make a few remarks:

(i) The Hartree-Fock method is of course not restricted to atoms; it is applicable to other problems such as the nucleons in atomic nuclei.

(ii) When applied to two-electron atoms, the starting point is the wave function (IX-49), where the symmetries in the space and the spin part are separable. For states symmetric in the spin part, the two functions $\phi_a$, $\phi_b$ are effectively orthogonal, since the spatial part is a determinant

$$\begin{vmatrix} \psi_a(1) & \psi_b(1) \\ \psi_a(2) & \psi_b(2) \end{vmatrix} \tag{IX-53}$$

and any part of $\psi_b$ which is not orthogonal to $\psi_a$ contributes nothing to the determinant.

For states antisymmetric in the spins, the two functions $\phi_a$ and $\phi_b$ do not have to be orthogonal.

(iii) For a three-electron or many-electron atom, it is not possible to have states antisymmetric in the spins of all the electrons, as pointed out at the end of Sec. 1. We shall treat the many-electron atom by the Hartree-Fock method in a later section.

### (4) Hylleraas' method

The use of one-electron product wave functions in the Ritz variational and the Hartree-Fock method suffers from the inadequate account of the correlation of the two electrons. To improve on this, Hylleraas (1928) introduced a very effective variational method by introducing, in the case of two-electron atoms, the following 6 coordinates

$$s = r_1 + r_2, \qquad t = r_1 - r_2, \qquad u = |\mathbf{r}_{12}| \tag{IX-54}$$

and

$$\vartheta, \phi, \psi,$$

which are the Euler angles to specify the orientation of the (rigid) triangle formed by the nucleus and the two electrons. For states with zero angular momentum, the wave function is independent of the Euler angles, and the wave function $\Psi(\mathbf{r}_1, \mathbf{r}_2)$ is a function of $s, t, u$ only. The wave function $\Psi(\mathbf{r}_1, \mathbf{r}_2)$ is then a non-separable function $\Psi(s, t, u)$.

The ranges of the variables are as follows

$$0 \leq t \leq u, \qquad 0 \leq u \leq s, \qquad 0 \leq s \leq \infty \qquad \text{(IX-55)}$$

and the volume element is

$$d\tau = (s^2 - t^2)u\, ds\, dt\, du. \qquad \text{(IX-56)}$$

We shall express all lengths in units of the Bohr radius $a = \hbar^2/me^2$, and energy in units of $2Rhc = e^2/a$, twice the ionization potential of hydrogen.

The Schrödinger equation in the form of the variational principle (III-38) is

$$\delta \int \left[ \sum_i \frac{\hbar^2}{2m}(\nabla_i \Psi)^2 - (E - V)\Psi^2 \right] d\tau = 0, \qquad \text{(IX-57)}$$

where

$$V = -Ze^2 \left( \frac{1}{r_1} + \frac{1}{r_2} \right) + \frac{e^2}{r_{12}}.$$

The integral in (IX-57) in the coordinates $s, t, u$ can be shown to be

$$M - L - NE = 0, \qquad \text{(IX-58)}$$

where $M, L, N$, the kinetic energy, the potential energy and the normalization part are

$$M = \int_0^\infty ds \int_0^s du \int_0^u dt \left\{ (s^2 - t^2)u \left[ \left( \frac{\partial \Psi}{\partial s} \right)^2 + \left( \frac{\partial \Psi}{\partial t} \right)^2 + \left( \frac{\partial \Psi}{\partial u} \right)^2 \right] \right.$$
$$\left. + 2s(u^2 - t^2)\frac{\partial \Psi}{\partial s}\frac{\partial \Psi}{\partial u} + 2(s^2 - u^2)t\frac{\partial \Psi}{\partial u}\frac{\partial \Psi}{\partial t} \right\}, \qquad \text{(IX-59a)}$$

$$L = \int_0^\infty ds \int_0^s du \int_0^u dt(4Zsu - s^2 + t^2)\Psi^2, \qquad \text{(IX-59b)}$$

$$N = \int_0^\infty ds \int_0^s dt \int_0^u du(s^2 - t^2)u\Psi^2. \qquad \text{(IX-59c)}$$

If all the lengths $r_1, r_2, r_{12}, s, t, u$ are changed in scale by a factor $k$, then (IX-58) becomes

$$k^2M - kL - NE = 0. \qquad \text{(IX-60)}$$

On varying the scale factor $k$ in (IX-57), one obtains

$$E = -\frac{1}{4NM}L^2. \tag{IX-61}$$

For the trial wave-function $\Psi(s, t, u)$, Hylleraas took the series

$$\Psi(s, t, u) = e^{-(1/2)s} \sum_{n,m,l} C_{n,2l,m} s^n u^m, t^{2l}, \tag{IX-62}$$

where odd powers of $t$ are excluded since for $1s^2\ {}^1S$, the space part $\Psi(s, t, u)$ must be symmetric in the two electrons.

The variations in the coefficients $C_{n,2l,m}$ lead to

$$\frac{\partial E}{\partial C_{n,2l,m}} = 0 \text{ for all } n,\ l,\ m,$$

which give an approximate value for the eigenvalue $E$.

The following values are some of the earlier results of Hylleraas for helium, using 3, 6 and 8 terms in the variational wave function (IX-62).

|  |  | $E$ in $Rhc$ |
|---|---|---|
| $\Psi = e^{-s/2}(1 + 0.08u + 0.01t^2)$, | | $-5.80488$ |
| $\Psi = e^{-s/2}(1 + 0.0972u + 0.0097t^2 - 0.0277s$ | | $-5.80648$ |
| $\quad + 0.0025s^2 - 0.0024u^2)$, | (IX-63) | |
| $\Psi = 8$ terms. | | $-5.80749$ |

It is important to note that even with only three terms in $\Psi$, the result $E = -5.80648Rhc$ is much closer to the observed value $E = -5.810$ than the Ritz method value $E = -5.695$. The great improvement is due to the fact that the $u$ term in $\Psi$ effectively takes account of the correlation effect of the two electrons.[c] This shows that the use of products of single-electron wave function is not a good approximation. Strictly speaking, the wave function $\Psi(\mathbf{r}_1, \mathbf{r}_2)$ is non-separable, and the single-electron, central field quantum numbers $n$, $l$, $m_l$ are not exact quantum numbers. It is good to be aware of this point, even though the quantum numbers $n_i$, $l_i$ have been used all the time.[d]

---

[c] The 8-term $\Psi$ gives $E = -5.80749\ Rhc$ which is $-198322$ cm$^{-1}$. This is in fact lower than the experimental value $-198298 \pm 6$ cm, a result seemingly contradictory to the theorem that all variational methods must necessarily approach the true eigenvalue from above, as shown in Sec. 2, (II-44). But when the calculated result is corrected for the relative motion of the electrons and the nucleus, and the relativistic effects, and for the Lamb shift, the theoretical value lies *above* the observed value. See references at the end of this chapter.

[d] The non-separable wave function $\Psi(\mathbf{r}_1, \mathbf{r}_2)$ can be expanded in terms of the complete set of the central field (hydrogenic) wave functions $\phi_{n,l,m}(r, \vartheta, \phi)$, in a series

$$\Psi(\mathbf{r}_1, \mathbf{r}_2) = \sum_{nlm, n'l'm'} C_{n,l,m;n',l',m'} \phi_{n,l,m}(\mathbf{r}_1)\phi_{n',l',m'}(\mathbf{r}_2) \tag{IX-64}$$

(to be properly antisymmetrized, together with the spins). It is the use of only one or two terms of the series that renders the approximation inadequate for the electron-electron correlations.

### 3.  Two-electron atom: configuration and $(L, S)$-coupling states $^{2S+1}L$

*(1)  L, S states, $e^2/r_{12}$ energies*

The Hamiltonian of a 2-electron atom is taken to be

$$H = \sum_{i=1}^{2} H_0(i) + \frac{e^2}{r_{12}} + \sum_{i=1}^{2} H_{s.o.}(i), \tag{IX-65}$$

where $H_{s.o.}(i)$ is the spin-orbit interaction of electron $i$ (VIII-11),

$$H_0(i) = -\frac{\hbar^2}{2\mu}\nabla_i^2 - Ze^2\frac{1}{r_i}, \tag{IX-66}$$

$$H_{s.o.}(i) = 2\mu_B^2\frac{Z}{r_i^3}(\mathbf{l}_i\cdot\mathbf{s}_i). \tag{IX-67}$$

In this section, we shall assume that

$$H_0 \gg \frac{e^2}{r_{12}} \gg H_{s.o.} \tag{IX-68}$$

In this case, one starts with the representation in which the basic kets are the two complete sets $\phi_{n,l,m}(i)$ of the two independent $H_0(1)$, $H_0(2)$, and

$$\Psi^0(\mathbf{r}_1,\mathbf{r}_2) = \phi_{n,l,m_l}(\mathbf{r}_1)\chi_{m_s}(1)\phi_{n',l',m_l'}(\mathbf{r}_2)\chi_{m_s'}(2)$$

to be properly antisymmetrized. With the nomenclature

| $l$ | 0 | 1 | 2 | 3 | 4 | |
|---|---|---|---|---|---|---|
| $n$ | | | | | | |
| 1 | 1s | | | | | |
| 2 | 2s | 2p | | | | |
| 3 | 3s | 3p | 3d | | | etc., |
| 4 | 4s | 4p | 4d | 4f | | |

$$\tag{IX-69}$$

the 2-electron *configurations* are, for example,

$$1s^2,\ 1s\,2s,\ 1s\,2p,\dots,2s^2,\ 2s\,2p,\ 2s\,3d,\dots 2p^2,\ 2p\,3s,\dots \tag{IX-70}$$

For the Hamiltonian

$$H_0 = H_0(1) + H_0(2), \tag{IX-71}$$

it is seen that the following angular momenta

$$\mathbf{L} = \mathbf{l}_1 + \mathbf{l}_2, \qquad \mathbf{S} = \mathbf{s}_1 + \mathbf{s}_2, \qquad \mathbf{J} = \mathbf{L} + \mathbf{S},$$
$$L_z = l_{1z} + l_{2z}, \qquad S_z = s_{1z} + s_{2z}, \qquad J_z = L_z + S_z, \tag{IX-72}$$

commute with each other and also with $H_0$, so that they are simultaneous diagonal matrices and their quantum numbers

$$L, \quad S, \quad J,$$
$$M, \quad M_s, \quad M_L, \tag{IX-73}$$

are all good quantum numbers. The spectroscopic nomenclature (I-89b) is

$$
\begin{array}{ccccccc}
L = & 0 & 1 & 2 & 3 & 4 & 5 \\
\text{State} & {}^{2S+1}\text{S} & {}^{2S+1}\text{P} & {}^{2S+1}\text{D} & {}^{2S+1}\text{F} & {}^{2S+1}\text{G} & {}^{2S+1}\text{H}
\end{array}
\tag{IX-74}
$$

This case is called the $\{L, S\}$-, or Russell-Saunders, coupling. The values of $J$ are

$$L - S \leq J \leq L + S, \qquad \text{if } S < L,$$
$$S - L \leq J \leq S + L, \qquad \text{if } L < S. \tag{IX-75}$$

Of the 6 quantum numbers in (IX-73), only 4 are independent. Thus we may use either the

$$(S, L, M_s, M_L)\text{-representation}, \tag{IX-76a}$$

or the

$$(S, L, J, M)\text{-representation}. \tag{IX-76b}$$

To obtain the ${}^{2S+1}L$ states for any electron configuration, let us give the following illustrative examples.

*(i)* $ns\,n'p\ {}^1P,\ {}^3P$

For the configuration ns n'p, there are 12 independent antisymmetric wave functions from the functions

$$\phi_{n,0,0}(1)\begin{cases}\chi_+(1)\\\chi_-(1)\end{cases}, \qquad \phi_{n',1,m_l}(2)\begin{cases}\chi_+(2)\\\chi_-(2)\end{cases}, \qquad m_l = -1, 0, 1. \tag{IX-77}$$

They are:

$$\Psi_1, \Psi_4, \Psi_7 = \frac{1}{\sqrt{2}}\begin{vmatrix}\phi_{n00}(1) & \phi_{n'11}(1)\\ \phi_{n00}(2) & \phi_{n'11}(2)\end{vmatrix} \times (\chi_1^s,\ \chi_0^s,\ \chi_{-1}^s), \tag{IX-77a}$$

$$\Psi_2, \Psi_5, \Psi_8 = \frac{1}{\sqrt{2}}\begin{vmatrix}\phi_{n00}(1) & \phi_{n'10}(1)\\ \phi_{n00}(2) & \phi_{n'10}(2)\end{vmatrix} \times (\chi_1^s,\ \chi_0^s,\ \chi_{-1}^s), \tag{IX-77b}$$

$$\Psi_3, \Psi_6, \Psi_9 = \frac{1}{\sqrt{2}}\begin{vmatrix}\phi_{n00}(1) & \phi_{n'1,-1}(1)\\ \phi_{n00}(2) & \phi_{n'1,-1}(2)\end{vmatrix} \times (\chi_1^s,\ \chi_0^s,\ \chi_{-1}^s), \tag{IX-77c}$$

$$\left.\begin{array}{c}\Psi_{10}\\ \Psi_{11}\\ \Psi_{12}\end{array}\right\} = \frac{1}{\sqrt{2}}\left\{\begin{array}{l}\phi_{n00}(1)\phi_{n'11}(2) + \phi_{n00}(2)\phi_{n'11}(1)\\ \phi_{n00}(1)\phi_{n'10}(2) + \phi_{n00}(1)\phi_{n'10}(2)\\ \phi_{n00}(1)\phi_{n'1,-1}(2) + \phi_{n00}(2)\phi_{n'1,-1}(1)\end{array}\right\}\chi_0^a, \tag{IX-77d}$$

where the $\chi_{m_s}^s$, $\chi_0^a$ are given in (IX-10a, b). The $M_L$, $M_s$ values of these 12 functions are

| | $\Psi_1$ | $\Psi_2$ | $\Psi_3$ | $\Psi_4$ | $\Psi_5$ | $\Psi_6$ | $\Psi_7$ | $\Psi_8$ | $\Psi_9$ | $\Psi_{10}$ | $\Psi_{11}$ | $\Psi_{12}$ | |
|---|---|---|---|---|---|---|---|---|---|---|---|---|---|
| $M_s$ | 1 | 1 | 1 | 0 | 0 | 0 | $-1$ | $-1$ | $-1$ | 0 | 0 | 0 | (IX-78) |
| $M_L$ | 1 | 0 | $-1$ | 1 | 0 | $-1$ | 1 | 0 | $-1$ | 1 | 0 | $-1$ | |

Let us now consider the Hamiltonian

$$H_0 = \sum_{i=1}^{2} H_0(i) + \frac{e^2}{r_{12}} \tag{IX-79}$$

of (IX-65), i.e., neglect the spin-orbit interactions. $H_0$ commutes with $L_z$ and $S_z$, so that $H_0$ is diagonal in $M_L$ and $M_s$. Also

$$\langle \chi^s | H_0 | \chi^a \rangle = 0. \tag{IX-80}$$

Hence

$$\langle \Psi_i | H_0 | \Psi_j \rangle = 0 \quad \text{for } i \neq j, \quad i, j = 1, \dots 12. \tag{IX-81}$$

For the $\Psi_1, \dots, \Psi_9$ states,

$$\langle \Psi_1 | H_0 | \Psi_1 \rangle = \cdots = \langle \Psi_9 | H_0 | \Psi_9 \rangle, \tag{IX-82}$$

so that the $\Psi_1, \dots, \Psi_9$ states are 9-fold degenerate:

$$^{2S+1}L_J = {}^3P_{0,1,2}, \qquad M = 2J + 1. \tag{IX-83}$$

Similarly $\Psi_{10}, \Psi_{11}, \Psi_{12}$ are 3-fold degenerate

$$\langle \Psi_{10} | H_0 | \Psi_{10} \rangle = \langle \Psi_{11} | H_0 | \Psi_{11} \rangle = \langle \Psi_{12} | H_0 | \Psi_{12} \rangle, \tag{IX-84}$$

$$^{2S+1}L_J = {}^1P_1, \qquad M = 2J + 1. \tag{IX-85}$$

The energy of (any one of the 9 states of) $^3P$ is, for simplicity, from $\Psi_5$,

$$\left\langle \Psi_5 \left| \frac{e^2}{r_{12}} \right| \Psi_5 \right\rangle = \left\langle n\text{s}, n'\text{p} \left| \frac{e^2}{r_{12}} \right| n\text{s}, n'\text{p} \right\rangle - \left\langle n\text{s}, n'\text{p} \left| \frac{e^2}{r_{12}} \right| n'\text{p}, n\text{s} \right\rangle$$

$$= F^{(0)}(n\text{s}, n'\text{p}) - \frac{1}{3} G^{(0)}(n\text{s}, n'\text{p}), \tag{IX-86}$$

where the $F^{(0)}$, $G^{(0)}$ are the integrals defined in (IX-32a, b). For $^1P_1$ one may use $\Psi_{10}$, and

$$\left\langle \Psi_{10} \left| \frac{e^2}{r_{12}} \right| \Psi_{10} \right\rangle = F^{(0)}(n\text{s}, n'\text{p}) + \frac{1}{3} G^{(0)}(n\text{s}, n'\text{p}). \tag{IX-87}$$

The energies for $n\text{s} \, n'\text{p} \, ^3P, \, ^1P$ are finally given by

$$\left.\begin{array}{l} nsn'\mathrm{p}\ ^1\mathrm{P} \\ nsn'\mathrm{p}\ ^3\mathrm{P} \end{array}\right\} : E = E_0(ns) + E_0(n'\mathrm{p}) + F^{(0)}(ns, n'\mathrm{p}) + \begin{cases} +\tfrac{1}{3}G^0(ns, n'\mathrm{p}), \\ -\tfrac{1}{3}G^0(ns, n'\mathrm{p}). \end{cases} \quad \text{(IX-88)}$$

The integrals $F^{(0)}$, $G^{(0)}$ can be evaluated if the radial wave functions $R_{n,0}(r)$, $R_{n',1}(r)$ are known.[e]

### (ii) $np^2$, $^3P$, $^1D$, $^1S$

With a configuration $npn'\mathrm{p}$, there will be $6 \times 6 = 36$ independent states by combining the $\phi_{n,1,m_l}(1)\chi_{m_s}(1)$ with $\phi_{n',1,m_{l'}}(2)\chi_{m_{s'}}(2)$. If $n' = n$ ($n\mathrm{p}$, $n\mathrm{p}$ are called equivalent electrons), the Pauli principle allows only $6!/2!4! = 15$ states. The 15 antisymmetric wave functions are determinental functions constructed in a way similar to (IX-77). The following table give the $(n, l, m_l, m_s)$-representation of the 15 states.

|  | $m_l$ | $m_l'$ | $m_s$ | $m_s'$ | $M_L$ | $M_s$ | $M$ | state |
|---|---|---|---|---|---|---|---|---|
| $\Psi_1$ | 1 | 1 | + | − | 2 | 0 | 2 | $^1D$ |
| $\Psi_2$ | 1 | 0 | + | + | 1 | 1 | 2 | $^3P$ |
| $\Psi_3$ | 1 | 0 | + | − | 1 | 0 | $\left.1\right\}$ | $^2P + {}^3D$ |
| $\Psi_4$ | 1 | 0 | − | + | 1 | 0 | $1$ |  |
| $\Psi_5$ | 1 | 0 | − | − | 1 | −1 | 0 | $^3P$ |
| $\Psi_6$ | 1 | −1 | + | + | 0 | 1 | 1 | $^3P$ |
| $\Psi_7$ | 1 | −1 | + | − | 0 | 0 | $\left.0\right]$ | $^3P + {}^1D + {}^1S$ |
| $\Psi_8$ | 1 | −1 | − | + | 0 | 0 | $0$ |  |
| $\Psi_9$ | 0 | 0 | + | − | 0 | 0 | $0$ |  |
| $\Psi_{10}$ | 1 | −1 | − | − | 0 | −1 | −1 | $^3P$ |
| $\Psi_{11}$ | −1 | 0 | + | + | −1 | 1 | 0 | $^3P$ |
| $\Psi_{12}$ | −1 | 0 | − | + | −1 | 0 | $\left.-1\right\}$ | $^3P + {}^1D$ |
| $\Psi_{13}$ | −1 | 0 | + | − | −1 | 0 | $-1$ |  |
| $\Psi_{14}$ | −1 | 0 | − | − | −1 | −1 | −2 | $^3P$ |
| $\Psi_{15}$ | −1 | −1 | + | − | −2 | 0 | −2 | $^1D$ |

$$\text{(IX-89)}$$

[e] For hydrogenic atoms, $R_{n,0}(r)$, $R_{n',1}(r)$ are known; but when the above theory is applied to the two-electron spectra of such atom as Mg, Ca, ..., the $R_{n,0}$, $R_{n',1}$ must be obtained by another approximate method, such as the Hartree-Fock, or the Thomas-Fermi statistical potential.

The Hamiltonian $H_0$ in (IX-79) is diagonal in $M_L$ and $M_s$. The matrix $H_0$ is as shown in the following figure

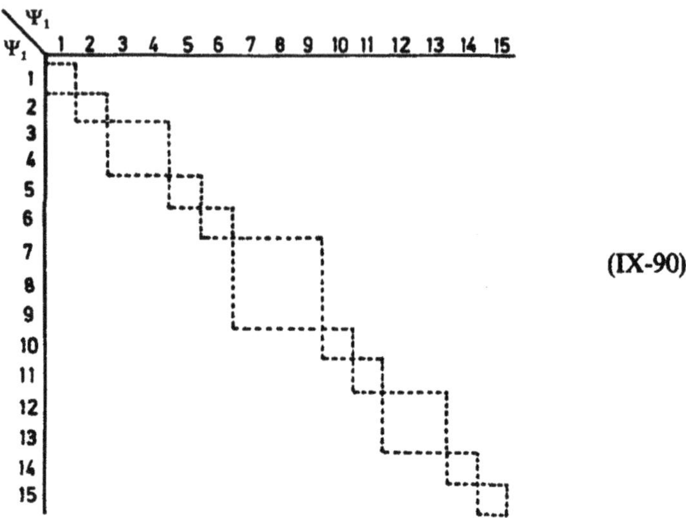

(IX-90)

The $^{2S+1}L$ states can be determined "by inspection" as follows. For $\Psi_1$, $M_L = 2$ so that $L$ is at least 2, i.e., a D state. There is no state $M_s > 0$ for $M_L = 2$. Hence it is a $^1$D state. For $\Psi_2$, $M_s = 1$ so that $S$ is at 1, i.e., a triplet state. There is no state $M_L > 1$ for $M_s = 1$. Hence $\Psi_2$ is a $^3$P state. The $\Psi_3$, $\Psi_4$ have the same $M_L$, $M_s$. The quadratic block in (IX-90) must correspond to a combination $^3$P + $^1$D. Similarly, $\Psi_7$, $\Psi_8$, $\Psi_9$ must be linear combinations of $^3$P, $^1$D, $^1$S.

For $^1$D,

$$\left\langle \Psi_1 \left| \frac{e^2}{r_{12}} \right| \Psi_1 \right\rangle = F^{(0)}(n\text{p}, n\text{p}) + \frac{1}{25} F^{(2)}(n\text{p}, n\text{p}). \qquad \text{(IX-91)}$$

For $^3$P,

$$\left\langle \Psi_2 \left| \frac{e^2}{r_{12}} \right| \Psi_2 \right\rangle = F^{(0)}(n\text{p}, n\text{p}) - \frac{2}{25} F^{(2)}(n\text{p}, n\text{p}) - \frac{3}{25} G^{(0)}(n\text{p}, n\text{p}). \qquad \text{(IX-92)}$$

For equivalent electrons $n\text{p}^2$, as seen from (IX-32a, b),

$$G^{(2)}(n\text{p}, n\text{p}) = F^{(2)}(n\text{p}, n\text{p}). \qquad \text{(IX-93)}$$

For $^1$S, since $^1$S does not appear in the matrix (IX-90) except in the combination $\Psi_7 + \Psi_8 + \Psi_9$, we proceed by writing down the cubic block in (IX-90) for $e^2/r_{12}$.

|              | $\Psi_7$ | $\Psi_8$ | $\Psi_9$ |
|--------------|----------|----------|----------|
| $\Psi_7$ | $F^{(0)} + \frac{1}{25}F^{(2)}$ | $-\frac{6}{25}G^{(2)}$ | $\frac{3}{25}G^{(2)}$ |
| $\Psi_8$ | $-\frac{6}{25}G^{(2)}$ | $F^{(0)} + \frac{1}{25}F^{(2)}$ | $-\frac{3}{25}G^{(2)}$ |
| $\Psi_9$ | $\frac{3}{25}G^{(2)}$ | $-\frac{3}{25}G^{(2)}$ | $F^{(0)} + \frac{4}{25}F^{(2)}$ |

$$\text{(IX-94)}$$

where $G^{(2)} = F^{(2)}$ by (IX-93). Now by the theorem for the trace,[f] the sum of the diagonal elements is invariant under any similarity transformation, in particular the one that diagonalizes (IX-94). Hence the sum below is invariant:

$$\left\langle {}^3\mathrm{P}\left|\frac{e^2}{r_{12}}\right|{}^3P\right\rangle + \left\langle {}^1\mathrm{D}\left|\frac{e^2}{r_{12}}\right|{}^1\mathrm{D}\right\rangle + \left\langle {}^1\mathrm{S}\left|\frac{e^2}{r_{12}}\right|{}^1\mathrm{S}\right\rangle$$

$$= F^{(0)} + \frac{1}{25}F^{(2)} + F^{(0)} + \frac{1}{25}F^{(2)} + F^{(0)} + \frac{4}{25}F^{(2)}. \qquad \text{(IX-95)}$$

From (IX-91) and (IX-92), one obtains for

$${}^1\mathrm{S}: \qquad \left\langle {}^1\mathrm{S}\left|\frac{e^2}{r_{12}}\right|{}^1\mathrm{S}\right\rangle = F^{(0)}(n\mathrm{p}, n\mathrm{p}) + \frac{7}{25}F^{(2)}(n\mathrm{p}, n\mathrm{p}) + \frac{3}{25}G^{(2)}(n\mathrm{p}, n\mathrm{p}),$$

$$(G^{(2)} = F^{(2)}). \qquad \text{(IX-96)}$$

The energy of $H_0$ in (IX-79) is

$$E({}^1\mathrm{S}) = 2E_0(n\mathrm{p}) + F^{(0)} + \tfrac{2}{5}F^{(2)},$$

$$E({}^1\mathrm{D}) = 2E_0(n\mathrm{p}) + F^{(0)} + \tfrac{1}{25}F^{(2)}, \qquad \text{(IX-97)}$$

$$E({}^3\mathrm{P}) = 2E_0(n\mathrm{p}) + F^{(0)} - \tfrac{1}{5}F^{(2)},$$

so that, under the condition (IX-68), one has the relation

$$\frac{E({}^1\mathrm{S}) - E({}^1\mathrm{D})}{E({}^1\mathrm{D}) - E({}^3\mathrm{P})} = \frac{3}{2}, \qquad \text{independent of } F^{(0)}, F^{(2)}.$$

The linear combinations of $\Psi_7$, $\Psi_8$, $\Psi_9$ for ${}^1\mathrm{S}$, ${}^1\mathrm{D}$ and ${}^3\mathrm{P}$ are:

$$\Psi({}^1\mathrm{S}) = \frac{1}{\sqrt{3}}(\Psi_7 - \Psi_8 + \Psi_9),$$

$$= \frac{1}{\sqrt{3}}\{\phi_1(r_1)\phi_{-1}(r_2) + \phi_1(r_2)\phi_{-1}(r_1) + \phi_0(r_1)\phi_0(r_2)\}\chi_0^a,$$

---

[f] Theorem 11 of Chap. 2, Sec. 2.

$$\Psi(^1D) = \frac{1}{\sqrt{6}}(\Psi_7 - \Psi_8 - 2\Psi_9),$$

$$= \frac{1}{\sqrt{6}}\{\phi_1(r_1)\phi_{-1}(r_2) + \phi_1(r_1)\phi_{-1}(r_2) - 2\phi_0(r_1)\phi_0(r_2)\}\chi_0^a,$$

(IX-98)

$$\Psi(^3P) = \frac{1}{\sqrt{2}}(\Psi_7 + \Psi_8),$$

$$= \frac{1}{\sqrt{2}}\{\phi_1(r_1)\phi_{-1}(r_2) - \phi_1(r_2)\phi_{-1}(r_1)\}\chi_0^s,$$

where the subscript in $\phi$ is $m_l$.

It is seen that for $^3P$, the wave function is symmetric (antisymmetric) in the spin $(r_1, r_2)$ of the two electrons.

### (iii) np n'p

In an entirely similar but slightly lengthier study, one finds that the configuration $np\, n'p$ gives rise to $^3D$, $^1D$, $^3P$, $^1P$, $^3S$, $^1S$ states—$6 \times 6 = 36$ states.

### (iv) $np^3$

For three equivalent p electrons, the Pauli principle allows $6!/3!3! = 20$ states. They can be shown to be $^2P$, $^2D$, $^4S$. Their energies are

$$E(^2P) = 3E_0(np) + 3F^{(0)}(np, np),$$

$$E(^2D) = 3E_0(np) + 3F^{(0)}(np, np) - \tfrac{6}{25}F^{(2)}(np, np),$$

(IX-99)

$$E(^4S) = 3E_0(np) + 3F^{(0)}(np, np) - \tfrac{15}{25}F^{(2)}(np, np).$$

The wave function of $^4S$ is totally symmetric in the spins of the three electrons, but for $^2D$, the wave function

$$\begin{vmatrix} \phi_1(1)\chi^+(1) & \phi_1(1)\chi^-(1) & \phi_0(1)\chi^+(1) \\ \phi_1(2)\chi^+(2) & \phi_1(2)\chi^-(2) & \phi_0(2)\chi^+(2) \\ \phi_1(3)\chi^+(3) & \phi_1(3)\chi^-(3) & \phi_0(3)\chi^+(3) \end{vmatrix}$$

(IX-100)

does not have definite symmetry in the spins of the three electrons. A similar result holds for $^2D$.

### (2) Spin-orbit interactions; multiplet structure

After one has found the $^{2S+1}L$ states of an electron configuration, one next seeks the energies due to the spin-orbit interactions $H_{s.o.}(i)$ in (IX-65). Now $H$ in (IX-65)

$$H = \sum_{i=1}^{2} H_0(i) + \frac{e^2}{r_{12}} + \sum_{i=1}^{2} H_{s.o.}(i)$$

commutes only with $\mathbf{J}^2$ and $J_z$, but does not commute with $\mathbf{L}^2$, $\mathbf{S}^2$, $L_z$, $S_z$ of (IX-72). For this reason $H$ is not diagonal in the quantum numbers $L$, $S$, $M_L$, $M_s$, but is diagonal only in $J$ and $M$ of (IX-73), i.e., in general, the nonvanishing matrix elements of $\sum_i H_{s.o.}(i)$ are

$$\left\langle n, n'; S, L, J, M, \left| \sum_i H_{s.o.}(i) \right| n'', n'''; S', L', J, M \right\rangle. \qquad \text{(IX-101)}$$

The non-diagonal elements in $n$, $n'$, $S$, $L$ contribute only in the second order and will be neglected here. The diagonal elements are

$$\left\langle \gamma, J, M \left| \sum_i H_{s.o.}(i) \right| \gamma, J, M \right\rangle, \qquad \text{(IX-101a)}$$

where $\gamma$ denotes $n$, $n'$, $S$, $L$ and other quantum numbers. In the $(J, M)$-representation, $(\mathbf{L} \cdot \mathbf{S})$ is diagonal in $J$, $M$, and[g]

$$\langle S, L, J, M | \mathbf{L} \cdot \mathbf{S} | S, L, J, M \rangle = \tfrac{1}{2}\{J(J+1) - L(L+1) - S(S+1)\}. \tag{IX-102}$$

We now define a parameter $\zeta(\gamma, S, L)$ by

$$\left\langle \gamma, J, M \left| \sum_i H_{s.o.}(i) \right| \gamma, J, M \right\rangle = \zeta(\gamma, S, L) \langle S, L, J, M | \mathbf{L} \cdot \mathbf{S} | S, L, J, M \rangle$$

$$= \zeta(J, S, L)\frac{1}{2}\{J(J+1) - L(L+1) - S(S+1)\}. \tag{IX-103}$$

Now in (IX-90), the representation used is the $(M_L, M_s)$-representation. Hence we need the matrix elements of $\sum_i H_{s.o.}(i)$ in the $(M_L, M_s)$-representation. One performs a unitary transformation and

$$\left\langle \gamma, S, L, M_s, M_L \left| \sum_i H_{s.o.}(i) \right| \gamma, S, L, M_s, M_L \right\rangle$$

$$= \sum_{J,M} \langle S, L, M_s, M_L | U^{-1} | S, L, J, M \rangle \left\langle \gamma, J, M \left| \sum_i H_{s.o.}(i) \right| \gamma, J, M \right\rangle$$

$$\times \langle S, L, J, M | U | S, L, M_s, M_L \rangle$$

$$= \zeta(\gamma, S, L) \langle S, L, M_s, M_L | \mathbf{L} \cdot \mathbf{S} | S, L, M_s, M_L \rangle$$

$$= \zeta(\gamma, S, L) M_L M_s.\text{[h]} \tag{IX-104}$$

---

[g] See Chap. 8, Sec. 1, (1).
[h] See (VIII-14c).

The next step is the evaluation of the parameter $\zeta(\gamma, S, L)$ introduced in (IX-103).

On going back to the $(n, l, m_l, m_s)$-representation in Table (IX-89) as an example, it is seen that for $M_L = 1$, $M_s = 0$, there are two states $^3P$ and $^1D_1$, i.e., for a given $(M_L, M_s)$, there may be more than one $(L, S)$ states. Now one wants the diagonal elements of $\sum_i H_{\text{s.o.}}(i)$ in this $(m_l, m_s)$-representation,

$$\left\langle m_l, m_s \left| \sum_i H_{\text{s.o.}}(i) \right| m_l, m_s \right\rangle = \sum_{i=1}^{2} m_{l_i} m_{s_i} \xi_{n_i l_i}, \qquad \text{(IX-105)}$$

where $\xi_{n_1 l_1}$ is given by (VIII-25a). (See (VIII-14c)).

On calculates the trace of (IX-105), for a given $(M_L, M_s)$, over the various $(L, S)$ states.[i] By the theorem of the trace,[j] the trace is independent of the representation, so that this quantity is equal to the trace of (IX-104) for a given $(M_L, M_s)$ and various $(L, S)$ states.

To make the above method clearer, let us illustrate it by an example. Take the $np\,n'p$ configuration and construct a part of the table similar to (IX-89).

|          | $m_l$ | $m_l'$ | $m_s$ | $m_s'$ | $M_L$ | $M_s$ | $M$ | State |
|----------|-------|--------|-------|--------|-------|-------|-----|-------|
| $\Psi_1$ | 1     | 1      | +     | +      | 2     | 1     | 3   | $^3D$ |
| $\Psi_2$ | 1     | 0      | +     | +      | 1     | 1     | 2   | $^3D, ^3P$ |
| $\Psi_3$ | 0     | 1      | +     | +      | 1     | 1     | 2   | $^3D, ^3P$ |

$$\text{(IX-106)}$$

For $\Psi_1$, $^3D$, one has, from (IX-104) and (IX-105),

$$M_L = 2, M_s = 1, \qquad m_l = 1, m_s = \tfrac{1}{2}, m_l' = 1, m_s' = \tfrac{1}{2},$$

$$2 \cdot 1\zeta(\gamma\,^3D) = \tfrac{1}{2}\xi_{np} + \tfrac{1}{2}\xi_{n'p}. \qquad \text{(IX-107)}$$

For $\Psi_2$, $\Psi_3$,

$$M_L = 1, M_s = 1, \qquad \begin{cases} m_l = 1, m_s = \tfrac{1}{2}, m_l' = 0, m_s' = \tfrac{1}{2}, \\ m_l = 0, m_s = \tfrac{1}{2}, m_l' = 1, m_s' = \tfrac{1}{2}, \end{cases}$$

$$\zeta(\gamma\,^3D) + \zeta(\gamma\,^3P) = \tfrac{1}{2}\xi_{np} + \tfrac{1}{2}\xi_{n'p}. \qquad \text{(IX-108)}$$

From (IX-107), (IX-108), one obtains

$$\zeta(\gamma\,^3D) = \zeta(\gamma\,^3P) = \tfrac{1}{4}\xi_{np} + \tfrac{1}{4}\xi_{n'p}. \qquad \text{(IX-109)}$$

From (IX-103), one obtains for the "fine-structure" due to spin-orbit interactions, the relation

---

[i] In the example above, for $(M_L, M_s) = (1, 0)$ in Table (IX-89), the trace is taken over the submatrix $\langle \Psi_3, \Psi_4 || \Psi_3, \Psi_4 \rangle$, i.e., over the $^3P$ and $^1D$ states.
[j] Theorem 11, Chap. 2, Sec. 2.

$$E_{\text{s.o.}}(J) \equiv \zeta(\gamma, S, L)\tfrac{1}{2}\{J(J+1) - L(L+1) - S(S+1)\}$$

and the successive intervals

$$\Delta E_{\text{s.o.}}(J) = E_{\text{s.o.}}(J) - E_{\text{s.o.}}(J-1)$$

$$= \zeta(\gamma, S, L)J, \tag{IX-110}$$

which is the empirical *interval rule* found by Landé.

Applying this relation to the $^3P_{0,1,2}$ and $^3D_{1,2,3}$, one finds

$$^3P_{0,1,2}: \quad \frac{E_{\text{s.o.}}(^3P_2) - E_{\text{s.o.}}(^3P_1)}{E_{\text{s.o.}}(^3P_1) - E_{\text{s.o.}}(^3P_0)} = \frac{2}{1}, \tag{IX-111a}$$

$$^3D_{1,2,3}: \quad \frac{E_{\text{s.o.}}(^3D_3) - E_{\text{s.o.}}(^3D_2)}{E_{\text{s.o.}}(^3D_2) - E_{\text{s.o.}}(^3D_1)} = \frac{3}{2}. \tag{IX-111b}$$

These relations are obtained for $\{L, S\}$-coupling, and deviations of the observed fine structure in a spectrum from these relations will indicate deviations from strict $\{L, S\}$-coupling.[k]

## 4. Two-electron atom: $\{j, j\}$-coupling

If in the Hamiltonian (IX-65), the relation (IX-68) is replaced by

$$H_{\text{s.o.}}(i) \gg \frac{e^2}{r_{12}}, \tag{IX-112}$$

a situation approximately achieved for large values of $Z$, (see (VIII-25)), one starts with

$$H_0 = -\frac{\hbar^2}{2\mu}(\nabla_1^2 + \nabla_2^2) - Ze^2\left(\frac{1}{r_1} + \frac{1}{r_2}\right) + \sum_i^2 H_{\text{s.o.}}(i) \tag{IX-113}$$

as the zeroth-order Hamiltonian and treats

$$V = \frac{e^2}{r_{12}} \tag{IX-114}$$

as a perturbation. The spin-orbit interaction

$$H_{\text{s.o.}}(i) = 2\mu_B^2 \frac{Z}{r_i}(\mathbf{l}_i \cdot \mathbf{s}_i) \tag{IX-115}$$

couples the orbital and spin angular momentum into a resultant

$$\mathbf{j}_1 = \mathbf{l}_1 + \mathbf{s}_i \tag{IX-116}$$

---

[k] Deviations from (IX-111a, b) may be due to the so-called *configuration interactions* (second order perturbations from non-diagonal elements of (IX-101)).

354



and the $\mathbf{j}_i$ of the two electrons then form a total angular momentum

$$\mathbf{J} = \mathbf{j}_1 + \mathbf{j}_2. \qquad \text{(IX-117)}$$

This coupling is called $\{j,j\}$-coupling.

The Hamiltonian $H_0$ of (IX-113) commutes only with $\mathbf{J}^2$ and $J_z$, but not with $\mathbf{L}^2$, $\mathbf{S}^2$, $L_z$, $S_z$, so that only $H_0$, $\mathbf{J}^2$, $J_z$ are simultaneously diagonal. Thus in the $\{j,j\}$-coupling $L$ and $S$ do not have exact meaning and the $^{2S+1}L$ spectroscopic nomenclature, such as $^3P$, $^1D$, $^1S$, also do not have exact meaning. Thus the $np^2$ configuration has, in the case of $\{L,S\}$-coupling, the $^3P_{0,1,2}$, $^1D_2$, $^1S_0$ states; but in the case of $\{j,j\}$-coupling, one may only say, strictly speaking, that there are two states with $J = 2$: one state with $J = 1$ and two states with $J = 0$.

Similarly, for the $npn'p$ configuration, the states in the two coupling schemes are as follows.

$$
\begin{array}{cc}
\{L,S\} & \{j,j\} \\[4pt]
^3D_{1,2,3} \quad ^3P_{0,1,2} \quad ^3S_1 & J = 3(1) \\[4pt]
^1D_2, \quad\quad ^1P_1, \quad\quad ^1S_0 & J = 2(3) \qquad \text{(IX-118)} \\[4pt]
& J = 1(4) \\[4pt]
& J = 0(2)
\end{array}
$$

The number in parentheses indicates the number of times the state $J$ appears; thus in either coupling, levels with $J = 2$ appear 3 times, etc. The total number of states for $np\,n'p$ is 36 in either coupling. The following table gives the $(j, m)$, $(j', m')$ values for the $np\,n'p$ configuration. Those states allowed by the Pauli principle for $np^2$ are marked by an asterisk $*$.

| $M$ | $j, j' = 3/2, 3/2$ | $(m, m')$<br>$3/2, \tfrac12$ | $\tfrac12, 3/2$ | $\tfrac12, \tfrac12$ | |
|---|---|---|---|---|---|
| 3 | $(3/2, 3/2)$ | | | | |
| 2 | $(3/2, \tfrac12)^*,\ (\tfrac12, 3/2)$ | $(3/2, \tfrac12)^*$ | $(\tfrac12, 3/2)$ | | |
| 1 | $(\tfrac12, \tfrac12)$<br>$(3/2, -\tfrac12)^*,\ (-\tfrac12, 3/2)$ | $(\tfrac12, \tfrac12)^*$<br>$(3/2, -\tfrac12)^*$ | $(\tfrac12, \tfrac12)$<br>$(-\tfrac12, 3/2)$ | $(\tfrac12, \tfrac12)$ | |
| 0 | $(3/2, -3/2)^*,\ (\tfrac12, -\tfrac12)^*$<br>$(-3/2, 3/2),\ (-\tfrac12, \tfrac12)$ | $(\tfrac12, -\tfrac12)^*$<br>$(-\tfrac12, \tfrac12)^*$ | $(-\tfrac12, \tfrac12)$<br>$(\tfrac12, -\tfrac12)$ | $(\tfrac12, -\tfrac12)^*$<br>$(-\tfrac12, \tfrac12)$ | (IX-119) |
| $-1$ | $(-3/2, \tfrac12),\ (\tfrac12, -3/2)^*$<br>$(-\tfrac12, -\tfrac12)$ | $(-3/2, \tfrac12)^*$<br>$(-\tfrac12, -\tfrac12)^*$ | $(\tfrac12, -3/2)$<br>$(-\tfrac12, -\tfrac12)$ | $(-\tfrac12, -\tfrac12)$ | |
| $-2$ | $(-3/2, -\tfrac12)^*,\ (-\tfrac12, -3/2)$ | $(-3/2, \tfrac12)^*$ | $(-\tfrac12, -3/2)$ | | |
| $-3$ | $(-3/2, -3/2)$ | | | | |
| $J =$ | $3, 2^*, 1, 0^*$ | $2^*, 1^*$ | $2, 1$ | $1, 0^*$ | |

To obtain the eigenvalues of $H_0$ of (IX-113), the appropriate representation is the $(n, l, j, m)$-representation. Now

$$
\begin{aligned}
E_{\text{s.o.}} &= \sum_i \langle n_i, l_i, j_i, m_i | H_{\text{s.o.}}(i) | n_i, l_i, j_i, m_i \rangle \\
&= \sum_i \xi_{n_i l_i} \langle n_i, l_i, j_i, m_i | (\mathbf{l}_i \cdot \mathbf{s}_i) | n_i, l_i, j_i, m_i \rangle \\
&= \sum_i \frac{1}{2} \{ j_i(j_i + 1) - l_i(l_i + 1) - s_i(s_i + 1) \} \xi_{n_i l_i}.
\end{aligned}
\tag{IX-120}
$$

This is independent of the quantum number $J$, so that the spin-orbit energy is the same for different $J$ values in (IX-118).

The following table gives the values of $E_{\text{s.o.}}$ for various $j, j'$ values for two non-equivalent p electrons $np$, $n'\text{p}$, and for $np^2$ in which case the Pauli principle excludes certain states.

| $j$ | $j'$ | $np\ n'\text{p}$ | | | $np^2$ | | |
|---|---|---|---|---|---|---|---|
| | | $J$ | degeneracy | $E_{\text{s.o.}}$ | $J$ | degeneracy | $E_{\text{s.o.}}$ |
| $\frac{3}{2}$ | $\frac{3}{2}$ | 3, 2, 1, 0 | (16) | $\frac{1}{2}\xi_{np} + \frac{1}{2}\xi_{n'p}$ | 2, 0 | (6) | $\xi_{np}$ |
| $\frac{3}{2}$ | $\frac{1}{2}$ | 2, 1 | (8) | $\frac{1}{2}\xi_{np} - \xi_{n'p}$ | 2, 1 | (8) | $-\frac{1}{2}\xi_{np}$ |
| $\frac{1}{2}$ | $\frac{3}{2}$ | 2, 1 | (8) | $-\xi_{np} + \frac{1}{2}\xi_{n'p}$ | | | |
| $\frac{1}{2}$ | $\frac{1}{2}$ | 1, 0 | (4) | $-\xi_{np} - \xi_{n'p}$ | 0 | 1 | $-2\xi_{np}$ |

$$\tag{IX-121}$$

The energy of $np\ n'\text{p}$ is then

$$
E = E^0(np) + E^0(n'\text{p}) + E_{\text{s.o.}},
\tag{IX-122}
$$

where $E^0(np)$, $E^0(n'\text{p})$ are the one-electron energies.

The next step is to calculate the Coulomb interaction $e^2/r_{12}$ as a perturbation, in the $(n, l, j, m)$-representation, which has just been used in evaluating $H_{\text{s.o.}}$ in (IX-120). Let us use the abbreviation

$$
|k\rangle \text{ for } \quad |n, l, j, m\rangle \quad (\phi_{n,l,j,m}(\mathbf{r}))
$$

and

$$
|t\rangle \text{ for } \quad |n', l', j', m'\rangle \quad (\phi_{n',l',j',m'}(\mathbf{r})).
$$

$$\tag{IX-123}$$

Then an antisymmetrized function (including the spin) is

$$
\Psi(1, 2) = \frac{1}{\sqrt{2}} \begin{vmatrix} \phi_{nljm}(1) & \phi_{n'l'j'm'}(1) \\ \phi_{nljm}(2) & \phi_{n'l'j'm'}(2) \end{vmatrix}
\tag{IX-124}
$$

and

$$
\left\langle \Psi(1,2) \left| \frac{e^2}{r_{12}} \right| \Psi(1,2) \right\rangle = \left\langle k, t \left| \frac{e^2}{r_{12}} \right| k, t \right\rangle, \ -\left\langle k, t \left| \frac{e^2}{r_{12}} \right| t, k \right\rangle,
\tag{IX-125}
$$

$$
\equiv T(n, l, j, m; n', l', j', m').
\tag{IX-125a}
$$

The $|n, l, j, m\rangle$ can be transformed into the $|n, l, m_l, m_s\rangle$ representation by means of (VIII-31a, b),

for $j = l + \frac{1}{2}$,

$$|l, j, m\rangle = \sqrt{\frac{l + m + \frac{1}{2}}{2l + 1}}\,|l, m - \tfrac{1}{2}, \tfrac{1}{2}\rangle - \sqrt{\frac{l - m + \frac{1}{2}}{2l + 1}}\,|l, m + \tfrac{1}{2}, -\tfrac{1}{2}\rangle,$$

for $j = l - \frac{1}{2}$, $\qquad\qquad\qquad\qquad\qquad\qquad\qquad\qquad\qquad$ (IX-126)

$$|l, j, m\rangle = \sqrt{\frac{l - m + \frac{1}{2}}{2l + 1}}\,|l, m - \tfrac{1}{2}, \tfrac{1}{2}\rangle + \sqrt{\frac{l + m + \frac{1}{2}}{2l + 1}}\,|l, m + \tfrac{1}{2}, -\tfrac{1}{2}\rangle$$

and the $T(n, l, j, m; n', l', j', m')$ can then be expressed in terms of the $F^{(k)}(nl; n'l')$, $G^{(k)}(nl; n'l')$ integrals of (IX-32a, b).

The following table gives the $T(nljm; n'l'j'm')$ for a few configurations $ns\,n's$, $ns\,n'p$, $np\,n'p$ as examples.

| $l$ | $l'$ | $j$ | $m$ | $j'$ | $m'$ | $F^{(0)}$ | $-G^0$ | $-\frac{1}{9}G^{(1)}$ | $-\frac{1}{25}G^{(2)}$ |
|---|---|---|---|---|---|---|---|---|---|
| s | s | $\frac{1}{2}$ | $\pm\frac{1}{2}$ | $\frac{1}{2}$ | $\pm\frac{1}{2}$ | 1 | 1 | | |
|   |   |   |   |   | $\mp\frac{1}{2}$ | 1 | 0 | | |
| s | p | $\frac{1}{2}$ | $\pm\frac{1}{2}$ | $\frac{3}{2}$ | $\pm\frac{3}{2}$ | 1 | | 3 | |
|   |   |   |   |   | $\pm\frac{1}{2}$ | 1 | | 2 | |
|   |   |   |   |   | $\mp\frac{3}{2}$ | 1 | | 0 | |
|   |   |   |   |   | $\mp\frac{1}{2}$ | 1 | | 1 | |
|   |   |   |   | $\frac{1}{2}$ | $\pm\frac{1}{2}$ | 1 | | 1 | |
|   |   |   |   |   | $\mp\frac{1}{2}$ | 1 | | 2 | |

| $l$ | $l'$ | $j$ | $m$ | $j'$ | $m'$ | $F^{(0)}$ | $\frac{1}{25}F^{(2)}$ | $-G^{(0)}$ | $-\frac{1}{25}G^{(2)}$ |
|---|---|---|---|---|---|---|---|---|---|
| p | p | $\frac{3}{2}$ | $\pm\frac{3}{2}$ | $\frac{3}{2}$ | $\pm\frac{3}{2}$ | 1 | 1 | 1 | 1 |
|   |   |   |   |   | $+\frac{1}{2}$ | 1 | $-1$ | 0 | 2 |
|   |   |   | $\pm\frac{1}{2}$ |   | $\pm\frac{1}{2}$ | 1 | 1 | 1 | 1 |
|   |   |   | $\pm\frac{3}{2}$ |   | $\mp\frac{3}{2}$ | 1 | 1 | 0 | 0 |
|   |   |   |   |   | $\mp\frac{1}{2}$ | 1 | $-1$ | 0 | 2 |
|   |   |   | $\pm\frac{1}{2}$ |   | $\mp\frac{1}{2}$ | 1 | 1 | 0 | 0 |
|   |   | $\frac{3}{2}$ | $\pm\frac{3}{2}$ | $\frac{1}{2}$ | $\pm\frac{1}{2}$ | 1 | | | 1 |
|   |   |   | $\pm\frac{1}{2}$ |   | $\pm\frac{1}{2}$ | 1 | | | 2 |
|   |   |   | $\pm\frac{3}{2}$ |   | $\mp\frac{1}{2}$ | 1 | | | 4 |
|   |   |   | $\pm\frac{1}{2}$ |   | $\mp\frac{1}{2}$ | 1 | | | 3 |
|   |   | $\frac{1}{2}$ | $\pm\frac{1}{2}$ | $\frac{1}{2}$ | $\pm\frac{1}{2}$ | 1 | | 1 | |
|   |   |   | $\pm\frac{1}{2}$ |   | $\mp\frac{1}{2}$ | 1 | | 0 | |

$$\text{(IX-127a)}$$

To obtain the matrix element

$$\left\langle J \left| \frac{e^2}{r_{12}} \right| J \right\rangle,$$

one constructs the matrix of $e^2/r_{12}$ with respect to the wave functions in tables such as (IX-119), and refer to tables such as (IX-127). The calculation is straightforward. The following table gives as examples the values of $\langle J|(e^2/r_{12})|J\rangle$ for states of different $J$ for $np\,n'p$.

| $j$ | $j'$ | $J$ | $\left\langle J\left\vert\dfrac{e^2}{r_{12}}\right\vert J\right\rangle$ |
|---|---|---|---|
| $\tfrac{3}{2}$ | $\tfrac{3}{2}$ | 3 | $F^{(0)} + \tfrac{1}{25}F^{(2)} - G^{(0)} - \tfrac{1}{25}G^{(2)}$ |
|  |  | 2 | $F^{(0)} - \tfrac{1}{2}F^{(2)} + G^{(0)} - \tfrac{3}{25}G^{(2)}$ |
|  |  | 1 | $F^{(0)} + \tfrac{1}{25}F^{(2)} - G^{(0)} - \tfrac{1}{25}G^{(2)}$ |
|  |  | 0 | $F^{(0)} + \tfrac{5}{25}F^{(2)} + G^{(0)} + \tfrac{5}{25}G^{(2)}$ |
| $\tfrac{3}{2}$ | $\tfrac{1}{2}$ | 2 | $F^{(0)} \qquad\qquad\qquad\quad -\tfrac{1}{25}G^{(2)}$ |
|  |  | 1 | $F^{(0)} \qquad\qquad\qquad\quad -\tfrac{5}{25}G^{(2)}$ |
| $\tfrac{1}{2}$ | $\tfrac{3}{2}$ | 2 | $F^{(0)} \qquad\qquad\qquad\quad -\tfrac{1}{25}G^{(2)}$ |
|  |  | 1 | $F^{(0)} \qquad\qquad\qquad\quad -\tfrac{5}{25}G^{(2)}$ |
| $\tfrac{1}{2}$ | $\tfrac{1}{2}$ | 1 | $F^{(0)} \qquad\quad - G^{(0)}$ |
|  |  | 0 | $F^{(0)} \qquad F \; + G^{(0)}$ |

$$(\text{IX-127b})$$

For further details, the reader is referred to Condon and Shortley's *Theory of Atomic Spectra*.

## 5.  Two-electron atom: intermediate coupling

If in the Hamiltonian (IX-65), $e^2/r_{12}$ and $\sum_i H_{\text{s.o.}}(i)$ are of the same magnitude, one may start with the $(S, L, J, M)$-representation in which

$$H_0 = \sum_{i=1}^{2} H_0(i) + \frac{e^2}{r_{12}}$$

is diagonal (although degenerate in $J$ and $M$), and it is necessary only to calculate the non-diagonal elements of[l] $\sum_i H_{\text{s.o.}}(i)$,

$$\left\langle \gamma JM \left\vert \sum_i H_{\text{s.o.}}(i) \right\vert \gamma J'M' \right\rangle. \tag{IX-128}$$

To calculate these, it is convenient to start with the $(m_l, m_s)$-representation[m]

$$\langle m_{l_i}, m_{s_i}|H_{\text{s.o.}}(i)|m_{l_i}, m_{s_i}\rangle = m_{l_i} m_{s_i}\zeta_{n_i l_i},$$
$$\langle m_{l_i}, m_{s_i}|H_{\text{s.o.}}(i)|m'_{l_i}, m'_{s_i}\rangle = \langle m_{l_i}, m_{s_i}|\mathbf{l}_i\cdot\mathbf{s}_i|m'_{l_i}, m'_{s_i}\rangle\zeta_{n_i l_i} \tag{IX-129}$$

and from this representation to transform to the $|\gamma JM\rangle$-representation

---

[l] $\gamma \equiv n, n', S, L$ and other quantum numbers. See (IX-101a).
[m] For the $(m_l, m_s)$-representation, see (IX-105), (VIII-14a, b, c).

$$\left\langle \gamma JM \left| \sum_i H_{\text{s.o.}}(i) \right| \gamma J'M' \right\rangle$$

$$= \sum \langle \gamma JM | U^{-1} | m_l m_s \rangle \left\langle m_l m_s \left| \sum_i H_{\text{s.o.}}(i) \right| m_l' m_s' \right\rangle \langle m_l' m_s' | U | \gamma J'M' \rangle,$$

$$(\text{IX-130})$$

where $m_l$, $m_s$, $m_l'$, $m_s'$ stand for $m_{l_i}$, $m_{s_i}$, $m_{l_i}'$, $m_{s_i}'$, and the summation $\sum$ is over all $m_l$, $m_s$, $m_l'$, $m_s'$.

The calculation is straightforward but lengthy in general. To illustrate the theory, take the simplest configuration $ns\,n'l$ whose states, in the limiting case of $\{L,S\}$-coupling, are $^1L$, $^3L$, but have the exact quantum numbers $J = l + 1$, $l, l, l - 1$. In the $|\gamma JM\rangle$- or $|SLJM\rangle$-representation, the matrix of $\sum_i H_{\text{s.o.}}(i)$ is

| $J \backslash J'$ | $l+1$ | $l$ | $l$ | $l-1$ |
|---|---|---|---|---|
| $l+1$ | $l$ | | | |
| $l$ | | $-\dfrac{1}{\sqrt{l(l+1)}}$ | $\sqrt{l(l+1)}$ | |
| $l$ | | $\sqrt{l(l+1)}$ | $0$ | |
| $l-1$ | | | | $-(l+1)$ |

$$\times\ \tfrac{1}{2}\xi_{n,l}. \qquad (\text{IX-131})$$

The matrix of $e^2/r_{12}$ in the same representation is given by an extension of (IX-86, 87) for $ns\,n'p$ $^1P$, $^3P$ to the present $ns\,nl$,

| | $^3L$ $\;l+1$ | $^3L$ $\;l$ | $^1L$ $\;l$ | $^3L$ $\;l-1$ |
|---|---|---|---|---|
| $^3L \quad l+1$ | $F^{(0)} - \dfrac{1}{2l+1} G^{(l)}$ | | | |
| $^3L \quad l$ | | $F^{(0)} - \dfrac{1}{2l+1} G^{(l)}$ | | |
| $^1L \quad l$ | | | $F^{(0)} + \dfrac{1}{2l+1} G^{(l)}$ | |
| $^3L \quad l-1$ | | | | $F^{(0)} - \dfrac{1}{2l+1} G^{(l)}$ |

$$(\text{IX-132})$$

where

$$F^{(0)} = F^{(0)}\left( ns,n'l \left| \frac{e^2}{r_{12}} \right| ns,n'l \right), \qquad G^{(l)} = G^{(l)}\left( ns,n'l \left| \frac{e^2}{r_{12}} \right| n'l,ns \right).$$

The eigenvalues of $(e^2/r_{12}) + \sum_i H_{\text{s.o.}}(i)$ are (using the approximate $L$, $S$ nomenclature)

$$^3L_{l+1}: F^{(0)} - \frac{1}{2l+1} G^{(l)} + \frac{1}{2} l\xi_{n,l},$$

$$\begin{matrix} ^3L_l \\ ^1L_l \end{matrix}: F^{(0)} - \frac{1}{4}\xi_{n,l} \mp \left[ \left( \frac{1}{2l+1} G^{(l)} + \frac{1}{4}\xi_{n,l} \right)^2 + \frac{1}{4} l(l+1)\xi_{n,l}^2 \right]^{1/2}, \qquad \text{(IX-133)}$$

$$^3L_{l-1}: F^{(0)} - \frac{1}{2l+1} G^{(l)} - \frac{1}{2}(l+1)\xi_{n,l}.$$

The wave functions of the two states $\begin{Bmatrix} ^3L_l \\ ^1L_l \end{Bmatrix}$ are

$$\begin{matrix} \Psi_1(J=l) \\ \Psi_2(J=l) \end{matrix} \Bigr\} = \left\{ \begin{matrix} \frac{1}{2}\sqrt{l(l+1)}\xi_{n,l} \\[2mm] \frac{1}{2l+1} G^{(l)} + \frac{1}{4}\xi_{n,l} \mp \sqrt{\phantom{xxxx}} \end{matrix} \right\} \Psi(^3L) + \Psi(^1L), \quad \text{(IX-134)}$$

where $\Psi(^3L)$, $\Psi(^1L)$ are the exact $\{L,S\}$ wave functions, and the radicand stands for the last term of (IX-133). In the limit $\xi_{n,l} = 0$, $\Psi_1$, $\Psi_2$ are these $\{L,S\}$-wave functions.

For other examples, such as $np^3$ $^4S$, $^2P$, $^2D$ (the $J$ values are $\frac{5}{2}, \frac{3}{2}, \frac{3}{2}, \frac{3}{2}, \frac{1}{2}$), the reader is referred to Condon and Shortley's *Theory of Atomic Spectra*.

## 6. Configuration interactions

In Sec. 3, (1), in the calculation of the energies of the various states $^3P$, $^1D$, $^1S$ of the $np^3$ configuration, the matrix elements of $e^2/r_{12}$ of other configurations, such as

$$\left\langle np, np \left| \frac{e^2}{r_{12}} \right| np, n'p \right\rangle,$$

$$\left\langle np, np \left| \frac{e^2}{r_{12}} \right| n'l', n''l'' \right\rangle \qquad \text{(IX-135)}$$

have been entirely left out of consideration. Such matrix elements joining states of different configurations play a role in the second-order perturbation calculation; they are said to represent "configuration interactions". They are significant when the matrix elements have large values or when the energy differences between the two configurations involved are small.

The concept of configuration interaction is a general one, not restricted to atomic systems. It is there whenever one starts with an approximate representation for the Hamiltonian of a system, for example, in a molecule, or in the independent-nucleon model of an atomic nucleus.

In an atomic system, the conditions for the presence of configuration interaction between states A and B from two configurations, arising from $e^2/r_{12}$, are:

(i)  A and B have the same parity,

(ii)  A and B have the same total angular momentum $J$,

(iii)  in the case of $\{L, S\}$-coupling, A and B have the same $S$, and the same $L$.

Configuration interaction may arise from the spin-orbit interaction

$$\sum_i H_{\text{s.o.}}(i) = \sum_i (\mathbf{l}_i \cdot \mathbf{s}_i)\xi_{n_i l_i}(r_i),$$

which couples states with different $J$ and $M$ as well as different[n] $n$, $n'$.

As notable examples of configuration interactions in atomic systems, we shall discuss the related phenomena of autoionization in atomic spectra and Auger effect in X-rays.


## *(1)  Autoionization*

For clarity and definiteness, let us take the two-electron atoms $H^-$, $He$, $Li^+$, $Be^{++}$, ... We shall use the nomenclature which, strictly speaking, is only accurate in the central-field, independent-electron, $\{L, S\}$-coupling approxi-

---

[n] One example of configuration interaction through $H_{\text{s.o.}}$ is Fermi's explanation of the deviation from the value 2 of the ratio of the two components of the principal doublets s $^2S_{1/2}$—$np\ ^2P_{1/2, 3/2}$ of Cs atom. From Chap. 8, Sec. 1, (2), it is known $(\mathbf{l} \cdot \mathbf{s})$ is diagonal in $j$, $l$, $s$, $m$ and from (VIII-33), one obtains

$$\begin{aligned}
\langle n\,^2P_{3/2}|H_{\text{s.o.}}|n'\,^2P_{3/2}\rangle &= \tfrac{1}{2}\xi_{np, n'p}, \\
\langle n\,^2P_{1/2}|H_{\text{s.o.}}|n'\,^2P_{1/2}\rangle &= -\xi_{np, n'p},
\end{aligned} \tag{IX-136}$$

where

$$\xi_{np, n'p} = 2\mu_B^2 Z \int \Psi_{np}^*(r)\frac{1}{r^3}\Psi_{n'p}(r)\,dr. \tag{IX-136a}$$

The wave functions of $np\ ^2P_{1/2, 3/2}$ are then

$$\psi(np\,^2P_{1/2}) = \psi^0(np\,^2P_{3/2}) + \frac{1}{2}\sum_{n'} \frac{\xi_{np, n'p}}{E_n^0 - E_{n'}^0}\psi^0(n'p\,^2P_{3/2}),$$

$$\psi(np\,^2P_{1/2}) = \psi^0(np\,^2P_{1/2}) - \sum_{n'} \frac{\xi_{np, n'p}}{E_n^0 - E_{n'}^0}\psi^0(n'p\,^2P_{1/2}). \tag{IX-137}$$

The ratio of the intensities of $^2S_{1/2}$—$n\ ^2P_{3/2}$ and $^2S_{1/2}$—$n\ ^2P_{1/2}$ is the ratio of the squares of the dipole moments

$$2\frac{\left|\langle s|r|np\rangle + \dfrac{1}{2}\sum_{n'}\dfrac{\xi_{np, np'}}{E_n^0 - E_{n'}^0}\langle s|r|n'p\rangle\right|^2}{\left|\langle s|r|np\rangle - \sum_{n'}\dfrac{\xi_{np, np'}}{E_n^0 - E_{n'}^0}\langle s|r|n'p\rangle\right|^2}, \tag{IX-138}$$

which is different from 2. See E. Fermi, *Z. Phys.* **59** 680 (1929).

mation. Thus the energy spectrum consists of configurations $nln'l'$, as shown in the following.

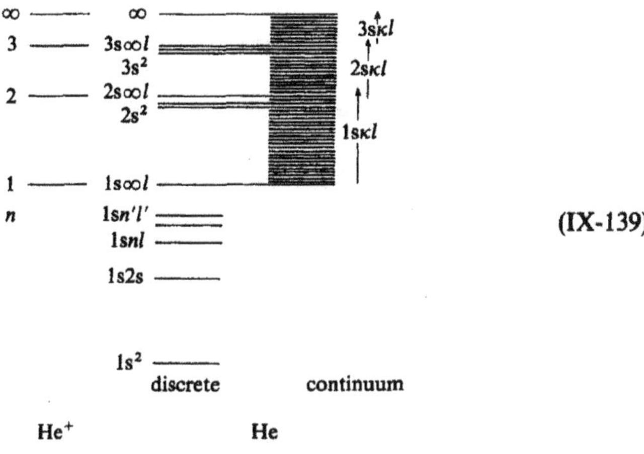



Energy levels of He$^+$ and He.

$$ \text{(IX-139)} $$

Starting from the ground state $1s^2\,{}^1S$, there are the infinitely many states described by $1s\,nl\,{}^3L$, ${}^1L$, $n \geq 2$, $l = 0, 1, 2, \ldots$ The limit is $1s\infty 1 = 1s\,{}^2S$, which is the ground state of He$^+$. Above $1s\,{}^2S$ and below $2s\,{}^2S$ ($2p\,{}^2P$), there is the continuum describable as $1s\,\kappa l$, $l = 0, 1, 2, \ldots$, being in the continuum, with energy

$$ E(1s\,\kappa l) = E^0(1s\,{}^2S) + \frac{\kappa^2\hbar^2}{2m}, \qquad \text{(IX-140)} $$

with one electron in $1s$ state of He$^+$ and a second electron with positive energy $\kappa^2\hbar^2/2m$ in the field of the He$^+$($1s$) ion. The continuum extends to all positive energies, but at and above He$^+$($2s$) (which is degenerate with He$^+$($2p$)), there is superposed a continuum with energy

$$ E(2s\,\kappa l) = E^0(2s\,{}^2S) + \frac{\kappa^2\hbar^2}{2m}, \qquad \text{(IX-141)} $$

etc.

Now between He$^+$($1s\,{}^2S$) and He$^+$($2s\,{}^2S$), there are states with configurations such as $2s^2\,{}^1S$, $2s\,np\,{}^3P$, ${}^1P$, $2p^2\,{}^3P$, ${}^1D$, ${}^1S$; $2s\,nd\,{}^3D$, ${}^1D$, $\ldots$, which approach, as $n$ increases, the limit He$^+$($3s\,{}^2S$) (or He$^+$($3p$), or He$^+$($3d$)). These energies of these doubly excited states $2s^2$, $2s\,nl$, $2p^2$, etc., can be estimated by some variational methods (such as the modified Ritz method of Chap. 8, Sec. 2, (2).); they are all imbedded in the continuum, i.e., a state such as $2s^2\,{}^1S$ is degenerate with a singlet $1s\,\kappa s\,{}^1S$ state of the continuum. For a radiationless transition from $2s^2\,{}^1S$ to $1s\,\kappa s\,{}^1S$, in which one of the $2s$ electrons drops to the 1s state of He$^+$ and the other 2s electron leaves the atom with a positive kinetic energy $\kappa^2\hbar^2/2m$ given by the energy relation

$$E(2s^2\,{}^1S) = E(1s\,{}^2S) + \frac{\hbar^2\kappa^2}{2m}, \qquad \text{(IX-142)}$$

the transition probability per unit time is given by (VII-40)

$$P = \frac{2\pi}{\hbar}\left|\left\langle 2s^2\,{}^1S\left|\frac{e^2}{r_{12}}\right|1s\,\kappa s\,{}^1S\right\rangle\right|^2 \rho(E_\kappa). \qquad \text{(IX-143)}$$

This transition determines the lifetime $\tau$ of the $1s^2\,{}^1S$ state (VII-163)

$$\tau \simeq \frac{1}{P}$$

and this lifetime $\tau$ causes a width $\Delta E$ of the level $1s^2\,{}^1S$ according to the uncertainty principle,

$$\Delta E \simeq \frac{h}{2\pi\tau}$$

$$= \frac{h}{2\pi}P. \qquad \text{(IX-144)}$$

Calculations based on this theory lead to P of the order $10^{13} - 10^{14}$/sec. for the doubly excited states such as $2s^2\,{}^1S$, $2s2p\,{}^3P$, ${}^1P$, ..., and a width of the order

$$\Delta E \simeq 0.006 - 0.06\ \text{eV}. \qquad \text{(IX-145)}$$

There is very scanty available experimental information on these doubly excited states in helium. Only two spectral lines at $\lambda = 320.4$ and $\lambda = 357.5$ (in Ångströms) have been observed in the extreme ultraviolet. The former has been tentatively identified as the transition

$$1s\,2p\,{}^3P - 2p^2\,{}^3P \qquad \text{(IX-146a)}$$

on the basis of the calculated position of $2p^2\,{}^3P$. If this identification is correct, it furnishes no check on the theory of autoionization above, as the state $2p^2$ ${}^3P$ is not subject to autoionization because of the absence of a state $1s\,\kappa l$ in the continuum that is of even parity, with $L = 1$. The $\lambda = 357.5$ line may be tentatively identified as

$$1s\,4s\,{}^3S - 2s\,2p\,{}^3P,$$
$$1s\,3d\,{}^3D - 2s\,2p\,{}^3P, \qquad \text{(IX-146b)}$$

but the line $\lambda = 357.5$ does not show such a width as $\Delta E$ in (IX-145).

The existence of states lying above the ionization limit in an atom is known in many atoms (such as alkaline earths Hg, Cu, Be, Zn, Cd and others). Such states are ascribed to electron configurations with two electrons excited— states having energies greater than necessary to (singly) ionize the atom. Their identification is made on the classification of observed spectral lines (similar to (IX-146a, b) above), sometimes on the basis of their widths as evidence of autoionization. In a general way, the theory of the observed spectral features is satisfactory.

The problem of autoionization, however, is still of interest from the theoretical side, on the following considerations. Let us take the helium atom. The electron-nucleus and the electron-electron interactions are of the same order, so that treating the latter $e^2/r_{12}$ as a "perturbation" of a central field, independent-electron system as in (IX-143) is not satisfactory. The method of Hylleraas (Sec. 2, (4) above) shows clearly the importance of the electron-electron correlation effect as represented by the nonseparability of the wave functions into products of one-electron wave functions. Strictly speaking, the problem of autoionization is the problem of the nature of the spectrum of eigenvalues and eigenfunctions of the Hamiltonian

$$H = -\frac{\hbar^2}{2\mu}(\nabla_1^2 + \nabla_2^2) - Ze^2\left(\frac{1}{r_1} + \frac{1}{r_2}\right) + \frac{e^2}{r_{12}}, \qquad \text{(IX-147)}$$

specifically, the nature of the spectrum of eigenfunctions at energies corresponding to the "doubly excited states" based on the independent-electron representation.

As an alternative to the perturbation treatment (IX-143), many authors treat the autoionization problem by the method of the scattering theory. The problem is formulated as follows: An electron of energy $E_0 = (\hbar^2 k^2/2m)$ is incident on a helium ion in the ground state $\text{He}^+$ (is $^2S$). As long as $E_0$ is not high enough for the system $\text{He}^+(1s\,^2S)$ + electron to reach the lowest of the "doubly-excited states" such as $2s^2\,^1S$, $2p^2\,^3P$, $2s\,2p\,^3P$, the scattering is elastic. But as $E_0 = (\hbar^2 k^2/2m)$ is such that

$$E(\text{He}^+\,1s\,^2S) + \frac{\hbar^2 k^2}{2m} = E(\text{He}\,2s^2\,^1S), \qquad \text{(IX-148)}$$

there is a change in the cross-section $\sigma$ in the scattering at this energy. Hence the problem is to formulate the scattering theory for "resonances" at certain values of the energy $\hbar^2 k^2/2m$ of the incident electron.

This problem, however, also entails some specification of the states at which resonances take place, and if the states are taken to be $2s^2$, $2p^2$, $2s\,2p$, etc., then one essentially is doing something not basically different from the perturbation theory approach. It seems that a satisfactory treatment of the autoionization in a system represented by the Hamiltonian $H$ in (IX-147) free from the central-field, independent-electron picture is still wanting.[o]

## (2) Auger effect

The Auger effect in X-rays can be briefly described as the following process. A K-electron has been ejected from an atom, say, by electron impact. An electron in the L shell (2p) drops into the "hole". Instead of emitting radiation 1s $^2$S $-$ 2p $^2$P, this energy is transferred to another electron, say, another L or an M electron ejecting it out of the atom as an "Auger electron". This radiationless transition is thus similar to autoionization for the optical electrons, and the usual theory is the perturbation theory represented by (IX-143), which in the present case becomes, say,

$$P = \frac{2\pi}{\hbar}\left|\left\langle 2s\,2p\left|\frac{e^2}{r_{12}}\right|1s\,\kappa p\right\rangle\right|^2 \rho(E_{\kappa p}),\qquad\text{(IX-149)}$$

where $\kappa p$ is an ejected (Auger) electron in a p state. In the case of X-ray electrons, the electron-electron interaction $e^2/r_{12}$ is "small" compared with the electron-nucleus interaction $(Z - \sigma)(e^2/r)$, where $\sigma\ (\simeq 0.3)$ the screening is small compared with $Z$, so that central-field, independent-electron approximation is a good one. Theoretical calculations have been carried out, some with relativistic approximations, and the relative intensities of different Auger transitions seem in good agreement with observed data. (See references at the end of this chapter.)

## 7.  Many-electron atoms

For the energy of a many-electron atom, Hylleraas' method of using the electron-electron distances $r_{ij}$ as coordinates in nonseparable wave functions $\Psi(r_1, r_2, \ldots, r_N)$ is too lengthy to be practicable. The simplest method is the modified Ritz method, using antisymmetrized product wave functions in the determinental form, with variational parameters. In the same category is the Hartree-Fock method which in the earlier years was limited by the tedious numerical calculations involved but is nowadays made practicable by computers. The basic theories of these methods have already been given in the preceding sections for the two-electron atoms. We shall give a brief discussion of a few topics in the present section.

[o] A treatment of the autoionization problem has been proposed (K. T. Chung, *Phys. Rev.* **A6**, 1809 (1972)) in which all states containing one electron in a 1s state are removed by means of projection operators in order to leave a Hamiltonian having only doubly excited states as eigenstates. The theory, however, is only formal; on account of the nonseparability of $\Psi(r_1, r_2)$ (the ground state 1s$^2$ $^1$S for example), it is not possible to remove all the states whose usual approximate notations are "1s $nl$" by means of a projection operator on a 1s electron.

### (1) Slater's method ($\{L, S\}$-coupling)

Let the individual electron wave functions be

$$\psi_a(i) = \phi_{nlm_l}(r_i)\chi_{m_s}(i), \qquad i = 1, 2, \ldots, N. \tag{IX-150}$$

Thus $a$ stands for the quantum numbers $n_a$, $l_a$, $m_{la}$, $m_{sa}$. The wave function of the $N$ electron system is approximated by the antisymmetrized combination of products

$$\Psi(1, 2, \ldots, N) = \sum (-1)^P \psi_a(1)\psi_b(2)\ldots\psi_n(N), \tag{IX-151}$$

where $(-1)^P$ stands for the sign $+(-)$ for an even(odd) number of interchanges of two electrons. This is equivalent to the determinent

$$\Psi(1, 2, \ldots N) = \frac{1}{N!}\begin{vmatrix} \psi_a(1) & \psi_b(1) & \ldots & \psi_n(1) \\ \psi_a(2) & \psi_b(2) & \ldots & \psi_n(2) \\ \ldots & \ldots & \ldots & \ldots \\ \ldots & \ldots & \ldots & \ldots \\ \psi_a(N) & \psi_b(N) & \ldots & \psi_n(N) \end{vmatrix}. \tag{IX-152}$$

It is seen that this $\Psi$ is antisymmetric with respect to the interchange of any two electrons, and that $\Psi$ vanishes whenever two electrons have the same set of quantum numbers, say, $a = b$. Thus the requirement of antisymmetry is equivalent to Pauli's exclusion principle discovered before quantum mechanics.

### (i) Energies of an atom

Let the Hamiltonian be

$$H = \sum_{i=1}^{N} H_0(i) + \sum_{1 \leq i < j}^{N} \frac{e^2}{r_{ij}}, \tag{IX-153}$$

$$H_0(i) = -\frac{\hbar^2}{2\mu}\nabla_i^2 - Ze^2\frac{1}{r_i} \tag{IX-153a}$$

and let

$$E = \int \ldots \int \Psi^*(1, 2, \ldots, N) H \Psi(1, 2, \ldots, N)\, d\tau_1 \ldots d\tau_N. \tag{IX-154}$$

The single-electron part $\sum H_0(i)$ can be calculated very simply. If the $\psi_a$, $\psi_b$, $\ldots$ in (IX-152) are orthonormal and

$$(H_0(i) - E_a^0)\psi_a = (H_0(i) - E_a^0)\phi_a(r)\chi_{m_s}$$

$$= 0, \tag{IX-155}$$

and

$$\int \psi_a^*(i)\psi_b(i)\,d\tau_i = \delta_{ab}, \tag{IX-156}$$

then

$$\int \cdots \int \Psi^* \left( \sum_i H_0(i) \right) \Psi \, d\tau_1 \ldots d\tau_N = E_a^0 + E_b^0 + \cdots + E_n^0. \tag{IX-157}$$

The $e^2/r_{ij}$ part of (IX-154) can be expressed in terms of the $F^{(k)}(nl;n'l')$, $G^{(k)}(nl;n'l')$ integrals of (IX-32a, b). Take a typical term in $e^2/r_{ij}$ in (IX-154), such as

$$\int \cdots \int \psi_a^*(1)\psi_b^*(2)\psi_p^*(3)\ldots\psi_n^*(N)\frac{e^2}{r_{12}}\psi_c(1)\psi_d(2)\psi_p(3)\ldots\psi_n(N)\,d\tau_1 \ldots d\tau_N$$

$$= \left\langle a,b \left| \frac{e^2}{r_{12}} \right| c,d \right\rangle. \tag{IX-158}$$

The spin part of the integral $\langle a,b|(e^2/r_{12})|c,d\rangle$: If we denote by $m_s^a, m_s^b, m_s^c, m_s^d$ the $m_s$ values of $\chi$ in $\psi_a, \psi_b, \psi_c, \psi_d$ then $\langle a,b|(e^2/r_{12})|c,d\rangle$ is seen to have the factor

$$\delta_{m_s^a,m_s^c}\delta_{m_s^b,m_s^d} = \begin{cases} 1, & m_s^a = m_s^c, \ m_s^b = m_s^d, \\ 0, & \text{otherwise.} \end{cases} \tag{IX-159}$$

On expanding $1/r_{12}$ as in (IX-28), from the integrals over $\phi_1$ and $\phi_2$, one obtains the relations for $\langle a,b|(1/r_{12})|c,d\rangle$ not to vanish,

$$m_l = m_l^a - m_l^c = m_l^d - m_l^b, \tag{IX-160a}$$

or

$$m_l^a + m_l^b = m_l^c + m_l^d. \tag{IX-160}$$

On combining (IX-159) and (IX-160), it is seen that $e^2/r_{12}$ is diagonal in

$$M_s = \sum_{i=1}^{N} m_{s_i}, \qquad M_L = \sum_{i=1}^{N} m_{l_i}, \qquad M = M_s + M_L. \tag{IX-161}$$

The integrations over $\vartheta_1, \vartheta_2$ proceed as in (IX-29)–(IX-35).

### (ii) Energy of a closed shell

As an example, let us find the energy of the six p electrons in a closed $p^6$ subshell. The 6 electrons have all different $(m_l, m_s)$ quantum numbers according to the Pauli principle. The $\psi_a, \psi_b, \ldots, \psi_f$ wave functions are as follows.

$$\begin{array}{c|cccccc} & a & b & c & d & e & f \\ \hline m & 1 & 0 & -1 & 1 & 0 & -1 \\ m_s & \tfrac{1}{2} & \tfrac{1}{2} & \tfrac{1}{2} & -\tfrac{1}{2} & -\tfrac{1}{2} & -\tfrac{1}{2} \end{array} \qquad (\text{IX-162})$$

One obtains for the nonvanishing matrix elements:

$$\left\langle a,b \left| \frac{e^2}{r_{12}} \right| a,b \right\rangle = \left\langle a,e \left| \frac{e^2}{r_{12}} \right| a,e \right\rangle = \left\langle b,c \left| \frac{e^2}{r_{12}} \right| b,c \right\rangle$$

$$= \left\langle b,d \left| \frac{e^2}{r_{12}} \right| b,d \right\rangle = \left\langle b,f \left| \frac{e^2}{r_{12}} \right| b,f \right\rangle = \left\langle e,f \left| \frac{e^2}{r_{12}} \right| e,f \right\rangle$$

$$= \left\langle c,f \left| \frac{e^2}{r_{12}} \right| c,f \right\rangle = \left\langle d,c \left| \frac{e^2}{r_{12}} \right| d,c \right\rangle = F^{(0)} - \frac{2}{25}F^{(2)}, \qquad (\text{IX-163a})$$

$$\left\langle a,c \left| \frac{e^2}{r_{12}} \right| a,c \right\rangle = \left\langle a,d \left| \frac{e^2}{r_{12}} \right| a,d \right\rangle = \left\langle a,f \left| \frac{e^2}{r_{12}} \right| a,f \right\rangle$$

$$= \left\langle c,d \left| \frac{e^2}{r_{12}} \right| c,d \right\rangle = \left\langle c,f \left| \frac{e^2}{r_{12}} \right| c,f \right\rangle = \left\langle d,f \left| \frac{e^2}{r_{12}} \right| d,f \right\rangle$$

$$= F^{(0)} + \frac{1}{25}F^{(2)}, \qquad (\text{IX-163b})$$

$$\left\langle b,e \left| \frac{e^2}{r_{12}} \right| b,e \right\rangle = F^{(0)} + \frac{4}{25}F^{(2)}, \qquad (\text{IX-163c})$$

$$\left\langle a,b \left| \frac{e^2}{r_{12}} \right| b,a \right\rangle = \left\langle b,c \left| \frac{e^2}{r_{12}} \right| c,b \right\rangle = \left\langle d,e \left| \frac{e^2}{r_{12}} \right| e,d \right\rangle$$

$$= \left\langle e,f \left| \frac{e^2}{r_{12}} \right| f,e \right\rangle = \frac{3}{25}G^{(2)}, \quad (G^{(2)} = F^{(2)}) \qquad (\text{IX-163d})$$

$$\left\langle a,c \left| \frac{e^2}{r_{12}} \right| c,a \right\rangle = \left\langle d,f \left| \frac{e^2}{r_{12}} \right| f,d \right\rangle = \frac{6}{25}G^{(2)}. \qquad (\text{IX-163e})$$

The energy of $np^6\ {}^1S$ is the sum

$$E(np^6\ {}^1S) = 6E^0(np) + 15F^{(0)}(np; np) - \tfrac{18}{25}F^{(2)}(np, np). \qquad (\text{IX-164})$$

*(iii)  Interaction between a (valence) electron with a closed shell*

Let the quantum numbers of the (valence) electron be $n'l'm_l'm_s'$, and those of the electrons in a closed shell be $n,\ l,\ m_l,\ m_s$.

The interaction energy between $(n',\ l',\ m_l',\ m_s')$ with all the $2(2l+1)$ electrons in the closed shell is

$$J - K = \sum_{m_l=-l}^{l} \sum_{m_s=-1/2}^{1/2} \left\{ \left\langle nlm_lm_s; n'l'm_l'm_s' \left| \frac{e^2}{r_{ij}} \right| nlm_lm_s; n'l'm_l'm_s' \right\rangle \right.$$

$$\left. - \left\langle nlm_lm_s; n'l'm_l'm_s' \left| \frac{e^2}{r_{ij}} \right| n'l'm_l'm_s'; nlm_lm_s \right\rangle \right\}. \qquad \text{(IX-165)}$$

The "direct" integral $J$ and the "exchange" integral $K$ can be shown to be[p]

[p] One expands

$$\frac{1}{r_{ij}} = \sum_{k=0}^{\infty} \frac{r_<^k}{r_>^{k+1}} P_k(\cos \omega), \qquad \text{(IX-167)}$$

where $\omega$ is the angle between $\mathbf{r}_i$ and $\mathbf{r}_j$. The angular part of the integral $J$ contains

$$\sum_{m=-l}^{l} [\Theta_{lm_l}(\vartheta_i)]^2 \Phi_{m_l}^*(\phi_i) \Phi_{m_l}(\phi_i) P_k(\cos \omega) [\Theta_{l'm_l'}(\vartheta_j)]^2 \Phi_{m_l'}^*(\phi_j)$$

$$\times \Phi_{m_l'}(\phi_j) \, d\cos\vartheta_i \, d\cos\vartheta_j \, d\phi_i \, d\phi_j.$$

By means of the relation (III-114)

$$\sum_{m=-l}^{l} [\Theta_{lm_l}(\vartheta_i)]^2 \Phi_{m_l}^*(\phi_i) \Phi_{m_l}(\phi_i) = \frac{2l+1}{4\pi}, \qquad \text{(IX-168)}$$

and choosing $\mathbf{r}_j$ as the polar axis for $\vartheta_i$, $\phi_i$ so that $P_k(\cos\omega)$ becomes $P_k(\cos\vartheta_i)$, one obtains for the angular part of $J$ above

$$\frac{2l+1}{4\pi} [\Theta_{l'm_l'}(\vartheta_j)]^2 \sum_{k=0}^{\infty} \frac{r_<^k}{r_>^{k+1}} P_k(\cos\vartheta_i) \cdot \frac{1}{2\pi} d\cos\vartheta_i \, d\cos\vartheta_j \, d\phi_i \, d\phi_j.$$

On integrating over $\vartheta_i$, $\vartheta_j$, $\phi_i$, $\phi_j$ it is seen that only the term $k = 0$ does not vanish. One gets

$$J = 2(2l+1) \iint [R_{n,l}(r_i)]^2 \frac{1}{r_>} [R_{n'l'}(r_j)]^2 r_i^2 \, dr_i r_j^2 \, dr_j$$

$$= 2(2l+1) F^{(0)}(nl; n'l'),$$

the factor 2 coming from the sum $\sum m_s$.

The "exchange" integral $K$ can be similarly evaluated. The angular part is, from (IX-167),

$$\sum_{m=-l}^{l} \Theta_{lm_l}(\vartheta_i) \Theta_{l'm_l'}(\vartheta_i) \Phi_{m_l}^*(\phi_i) \Phi_{m_l'}(\phi_i) P_k(\cos\omega) \Theta_{lm_l}(\vartheta_j) \Theta_{l'm_l'}(\vartheta_j)$$

$$\times \Phi_{m_l}(\phi_j) \Phi_{m_l'}^*(\phi_j).$$

By means of the addition theorem

$$\sum_{m=-l}^{l} \Theta_{lm}(i) \Theta_{lm}(j) \Phi_{m_l}^*(i) \Phi_{m_l}(j) = \frac{2l+1}{4\pi} P_l(\cos\omega),$$

this becomes

$$\frac{2l+1}{4\pi} \Theta_{l'm_l'}(\vartheta_i) \Theta_{l'm_l'}(\vartheta_j) \Phi_{m_l'}(\phi_i) \Phi_{m_l'}^*(\phi_j) P_k(\cos\omega) P_l(\cos\omega). \qquad \text{(IX-169)}$$

One expands $P_k(\cos\omega) P_l(\cos\omega)$ in terms of the complete set $P_l(\cos\omega)$,

$$P_l(x) P_k(x) = \sum_{\lambda=0}^{\infty} \frac{2\lambda+1}{2} C_{\lambda lk} P_\lambda(x), \qquad \text{(IX-170)}$$

where

$$C_{\lambda\mu\nu} = \int_{-1}^{1} P_\lambda(x) P_\mu(x) P_\nu(x) \, dx. \qquad \text{(IX-171)}$$

$$J = 2(2l + 1)F^{(0)}(nl; n'l') \qquad (n'l' \text{ for a valence electron}),$$

$$K = (2l + 1) \sum_{k=0} C_{l'lk} G^{(k)}(nl; n'l').  \qquad \text{(IX-166)}$$

It is important to note that the interaction of an electron $(n', l')$ with all the electrons in a closed $(n, l)$-shell

$$J - K = 2(2l + 1)\left[ F^{(0)}(nl; n'l') - \frac{1}{4}\sum_k C_{ll'k} G^{(k)}(nl; n'l') \right]  \qquad \text{(IX-172)}$$

is independent of the $m_l'$, $m_s'$ of the electron $(n', l')$, i.e., to the electron $(n', l')$, the closed shells (and the nucleus) form a centrally symmetric field. This result is of general importance, and in particular justifies an assumption made in the Hartree-Fock method that the closed shells act as a centrally symmetric field.

*(iv) Interactions between electrons in two closed shells $(n, l)$ and $(n', l')$*

From (IX-172), it immediately follows that the total interaction energy between the electrons of two closed $(n, l)$, $(n', l')$ shells is

$$J - K = 4(2l + 1)(2l' + 1)\left[ F^{(0)}(nl; n'l') - \frac{1}{4}\sum_k C_{ll'k} G^{(k)}(nl; n'l') \right].  \qquad \text{(IX-173)}$$

*(v) Interactions among electrons in a closed $(n, l)$ shell*

This can be shown to be

$$J - K = 2(2l + 1)^2\left[ F^{(0)}(nl; nl) - \frac{1}{4}\sum_{k=0} C_{llk} F^{(k)}(nl; nl) \right],  \qquad \text{(IX-174)}$$

---

The angular part (IX-169) becomes

$$\frac{2l + 1}{4\pi} \cdot \frac{2\lambda + 1}{2} C_{\lambda lk} \Theta_{l'm_i'}(i)\Phi_{m_i'}(i) P_\lambda(\cos \omega)\Theta_{l'm_i'}(j)\Phi_{m_i'}^*(j).$$

Expanding $P_\lambda(\cos \omega)$ again, this becomes

$$\sum_{m=-\lambda}^{\lambda} \frac{2l + 1}{4\pi} C_{\lambda lk} \Theta_{l'm_i'}(i)\Theta_{\lambda m}(i)\Phi_{m_i'}(i)\Phi_m^*(i)\Theta_{l'm_i'}(j)\Theta_{\lambda m}(j)\Phi_{m_i'}^*(j)\Phi_m(j);$$

when integrating over $\vartheta_j$, $\phi_j$ (or $\vartheta_i$, $\phi_i$), only the term

$$m = m_l', \qquad \lambda = l'$$

does not vanish. Hence

$$K = \frac{2l + 1}{2} \sum_{k=0}^{\infty} C_{l'lk} \iint R_{nl}(i)R_{n'l'}(i)\frac{r_<^k}{r_>^{k+1}} R_{nl}(j)R_{n'l'}(j)r_i^2\, dr_i r_j^2\, dr_j$$

$$= \frac{2l + 1}{2} \sum_{k=0}^{\infty} C_{l'lk} G^{(k)}(n'l'; n, l).$$

which is in agreement with (IX-164) obtained for the $p^6$ shell from direct calculations.

### (2) Hartree-Fock method

The principle of the method has been described in Chap. 9, Sec. 2, (3). For an $N$-electron atom, one starts with (IX-50a, b) in which $\Psi$ is the determinental wave function (IX-152), and carries out the variational calculation by making independent $\delta\phi_a$, $\delta\phi_b$, $\delta\phi_c$, ... The general procedure is sufficiently clear, and we shall illustrate it by two examples.

### (i) The closed shell $1s^2\,2s^2\,2p^6\,{}^1S$ of Ne, Na$^+$, Mg$^{++}$

For the configuration $1s^2\,2s^2\,2p^6$, let the radial wave functions be denoted by

$$1s: \frac{1}{r}R_1(r); \qquad 2s: \frac{1}{r}R_2(r); \qquad 2p: \frac{1}{r}R_3(r), \qquad \text{(IX-175)}$$

We shall use the atomic units: lengths in units of the Bohr radius $a = \hbar^2/me^2$, energies in units of the ionization potential of hydrogen $e^2/2a$. The differential-integral equations are

$$\frac{d^2R_1}{dr^2} + \left[ E_1 + \frac{2Z}{r_1} - V(r) + F_0^{11}(r) \right] R_1$$
$$= -F_0^{21}(r)R_2 - 3F_1^{33}(r)R_3, \qquad \text{(IX-176)}$$

$$\frac{d^2R_2}{dr^2} + \left[ E_2 + \frac{2Z}{r} - V(r) + F_0^{22}(r) \right] R_2$$
$$= -F_0^{21}(r)R_1 - 3F_1^{32}(r)R_3, \qquad \text{(IX-177)}$$

$$\frac{d^2R_3}{dr^2} + \left[ E_3 + \frac{2Z}{r} - \frac{2}{r^2} - V(r) + F_0^{33}(r) + 2F_2^{33}(r) \right] R_3(r)$$
$$= -F_0^{31}(r)R_1 - F_1^{31}(r)R_2, \qquad \text{(IX-178)}$$

where

$$V(r) = 2F_0^{11}(r) + 2F_0^{22}(r) + 6F_0^{33}(r) \qquad \text{(IX-179)}$$

and

$$F_k^{ij}(r) = \frac{2}{2k+1}\left\{ \frac{1}{r^{k+1}} \int_0^r \rho^k R_i(\rho)R_j(\rho)\,d\rho + r^k \int_r^\infty \frac{1}{\rho^{k+1}} R_i(\rho)R_j(\rho)\,d\rho \right\},$$
$$\text{(IX-180)}$$

$F_k^{ij}(r)$ satisfying the following equation

$$\frac{d}{dr}\left(r^2\frac{dF_k^{ij}}{dr}\right) - k(k+1)F_k^{ij} = -2R_iR_j(r). \qquad \text{(IX-181)}$$

The meaning of the various terms in these equation is rather clear. $V(r)$ of (IX-179) is the potential energy of a "test" electron due to the $1s^2\,2s^2\,2p^6$ electrons, so that in (IX-176),

$$V(r) - F_0^{11}(r) \qquad \text{(IX-182)}$$

is the potential energy of a 1s electron due to the repulsion by the $1s^2\,2s^2\,2p^6$ electrons. In (IX-178), similarly,

$$V(r) - F_0^{33}(r) - F_2^{33}(r) \qquad \text{(IX-183)}$$

is the potential energy of a 2p electron due to the repulsion by the other $1s^2\,2s^2\,2p^5$ electrons.[q]

The terms on the right-hand side in (IX-176)–(IX-178) are the exchange terms. Thus $F_0^{21}(r)R_2(r)$ in (IX-176) is the exchange energy between a 1s electron with *one* of the $2s^2$ electrons on account of the orthogonality of the spin wave functions; $3F_1^{33}(r)R_3(r)$ is the exchange energy between a 1s electron with *three* of the $2p^6$ electrons, etc.

On such considerations, it is possible to write down the equations (IX-176)–(IX-178) without having to actually go through the variational calculations.

*(ii)  A valence electron n, l outside $1s^2\,2s^2\,2p^6$ closed shell*

Consider Na, or Mg$^+$. Let the radial wave function of the valence electron be denoted by

$$n,\,l: \quad \frac{1}{r}R_4(r).$$

---

[q] Let us for a moment take the "no. 6" p electron to have the quantum numbers $m_l,\ m_s$. The interaction between this p electron with the other 5 p electrons depends on the value $(m_l, m_s)$. Calculation shows that

$$\begin{aligned}
m &= \pm 1, & J &= 5F_0^{33} - \tfrac{1}{5}F_2^{33}, & K &= \tfrac{9}{5}F_2^{33}, \\
m &= 0, & J &= 5F_0^{33} - \tfrac{4}{5}F_2^{33}, & K &= \tfrac{6}{5}F_2^{33}.
\end{aligned} \qquad \text{(IX-184)}$$

It is physically natural, and mathematically simpler, to average over the values $m = -1, 0, 1$ for the "number 6" p electron. Thus

$$\begin{aligned}
J - K &= 5F_0^{33} - \tfrac{2}{5}F_2^{33} - \tfrac{8}{5}F_2^{33} \\
&= 5F_0^{33} - 2F_0^{33} \qquad \text{(IX-185)}
\end{aligned}$$

and the average interaction energy of one p electron with the other $1s^2\,2s^2\,2p^5$ electrons is

$$2F_0^{11} + 2F_0^{22} + 5F_0^{33} - 2F_0^{33} = V(r) - F_0^{33} - 2F_0^{33}, \text{ which is (IX-183)}.$$

For simplicity, we shall make an approximation by treating the valence electron as moving in the field of the closed-shell electrons and neglect the effect of the valence electron on the closed-shell electrons themselves.

On this approximation, the closed-shell electrons are given by (IX-176)–(IX-178), and the equation for the valence electron $(n, l)$

$$\frac{d^2 R_4}{dr^2} + \left[ E_4 + \frac{2Z}{r} - \frac{l(l+1)}{r^2} - V(r) \right] R_4$$
$$= -F^{14}(r)R_1(r) - F^{24}(r)R_2(r) - 3\left[ \frac{l}{2l+1} F_{l-1}^{34} + \frac{l+1}{2l+1} F_{l+1}^{34} \right]. \quad \text{(IX-186)}$$

This can be simplied if the $(n, l)$ electron is an s electron (such as 3s). The right-hand side becomes

$$-F_0^{14}(r)R_1(r) - F_0^{24}(r)R_2(r) - 3F_1^{34}(r)R_3(r). \quad \text{(IX-187)}$$

As shown in subsection 7, (1), (iii) above, the field due to closed-shells is centrally symmetric so that for a valence electron, the quantum number $m_l$ of the valence electron does not appear in (IX-186).

The Hartree-Fock method of Secs. 7, 2(3) and 7(2) gives the best approximate energy of the ground state of an atom, subject to the limitation of the use of product single-electron wave functions, which do not take proper account of the electron-electron correlation effect. This neglect can in principle be remedied by using, instead of a single determinant wave function (IX-152), a sum of determinants corresponding to configurations having the same symmetry properties (with respect to rotation, spin) as the ground state. The use of a sum of a large number, not to say a "complete set set", of determinantal wave functions in the variational problem is however not very practicable, even with the help of computers.

*(3) Selection rules for many-electron atoms*

*(i) Parity*

For electric dipole transitions, the selection rule is:

$$\text{even} \leftrightarrow \text{odd}. \quad \text{(IX-188)}$$

For electric quadrupole and magnetic dipole transitions, the selection rule is

$$\text{even} \leftrightarrow \text{even},$$
$$\text{odd} \leftrightarrow \text{odd}. \quad \text{(IX-189)}$$

In the single-electron wave-function product approximation (the determinental $\Psi(r_1, \ldots, r_N)$ in (IX-152)), we have the properties (III-118)-(III-120) of the one-electron, central field wave functions, so that the

$$P\Psi(\mathbf{r}_1,\mathbf{r}_2,\ldots,\mathbf{r}_N) = P\phi_a(\mathbf{r}_1)\phi_b(\mathbf{r}_2)\ldots\phi_n(\mathbf{r}_N)$$
$$= (-1)^{l_a+l_b+\cdots l_n}\Psi(\mathbf{r}_1,\ldots,\mathbf{r}_N),$$

i.e.,

$$\Psi \text{ is even(odd) for } \sum_{i=1}^{N} l_i = \text{even(odd)}, \tag{IX-190}$$

and the selection for electric dipole transitions is

$$\Delta\left(\sum_{i=1}^{N} l\right) = \text{odd integer.} \tag{IX-191}$$

### (ii) $J, M$

The method of obtaining (VIII-37, 38) in Chap. 8, Sec. 2, can be extended. The results are

$$\Delta M = 0, \pm 1,$$
$$\Delta J = 0, \pm 1 \qquad (\text{excluding } 0 \leftrightarrow 0). \tag{IX-192}$$

### (iii) $L, S, M_L, M_s$

In the limiting case of $\{L, S\}$-coupling, the selection rules are

$$\Delta L = 0, \pm 1,$$
$$\Delta S = 0,$$
$$\Delta M_L = 0, \pm 1, \tag{IX-193}$$
$$\Delta M_s = 0.$$

### (4) Configuration interactions

The general theory of configuration interaction has been given in Sec. 6 (the preceding section). We shall briefly discuss two types of configuration interaction here.

### (i) Perturbations in series spectra-anomalous Rydberg corrections

Empirically, the term values of the various series (S, P, D, ...) are well represented by a Rydberg-Ritz formula

$$E_n = -\frac{Z^2 Rhc}{(n - \Delta_n)^2}, \qquad n = \text{integers}, \tag{IX-194}$$

where $\Delta_n$, the Rydberg corrections, also called the "quantum defects", usually show a small variation with $n$, i.e.,

$$\Delta_n = \mu - \alpha E_n. \tag{IX-195}$$

It is found empirically, however, that in many spectral series, the $\Delta_n$'s show, in the neighborhood of a certain $n$, an anomalous variation with $n$. Table (IX-196), giving the empirical term values and $\Delta_n$ for the 3s $nd$ $^1$D series of Mg is an example. The variation of $\Delta_n$ with $n$ is such that between $n = 7$ and $n = 8$, the behavior of $\Delta_n$ is like that of the index of refraction showing the anomalous dispersion near an absorption line in the spectrum.

| n | 3snd | | $^1$D | $\Delta$ |
|---|---|---|---|---|
| | $^3$D | $^1$D | | |
| 13 | 61095 | 60956 | | |
| 12 | 60685 | 60827 | 60956 | $-0.403$ |
| 11 | 60734 | 60658 | 60827 | $-0.412$ |
| 10 | 60534 | 60135 | 60658 | $-0.421$ |
| 9 | 60263 | 60127 | 00136 | $-0.430$ |
| 8 | 59880 | 59690 | 60427 | $-0.436$ |
| ($3p^2\,^1$D | | | 59690) | |
| 7 | 59317 | 59041 | 59041 | 0.538 |
| 6 | 58443 | 58023 | 58023 | 0.513 |
| 5 | 56968 | 56308 | 56308 | 0.475 |
| 4 | 54192 | 53135 | 53134 | 0.413 |
| 3 | 47957 | 46403 | 16403 | 0.319 |

$$\text{(IX-196)}$$

The theory of such anomalous Rydberg corrections is the configuration interaction. For the sake of definiteness, take the case of the 3s $nd$ $^1$D series Mg. Let $|k\rangle$ stand for the $2p^2$ $^1$D state of Mg, which lies between the $n = 7$ and $n = 8$ levels of the 3s $nd$ $^1$D series. The 2s $nd$ $^1$D and the $2p^2$ $^1$D states have the same parity and $J$ value and are in the position to have configuration interaction. According to perturbation theory, the energy $E_n$ of the state 2s $nd$ $^1$D near $2p^2$ $^1$D is then

$$E_n = -\frac{Z^2 Rhc}{(n - \Delta_n)^2} + \frac{\left|\left\langle n\left|\dfrac{e^2}{r_{12}}\right|k\right\rangle\right|^2}{E_n^0 - \varepsilon_n}, \tag{IX-197}$$

where $\varepsilon_n$ is the energy of the unperturbed (as distinct from the observed value 59690 cm$^{-1}$ in (IX-196)) $2p^2$ $^1$D state. Equation (IX-197) can be reexpressed in the form

$$E_n = -\frac{Z^2 Rhc}{(n - \Delta_n)^2}\left\{1 - \frac{(n - \Delta_n)^2}{Z^2 Rhc}\frac{\left|\left\langle n\left|\dfrac{e^2}{r_{12}}\right|k\right\rangle\right|^2}{E_n^0 - \varepsilon_n}\right\}$$

$$\cong \frac{-ZRhc}{\left[n - \Delta_n + \dfrac{\beta_{nk}}{E_n^0 - \Delta_n}\right]^2}, \tag{IX-198}$$

where

$$\beta_{nk} = \frac{\left|\left\langle n\left|\dfrac{e^2}{r_{12}}\right|k\right\rangle\right|^2}{2Z^2 Rhc}(n - \Delta_n)^2. \tag{IX-199}$$

From (IX-198), one obtains for the Rydberg correction $\Delta_n^*$ of the perturbed series

$$\Delta_n^* = \Delta_n - \frac{\beta_{nk}}{E_n^0 - \varepsilon_n}, \tag{IX-200}$$

which exhibits the anomalous behavior near $E_n^0 \simeq \varepsilon_n$.

There are numerous cases of such anomalous $\Delta_n$ in the spectra of many atoms. The reader is referred to a review in an article by the author.

### (ii) Inverted doublets of alkali metal atoms

In the "one-electron" configuration of alkali metal atoms, the doublet states $(^2P_{1/2,\,3/2}, {}^2D_{3/2,\,5/2}, {}^2F_{5/2,\,7/2})$ are normal in the sense that the level with lower $J$ lies below that of high $J$ value. From the theory of Chap. 8, (VIII-28), we have seen that the spin-orbit interactions lead to these doublets

$$E_{\text{s.o.}} = \left\{\begin{array}{c} \dfrac{l}{2} \\[2mm] -\dfrac{l+1}{2} \end{array}\right\}\xi_{n,l}, \qquad j = \left\{\begin{array}{c} l + \dfrac{1}{2} \\[2mm] l - \dfrac{1}{2} \end{array}\right\} \tag{IX-201}$$

so that

$$^2L_{l+(1/2)} - {}^2L_{l-(1/2)} = \frac{2l + 1}{2}\xi_{n,l} > 0. \tag{IX-202}$$

But in the $^2D$ of the K atom, certain $^2D$, $^2F$ of Rb, and the $^2F$ of Cs atoms, the doublets are inverted.

The explanation of this inversion is found in the configuration interaction between the configuration for the doublet $^2L_{l-(1/2),\,l+(1/2)}$ in question and another configuration involving the excitation of one p electron from the closed shell. As an example, take the $nd$ $^2D$ of the K atom. When a p electron is excited from the $3p^6$ closed shell to 4p, the resulting configuration $1s^2$ $2s^2$ $2p^6$ $3s^2$ $3p^5$ $4s$ $4p$ has a $^2D$ state, which by virtue of $3p^5$ having one "hole" in the closed $3p^6$ shell, is itself inverted. The "repulsion" between $nd$ $^2D_{5/2}$ and $5p^5$ $4s$ $4p$ $^2D_{5/2}$ pushes $nd$ $^2D_{5/2}$ downward by an amount greater than that of the downward shift of $nd$ $^2D_{3/2}$ due to the "repulsion" by $5p^5$ $4s$ $4p$ $^2D_{5/2,\,3/2}$. This is shown in the following figure.[r]

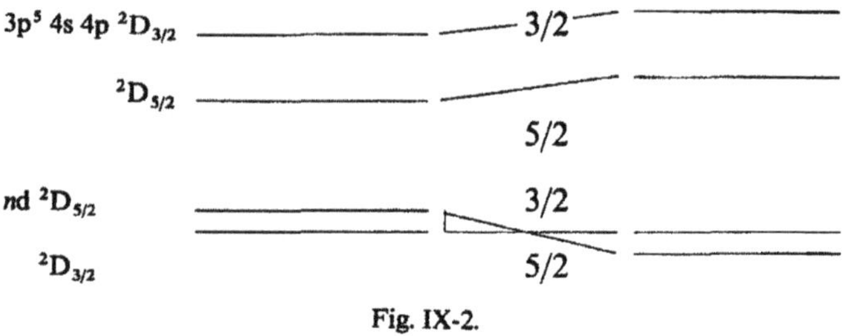


Fig. IX-2.

The theory, due to White, is plausible, but no theoretical calculation seems to have been carried out.

## 8. The Thomas-Fermi potential and many-electron atoms

### (1) The statistical potential

For many purposes it is desirable to have an approximate but conveniently simple potential for an electron in a neutral atom or a positive ion. For example, let us imagine the feeding of the electrons, one at a time, into the field of an initially bare nucleus, and consider the successive build-up of the electron configurations. Such considerations are useful in a qualitative understanding of the actual structure of the periodic table of chemical elements. For this and other problems, the statistical potential of L. H. Thomas (1927) and E. Fermi (1928) has proved very useful.

The basic idea in the Thomas-Fermi theory is that the electrons in an atom (especially, in the positive ion of a heavy atom) are in a degenerate state in the sense of Fermi-Dirac statistics. Consider a positive ion consisting of a nucleus of charge $Ze$ and $Z-z$ electrons, so that the ion has a positive net charge $ze$.

---

[r] For the "repulsion", see the formula (VI-11) for second order perturbation. A state $E_n$ is "pushed" upward by a "perturbing state" $E_l$ when $E_n > E_l$; similarly $E_l$ is "pushed" downward by $E_n$, when $E_n > E_l$.

The inverted doublet $3p^5$ $4s$ $4p$ $^2D$ has a large doublet separation $\Delta E_{s.o.}$ because 3p has larger value for the effective charge $Z_{eff}$ that determines $-\Delta E_{s.o.}$.

Let $U(r)$ be the potential at a distance $r$ from the nucleus. The energy of an electron must be negative, i.e.,

$$\frac{p^2}{2m} - eU = -ev, \tag{IX-203}$$

where $v$ is a positive constant to be determined. The electron number density, according to the statistical mechanics of Fermi and Dirac, is

$$n(r) \simeq \frac{8\pi}{3h^3}(2me)^{3/2}(U - v)^{3/2}. \tag{IX-204}$$

The assumption is made that $U$ and $n$ satisfy the Poisson equation

$$\nabla^2 U = -4\pi(-ne)$$
$$= C(U - v)^{3/2}, \tag{IX-205}$$

$$C = \frac{32\pi^2 e}{3h^3}(2me)^{3/2}. \tag{IX-206}$$

$U(r)$ satisfies the condition

$$\lim_{r\to 0} rU = Ze. \tag{IX-207}$$

We shall assume that the electron charge distribution $n(r)$ has a boundary at $r = r_0$ so that

$$n(r) = 0 \quad \text{for } r \geq r_0, \tag{IX-208}$$

$$4\pi \int_0^{r_0} n(r)r^2\, dr = Z - z, \tag{IX-209}$$

$$U(r) = \begin{cases} v & \text{at } r = r_0, \\ \dfrac{ze}{r}, & r_0 \leq r. \end{cases} \tag{IX-210}$$

The Poisson equation is then

$$\nabla^2 U = C\left(U - \frac{ze}{r_0}\right)^{3/2}, \quad r < r_0, \tag{IX-211}$$

$$\nabla^2 U = 0, \qquad\qquad r_0 < r.$$

Let us introduce a function $\phi(x) = \phi(\mu r)$ defined by

$$\frac{Ze}{r}\phi(x) = U(r) - \frac{ze}{r_0}, \tag{IX-212}$$

where

$$\mu = \left(\frac{128Z}{9\pi^2}\right)^{1/3} \frac{1}{a}, \qquad a = \frac{h^2}{4\pi^2 m e^2}. \tag{IX-213}$$

The Poisson equation then takes the form

$$\frac{d^2\phi}{dx^2} = \frac{\phi^{3/2}}{\sqrt{x}} \quad \text{for } x < x_0 = \mu r_0,$$

$$= 0 \qquad \text{for } x_0 < x. \tag{IX-214}$$

These equations are to be solved subject to the conditions (IX-207), (IX-210), (IX-209),

$$\phi(0) = 1, \tag{IX-207a}$$

$$\phi(x_0) = 0, \tag{IX-210a}$$

$$\int_0^{x_0} \phi^{3/2}\sqrt{x}\, dx = 1 - \frac{z}{Z}. \tag{IX-209a}$$

The last two equations lead to

$$-x_0\left(\frac{d\phi}{dx}\right)_{x_0} = \frac{z}{Z}. \tag{IX-209b}$$

This relation determines $x_0 = \mu r_0$ for a given $z/Z$.

The family of solutions of (IX-214) for various values of the parameter $z/Z$ can be used for ions of various degrees of ionization of any atom. In particular, for neutral atoms, $z = 0$, the boundary conditions are

$$\phi(0) = 1, \qquad \phi(\infty) = 0. \tag{IX-215}$$

The function $\phi(x)$ for neutral atoms has been obtained by various authors by numerical integration. See, for example, P. Gombás, *Die Statistische Theorie des Atoms und Ihre Anwendungen* (Vienna, 1949).

For positive ions, $\phi$ has been obtained for a few values of $z/Z$ by E. Fermi, *Mem. della reale Acad. d'Italia I*, (1930), and approximate solutions have been obtained by A. Sommerfeld, *Z. Phys.* **78**, 283 (1932).

## (2) Energy of an electron in an atom or ion

As an application of the statistical potential to atomic problems, let us consider the motion of an electron of quantum numbers $n, l$ in a neutral atom

$Z$. The electron $n, l$ is moving in the field of the nucleus and the other $Z - 1$ electrons, i.e.,

$$U(r) = \frac{Ze}{r}\phi(\mu r) + \frac{e}{r_0} \quad \text{for } r < r_0,$$

$$= \frac{e}{r} \qquad\qquad \text{for } r_0 < r, \tag{IX-216}$$

where $\phi(\mu r)$ is the function satisfying (IX-214) for $z = 1$. The potential energy of an electron is $V(r) = -eU(r)$.

Let $R(\rho)$ be the radial wave function and

$$\psi(\rho) = \rho R(\rho), \qquad \rho = \frac{r}{a}, \qquad a = \frac{\hbar^2}{me^2}. \tag{IX-217}$$

Let energy be expressed in units of rydbergs. The Schrödinger equation for $\psi(\rho)$ is

$$\frac{d^2\psi}{d\rho^2} + [E - \mathscr{W}(\rho)]\psi = 0, \tag{IX-218}$$

where the $\mathscr{W}(\rho)$ is the "effective" potential energy

$$\mathscr{W}(\rho) = V(\rho) + \frac{l(l + 1)}{\rho^2}, \tag{IX-219}$$

$$V(\rho) = -\frac{2Z}{\rho}\phi(\mu_0\rho) - \frac{2}{\rho_0}, \quad \rho < \rho_0,$$

$$= -\frac{2}{\rho}, \qquad\qquad \rho_0 < \rho.$$

$\phi(\mu_0\rho) = \phi(\mu r) = \phi(x)$ is given by the solution of (IX-214) subject to the conditons (IX-207a), (IX-210a), (IX-209b).

$\mathscr{W}(\rho)$ has the following properties: for $l \neq 0$, $\mathscr{W}(\rho)$ is positive for very small $\rho$ on account of the $l(l + 1)/\rho^2$ term; it becomes negative as $\rho$ increases by virtue of the negative $-(2Z/\rho)\phi(\mu_0\rho)$ term; $V(\rho)$ reaches a minimum and increases on account of the positive term $l(l + 1)/\rho^2$, and at large $\rho$, $\mathscr{W}(\rho)$ becomes Coulombic, i.e., $-2/\rho$.

For $l = 3$, i.e., for an $f$ electron, and for large $Z$, say $Z > 80$, as $\rho$ increases from very small values, $\mathscr{W}(\rho)$ first decreases to a deep minimum, rises to a maximum with $\mathscr{W}_{\max} > 0$, decreases again to negative values, reaches a flat minimum and behaves as $-2/\rho$ at large $\rho$. This is shown qualitatively in the following figure

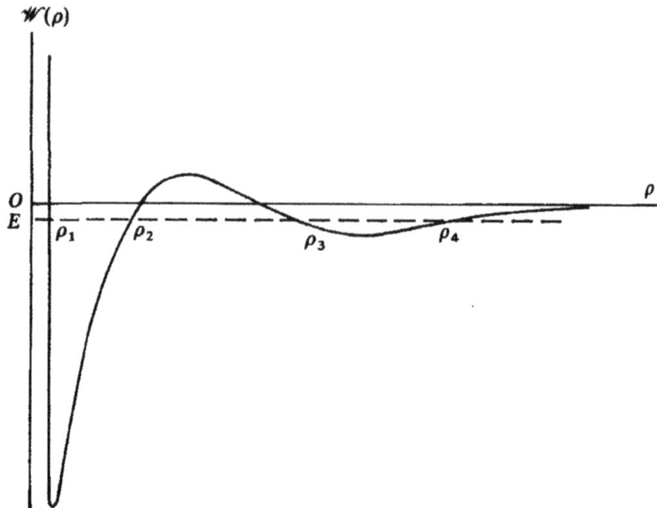

Fig. IX.3. Effective potential (qualitative) for an $f$ electron ($l = 2$) in a heavy atomic ion such as $U^+$.

The problem is to obtain the eigenvalues $E$ of Eq. (IX-218) corresponding to the situation shown in Fig. IX.3; namely, for negative $E$, there are four roots $P_1$, $P_2$, $P_3$, $P_4$ in the expression

$$E - \mathscr{W}(\rho) = E - V(\rho) - \frac{l(l + 1)}{\rho^2} = 0. \qquad \text{(IX-220)}$$

### (3) WKB method of solving Eq. (IX-218)

An approximate method to obtain $E$ is the WKB (or, Wentzel-Kramers-Brillouin-Jeffreys) method. A treatment of this method can be conveniently found in many books. We shall give only a brief sketch of the method here.

Let us write Eq. (IX-218) in the form

$$\frac{d^2\psi}{d\rho^2} + Q^2\psi = 0, \qquad \text{(IX-221)}$$

with

$$Q^2 = E - V(\rho) - \frac{l(l + 1)}{\rho^2}. \qquad \text{(IX-222)}$$

The starting point is to compare this equation with that for the eikonal in the classical theory (see Chap. 3, (III-39)) and express $\psi$ in the form

$$\psi = \exp\left(\frac{i}{\hbar} S\right), \qquad \text{(IX-223)}$$

with an expansion of the function $S$ as a series in powers of $\hbar$. To the first approximation, $\psi$ can readily be found.

Consider the regions $\rho < \rho_1$ and $\rho_1 < \rho$ in Fig. IX.3:

(i) $\rho < \rho_1, Q^2 < 0,$

$$\psi_1 = \frac{A}{\sqrt{|Q|}}\exp\left[-\int_\rho^{\rho_1}|Q|\,d\rho\right];\qquad\qquad\text{(IX-224)}$$

(ii) $\rho_1 < \rho, Q^2 > 0,$

$$\psi_{II} = \frac{B}{\sqrt{Q}}\cos\left[\int_{\rho_1}^\rho Q\,d\rho + \delta\right].\qquad\qquad\text{(IX-225)}$$

At $\rho = \rho_1$, $Q^2 = 0$ and the above approximate solutions diverge. One can, however, obtain more accurate solutions in the neighborhood of $\rho_1$ in terms of Bessel functions of order $\pm\frac{1}{3}$. In the WKB treatment, we shall not be concerned with these functions valid at $\rho_1$, but rather with the smooth joining of the two functions $\psi_1$ and $\psi_{II}$ in passing from $\rho < \rho_1$ to $\rho_1 < \rho$. The study of the behavior of the solutions at the singular point $\rho_1$ leads to the "connection formulas"

$$\frac{A}{\sqrt{|Q|}}\exp\left[-\int_\rho^{\rho_1}|Q|\,d\rho\right]\leftrightarrow\frac{2A}{\sqrt{Q}}\cos\left[\int_{\rho_1}^\rho Q\,d\rho - \frac{\pi}{4}\right],\qquad\text{(IX-226)}$$
$$\qquad(\rho < \rho_1)\qquad\qquad\qquad\qquad(\rho_1 < \rho)$$

$$\frac{2B}{\sqrt{Q}}\cos\left[\int_\rho^{\rho_2}Q\,d\rho - \frac{\pi}{4}\right]\leftrightarrow\frac{B}{\sqrt{|Q|}}\exp\left[-\int_{\rho_2}^\rho|Q|\,d\rho\right],\qquad\text{(IX-227)}$$
$$\qquad(\rho < \rho_2, Q^2 > 0)\qquad\qquad\qquad(\rho_2 < \rho, Q^2 < 0)$$

$$\frac{C}{\sqrt{Q}}\cos\left[\int_\rho^{\rho_2}Q\,d\rho + \frac{\pi}{4}\right]\leftrightarrow\frac{C}{\sqrt{|Q|}}\exp\left[\int_{\rho_2}^\rho|Q|\,d\rho\right].\qquad\text{(IX-228)}$$
$$\qquad(\rho < \rho_2, Q^2 > 0)\qquad\qquad\qquad(\rho_2 < \rho, Q^2 < 0)$$

Application of these relations to the eigenvalue problem as depicted in Fig. IX.3 leads to the condition

$$\int_{\rho_3}^{\rho_4}Q\,d\rho = (n - l - \tfrac{1}{2})\pi - \arctan\left[\frac{1}{4}\exp\left\{-2\int_{\rho_2}^{\rho_3}|Q|\,d\rho\right\}\tan\int_{\rho_1}^{\rho_2}Q\,d\rho\right],$$
$$\text{(IX-229)}$$

where, instead of (IX-222), $Q^2$ is given by

$$Q^2 = E - V(\rho) - (l + \tfrac{1}{2})^2/\rho^2,\qquad\qquad\text{(IX-230)}$$

$n$ being the principal quantum number.

Equation (IX-229) is the generalized Sommerfeld phase integral quantum condition for the case where the potential energy $V(r)$ has two valleys, corresponding to two regions of possible classical motion separated by a potential barrier.[s]

### (4) f states in the actinium series of atoms

In Chap. 1, Sec. 11, we saw how the Pauli principle accounts for the periodic table of chemical elements. There are questions which are not immediately answered by the Pauli principle alone, as for instance, the order in which the various subshells are filled. For example, the 4s shell is filled before the 3d; the 5s before the 4d; the 5s, 5p shells before the 4f, etc. Such questions, in principle, can be answered from calculations of the relative stabilities of the competitive configurations, as for example, the $1s^2 2s^2 2p^6 3s^2 3p^6 4s$ and the $1s^2 2s^2 2p^6 3s^2 3p^6 3d$ configurations in the case of K $(Z = 19)$, etc. It is on such considerations that the actual "build-up principle" of the periodic table is to be understood.

For radon, $Z = 86$, the electron configuration is $1s^2 2s^2 2p^6 3s^2 3p^6 3d^{10} 4s^2 4p^6 4d^{10} 4f^{14} 5s^2 5p^6 5d^{10} 6s^2 6p^6$. For Fr $(Z = 87)$, radon core + 7s; Ra $(Z = 88)$, radon core + $7s^2$; Ac $(Z = 89)$, radon core + $7s^2 6d$; Th $(Z = 90)$, radon core + $7s^2 6d^2$.

Referring back to the rare earths atoms, one sees that with increasing $Z$, the 4f state gains in stability relative to 5d, so that the 4f subshell is filled before the 5d subshell. In the radon core, the 5f subshell is left vacant even when the 5d, 6s, 6p subshells have been filled. The question arises whether, with increasing $Z$, the 5f state gains in stability relative to 6d, i.e., whether there will begin, in the neighborhood of uranium, a series of 14 elements of

---

[s] If (IX-221) is written in the form

$$\frac{d^2\psi}{dx^2} + \frac{1}{\hbar^2} P^2 \psi = 0,$$

$$P^2 = 2m(E - V),$$

all in ordinary units, then (IX-229) has the form

$$\frac{1}{\hbar}\int_{x_3}^{x_4} P\,dx = (n + \tfrac{1}{2})\pi - \arctan\left[\frac{1}{4}\exp\left\{-\frac{2}{\hbar}\int_{x_2}^{x_3}|P|\,dx\right\}\tan\frac{1}{\hbar}\int_{x_1}^{x_2}P\,dx\right]. \qquad \text{(IX-229a)}$$

This reduces, for the case of only one region of classical motion, to the Sommerfeld condition

$$\oint P\,dx = (n + \tfrac{1}{2})h.$$

The justification for having $(l + \tfrac{1}{2})^2$ instead of $l(l + 1)$ in (IX-230) comes from the theory of the WKB method itself, and is not artificial. See R. E. Langer, *Phys. Rev.* **51**, 669 (1937).

The result (IX-229) was obtained by T. Y. Wu [*Phys. Rev.* **44**, 727 (1933)]. The study of the WKB method in the second approximation was carried out by N. Fröman, *Ark. f. Fysik* **32**, 79 (1966). She obtained a result which reduces to (IX-229a) when proper approximations are made.

the 5f subshell having similar chemical properties, analogous to the case of the rare earths series.

Such considerations led Goudsmit and Wu, in 1932, to investigate the relative stabilities of the 4f and 6d electrons in the uranium atom and ions U, $U^+$, $U^{2+}$, $U^{3+}$, $U^{4+}$, $U^{5+}$. The idea was to see the relative energies of the 5f and 6d states if electrons are fed one at a time into a highly ionized atom.

When the statistical potential of Fermi (IX-214, 207a, 210a, 209b) was employed in the calculations, the two-valley potential in Fig. IX.3 was discovered and the relation (IX-229) was derived. Calculations showed that indeed somewhere around $Z = 92$, the 5f state begins to be lower than 6d, and suggested the beginning of a series of 14 elements having similar chemical properties corresponding to the filling up of the 5f subshell. (See. T. Y. Wu and S. Goudsmit, *Phys. Rev.* **43**, 496 (1933); similar studies were made by M. Goeppert-Mayer during the Second World War).

The trans-uranium radioactive elements ($Z = 93$ to $104$) artificially produced since World War II seem to bear out the idea of a series of 14 elements.

### *(5) The Rydberg correction to the f states of a heavy atom*

With the use of the statistical potential of Fermi, Eq. (IX-219), one can calculate the energies of an *nf* electron in a (heavy) atom. The "effective potential" $\mathscr{W}(\rho)$ for $l = 3$ is as shown in Fig. IX.3. For the region $\rho_3 < \rho < \rho_4$ of possible classical motion, the field is Coulombic,

$$V(\rho) = -\frac{2}{\rho}, \tag{IX-231}$$

so that the integral $\int_{\rho_3}^{\rho_4} Q \, d\rho$ can be readily calculated,

$$\int_{\rho_3}^{\rho_4} \left[ E - \frac{2}{\rho} - \frac{(l + \frac{1}{2})^2}{\rho^2} \right]^{1/2} d\rho = \frac{\pi}{\sqrt{-E}} - (l + \tfrac{1}{2})\pi. \tag{IX-232}$$

Let

$$\pi\Delta = \arctan\left[ \frac{1}{4}\exp\left( -2 \int_{\rho_2}^{\rho_3} |Q| \, d\rho \right) \tan \int_{\rho_1}^{\rho_2} Q \, d\rho \right]. \tag{IX-233}$$

The integrals $\int_{\rho_2}^{\rho_3} |Q| \, d\rho$, $\int_{\rho_1}^{\rho_2} Q \, d\rho$ are slowly varying with the energy $E$; the latter varies from about $1.0\pi$ to $1.2\pi$, and the exponential factor from 0.0027 to 0.0025 for atoms from Au ($Z = 79$) to U ($Z = 92$). Thus for all *nf* states of these atoms,

$$\Delta \simeq 1.00. \tag{IX-234}$$

On combining (IX-229), (IX-232) and (IX-233), one obtains for the energy $E$

of an $nf$ electron

$$E = -\frac{1}{(n - \Delta)^2} \text{ in rydbergs,} \qquad \text{(IX-235)}$$

and from the theoretically calculated value of $\Delta$ above,

$$E = -\frac{1}{(n - 1)^2}. \qquad \text{(IX-236)}$$

This shows that the rydberg correction (also called the quantum defect) of the f spectroscopic terms of heavy atoms has the value 1. It is important to point out that it is this closeness of $\Delta$ to unity that has led to the assignment of principal quantum numbers less by one unit to the empirical spectroscopic f terms for heavy atoms (for example, F. Hund, *Linianspektren und Periodische System der Elemente* (Springer, Berlin, 1930)). With the correct theoretical expression (IX-235), the empirical values of $\Delta$ are as follows

| $Z$ | 79 | 80 | 81 | 82 | 86 |
|---|---|---|---|---|---|
| atom | Au | Hg | Tl | Pb | Rn |
| $\Delta$ | 1.03 | 1.03 | 1.02 | 1.04 | 1.02 |

They are in good agreement with the value $\Delta = 1$ from the theory.

### References

W. Heisenberg, *Z. Phys.* **39**, 499 (1926) discusses the concept of exchange energy in the helium atom.

J. C. Slater, *Phys. Rev.* **34**, 1293 (1929) introduced the determinantal form for many-electron wave functions, used previously by

P. A. M. Dirac, *Proc. Roy. Soc.* London **A 122**, 661 (1926).

D. R. Hartree, *Proc. Camb. Phil. Soc.* **24**, 89 (1928) covers the self-consistent field method without exchange.

J. C. Slater, *Phys. Rev.* **35**, 210 (1930), and

V. Fock, *Z. Phys.* **61**, 126 (1930) deal with the self-consistent field method with exchange.

L. Gaunt, *Trans. Roy. Soc.* London **A 228**, 151 (1929) treats the integrals (IX-30) containing three associated Legendre polynomials. See also

E. U. Condon and G. H. Shortley, *Theory of Atomic Spectra* (Cambridge Univ. Press, Cambridge, 1934) for tables of coefficients $a^k(l, m; l', m')$ and $b^k(l, m; l', m')$ of (IX-31a), (IX-31b).

E. Hylleraas, *Z. Phys.* **48**, 469 (1928); **54**, 347 (1929).

T. Kinoshita, *Phys. Rev.* **105**, 1490 (1957) applies a 39-parameter Hylleraas method calculation for helium.

H. Bethe and E. E. Salpeter, *Quantum Mechanics of One- and Two-Electron Atoms* (Springer, Heidelberg, 1957) gives a review of the corrections to the non-relativistic calculation of the $1s^2\,{}^1S$ of helium-like atoms.

T. Y. Wu, *Phys. Rev.* **66**, 291 (1944), discusses the problem of autoionization of doubly excited states of helium and identification of the lines $\lambda = 320.4$ and $\lambda = 357.5$.

Condon and Shortley, *loc. cit*, Chap. 15, Sec. 3, also discusses autoionization.

E. Ramberg and F. Richtmyer, *Phys. Rev.* **51**, 913 (1937) presents calculations on the Auger effect and radiative transition probabilities.

T. Y. Wu, "Configuration Interaction in Atomic States" in *Mem. Vol. to Chiang Kai-shek*, Acad. Sinica (1976).

H. E. White, *Phys. Rev.* **40**, 316 (1932) gives a theory of inverted alkali atom doublets. See also M. Phillips, *Phys. Rev.* **44**, 644 (1933).

L. H. Thomas, *Proc. Camb. Phil. Soc.* **23**, 542 (1926).

E. Fermi, *Acc. Lencei* **6**, 602 (1927) discusses neutral atoms, while E. Fermi, *Mem. della reale Acad. d'Italia I* (1930) is on positive ions.

A. Sommerfeld, *Z. Phys.* **78**, 283 (1932) gives approximate forms of $\phi(x)$ for neutral atoms and positive ions.

P. Gombás, *Die Statistiche Theorie des Atoms und ihre Anwendungen* (Springer, Vienna, 1949) gives the values of the Thomas-Fermi function $\phi(x)$ and references to the literature.

G. Wentzel, *Z. Phys.* **38**, 518 (1926).

H. A. Kramers, *Z. Phys.* **39**, 828 (1926).

L. Brillouin, *Comptes Rendus* **183**, 24 (1926); *J. de Physique* **I**, 353 (1926).

H. Jeffreys, *Proc. London Math. Soc.* (2) **23**, 420 (1923), gives the method now known as the WKB method before its application to the Schrödinger equation.

E. C. Kemble, *The Fundamental Principles of Quantum Mechanics* (McGraw-Hill, New York, 1937; Dover, New York, 1958). Chapter III, Sec. 21, gives a detailed treatment of the WKB method.

Fröman, *Arkiv f. Fysik* **31**, 381, 445; **32**, 79, 541 (1966) present detailed analysis of the WKB theory, including the second approximation.

T. Y. Wu, *Phys. Rev.* **44**, 727 (1933) is on the quantum condition (IX-229), and its application to the problem of the f states of heavy atoms.

T. Y. Wu and S. Goudsmit, *Phys. Rev.* **43**, 496 (1933) uses the Fermi potential and the quantum condition (IX-229) to study the energies of a 5f and a 6d electron in $U^+, \ldots, U^{5+}$ and the possibility of a series of 14 elements in the filling up of the 5f subshell starting with uranium.

Chapter 10

# Quantum Mechanics of Molecules

## 1. Introduction

A stable molecule in the normal state may be viewed either as a system of electrons in the field of the atomic nuclei, or as a number of atomic nuclei imbedded in a distribution of electrons. Just as in atoms, the electrons in a molecule can exist in different energy states. These states are quantized, as in the case of atoms. This was established directly by experiments on the excitation by electron impact—the Franck-Hertz experiment, or by the analysis of the electronic spectra of molecules.

Now, in any given electronic state of a stable molecule, the atomic nuclei are capable of executing small oscillations about their equilibrium position. By virtue of the very much larger masses of the atomic nuclei compared with that of the electron, one may picture these vibrational motions as those of the atomic ions (formed by the nuclei with their more tightly bound electrons) in the (time) average field of the electrons that are responsible for the formation of the stable molecular state. To a first approximation, the vibrational motions of a molecule are describable as the normal vibrations of a system of particles from their equilibrium position in classical dynamics. The vibrational states are quantized.

Now a non-vibrating molecule certainly can rotate as a whole, the rotational motion being also quantized. To a "zeroth-order" approximation, one regards the electronic, vibrational and rotational motion as separate and independent, and their energies are additive[a]

$$E = E_n + E_v + E_r.$$

Here $n, v, r$ denote the totality of electronic, vibrational and rotational quantum numbers. $E_n$ corresponds to the electronic states and is of the order of a few electron volts (eV); $E_v$, depending on the molecule and the modes of vibrations, ranges from 0.4 eV for such light chemical bonds as CH (in $C_2H_2$, say) to 0.03–0.10 eV in $CCl_4$. $E_r$ depends again on the molecule and is, for the

---

[a] There is certainly the translational energy $E_t$ which is, for a molecule in a box of macroscopic dimensions, practically continuous. We shall leave out this part.

lower rotational states, much smaller than $E_v$. Thus, generally,

$$E_n \gg E_v \gg E_r \gg E_s,$$

where $E_s$ is the energy arising from the nuclear spins.

Transitions among electronic states correspond to atomic transitions and the spectra lie usually in the ultraviolet and the visible region. Transitions between two vibrational states usually lie in the infrared region ($3\mu - 30\mu$, or, $3000 \text{ cm}^{-1} - 300 \text{ cm}^{-1}$). Transitions among rotational states usually lie in the far infared, $\lambda > 50\mu$, say. But simultaneous changes in the electronic, vibrational and rotational states can take place and these give rise to spectral lines

$$h\nu = (E_{n'} - E_n) + (E_{v'} - E_v) + (E_{r'} - E_r).$$

Given $(n' - n)$, various $(v', v)$ and $(r', r)$ transitions give rise to a "band" centered around $E_{n'} - E_n$ of the electronic part of the transition, usually in the visible or the ultraviolet region of the spectrum.

For a given vibrational transition between two states of the same electronic states, the various rotational transitons give rise to

$$h\nu = (E_{v'} - E_v) + (E_{r'} - E_r),$$

which form a vibration-rotational band in the (near) infrared.

Of course the transitions are governed by selection rules for the quantum numbers $n$, $v$ and $r$, as determined by the symmetry properties of the wave functions.

*In principle*, the energy of a molecule in its electronic, vibrational and rotational states can be calculated from first principles in quantum mechanics. In practice, the calculation of even the electronic energies alone of a molecule in general is exceedingly lengthy, if not impraticable. For example, the calculation of the electronic states of the hydrogen molecule $H_2$ is more difficult than that for the helium atom by more than an order of magnitude.

Also *in principle*, the simultaneous electronic, vibrational and rotational motion of a molecule can be treated in quantum mechanics. But the complexity of the problem, as mentioned in the last paragraph, is here further greatly enhanced by the "couplings" among the idealized separate motions. For example, we may mention the change of the moment of inertia of a molecule due to the vibrational motion; and in turn, the change in the frequency (energy) of an anharmonic vibration due to the rotational stretching of a molecule. Such "couplings" can, however, be treated in an approximate (perturbation) theory.

*In principle*, such questions as the formation of molecules from atoms and their geometrical configurations—the theory of directed valence—can all be answered in quantum mechanics. The electron spin plays a fundamental role in the theory of molecular formation through its part in the symmetry of the electronic wave functions (in a somewhat similar way to its part in the atoms). Finally the spins of the atomic nuclei also play a role both through their effect on the symmetry of such molecules as $H_2$, $N_2$, $CH_4$ possessing elements of symmetry, and through their contribution to the magnetic moment and angular momentum of the molecule, which become important when one is interested in the fine structure of the energy states in the microwave spectra of the molecule.

A detailed treatment of even one of the above mentioned topics is not possible in the present book. However, a brief account of the basic ideas by means of simple examples will be given below.

## 2.  The electronic, vibrational and rotational motion of a molecule

We shall bring out the essential features of the theory by means of the simplest of molecules, namely, the hydrogen molecular ion $H_2^+$.

We shall first separate off the center-of-mass motion of the M-m-M system, and employ the following coordinates:

for the M-M, the separation $\rho$ and the polar angles $\vartheta$, $\varphi$;
for the electron m, the elliptic coordinates $\xi$ and $\eta$ with respect to M-M, and
  the angle $\psi$ around the line M-M.

As usual

$$\xi = \frac{r_1 + r_2}{\rho}, \qquad \eta = \frac{r_1 - r_2}{\rho}. \tag{X-1}$$

Here all lengths are in units of

$$\bar{a} = \frac{m}{\mu} a_B, \qquad \mu = \frac{mM}{m + M}, \tag{X-2a}$$

and energies in units of

$$\varepsilon = \frac{\mu}{m} \quad \text{Rydberg.} \tag{X-2b}$$

A calculation gives the following expression for the Hamiltonian

$$H = -\frac{\hbar^2}{2m} \nabla_m^2 - \frac{\hbar^2}{2M}(\nabla_{M_1}^2 + \nabla_{M_2}^2) + \frac{e^2}{|\mathbf{r}_{M_1} - \mathbf{r}_{M_2}|}$$

$$- \frac{e^2}{|\mathbf{r}_m - \mathbf{r}_{M_1}|} - \frac{e^2}{|\mathbf{r}_m - \mathbf{r}_{M_2}|}, \tag{X-3}$$

in the coordinates $\xi$, $\eta$, $\psi$, $\rho$, $\vartheta$, $\varphi$ and

$$H = H_e + H'_e + H_{ev} + H_{er} + H_v + H_r, \tag{X-4}$$

$$H_e = -\frac{4}{\rho^2(\xi^2 - \eta^2)}[X + 2\rho\xi - \tfrac{1}{2}\rho(\xi^2 - \eta^2)], \tag{X-4a}$$

$$H'_e = -\frac{4}{M\rho^2}\left[\frac{(\xi^2 + \eta^2 - 2)}{\xi^2 - \eta^2}X - 2Y - 2\frac{\partial^2}{\partial\psi^2}\right], \tag{X-4b}$$

$$H_{ev} = \frac{4\mu}{M}Y\frac{1}{\rho}\frac{\partial}{\partial\rho}, \tag{X-4c}$$

$$H_{er} = -\frac{4\mu}{M\rho^2}\left[\frac{1}{\sin^2\vartheta}\frac{\partial^2}{\partial\psi^2} - 2\frac{\cot\vartheta}{\sin\vartheta}\frac{\partial^2}{\partial\varphi\partial\psi}\right.$$

$$+ \frac{2\xi\eta}{\sqrt{(\xi^2 - 1)(1 - \eta^2)}}\left\{\sin\psi\frac{\partial}{\partial\vartheta} - \frac{\cos\psi}{\sin\vartheta}\frac{\partial}{\partial\varphi} + \cot\vartheta\cos\psi\frac{\partial}{\partial\psi}\right\}\frac{\partial}{\partial\psi}$$

$$- \frac{2\sqrt{(\xi^2 - 1)(1 - \eta^2)}}{\xi^2 - \eta^2}\left\{\cos\psi\frac{\partial}{\partial\vartheta} + \frac{\sin\psi}{\sin\vartheta}\frac{\partial}{\partial\varphi}\right.$$

$$\left.\left. - \cot\vartheta\sin\psi\frac{\partial}{\partial\psi}\right\}\left(\eta\frac{\partial}{\partial\xi} - \xi\frac{\partial}{\partial\eta}\right)\right], \tag{X-4d}$$

$$H_v = -\frac{2\mu}{M}\left[\frac{\partial^2}{\partial\rho^2} + \frac{2}{\rho}\frac{\partial}{\partial\rho}\right], \tag{X-4e}$$

$$H_r = -\frac{2\mu}{M\rho^2}\left[\frac{\partial^2}{\partial\vartheta^2} + \cot\vartheta\frac{\partial}{\partial\vartheta} + \frac{1}{\sin^2\vartheta}\frac{\partial^2}{\partial\varphi^2}\right], \tag{X-4f}$$

$$X = \frac{\partial}{\partial\xi}(\xi^2 - 1)\frac{\partial}{\partial\xi} + \frac{\partial}{\partial\eta}(1 - \eta^2)\frac{\partial}{\partial\eta} + \left(\frac{1}{\xi^2 - 1} + \frac{1}{1 - \eta^2}\right)\frac{\partial^2}{\partial\psi^2}, \tag{X-4g}$$

$$Y = \frac{1}{\xi^2 - \eta^2}\left[\xi(\xi^2 - 1)\frac{\partial}{\partial\xi} + \eta(1 - \eta^2)\frac{\partial}{\partial\eta}\right]. \tag{X-4h}$$

The meaning of the various terms is as follows: $H_e$ is the Hamiltonian of $H_2^+$ for $M \to \infty$; $H'_e$ is the correction for the relative motion of m and M-M; $H_v$ is the kinetic energy of the vibration M-M; $H_r$ is the kinetic energy of the rotator M-M; $H_{ev}$, $H_{er}$ are the coupling between the electronic motion and the vibrational and rotational motion, respectively. Note that all the terms except $H_e$ are of order $\mu/M = m/(m + M)$.

From Eqs. (X-4)–(X-4h), it is clear that the Schrödinger equation

$$(H - E)\Psi = 0 \tag{X-5}$$

cannot be solved in an exact way. An approximate method in the general case has been given by Born and Oppenheimer. Instead of following their treatment, we shall illustrate the essential results by means of the simple molecular system $H_2^+$ as given by (X-3).[b]

## (1) $H_2^+$ electronic states with fixed nuclei

The lowest order approximation to (X-5) is

$$(H_e - V_n(\rho))\Psi_n(\xi, \eta, \psi; \rho) = 0, \qquad (X\text{-}6)$$

where the internuclear distance $\rho$ appears as a fixed parameter. $V_n(\rho)$ is the eigenvalue for the electronic state specified by the quantum numbers denoted collectively by $n$.

The states having zero total angular momentum are those given by putting $\partial\Psi/\partial\vartheta = \partial\Psi/\partial\rho = \partial\Psi/\partial\psi = 0$ in $H$, and (X-6) becomes

$$\left\{-\frac{4}{\rho^2(\xi^2 - \eta^2)}\left[\frac{\partial}{\partial\xi}(\xi^2 - 1)\frac{\partial}{\partial\xi} + \frac{\partial}{\partial\eta}(1 - \eta^2)\frac{\partial}{\partial\eta} + 2\rho\xi - \frac{1}{2}\rho(\xi^2 - \eta^2)\right]\right.$$

$$\left. - V_n(\rho)\right\}\Psi_n(\xi, \eta; \rho) = 0. \qquad (X\text{-}7)$$

The system, being symmetric with respect to the center of the line M-M, has two solutions $\Psi_g$, $\Psi_u$ which are even and odd, respectively, with respect to the interchange of the two identical nuclei. They are called the $1s\sigma g$, $1s\sigma u$ electron orbitals, $\sigma$ signifying zero-angular momentum component along the line M-M. The eigenvalues[c] $V(\rho)$ obey

$$V_g(\infty) = V_u(\infty) = 0. \qquad (X\text{-}8)$$

Equation (X-7) is separable and exact solutions for the lowest state $V_g(\rho)$ and $\Psi_g(\xi, \eta; \rho)$ (and some excited states) are known (Bates *et al.*). The lowest $V_g(\rho)$ is $-1.20$ ($\mu/m$ Rydberg) at the equilibrium $\rho = 2.0$ ($m/\mu$ Bohr radius). The antisymmetric $\Psi_u$ lead to repulsive $V_u(g)$ for all $\rho$ so that there are no stable states $\Psi_u(\xi, \eta; \rho)$.[d]

---

[b] The classic paper by M. Born and R. Oppenheimer, *Annalen d. Phys.* **84**, 457 (1927), is often quoted. The treatment of $H_2^+$ in (X-3), (X-4a)–(X-4h) is taken from T. Y. Wu, R. L. Rosenberg and H. Sandstrom, *Nucl. Phys.* **16**, 432 (1960). There the more general HD$^+$ system is treated.

[c] The eigenvalues $V(\rho)$ contain the term $2/\rho$ which is the electrostatic energy between the two nuclei.

[d] This can be seen very readily in a qualitative manner by approximating the "molecular orbital" $\Psi(\xi, \eta; \rho)$ by a linear combination of "atomic orbitals" (LCAO)

$$\left.\begin{array}{c}\Psi_g \\ \Psi_e\end{array}\right\} = \binom{N_g}{N_u}(u(r_A) \pm u(r_B)), \qquad (X\text{-}9)$$

where $u(r_A)$ is an atomic wave function, say, 1s, centred at nucleus A, $u(r_B)$ is another centred at nucleus B. It is seen that $\Psi_u(\rho)$ is zero at the center of A $-$ B, so that the probability of the electron being in the region between the two nuclei is small, leading to a small binding effect. $\Psi_g(\rho)$ has non-vanishing values near the center of A $-$ B.

### (2) Nuclear vibration

For simplicity, consider only states of zero total angular momentum. The Schrödinger equation is, from (X-4a, b, c, e, g, h), for $\Psi_g$ states,

$$(H_e + H_e' + H_{ev} + H_v - E)\Psi(\xi, \eta; \rho) = 0. \tag{X-10}$$

On expanding

$$\Psi(\xi, \eta; \rho) = \sum_n \Psi_n(\xi, \eta; \rho)\Phi_n(\rho); \tag{X-11}$$

multiplying by $\Psi_0(\xi, \eta; \rho)\, dr$, $dr = \tfrac{1}{8}\rho^3(\xi^2 - \eta^2)\, d\xi\, d\eta$, and integrating over $r$,

$$1 \le \xi < \infty, \qquad -1 \le \eta \le 1,$$

one obtains

$$\left\{ -\frac{2\mu}{M}\left[ \frac{\partial^2}{\partial\rho^2} + \frac{2}{\rho}\frac{\partial}{\partial\rho} \right] + V_0(\rho) + \Delta V_0(\rho) - E \right\}\Phi_0(\rho) = 0, \tag{X-12}$$

where $V_0(\rho)$ is the eigenvalue in (X-7), namely,

$$V_0(\rho) = -\frac{\rho}{2}\iint \Psi_0^*(\xi, \eta; \rho)\left[ \frac{\partial}{\partial\xi}(\xi^2 - 1)\frac{\partial}{\partial\xi} + \frac{\partial}{\partial\eta}(1 - \eta^2)\frac{\partial}{\partial\eta} \right.$$
$$\left. + 2\rho\xi - \frac{\rho}{2}(\xi^2 - \eta^2) \right]\Psi_0(\xi, \eta; \rho)\, d\xi\, d\eta. \tag{X-13}$$

$\Delta V_0(\rho)$ is the correction to the eigenvalue $V_0(\rho)$,

$$\Delta V_0(\rho) = \frac{2\mu}{M}\int \frac{\xi^2 + \eta^2 - 2}{\xi^2 - \eta^2}\left[ (\xi^2 - 1)\left(\frac{\partial\Psi_0}{\partial\xi}\right)^2 + (1 - \eta^2)\left(\frac{\partial\Psi_0}{\partial\eta}\right)^2 \right] dr$$
$$+ \frac{2\mu}{M}\int \left(\frac{\partial\Psi_0}{\partial\rho}\right)^2 dr - \frac{6\mu}{M\rho^2}\left( 1 + \frac{d\ln\Phi_0}{d\ln\rho} \right). \tag{X-14}$$

The first integral and the term $-6\mu/M\rho^2$ come from $H_e'$, the second integral comes from $H_v$; the last term in $d\ln\Phi_0/d\ln\rho$ from $H_{ev}$. They are all of order $\mu/M$.[e]

The meaning of (X-12) is clear: it is the equation of vibrational motion (of the coordinate $\rho$) in the "potential" $V_0(\rho) + \Delta V_0(\rho)$. $V_0(\rho)$ is the potential in the static nuclei approximation; $\Delta V_0(\rho)$ is the correction due to relative motion of the electron and the two nuclei, and to the coupling between the electronic and the nuclear motion $H_{ev}$.

---

[e] The integrals in (X-14) have been calculated from the exact $\Psi_0(\xi, \eta; \rho)$ by Dalgarno and McCarroll and from the approximate $\Psi_0(\xi, \eta; \rho)$ of Guillemin-Zener by Wu *et al.*

### *(3) Rotational motion*

For the "pure" rotational motion (i.e., ignoring the electronic motion and setting the internuclear distance $\rho = $ a constant), the Schrödinger equation is, from (X-4f),

$$\left\{ -\frac{2\mu}{M\rho^2}\left[ \frac{1}{\sin\vartheta}\frac{\partial}{\partial\vartheta}\left(\sin\vartheta\frac{\partial}{\partial\vartheta}\right) + \frac{1}{\sin^2\vartheta}\frac{\partial^2}{\partial\varphi^2}\right] - E_r \right\}\chi(\vartheta,\varphi) = 0. \quad \text{(X-15)}$$

The exact eigenvalues and eigenfunctions are

$$E_J = \frac{2\mu}{M\rho^2}J(J+1)\left(\text{in }\frac{\mu}{m_e}\text{ Rydberg}\right)^{\text{f}},$$

$$\chi(\vartheta,\varphi) = \left[\frac{1}{2\pi}\frac{2J+1}{2}\frac{(J-|m|)!}{(J+|m|)!}\right]^{1/2} e^{im\varphi}P_J^m(\cos\vartheta), \quad \text{(X-16)}$$

where $-J \le m \le J$, $J = 0, 1, 2, \ldots$, $J(J+1)\hbar^2$ being the square of the total angular momentum of M-M. The state $J$ is $(2J+1)$-fold degenerate.

On account of the complexity of the calculations, we shall not deal with the problem of the coupling of the vibrational and the rotational motion here, but instead we shall bring out the essential nature of the problem by another method in the following.

The selection rule for dipole transition is determined by the matrix element of the electric moment $\mathcal{M}_0$ whose components $\mathcal{M}_x$, $\mathcal{M}_y$, $\mathcal{M}_z$ are

$$\mathcal{M}_z = \mathcal{M}_0\cos\vartheta,$$

$$\mathcal{M}_x \pm i\mathcal{M}_y = \mathcal{M}_0\sin\vartheta\, e^{\pm i\varphi}.$$

For homopolar molecules, such as $H_2$, $O_2$ (or symmetrical linear molecules $CO_2$, $C_2H_2$), the permanent moment $\mathcal{M}_0 = 0$, and there is no pure rotational absorption spectrum in the infrared. For molecules such as HCl, HCN, $\mathcal{M}_0 \ne 0$, and the selection rules are given by the non-vanishing matrix elements $\langle J, m|\cos\vartheta|J', m'\rangle$, $\langle J, m|\sin\vartheta\, e^{\pm i\varphi}|J', m'\rangle$,

$$\Delta J = J - J' = \pm 1 \quad \text{for } \mathcal{M}_x, \mathcal{M}_y, \mathcal{M}_z,$$

$$\Delta m = 0 \quad \text{for } \mathcal{M}_z,$$

$$\Delta m = \pm 1 \quad \text{for } \mathcal{M}_x, \mathcal{M}_y.$$

See (III-122) of Chapter 3.

---

f In cgs units, this is

$$E_J = \frac{\hbar^2}{2I}J(J+1), \quad \text{(X-17)}$$

where $I = \frac{1}{2}MR^2$ is the moment of inertia of M-M.

### 3. The hydrogen molecule—the theory of Heitler and London

The first treatment of the problem of the formation of valence bonds in (homopolar) molecules is the work of Heitler and London (1927), on the hydrogen molecule. Although in its original form it involves many approximations, it has brought out certain essential features, such as the importance of the role played by the electron spin through its effect on the symmetry and hence on the stability of valence bonds, and the (technical) concept of "overlap" in "exchange integrals" leading to an *anschaulich* understanding of directed valence, etc.

We shall take the static nuclei approximation so that, for the electronic motion, the Hamiltonian is

$$H = -\frac{\hbar^2}{2m}(\nabla_1^2 + \nabla_2^2) - e^2\left(\frac{1}{r_{1A}} + \frac{1}{r_{2B}} + \frac{1}{r_{1B}} + \frac{1}{r_{2A}} - \frac{1}{r_{AB}} - \frac{1}{r_{12}}\right), \quad \text{(X-18)}$$

where the subscripts 1, 2 denote the electrons and A, B the nuclei, $r_{1A}$ is the distance between electron 1 and nucleus A, etc.

Let $u_a(r_{1A})$ be an atomic orbital (wave function) of electron 1, centred around A, $a$ being the quantum numbers of the one-electron $u(r_{1A})$. Let $\chi^s$, $\chi^a$ be the symmetric and antisymmetric spin wave functions

$$\chi^s = \begin{cases} \alpha_1\alpha_1, \\ \dfrac{1}{\sqrt{2}}(\alpha_1\beta_2 + \alpha_2\beta_1), \\ \beta_1\beta_2. \end{cases} \qquad \chi^a = \frac{1}{\sqrt{2}}(\alpha_1\beta_2 - \alpha_2\beta_1), \qquad \text{(X-19)}$$

The wave functions of the two electrons are assumed to be representable by the approximate form

$$\left.\begin{array}{c} {}^1\Psi_{(1,2)} \\ {}^3\Psi_{(1,2)} \end{array}\right\} = \left\{\begin{array}{c} N_1 \\ N_3 \end{array}\right\} [u_a(r_{1A})u_b(r_{2B}) \pm u_a(r_{2A})u_b(r_{1B})]\begin{pmatrix} \chi^a \\ \chi^s \end{pmatrix}, \qquad \text{(X-20)}$$

where

$$N_1^2 = \frac{1}{2(1 + \Delta^2)}, \qquad N_3^2 = \frac{1}{2(1 - \Delta^2)}, \qquad \text{(X-21)}$$

$$\Delta = \int u_a^*(r_{1A})u_b(r_{1B})\,dr_1. \qquad \text{(X-22)}$$

$\Delta$ is thus a measure of the overlap of an electron's atomic wave function centred at the two nuclei. $\Delta$ obviously vanishes when the two nuclei are very far apart. To simplify notations in the following, we shall denote $u_a$, $u_b$ by $u$, $v$ respectively.

For simplicity, consider states with zero total angular momentum, which in molecular spectroscopic notations are denoted by $\Sigma$ (analogue of S atomic states). The energy of the $^3\Sigma$ state is

$$E(^3\Sigma) = \iint {}^3\Psi^*(1,2)H\, {}^3\Psi(1,2)\, dr_1\, dr_2.$$

$$= \frac{1}{2(1-\Delta^2)}2\left[\iint u^*(r_{1A})v^*(r_{2B})H\, u(r_{1A})v(r_{2B})\, dr_1\, dr_2\right.$$

$$\left. - \iint u^*(r_{1A})v^*(r_{2B})H\, u(r_{2A})v(r_{1B})\, dr_1\, dr_2\right]. \qquad \text{(X-23)}$$

The first integral, called the direct integral, can be shown to be

$$E_a + E_b + 2J + 2J' + \frac{e^2}{R}, \qquad R = r_{AB}, \qquad \text{(X-24)}$$

where

$$E_a = \int u^*(r_{1A})\left[-\frac{\hbar^2}{2m}\nabla_1^2 - \frac{e^2}{r_{1A}}\right]u(r_{1A})\, dr_1,$$

$$E_b = \int v^*(r_{2B})\left[-\frac{\hbar^2}{2m}\nabla_2^2 - \frac{e^2}{r_{2B}}\right]v(r_{2B})\, dr_2, \qquad \text{(X-24a)}$$

$$J = -e^2\int \frac{u^*(r_{1A})u(r_{1A})}{r_{1B}}\, dr_1,$$

$$J' = e^2\iint \frac{|u(r_{1A})|^2|v(r_{2B})|^2}{r_{12}}\, dr_1\, dr_2.$$

The second integral in $E(^3\Sigma)$, called the exchange integral, is

$$(E_a + E_b)\Delta^2 + 2K\Delta + K' + \frac{e^2}{R}\Delta^2, \qquad \text{(X-25)}$$

where

$$K\Delta = -e^2\Delta\int \frac{v^*(r_{2B})u(r_{2A})}{r_{2B}}\, dr_2,$$

$$K' = e^2\iint \frac{u^*(r_{1A})v^*(r_{2B})u(r_{2A})v(r_{1B})}{r_{12}}\, dr_1\, dr_2, \qquad \text{(X-25a)}$$

$$\frac{e^2}{R}\Delta^2 = e^2\iint \frac{u^*(r_{1A})v^*(r_{2B})u(r_{2A})v(r_{1B})}{R}\, dr_1\, dr_2.$$

With these, one obtains from (X-23),

$$E(^3\Sigma) = E_a + E_b + \frac{e^2}{R} + \frac{1}{1 - \Delta^2}[2J + J' - 2K\Delta - K']. \qquad \text{(X-26)}$$

In a similar manner, one obtains for the singlet state $(^1\Sigma)$

$$E(^1\Sigma) = \iint {}^1\Psi^*(1,2)H\,{}^1\Psi(1,2)\,dr_1\,dr_2 \qquad \text{(X-27)}$$

$$= E_a + E_b + \frac{e^2}{R} + \frac{1}{1 + \Delta^2}[2J + J' + 2K\Delta + K']. \qquad \text{(X-28)}$$

All the integrals, involving two centers A and B, are most conveniently evaluated by expressing the atomic wave functions $u(r)$, $v(r)$ in elliptic coordinates.

For the lowest state $^1\Sigma$ of $H_2$, one takes

$$v(r_{1A}) = u(r_{1A}) = \frac{1}{\sqrt{\pi}}\left(\frac{Z}{a}\right)^{3/2}\exp\left(-\frac{Zr_{1A}}{a}\right), \qquad a = \frac{\hbar^2}{me^2}. \qquad \text{(X-29)}$$

One obtains

$$\Delta = \frac{1}{\pi}\left(\frac{Z}{a}\right)^3\frac{R^3}{8}\iiint \exp\left(-\frac{ZR}{a}\xi\right)(\xi^2 - \eta^2)\,d\xi\,d\eta\,d\psi$$

$$= e^{-\rho}(1 + \rho + \tfrac{1}{3}\rho^2), \qquad \rho = \frac{ZR}{a},$$

$$J = -\frac{e^2}{R}[1 - e^{-2\rho}(1 + \rho)],$$

$$J' = \frac{e^2}{R}\left[1 - e^{-2\rho}\left(1 + \frac{11}{8}\rho + \frac{3}{4}\rho^2 + \frac{1}{6}\rho^3\right)\right], \qquad \text{(X-30)}$$

$$K = -\frac{e^2}{R}e^{-\rho}(\rho + \rho^2),$$

$$K' = \frac{e^2}{5a}\left[e^{-2\rho}\left(\frac{25}{8} - \frac{23}{4}\rho - 3\rho^2 - \frac{1}{3}\rho^3\right) + \frac{6}{\rho}\{(\gamma + \ln\rho)\Delta^2 \right.$$

$$\left. - 2\Delta\Delta'E_i(-2\rho) + \Delta'^2 E_i(-4\rho)\}\right],$$

$$\Delta' = e^{\rho}(1 - \rho + \tfrac{1}{3}\rho^2),$$

$$E_i(-x) = -\int_x^\infty \frac{1}{t}e^{-t}\,dt,$$

$$\gamma = \text{Euler constant} = 0.5772\ldots$$

Calculations of the above theory of Heitler and London have been carried out by Sugiura (1928), with $Z = 1$. The result is

$$E(^3\Sigma) \text{ is positive for all } \rho \text{ (repulsive),}$$

$$E(^1\Sigma) \text{ has a minimum at } R = 0.8 \times 10^{-8} \text{ cm,}$$

and

$$E(^1\Sigma)_{\text{min.}} = -3.14 \text{ eV below the asymptotic value}$$

$$2E(1s\,^2S) \quad \text{as } R \to \infty. \tag{X-31}$$

The experimental value of the dissociation energy is

$$E(^1\Sigma) - 2E(1s\,^2S) = -4.72 \text{ eV} \quad \text{at } R = 0.74 \times 10^{-8} \text{ cm.} \tag{X-32}$$

The above simple theory has then been "improved" by:

(i)   a variational method (S. C. Wang, 1928) in which the nuclear charge $Z$ in (X-29) is regarded as a variational parameter;

(ii)  an approximate molecular orbital method, in which $^1\Psi(1,2)$ is represented by

$$^1\Psi(1,2) = N[u(r_{1A}) + u(r_{1B})][u(r_{2A}) + u(r_{2B})]; \tag{X-33}$$

(iii) reducing the "ionic effect" in (X-32), namely, the over-emphasis of both electrons being localized at each nucleus, by taking

$$^1\Psi(1,2) = N[u(r_{1A})u(r_{2B}) + u(r_{2A})u(r_{1B})$$
$$+ k\{u(r_{1A})u(r_{2A}) + u(r_{1B})u(r_{2B})\}] \tag{X-34}$$

and regarding $k$ as a variational parameter;

(iv)  using in (X-34), instead of the 1s spherically symmetric atomic orbital for $u$, the orbital

$$u(r_{1A}) = u_{1s}(r_{1A}) + \sigma u_{2p}(r_{1A}); \tag{X-35}$$

(v)   using wave functions of the kind employed by Hylleraas for the helium atom to take into account the electron-electron correlation (James and Coolidge, 1933).

The following table summarizes the various results.

$$E(^1\Sigma) - 2E(1s\,^2S)$$

| | |
|---|---|
| Heitler-London-Sugiura | $-3.14$ eV |
| Wang | $-3.76$ |
| molecular orbital (X-33) | $-3.47$ |
| Weinbaum (X-34) | $-4.00$ |

$$
\begin{array}{ll}
\text{Rosen (X-35)} & -4.02 \\
\text{Ionic term} + 2p \text{ term} & -4.10 \\
\text{James \& Coolidge} & -4.722 \\
\text{Experimental} & -4.72 \qquad\qquad (\text{X-36})
\end{array}
$$

We shall not be particularly emphasizing the numerical results of these and other calculations. The essential result of the theory, simple as it is, is that the formation of a stable valence bond arises from the overlap of the $u(r_{1A})$ and $v(r_{1B})$ orbitals centred at A and B, which determines the "electron exchange" between two electrons of the two atoms. This can be seen as follows: From (X-26) and (X-28), it is seen that the difference $E(^3\Sigma) - E(^1\Sigma)$ is, for $\Delta < 1$,

$$E(^3\Sigma) - E(^1\Sigma) \simeq -2(2K\Delta + K')$$

$$> 0. \qquad\qquad (\text{X-37})$$

$K\Delta$ represents the "exchange" effect and its absolute value $|K\Delta|$ increases as the overlap $\Delta$ increases. It is also seen that the pairing of the spins into a singlet state is essential for forming a stable covalent bond. One may say that the theory in even such a simple form has already laid the foundation of the theory of homopolar bond and of directed valence.

The theory of Heitler and London, as represented by the use of the wave function (X-20), is known as the method of atomic orbitals, whereas the method of using one-electron wave functions which are solutions of the two-nuclei system (of which the $H_2^+$ in (X-7) is an example and (X-33) is a rough approximation) is known as the method of molecular orbitals. This latter method is initially associated with the names of F. Hund, R. S. Mulliken G. Herzberg and others.

It is to be noted that the former method will be a good approximation for large internuclear separations, whereas the latter method will be a good approximation for small internuclear separation. The two methods may thus be regarded as approximating the actual problem for two limiting situations.

## 4. Coupling between vibrational and rotational motions

Let us take the simple diatomic system $H_2^+$ in (X-4), but neglect the coupling $H_{er}$ between the electronic and the rotational motion. The lowest order approximation is again taken to be (X-6), and equation (X-10), (X-11) are now replaced by

$$(H_e + H'_e + H_{ev} + H_v + H_r - E)\Psi(\xi, \eta, \psi; \rho, \vartheta, \varphi) = 0, \qquad (\text{X-38})$$

$$\Psi(\xi, \eta, \psi; \rho, \vartheta, \varphi) = \sum_n \Psi_n(\xi, \eta, \psi; \rho)\Phi_n(\rho, \vartheta, \varphi). \qquad (\text{X-39})$$

By a procedure similar to that employed from (X-10) to (X-12), one obtains

$$[H_v + H_r + U_0(\rho) - E]\Phi_0(\rho, \vartheta, \varphi) = 0, \qquad \text{(X-40)}$$

where $U_0(\rho)$ is given by an expression more complicated than the $V_0(\rho) + \Delta V_0(\rho)$ given in (X-13, 14). We shall now depart from the $H_2^+$ system, but start from (X-40) in which $U_0(\rho)$ is a given function.

For convenience, let us revert to the coordinates $R$, $\vartheta$, $\varphi$ in ordinary cgs units. Equation (X-40) is separable by setting

$$\Phi_0(R, \vartheta, \varphi) = f(R)\Theta(x)\phi(\varphi), \qquad x = \cos \vartheta, \qquad \text{(X-41)}$$

into

$$\frac{d^2\phi}{d\varphi^2} = -m^2\phi, \qquad \text{(X-42)}$$

$$\frac{d}{dx}\left[(1 - x^2)\frac{d\Theta}{dx}\right] + \left[J(J + 1) - \frac{m^2}{1 - x^2}\right]\Theta = 0, \qquad \text{(X-43)}$$

$$\frac{1}{R^2}\frac{d}{dR}\left(R^2\frac{df}{dR}\right) + \left\{\frac{M}{\hbar^2}[E - U(R)] - \frac{J(J + 1)}{R^2}\right\}f = 0, \qquad \text{(X-44)}$$

where

$$-J \leq m \leq J,$$

$$J = 0, 1, 2, \ldots,$$

$\phi(\varphi)\Theta(\cos \vartheta)$ is as given in (X-16).

$$E_J = \frac{\hbar^2}{MR^2}J(J + 1) \qquad \text{(X-45)}$$

is the rotational energy of the non-rigid system M-M, since $R$ is not a constant.

To solve the equation (X-45), one assumes for $U(R)$ a form which has the general characteristics of having a minimum at an equilibrium value $R_e$, being repulsive for $R < R_e$, attractive for $R_e < R$ and having an asymptotic value $U(\infty)$ such that

$$U(\infty) - U(R_e) = D \qquad \text{(X-46)}$$

corresponds to the dissociation energy $D - \frac{1}{2}hv$. The following potential, due to P. M. Morse,

$$U(R) = D(1 - y)^2, \qquad \text{(X-47)}$$

$$\ln y = -\alpha(R - R_e), \qquad \alpha = \text{a constant},$$

has these properties.

On using this $U(R)$ in (X-44), defining

$$A \equiv \frac{\hbar^2}{MR_e^2} J(J+1) = \frac{\hbar^2}{2I_e} J(J+1), \tag{X-48}$$

and

$$\psi(R) \equiv Rf(R),$$

one transforms (X-44) into the form

$$\frac{1}{y}\frac{d}{dy}\left(y\frac{d\psi}{dy}\right) + \frac{M}{\alpha^2\hbar^2}\left[\frac{E-D}{y^2} + \frac{2D}{y} - D - A\frac{1}{y^2}\left(\frac{R_e}{R}\right)^2\right]\psi(R) = 0. \tag{X-49}$$

On expanding $(R_e/R)^2$ about $R_e$, i.e., about $y = 1$, one obtains

$$\left(\frac{R_e}{R}\right)^2 = \left(1 - \frac{1}{b}\ln y\right)^{-2}, \qquad b = \alpha R_e,$$

$$= 1 + \frac{2}{b}(y-1) + \left\{-\frac{1}{b} + \frac{3}{b^2}\right\}(y-1)^2 + \cdots, \tag{X-50}$$

and neglecting terms beyond $(y-1)^2$, one obtains for (X-49)

$$\frac{1}{y}\left(y\frac{d\psi}{dy}\right) + \frac{M}{(\alpha\hbar)^2}\left\{\frac{E-D-C_0}{y^2} + \frac{2D-C_1}{y} - (D+C_2)\right\}\psi(R) = 0, \tag{X-51}$$

where

$$C_0 = \left(1 - \frac{3}{b} + \frac{3}{b^2}\right)A,$$

$$C_1 = \left(\frac{4}{b} - \frac{6}{b^2}\right)A,$$

$$C_2 = \left(1 - \frac{1}{b} + \frac{3}{b^2}\right)A.$$

Let

$$\eta \equiv \frac{\sqrt{M(D+C_2)}}{\alpha\hbar}, \qquad -\frac{(\alpha\hbar\beta)^2}{4M} \equiv E - D - C_0, \tag{X-52}$$

$$\psi(z) = e^{-z/2}z^{\beta/2}F(z), \qquad z \equiv 2\eta y.$$

Equation (X-51) then takes the form

$$z\frac{d^2F}{dz^2} + (\beta + 1 - z)\frac{dF}{dz} + \frac{1}{2}(2\eta - \beta - 1)F = 0. \tag{X-53}$$

The eigenvalues $E$ are given by

$$\tfrac{1}{2}(2\eta - \beta - 1) = v, \tag{X-54}$$

$$v = 0, 1, 2, \ldots,$$

and $F$ is then the hypergeometric function

$$F(-v, \beta + 1; z) = 1 - \frac{v}{\beta + 1} z + \frac{v(v-1)}{(\beta+1)(\beta+2)} \frac{z^2}{2!} + \cdots \tag{X-55}$$

Let

$$v \equiv \frac{\alpha}{\pi} \sqrt{\frac{D}{M}} \quad \text{(in cm}^{-1}\text{)},$$

$$x \equiv \frac{hv}{4D} = \frac{ha^2}{4\pi^2 Mv},$$

$$B_e \equiv \frac{h}{8\pi^2 I_e c}, \qquad I_e = \tfrac{1}{2}MR_e^2, \tag{X-56}$$

$$\zeta \equiv 6B_e x \left( \sqrt{\frac{B_e}{xv}} - \frac{B_e}{xv} \right) \quad \text{(in cm}^{-1}\text{)},$$

$$\zeta \equiv \frac{4}{v^2} B_e^2 \quad \text{(in cm}^{-1}\text{)}.$$

The eigenvalues $E$ of (X-44) are then given by the (approximate) expression, in cm$^{-1}$,

$$\frac{E}{h} = (v + \tfrac{1}{2})v - (v + \tfrac{1}{2})^2 xv + B_e J(J+1)$$

$$- \zeta(v + \tfrac{1}{2})J(J+1) - \xi J^2(J+1)^2. \tag{X-57}$$

The meaning of the various terms in (X-57) is as follows: The first term $(v + \tfrac{1}{2})hv$ is the vibrational energy and the second term $(v + \tfrac{1}{2})^2 xhv$ is the correction arising from the anharmonicity, i.e., the deviation of $U(R)$ from the harmonic potential in the neighborhood of $R_e$. These two terms, being independent of the rotational quantum number $J$, are the energy of a non-rotating oscillator. The third term $B_e J(J+1)h$ is the energy of rotation of a rigid (i.e., non-vibrating) rotator M-M. The last term $-\xi h J^2(J+1)^2$ is the correction to $B_e J(J+1)h$ arising from the change in the moment of inertia due to the rotational stretching of the M-M distance. The fourth term $-\zeta(v + \tfrac{1}{2})J(J+1)h$ represents the effects of the coupling between the vibrational and the rotational motion. The effects can be traced to two causes as follows:

(i) On account of the vibration, the moment of inertia $I = \frac{1}{2}MR^2$ is not a constant, but varies periodically. Thus the rotational energy $(h^2/8\pi^2 I)J(J + 1)$ also varies. To calculate this small correction to the rotational energy averaged over a period of vibration, one may regard the vibration as harmonic. Since

$$\frac{1}{(R_e - \Delta R)^2} > \frac{1}{(R_e + \Delta R)^2} \quad \text{for } \Delta R > 0,$$

it is seen that this correction is positive. This is represented by the second part of $\zeta$, and

$$\Delta E_J = 6B_e x \left(\frac{B_e}{xv}\right)(v + \tfrac{1}{2})J(J + 1).$$

(ii) The rotation stretches the equilibrium separation $R_e$ by $\Delta R_e$. If the vibration is harmonic, i.e.,

$$U(R) = \tfrac{1}{2}k(R - R_e)^2,$$

a change $\Delta R_e$ does not affect the frequency of the vibration. But if $U(R)$ is anharmonic about $R_e$, a change $\Delta R_e$ does affect the frequency, and this effect is related to the parameter $x$ which is a measure of the anharmonicity. This effect is represented by the first part of $\zeta$, i.e.,

$$\Delta E_v = -6B_e x \sqrt{\frac{B_e}{xv}}(v + \tfrac{1}{2})J(J + 1).$$

In other more complicated molecules, the calculations are necessarily much more lengthy, but the essential nature of the effects is the same.

## 5. Vibrational motion of a polyatomic molecule

Consider the vibrational motion of a molecule (in a given electronic state) alone by disregarding their coupling with the rotational motion. For this problem, one may start with the theory of small oscillations in classical dynamics. A molecule formed by $N$ atoms has $n = 3N - 6$ degrees of freedom for vibration,[8] and for small amplitudes one introduces $n$ normal coordinates $q_1, \ldots, q_n$ in terms of which the energy of the system can be expressed in the form

$$E = \tfrac{1}{2}\sum_i (\dot{q}_i^2 + \lambda_i q_i^2), \tag{X-58}$$

$$\lambda_i = (2\pi v_i)^2 = \omega_i^2, \qquad i = 1, 2, \ldots, n,$$

$v_i$ being the frequencies of the normal vibrations.

---

[8] For linear molecules, $n = 3N - 5$, as axial symmetry leads to a two-fold degeneracy for vibrations perpendicular to the axis.

The transcription into quantum mechanics can be carried out by transforming to the (dimensionless) coordinates $\xi_i$

$$\xi_i = 2\pi \sqrt{\frac{v_i}{h}} q_i, \qquad \eta_i = \frac{\partial T}{\partial \dot{\xi}_i}, \tag{X-59}$$

so that the Hamiltonian is

$$H = \frac{2\pi^2}{h} \sum_i v_i \eta_i^2 + \frac{h}{2} \sum_i v_i \xi_i^2. \tag{X-60}$$

The Schrödinger equation

$$(H - E)\Psi = 0,$$

then separates into $n$ equations

$$\frac{d^2 \psi_i(\xi_i)}{d\xi_i^2} + \left(\frac{2E_i}{hv_i} - \xi_i^2\right)\psi_i(\xi_i) = 0, \tag{X-61}$$

which are the equations of simple harmonic oscillators, with eigenvalues

$$E_i = (v_i + \tfrac{1}{2})hv_i, \qquad i = 1, 2, \ldots, n. \tag{X-62}$$

The selection rules for the emission and absorption spectrum (in the infrared) are given by the matrix elements of the electric moment $\mathcal{M}$ of the molecule. Consider for example the $x$ component $\mathcal{M}_x$ (the molecule being referred to a coordinate system in which the non-vibrating molecule is at rest). One develops $\mathcal{M}_x$ about the configuration of equilibrium,

$$\mathcal{M}_x = \mathcal{M}_x^0 + \sum_i \left(\frac{\partial \mathcal{M}_x}{\partial \xi_i}\right)_e \xi_i + \cdots \tag{X-63}$$

$\mathcal{M}^0$ is the permanent moment, and $(\partial \mathcal{M}/\partial \xi_i)_e \xi_i$ is the change in $\mathcal{M}$ due to the normal vibration $\xi_i$. The matrix element associated with a transition $(v_1, v_2, \ldots) \leftrightarrow (v_1', v_2', \ldots)$ is

$$\langle v_1', v_2', \ldots | \mathcal{M}_x^0 | v_1, v_2, \ldots \rangle + \sum_i \langle v_1', v_2', \ldots | \xi_i | v_1, v_2, \ldots \rangle \left(\frac{\partial \mathcal{M}_x}{\partial \xi_i}\right) + \cdots \tag{X-64}$$

The first term vanishes unless $v_i' = v_i$ for all $i$, so that the permanent moment $\mathcal{M}^0$ gives rise to no emission or absorption spectrum. The second sum in (X-64) consists of terms of the form

$$\left(\frac{\partial \mathcal{M}_x}{\partial \xi_i}\right)_e \langle v_j | \xi_j | v_j \pm 1 \rangle \delta_{v_i', v_i}, \quad v_i \neq v_j, \tag{X-65}$$

which lead to the emission or absorption of only the fundamental frequencies $v_j$, provided the corresponding $(\partial \mathcal{M}/\partial \xi_j)_e$ do not vanish.

The absence of harmonics such as $2v_j$, $3v_j$, ... and combination frequencies such as $v_i + v_j$ is a consequence of the assumption (approximation) in (X-58), or (X-60), that the small vibrations are harmonic. When anharmonic vibrations are considered, harmonic and combination frequencies do appear.

The question whether a given frequency $v_j$ appears or not in emission and absorption is answered by an examination of the coefficients

$$\left(\frac{\partial \mathcal{M}_x}{\partial \xi_j}\right)_e, \qquad \left(\frac{\partial \mathcal{M}_y}{\partial \xi_j}\right)_e, \qquad \left(\frac{\partial \mathcal{M}_z}{\partial \xi_j}\right)_e.$$

The vanishing or nonvanishing of these coefficients is completely determined by the symmetry of the molecule and the vibration $\xi_j$. This can be illustrated by a few simple examples.

$$
\begin{array}{llll}
CO_2 & & & \\
HCN & & & \text{(X-66)} \\
H_2O & & &
\end{array}
$$

$$v_1 \qquad\qquad v_2 \qquad\qquad v_3$$

|        | $v_1$ | $v_2$ | $v_3$ |
|--------|-------|-------|-------|
| $CO_2$ |  | $\dfrac{\partial \mathcal{M}_y}{\partial \xi_2}, \dfrac{\partial \mathcal{M}_z}{\partial \xi_2}$ | $\dfrac{\partial \mathcal{M}_x}{\partial \xi_3}$ |
| $HCN$ | $\dfrac{\partial \mathcal{M}_x}{\partial \xi_1}$ | $\dfrac{\partial \mathcal{M}_y}{\partial \xi_2}, \dfrac{\partial \mathcal{M}_z}{\partial \xi_2}$ | $\dfrac{\partial \mathcal{M}_x}{\partial \xi_3}$ |
| $H_2O$ | $\dfrac{\partial \mathcal{M}_y}{\partial \xi_1}$ | $\dfrac{\partial \mathcal{M}_y}{\partial \xi_2}$ | $\dfrac{\partial \mathcal{M}_x}{\partial \xi_3}$ |

$$\text{(X-67)}$$

The figures give the normal vibrations, and the table gives the nonvanishing coefficients for the various normal vibrations. (The $x$ axis is in the horizontal and the $y$ axis in a perpendicular direction, the $z$ axis is normal to the plane of the paper.)

The selection rules for the Raman effect of the vibrational spectrum are governed by the induced electric moment,

$$\mathcal{M}_x = \sum_{x}^{z} \alpha_{xk} \mathscr{E}_k,$$

where $\alpha_{xk}$ is the polarizability of the molecule, and the summation is over $k = x, y, z$. Here the coordinate axes are chosen along the symmetry axes, (or principal axes) of the non-vibrating molecule.

For a vibration $\xi_j$, the polarizibility tensor changes

$$\alpha_{xk} = \alpha_{xk}^0 + \left(\frac{\partial \alpha_{xk}}{\partial \xi_j}\right)_e \xi_j. \qquad \text{(X-68)}$$

Thus the matrix element of the electric moment associated with the transition $(v_1', v_2', \ldots) \leftrightarrow (v_1, v_2, \ldots)$ is given by

$$\mathscr{E}_k \left(\frac{\partial \alpha_{xk}}{\partial \xi_j}\right)_e \langle v_j | \xi_j | v_j \pm 1 \rangle \delta_{v_i', v_j}, \quad \text{for all } i \neq j, \qquad \text{(X-69)}$$

$$k = x, y, z, \qquad x = x, y, z.$$

Thus again, in this harmonic approximation, only the fundamental frequencies $\xi_j$ are expected in the Raman spectrum. Whether a given mode $\xi_j$ appears or not is determined by the coefficients $(\partial \alpha_{xy}/\partial \xi_j)_e$ for all $x$, $y$, and this is governed by the symmetry property of the molecule and its normal vibrations. The following table illustrates the selection rules by giving the nonvanishing coefficients,

| | $v_1$ | $v_2$ | $v_3$ | |
|---|---|---|---|---|
| $CO_2$ | $\left(\dfrac{\partial \alpha_{xx}}{\partial \xi_1}\right)_e \quad x = x, y, z$ | | | |
| HCN | $\left(\dfrac{\partial \alpha_{xx}}{\partial \xi_1}\right)_e$ | | $\left(\dfrac{\partial \alpha_{xx}}{\partial \xi_3}\right)_e$ | (X-70) |
| $H_2O$ | $\left(\dfrac{\partial \alpha_{xx}}{\partial \xi_1}\right)_e \quad x = x, y, z$ | $\left(\dfrac{\partial \alpha_{xx}}{\partial \xi_2}\right)_e \quad x = x, y, z$ | | |

The vanishing of such nondiagonal components such as $\alpha_{xy}$ is due to the choice of the symmetry axes as the coordinate axes.

It is seen that for a molecule possessing a center of symmetry such as $CO_2$, the selection rules for the dipole electric moment and the induced dipole moment are exactly complementary to each other. Thus in such cases the combined infrared spectrum and the Raman spectrum data are needed for obtaining the frequencies of all the normal vibrations.

This difference in the selection rules for the infrared and the Raman spectrum is thus very useful in the determination of the geometrical shape of a molecule. Thus the complementary nature of these selection rules in the case of $CO_2$, $CS_2$ rules out any nonlinear or unsymmetrical model. On the other hand, the absence of this complementary nature in the case of $H_2O$ rules out the linear symmetric model.

## 6.  Rotational motion of a symmetrical top molecule

In Sec. 2 (3), we treated the rotation of a linear molecule. Here we shall treat the rotation of a symmetrical top, i.e., a body having two equal principal moments of inertia, $I_A = I_B \neq I_C$. Such a model represents molecules having an axis of symmetry, such as $NH_3$, $CH_3Cl$, $C_2H_6$ (ethane), $H_3CCN$ (methyl cyanide) and others.

Let $\xi, \eta, \zeta$ be a set of rectangular axes fixed in the rotator, with $\zeta$ along the symmetry axis and the origin at the center of mass. Let $\vartheta, \psi, \varphi$ be the Euler angles. The Schrödinger equation of the rotational motion can be shown to be

$$\left\{ \frac{\partial^2}{\partial \vartheta^2} + \cot \vartheta \frac{\partial}{\partial \vartheta} + \left( \frac{I_A}{I_C} + \cot^2 \vartheta \right) \frac{\partial^2}{\partial \varphi^2} + \frac{1}{\sin^2 \vartheta} \frac{\partial^2}{\partial \psi^2} \right.$$

$$\left. - 2 \frac{\cot \vartheta}{\sin \vartheta} \frac{\partial^2}{\partial \psi \partial \varphi} + \frac{8\pi^2 I_A}{h^2} E \right\} \Psi(\vartheta, \psi, \varphi) = 0. \tag{X-71}$$

The angles $\psi$ and $\varphi$ are cyclic coordinates, and the equation is separable, by means of

$$\Psi(\vartheta, \psi, \varphi) = \Theta(\vartheta) e^{iM\psi} e^{iK\varphi}, \tag{X-72}$$

$M, K$ being positive or negative integers, and

$$\left\{ \frac{d^2}{d\vartheta^2} + \cot \vartheta \frac{d}{d\vartheta} - \left( \frac{M - K \cos \vartheta}{\sin \vartheta} \right)^2 + \sigma \right\} \Theta(\vartheta) = 0, \tag{X-73}$$

$$\sigma = \frac{8\pi^2 I_A}{h^2} E - \frac{I_A}{I_C} K^2. \tag{X-74}$$

The wave equation can be transformed by

$$s = |K + M|, \qquad d = |K - M|,$$

$$t = \sin^2 \frac{\vartheta}{2}, \qquad \Theta(\vartheta) = \sin^d \frac{\vartheta}{2} \cos^s \frac{\vartheta}{2} F(t),$$

into

$$(1 - t) \frac{d^2 F}{dt^2} + [\gamma^2 - (\alpha + \beta + 1)t] \frac{dF}{dt} - \alpha\beta F = 0, \tag{X-75}$$

where

$$\gamma = 1 + d, \qquad \alpha + \beta = 1 + d + s, \tag{X-76}$$

$$\alpha\beta = \frac{d + s}{2} \left( \frac{d + s}{2} + 1 \right) - \sigma - K^2.$$

The solution of (X-75) is the hypergeometric function

$$F(\alpha, \beta, \gamma, t) = 1 + \frac{\alpha \cdot \beta}{1 \cdot \gamma} t + \frac{\alpha(\alpha + 1)\beta(\beta + 1)}{1 \cdot 2 \cdot \gamma(\gamma + 1)} t^2 + \cdots \qquad \text{(X-77)}$$

The condition that $F$ terminate at a certain power, $t^p$, is that either $\alpha$ or $\beta$ be equal to $-p$. From (X-74) and (X-76) one then obtains the eigenvalue

$$E = \frac{h^2}{8\pi^2 I_A} \left\{ J(J + 1) + \left( \frac{I_A}{I_C} - 1 \right) K^2 \right\}, \qquad \text{(X-78)}$$

where

$$J = p + \tfrac{1}{2}(d + s). \qquad \text{(X-79)}$$

Now $M\hbar$, $K\hbar$ are the components of angular momentum along the fixed space $z$-axis and the figure $\zeta$-axis respectively, and from (X-79), it follows that

$$|K| \leq J, \qquad |M| \leq J. \qquad \text{(X-80)}$$

$J(J + 1)\hbar^2$ is the square of the total angular momentum. As $E$ is independent of $M$ and of the sign of $K$, it follows that a state $(J, K)$ is $2(2J + 1)$-fold degenerate if $K \neq 0$, and $(2J + 1)$-fold degenerate if $K = 0$.

The selection rules are obtained from the matrix elements $\langle J, M, K |$ $| J', M', K' \rangle$ of the components of the permanent moment $\mathcal{M}_0$ (assumed to be along the figure axis on symmetry considerations),[h]

$$\mathcal{M}_z = \mathcal{M}_0 \cos \vartheta,$$

$$\mathcal{M}_x \pm i\mathcal{M}_y = \mathcal{M}_0 \sin \vartheta \, e^{\mp i\psi}.$$

It can be shown that the matrix elements vanish unless

$$\Delta J = 0, \pm 1, \qquad \Delta K = 0, \qquad \Delta M = 0, \pm 1. \qquad \text{(X-81)}$$

The rotational spectrum (in the infrared) consists of equidistant lines (in this approximation of the rigid rotator)

$$\nu_r = \frac{h}{4\pi^2 I_A}(J + 1), \qquad J = 0, 1, 2, \ldots \qquad \text{(X-82)}$$

The Schrödinger equation of the rotation of an asymmetrical top (such as $H_2O$) for which the three principal moments of inertia are all different, is not solvable in an exact manner. The degeneracies in (X-78) are here removed, and the rotational spectrum is extremely complicated. Reference must be made to the literature for such molecules.

---

[h] A molecule such as $C_2H_6$ (ethane) has zero permanent moment and hence has no pure rotational absorption spectrum.

### 7. Nuclear spin in molecules

We shall consider the question of the effect of nuclear spin on the energy and state of a molecule.

The magnetic moment of a nucleus of mass $M$ and spin $I$ is $I(e\hbar/2Mc)$ which is $(m/M)I$ times that due to the electron spin, and is very small. Thus the effect of nuclear spin on the energy of a molecule is very small, compared with the electronic, vibrational and rotational energies. The effect, however, on the symmetry property of the states of molecule is not small, and is, in fact, very important in the case of molecules having high symmetries. We shall give the theory of the effect of nuclear spin for such molecules as $H_2$, $N_2$, $O_2$, $CO_2$ which are linear, have a center of symmetry, with two atoms at of nuclear spin $I$ at the ends.

We shall assume that the coupling of the nuclear magnetic moment with the electronic and nuclear motion is small, so that the wave function of the molecule may be expressed as

$$\Psi = \psi_e \psi_v \psi_r \psi_i, \tag{X-83}$$

where $\psi_e$, $\psi_v$, $\psi_r$, $\psi_i$ are the wave functions of electronic, vibrational, rotational motion and nuclear spins respectively. For such molecules as $H_2$, etc. mentioned above, we assume for $\psi_i(1, 2)$ the product form

$$\chi_{m_i}(1)\chi_{m_i'}(2),$$

where

$$-I \leq m_i \leq I, \qquad -I \leq m_i' \leq I. \tag{X-84}$$

There are $(2I + 1)^2$ combinations of which $(I + 1)(2I + 1)$ are symmetric and $I(2I + 1)$ antisymmetric with respect to the interchange of the two nuclei, namely,

symmetric:

$$\psi_i^s(1, 2) = \begin{cases} \chi_{m_i}(1)\chi_{m_i}(2), \\ \dfrac{1}{\sqrt{2}}[\chi_{m_i}(1)\chi_{m_i'}(2) + \chi_{m_i'}(1)\chi_{m_i}(2)], & m_i' \neq m_i, \end{cases} \tag{X-85a}$$

antisymmetric:

$$\psi_i^a(1, 2) = \frac{1}{\sqrt{2}}[\chi_{m_i}(1)\chi_{m_i'}(2) - \chi_{m_i'}(1)\chi_{m_i}(2)], \quad m_i' \neq m_i. \tag{X-85b}$$

Consider now the rotational wave function $\psi_r$. The interchange of the two nuclei 1, 2 is effected by the parity operation $P$, which acts on the rotational wave function (X-16) as shown in (III-120)

$$P\psi_r(\vartheta, \varphi) = (-1)^J \psi_r(\vartheta, \varphi),$$

i.e.,

$$\psi_r(\vartheta, \varphi) \text{ is } \begin{Bmatrix} \text{symm.} \\ \text{antisym.} \end{Bmatrix} \text{ according as } J \text{ is an } \begin{Bmatrix} \text{even} \\ \text{odd} \end{Bmatrix} \text{ integer.} \quad \text{(X-86)}$$

Consider then the total wave function $\Psi$ in (X-83) and let us confine ourselves to those states of which the $\psi_e \psi_v$ part is symmetric in the interchange of the two nuclei. For $I = 0, 1, 2, \ldots, \psi_r(\vartheta, \varphi)\psi_i(1, 2)$ must be symmetric in the two nuclei, i.e.,

$$\psi_r(\vartheta, \varphi)\psi_i(1, 2) \text{ is symm. for } \begin{cases} J = \text{odd, and } \psi_i = \psi_i^a, \text{ or} \\ J = \text{even, and } \psi_i = \psi_i^s. \end{cases} \quad \text{(X-87)}$$

From (X-85), it is seen that the rotational states have the following weights on account of the nuclear spins.

$$J \text{ odd: } I(2I + 1),$$
$$J \text{ even: } (I + 1)(2I + 1). \quad \text{(X-88)}$$

For $I = \frac{1}{2}, \frac{3}{2}, \ldots$, then $\psi_r(\vartheta, \varphi)\psi_i(1, 2)$ must be antisymmetric, i.e.,

$$\psi_r(\vartheta, \varphi)\psi_i(1, 2) \text{ is antisym. for } \begin{cases} J = \text{even, and } \psi_i = \psi_i^a, \text{ or} \\ J = \text{odd, and } \psi_i = \psi_i^s. \end{cases} \quad \text{(X-89)}$$

Let us apply the result (X-87), (X-89) to the $O_2$ molecule. For the oxygen nucleus, $I = 0$. The $\psi_e \psi_v$ in the ground state is symmetric in the two nuclei. Since $\psi_i(1, 2)$ is symmetric, it is seen from (X-87) that only the $J = 0, 2, 4, \ldots$ rotational states exist. The Raman spectrum of $O_2$ has been found (F. Rasetti, in 1930) to consist of lines $\Delta J = 0 \to 2, 2 \to 4, 4 \to 6, \ldots$ The spacing (in frequency) between two neighboring lines is

$$8 \cdot \frac{h}{4\pi^2 I},$$

in agreement with (X-94), showing the absence of the $\Delta J = 1 \to 3, 3 \to 5, \ldots$ lines.

For $N_2$, $I = 1$ and from (X-87), (88), it is seen that the $J = \begin{Bmatrix} \text{odd} \\ \text{even} \end{Bmatrix}$ states have the relative weights $\begin{Bmatrix} 1 \\ 2 \end{Bmatrix}$. The Raman spectrum (Rasetti, 1930) shows alternating intensities 2:1 for the $\Delta J$ lines $= 0 \to 2, 1 \to 3, 2 \to 4, 3 \to 5$, etc. It may be remembered that the value $I = 1$, as indicated by the relative intensities of spectral lines in a band, is a difficulty in the proton-and-electron model of atomic nuclei, which was removed only when the neutron was discovered by Chadwick (1932).

For the molecule $H_2$, $I = \frac{1}{2}$, we have from (X-89) and (X-88), two possible symmetries for $\psi_i$. If the nuclear spins of the two nuclei interact very weakly with each other, the two wave functions $\psi_i^s$, $\psi_i^a$ of (X-85) are then independent, and there will be two species of hydrogen molecules which will be called

> para $H_2$, with symmetric $\psi_r$ ($J$ = even integers) and $\psi_i^a$,
>
> ortho $H_2$, with antisymmetric $\psi_r$ ($J$ = odd integers) and $\psi_i^s$.
>
> (X-90)

But at temperatures not too low (i.e., not near the absolute zero), $H_2$ exists in a mixture with an ortho:para ratio of 3:1 in (X-88). The specific heat of the mixture is then

$$C_v = \tfrac{3}{4}C_v(\text{ortho}) + \tfrac{1}{4}C_v(\text{para}), \qquad \text{(X-91)}$$

where

$$C_v = \frac{d\bar{E}}{dT},$$

$$\bar{E}(\text{ortho}) = \frac{1}{Z_{\text{ort.}}} \sum_{\text{odd } J} (2J + 1)E_J \exp(-E_J/kT),$$

$$\bar{E}(\text{para}) = \frac{1}{Z_{\text{para}}} \sum_{\text{even } J} (2J + 1)E_J \exp(-E_J/kT),$$

$$\text{(X-92)}$$

$$Z_{\text{ort.}} = \sum_{\text{odd } J} (2J + 1)\exp(-E_J/kT),$$

$$Z_{\text{para}} = \sum_{\text{even } J} (2J + 1)\exp(-E_J/kT),$$

$$\text{(X-93)}$$

$$E_J = J(J + 1)\frac{\hbar^2}{2I}, \qquad \text{(X-94)}$$

$I = \tfrac{1}{2}MR^2$ being the moment of inertia of the $H_2$ molecule. The calculated results from these relations are in agreement with the observed data. The theory is due to Dennison (1927).

The spectral lines $\Delta J = 0 \to 1$, $1 \to 2$, $2 \to 3$, etc. have been found to have alternating intensities $1:3:1:3$, etc. It was from this ratio that the spin of the proton was determined.

The spectrum of the acetylene molecule $C_2H_2$ has the same alternating intensity ration $1:3$.


## References

M. Born and R. Oppenheimer, *Annalen d. Physik* **84**, 457 (1927), is the classic paper on the quantum mechanics of molecules (Sec. 2).

T. Y. Wu, R. L. Rosenberg and H. Sandstrom, *Nucl. Phys.* **16**, 432 (1960). Section 2 treats the $\mu$-mesic H-$\mu$-D ion.

D. R. Bates, K. Ladsham and A. L. Stewart, *Phil. Trans. Roy. Soc.* A **246**. 215 (1953); R. F. Wallis and H. Hulbert, Jr., *J. Chem. Phys.* **22**, 774 (1954), Section 2 gives exact solutions of $H_2^+$.

W. Heitler and F. London, *Z. Phys.* **44**, 455 (1927).

Y. Sugiura, *ibid.* **45**, 484 (1927).

S. C. Wang, *Phys. Rev.* **31**, 579 (1928).

H. M. James and A. S. Coolidge, *J. Chem. Phys.* **3**, 129 (1935) (Sec. 3).

M. Born and R. Oppenheimer, *loc. cit.* (Sec. 4).

C. Eckart, *Phys. Rev.* **46**, 383 (1934), **47**, 552 (1935).

J. H. Van Vleck, *ibid.* **47**, 487 (1935).

P. M. Morse, *Phys. Rev.* **34**, 57 (1929) (Sec. 4).

P. Pauling and E. B. Wilson, Jr., *Introduction to Quantum Mechanics* (McGraw-Hill, New York, 1935) (Secs. 3, 4).

C. L. Pekeris, *ibid.* **45**, 98 (1934) (Sec. 4).

A. Adel and D. M. Dennison, *ibid.* **44**, 99 (1933) (Sec. 5).

T. Y. Wu, *Vibrational Spectra and Structure of Polyatomic Molecules* (Univ. of Peking Press and Prentice-Hall, New Jersey, 1940); 2nd ed. (Ed. Bro's., Ann Arbor, Mich., 1946), (Secs. 4, 5, 6).

D. M. Dennison, *Proc. Roy. Soc.* A **115**, 483 (1927) gives the theory of the specific heat of hydrogen. For a discussion of this problem, see R. H. Fowler, *Statistical Mechanics* (Cambridge Univ. Press, Cambridge, 1936).

F. Rasetti, *Z. Phys.* **61**, 598 (1930) treats the rotational Raman spectra of $O_2$, $N_2$.

B. Stoicheff, *Adv. Spectroscopy* **1**, 91 (1959) discusses the Raman spectrum of $CO_2$.

G. Herzberg, *Molecular Spectra and Molecular Structure*, Vol. I, Diatomic Molecules (1950), Vol. II, Infrared and Raman Spectra of Polyatomic Molecules (1945), Vol. III, Electronic Spectra and Electronic Structure of Polyatomic Molecules (1966), (Van Nostrand-Rheinhold, New York).

# Name Index


Abraham, M. 79
Adel, A. 410
Auger, P. 364

Bailey, V. A. 135
Balmer, J. J. 65
Barkla, C. G. 42
Bates, D. R. 390, 410
Belinfante, F. J. 235, 258
Bethe, H. A. 188, 189, 384
Bohm, D. 258
Bohr, N. 18, 42, 43, 46–9, 52, 55, 59, 75, 82,
    134, 202, 209, 212, 213, 258,
Boltzmann, L. 1, 31
Born, M. 30, 45, 49, 52, 64, 82, 84, 85, 109,
    110, 113, 118, 126, 148, 191, 197, 198,
    205, 207, 209, 213, 274, 325, 390, 409, 410
Bose, S. N. 36, 37, 213, 239
Bothe, W. 42, 43
Bragg, W. H. 42
Bragg, W. L. 42
Brillouin, L. 380, 385
Bruns, H. 139
Burgers, J. M. 19, 48

Cayley, A. 109
Chadwick, J. 408
Chaudhuri, R. N. 135
Chung, K. T. 364
Collidge, A. S. 397, 410
Compton, A. H. 41, 42
Condon, E. U. 46, 273, 279, 357, 359, 384, 385
Conway, A. S. 47
Cross, W. G. 42, 43

Dauvillier, A. 135
Davisson, C. J. 43, 135
de Broglie, L. 30, 43, 83, 84, 131, 132, 133,
    135–6, 148–9, 191, 213, 258
Dalgarno, A. 391
Debye, P. 44, 45, 139
Dennison, D. M. 409, 410
Dirac, P. A. M. 30, 51, 54, 76, 79, 83, 84, 101,
    106, 109, 117, 126, 129, 130, 143, 191–3,
    199, 200, 205, 208–9, 213, 237, 239, 254,
    258, 274, 279, 293, 306, 315, 320, 328, 384

Eckart, C. 410
Ehrenfest, P. 19, 34, 40, 43, 75, 207
Einstein, A. 19, 37, 39, 40, 43, 44, 45, 52–5, 60,
    85, 131, 191, 202, 213, 239, 254, 255, 258

Elwert, G. 187
Epstein, P. 52
Estermann, I. 43, 135

Fano, U. 254
Fermat, P. de 131, 140
Fermi, E. 135, 213, 239, 311, 347, 360, 376,
    378, 385
Feynman, R. P. 306
Fock, V. 116, 148–9, 164, 207, 339–41, 347,
    370–2, 384
Foley, H. M. 263, 328
Fowler, R. H. 410
Franck, J. 47, 52, 64, 198
Frisch, R. 135
Fröman, N. 382, 385

Gaunt, L. 384
Gehrcke, E. 39
Geiger, W. 42, 43, 46, 48
Gerlach, W. 50, 52, 64, 76
Germer, L. H. 43, 135
Goeppert-Mayer, M. 383
Gombás, P. 378, 385
Gordon, W. 148, 149, 188
Goudsmit, S. A. 64, 73, 74, 79, 307, 383
Guillemin, V. 391

Hahn, O. 82
Hamilton, W. R. 1, 131, 132, 134–6
Hartree, D. R. 339, 340–1, 347, 370, 372,
    384
Heisenberg, W. 30, 54, 59, 62, 64, 83–5, 88,
    109, 110, 113, 118, 126, 144, 191–2, 200,
    208, 209, 213, 242, 254, 288, 384
Heitler, W. 393ff, 410
Hertz, G. 47, 52, 64
Hertz, H. 39, 132
Herzberg, G. 397, 410
Hofstadter, R. 42, 43
Hulbert, H. 410
Hund, F. 384, 397
Hylleraas, E. A. 341, 363, 364, 384

Jacobi, K. G. 137
James, H. M. 397, 410
Jammer, M. 31, 110
Jeans, J. H. 32, 33, 35, 40
Jeffreys, H. 380, 385
Jordan, P. 30, 84, 85, 109, 110, 113, 118, 126,
    191, 200, 213

Kallmann, H. 187
Kamerlingh-Onnes, H. 34
Kármán, Von, T. 45
Kayser, H. 65
Kelvin, Lord, 1, 51
Kemble, E. C. 152, 183, 385
Kennard, E. H. 109
Kikuchi, C. 135
Kinoshita, T. 384
Kirchhoff, G. 101
Klein, O. 83, 148, 149, 276
Kopfermann, H. 61, 288
Kossel, W. 58
Kramers, H. A. 42, 43, 54, 58–9, 61–2, 64, 86,
    288, 380, 385
Krishnan, K. S. 64
Kronig, R. de 75
Kuhn, W. 63–4, 117, 288
Kunsman, C. H. 135
Kusch, P. 328

Ladenburg, R. 54, 59, 60, 61, 288
Ladsham, K. 410
Lagrange, J. L. 5, 132
Lamb, Jr., W. E. 320–1, 328
Landé, A. 64, 67, 69, 72–4, 76, 80, 322
Landsberg, G. 64
Lang, W. 83
Langer, R. E. 382
Laporte, O. 52, 69, 72, 315
Laue, von, M. 42, 191
Lee, T. D. 25
Levi-Civita, T. 15
Lewis, G. N. 39
Lighthill, M. J. 110
London, F. 393ff, 410
Lorentz, H. A. 35, 41, 58, 70
Lummer, O. 31, 39

Main-Smith, J. D. 79, 82
Mandelstam, L. 64
Marsden, E. 46, 48
McCarroll, R. 391
McIntyre, J. A. 42, 43
Marney, M. C. 135
Marshall, L. 135
Maupertuis, P. de 131, 140
Mendeleev, D. I. 82
Millikan, R. A. 40
Montroll, E. W. 46
Morse, P. M. 398, 410
Moseley, H. G. J. 64, 82
Mulliken, R. S. 397

Neumann, J. von, 235, 245, 253, 258, 328
Newlands, J. A. R. 82

Odishaw, H. 46
Oppenheimer, J. R. 409, 410
Ohmura, T. 283, 306, 328

Paris, A. 31
Paschen, F. 31, 51, 70, 72, 77, 320, 322
Pasler, M. 183
Pauli, W. 73–5, 79, 81, 82, 114, 118, 209, 212,
    236, 307
Pauling, L. 410
Pekeris, C. L. 410
Phillips, M. 385
Planck, M. 31, 32, 33, 34, 35, 36, 41, 48, 191
Podolsky, B. 255, 258
Pringsheim, E. 31

Ramberg, E. 385
Raman, C. V. 62, 64, 274
Ramsauer, C. 135
Ramsey, N. F. 42, 43
Rayleigh, Lord 32, 33, 35, 64, 134, 259
Rassetti, F. 408, 410
Reid, A. 135
Retherford, R. C. 320, 328
Richtmyer, F. 385
Ritz, W. 66, 86, 87
Riviere, D. C. 254
Rosen, N. 254, 258, 397
Rosenberg, R. L. 390, 409
Rosenfeld, L. 209
Rubens, H. 34, 35
Rubinowicz, A. 58
Runge, J. 139
Rupp, E. 43, 135
Russell, H. N. 345
Rutherford, Lord 2, 46
Rydberg, J. R. 65

Salpeter, E. E. 188, 189, 384
Sandstrom, H. 390, 409
Schrödinger, E. 30, 45, 83–4, 135–6, 140–4,
    146, 148–9, 191, 213, 273–4, 306
Schwartz, L. 101, 110
Schwarz, H. A. 200, 203
Schwarzschild, K. 14, 52
Shortley, G. H. 273, 357, 359, 384, 385
Shull, E. G. 135
Simons, A. W. 42, 43
Smekal, A. 62, 64, 274
Slater, J. C. 42, 43, 340, 365, 384
Soddy, F. 2, 82
Sommerfeld, A. 48, 50–2, 58, 64, 68, 72, 80,
    82, 139, 320, 378, 385
Stande, O. 14
Stefan, J. 31
Stern, O. 43, 50, 52, 64, 76, 135
Stewart, A. L. 410
Stoichoff, B. 410
Stoner, E. C. 79, 80, 82
Sucher, J. 263
Sugiura, Y. 396, 410

Taube, G. E. 263
Thomas, L. H. 74–5, 79, 347, 376, 385

Thomas, W. 63, 64, 117, 288
Thomson, G. P. 43, 135
Thomson, J. J. 1, 64, 82
Townsend, J. S. 135

Uhlenbeck, G. E. 64, 73–4, 79, 307

Van Vleck, J. H. 53, 189, 410
Vigier, J. P. 258
Von Kármán, Th. 45
Von Laue, M. 42, 191
von Neumann, J. 245, 253, 258, 328

Wallis, R. F. 410
Wang, S. C. 396, 410
Watson G. N. 180
Weinbaum 396
Wentzel, G. 380, 385

Weyl, H. 183, 200, 245, 247, 254
White, H. E. 376, 385
Whittaker, E. T. 31, 180
Wien, W. 31, 33, 35, 38, 40
Wiener, N. 108, 109, 205
Wigner, E. P. 253, 328
Wilson, C. T. R. 42
Wilson, Jr., E. B. 410
Wilson, W. 48
Wu, C. S. 25
Wu, T. Y. 55, 187, 253, 254, 263, 279, 283,
    306, 328, 382, 383, 385, 390, 409, 410

Yang, C. N. 25
Yao, Y. T. 279

Zener, C. 391
Zinn, W. H. 135

# Subject Index


Action and angle variables 13–18, 110
Angular momentum
  commutation relations 118–119, 172
  eigenvalues 119, 120, 161
  matrices 118
  matrix elements 120–121, 161, 181
  operators, in wave mechanics 161
  spin 122
Anharmonic oscillator
  in matrix mechanics 125
  in wave mechanics 261
  Morse oscillator 398
Anomalous Zeeman effect 70–72, 76–77
Associated Laguerre polynomial 185
  integrals containing 186–188
Associated Legendre polynomial 169–172,
  175, 178
Auger effect 364
Autoionization 360–364

Balmer formula 65
Bohr-Kramers-Slater's statistical conservation
  of energy 42
Born approximation, 197, 293, 300, 325
Bose's derivation of Planck's law 40–41
Boundary condition 150
Bra 215

Canonical equations 5
  completely solvable 15
  in matrix mechanics 111, 113–114
  from Hamilton's principle 7, 27–29
Canonical transformation 5
  adiatatic invariants 19
  generator 8, 11
  sufficient condition 8, 9, 10
  Poincaré integral invariants 9, 10
Causality condition 297, 299
Central field problem
  eigenfunctions 157–160
Characteristic function 12, 138
Closure relation 145, 197, 220
Collision
  rearrangement, 289, 290–291
Commutation relation 111, 113, 129, 145,
  200, 210, 220, 233
Commuting variables, complete set of, 239
Compton effect 41–42
Complementarity principle 209, 212
  postulates 232

Configuration, atomic 344
Configuration interaction 359, 373–376
Connection formula, WKB method 381
Continuous wave functions 181–183
Copenhagen school 210, 254–257
Correspondence principle 55–58
  in frequency 56
  in selection rules 56-8, 86, 128
Correlation (electron-electron) effect 340-1
Coupling between vibration and rotational
  motion of molecules 389, 391, 400

de Broglie's idea 132–134
Degeneracy 14, 18, 50
Density matrix 247, 249
  trace 248, 250–251
  transformation 251
Density operator 247, 249, 250
Dirac's $\delta$ function 100
  representation of 105
  $\delta'$ function 103
Direct integral 334
Dispersion theory
  Classical 59–60
  multiply periodic system 42–48
  quantum theory 60–61
  quantum mechanical 286–288

Ehrenfest's theorem 207
Eikonal 139
Einstein's theory
  light quanta 37–40, 42–43
  philosophy toward quantum mechanics
    254–257
  Planck's law 40–41
  specific heat of solids 44
  transition coefficients $A$, $B$, 53, 283
Electronic energy of molecule 388, 390, 393ff
Exchange integral in He 334
Exchange integral in hydrogen molecule 394
Exclusion (Pauli) principle 79–82
Expectation value 192, 199, 210, 233–234
$f$ electron in actinium series 382ff
Fermat principle 131
Fine structure of $H$ spectrum
  Dirac's formula 320
  Lamb shift 320
  Sommerfeld's relativistic theory 50–51, 75,
    262–265, 318
  spin-orbit 73–74, 317

Franck-Hertz experiment 47
Frequency condition
  Bohr theory 46, 52, 55–56
  matrix mechanics 112
Fourier transform
  momentum representation 146–147

Geometrical (ray) optics 139
Green's function 193–195, 296
  stationary-state problem 196
  time-dependent problem 296
  propagator 200

Hamiltonian
  classical dynamics 6
  quantum mechanics 111, 141
Hamilton-Jacobi equation 10–14, 16, 137, 140
  condition for separability 15
Hamilton's principle 5, 26
Harmonic oscillator
  in matrix mechanics 114
  in Fock representation 116
  in quantum mechanics 154
  in Fock representation 164
  two-dimensional 176
  three-dimensional 177
Hermite polynomial 161–163
Hermitian operators (see self-adjoint operators) 150, 217
Hidden variable 235
Hydrogen atom
  classical dynamics 16–18
  continuous wave function 181
  discrete energy states 179
  discrete eigenfunctions 181
  normalization 183, 189
Hydrogen molecular ion 388ff
Hydrogen molecule 393ff
  Heitler-London theory 393ff

Intermediate coupling 357–359
Invariants
  adiabatic 19
  Poincaré integral 9, 10
  under space inversion, of Maxwell equations 23–25
  under time reversal, of Maxwell equations 23–25
Inverted doublet, in alkali atom 375

$\{j,j\}$ coupling 79, 353
  electron-electron interaction in, 355–357

Ket 215

Landé $g$-formula 72, 77
Laporte rule 69–70, 315, 372

Legendre polynomial 167–168
Legendre transformation 6
Liouville equation
  classical 253
  quantum mechanics 253
Lorentz relation 23
$\{L, S\}$-coupling 78, 344

Many-electron atom, energy
  Slater's determinantal wave function 365
  Hartree-Fock method 370–2
  closed shell 366
  between a valence electron and a closed shell 367
  between two closed shells 369
Matrix
  adjoint of 90
  algebra 88. 89
  continuous 100
  differentiation of 98–99
  eigenvalues of 93–94
  eigenvectors of 93–94
  hermitian 90–91, 94
  inverse of 90
  mechanics 110f
  orthogonal 91
  principal axes transformation 96
  similarity transformation 92
  simultaneous diagonalization of two matrices 97–98
  trace of 90
  transpose of 90
  unit continuous 101
  unitary transformation 95
Matrix mechanics 110ff
  equivalence with wave mechanics 144–5
Maupertuis principle 131–132
Maxwell's equations 21–24
Measurement, theory of 233, 239, 241
Morse potential 398
Multiply periodic system 14

Non-commutative operators, Dirac's theory 126
Noncrossing theorem 272, 324
Nuclear spin
  effect on molecular state 407ff

Observable 232
  complete set of commuting 239–240
Operator
  displacement 227–230
  hermitian 216–217
  integral 197, 301, 302
  projection 220
  self-adjoint 150, 217, 232, 233
  time translation 230, 236, 293
  unitary 219, 226

Parabolic coordinates 269, 326
Paschen-Back effect 70–72, 77, 322
Perturbation theory of nondegenerate systems
  matrix mechanics 123
  wave mechanics 259, 260
Perturbation theory for time-dependent
  Schrödinger equation 265, 275
  Dirac's method of variation of constants
    279–283
Pictures 242
  Heisenberg 242–244
  Schrödinger 242–244
  interaction 293
Planck's quantum theory 32–36
Physical (wave) optics 139
Polarizability tensor 278
Poisson bracket expression
  classical 9, 129
  quantum 111
Poynting vector 23
Principal function of Hamilton 9, 136
Probability
  in classical physics 25, 199
  in quantum mechanics 199
  in radioactive decay 26
Probability postulate 191, 233
Projection operator 294, 300–304

Quantization condition 46, 48–49, 83, 130,
    134, 148, 149
Quantum numbers
  $n$ 46–47; $n, k$ 52; $n, k, S, L, J$ 66; $L, S, J$
    77–79; $n, l, m_l, m_s$ 81; $n, l, j, m$ 81

Raman effect 274–8
  of $N_2$, $O_2$, 408
  relation with Stark effect 278
  Selection rules 403
Rayleigh-Jeans law 32
Ritz combination principle 66, 86–87
Representation
  matrix 103, 106
  $Q, P$ 220–225
  transformation of 107–109, 146, 191, 219
  $(j, m)$ 311, 313–314
  $(m_l, m_s)$ 311, 313–314
  Schrödinger representation 143
Representatives 216
Rotational energy 289
  of linear molecule 292, 399
  of symmetrical molecules 405
Rydberg correction 373, 383–384
  $f$ states of heavy atoms 384
Rydberg series 45

Selection rule, for
  $m$ 56, 316
  $S$ 67
  $J$ 68, 317–318, 373

$l(k)$ 318
  parity 69–70, 315, 372
  vibrational transitions in molecules 402
Scattering
  Born approximation 197, 293, 300, 325
  by a central field 194–196
  amplitude 196, 295, 298
  cross section 197
  integral equation, for 193–8
Schrödinger's initial ideas 135–141
Schrödinger time-dependent equation 142,
    235–236
  argument for first order equation 237
  symmetry and invariants 237
Schwarz inequality 203–204
Self-adjoint operator 150, 217, 226
  completeness of eigenfunctions 152–154
  orthogonality of eigenfunctions 151, 217
  reality of eigenvalues 152, 217
  simultaneous eigenstates of two self adjoint
    operators 218
Spatial quantization 50
Specific heat of ortho- and para-hydrogen 409
Specific heat of solid 44
  Debye, Born-Von Kármán theory 44
Spin, electron
  operators (Pauli) 307
  eigenvalues 308
Spin of two-electron system
  symmetric (triplet) state 331
  antisymmetric (singlet) state 331
Spin-orbit interaction 74, 309
  in $\{j,j\}$ coupling 355
  in $\{L, S\}$ coupling 350
Stark effect 261, 267, 269–271
  inverse Stark effect 274, 278–279
  relation to Raman effect 278
State
  vector 215, 232
  condition for pure state 249
  mixed state 246
  pure 246
Static-nuclei approximation 389, 391
Stationary state 46–48
Stefan-Boltzmann's law 31
Stern-Gerlach experiment 50, 76
Sturm-Liouville problem 150
Sum rule 63, 288

Thomas correction 75
Thomas-Fermi potential 376ff
Time concept, in special theory of relativity 3
  in quantum mechanics 4
Transformation
  canonical 8ff
  Fourier 146
  Representation 107, 146, 191, 219
  $(m, m_s)$ to $(j, m)$ 311, 313
  unitary 95, 219, 226, 227, 230

Transition probability
    amplitude 295, 296, 299, 303
    cross section 291–293
    Einstein's $A$ and $B$ coefficients 283–286
    Golden rules 281, 282, 304
Two-electron atom
    angular momenta 332
    energy, Coulomb integral 335
    energy, exchange integral 336
    energy, Hartree-Fock method 339, 341
    energy, Hylleraas method 341–343
    energy, perturbation method 336–337
    energy, Ritz method 337–339
Two unequal minima potential 379–380
    eigenvalue 381

Uncertainty principle 199, 200
    coordinate and momentum 201–5, 210
    energy and time 211, 304–6

Variational method 337, 339, 341
    theory 338

Vibrational energy of molecules 391, 400, 401–402
Vibration-rotation coupling in molecules 400
Velocity
    group 134
    phase 134

Wave mechanics
    equivalence with matrix mechanics 144–145
    initial ideas of de Broglie 131ff
    initial ideas of Schrödinger 136, 138ff
Wave packet 143, 202
Width of energy level 306
Wien's displacement law 31
Wien's radiation law 32
WKB method 380

Zeeman effect 57, 321, 323
Zero-point energy 36, 116, 156
    experimental evidence for, 36

9 789812 382863